Discrete Fourier Transform (DFT)

$$X[k] = \sum_{n=0}^{N-1} x[n] e^{-j2\pi \frac{k}{N} n} \qquad k = 0, 1, \ldots, N-1$$

Invers

$$x[n] = \frac{1}{N} \sum_{k=0}^{N-1} X[k] e^{j2\pi \frac{k}{N} n} \qquad n = 0, 1, \ldots, N-1$$

2D Discrete Fourier Transform (2D DFT)

$$X[i, k] = \sum_{m=0}^{M-1} \sum_{n=0}^{N-1} x[m, n] e^{-j2\pi \frac{i}{M} m} e^{-j2\pi \frac{k}{N} n} \qquad i = 0, \ldots, M-1; k = 0, \ldots, N-1$$

2D Inverse Discrete Fourier Transform (2D IDFT)

$$x[m, n] = \frac{1}{MN} \sum_{i=0}^{M-1} \sum_{k=0}^{N-1} X[i, k] e^{j2\pi \frac{i}{M} m} e^{j2\pi \frac{k}{N} n} \qquad m = 0, \ldots, M-1; n = 0, \ldots, N-1$$

2D Discrete Cosine Transform (2D DCT)

$$C[i, k] = 4 \sum_{m=0}^{N-1} \sum_{n=0}^{N-1} x[m, n] cos\left(\frac{(2m+1)i\pi}{2N}\right) cos\left(\frac{(2n+1)k\pi}{2N}\right) \qquad i, k = 0, \ldots, N-1$$

2D Inverse Discrete Cosine Transform (2D IDCT)

$$x[m, n] = \frac{1}{N^2} \sum_{i=0}^{N-1} \sum_{k=0}^{N-1} \beta[i]\beta[k] C[i, k] cos\left(\frac{(2m+1)i\pi}{2N}\right) cos\left(\frac{(2n+1)k\pi}{2N}\right) \quad m, n = 0, \ldots, N-1$$

$$\beta[p] = \begin{cases} \frac{1}{2} & p = 0 \\ 1 & p = 1, \ldots, N-1 \end{cases}$$

Discrete Wavelet Transform (DWT) Analysis

$$c_j[k] = \sum_{m=-\infty}^{\infty} c_{j+1}[m] h_0[m - 2k]$$

$$d_j[k] = \sum_{m=-\infty}^{\infty} c_{j+1}[m] h_1[m - 2k]$$

Discrete Wavelet Transform (DWT) Synthesis

$$c_{j+1}[k] = \sum_{m=-\infty}^{\infty} c_j[m] h_0[k - 2m] + \sum_{m=-\infty}^{\infty} d_j[m] h_1[k - 2m]$$

FUNDAMENTALS OF DIGITAL SIGNAL PROCESSING

Joyce Van de Vegte

Camosun College

Upper Saddle River, New Jersey
Columbus, Ohio

Library of Congress Cataloging-in-Publication Data

Van de Vegte, Joyce.
 Fundamentals of digital signal processing / Joyce Van de Vegte.
  p. cm.
 ISBN 0-13-016077-6
  1. Signal processing--Digital techniques. I. Title

TK5102.9 .V35 2001
621.382′2--dc21

2001021285

Editor in Chief: Stephen Helba
Assistant Vice President and Publisher: Charles E. Stewart, Jr.
Assistant Editor: Delia K. Uherec
Production Editor: Tricia L. Rawnsley
Design Coordinator: Robin Chukes
Production Coordination: Jeff Stiles, Carlisle Publishers Services
Text Designer: Carlisle Communications
Cover photo: Visual Edge Imaging
Cover Designer: Jeff Vanik
Production Manager: Matthew Ottenweller
Electronic Text Management: Karen L. Bretz

This book was set in Times New Roman by Carlisle Communications, Ltd., and was printed and bound by R. R. Donnelley & Sons Company. The cover was printed by Phoenix Color Corp.

Prentice-Hall International (UK) Limited, *London*
Prentice-Hall of Australia Pty. Limited, *Sydney*
Prentice-Hall Hispanoamericana, S.A., *Mexico*
Prentice-Hall of India Private Limited, *New Delhi*
Prentice-Hall Singapore Pte. Ltd.
Editora Prentice-Hall do Brasil, Ltda., *Rio de Janeiro*

10 9 8 7 6 5 4 3
ISBN: 0-13-016077-6

Like father, like daughter.

PREFACE

Digital signal processing (DSP) can no longer be considered the domain of graduate students and researchers. It now pervades the technology that we take for granted in our homes and offices, and its influence is growing. This book was written to create an accessible resource for college students, engineers, and computer scientists wanting to gain a working knowledge of the principles, applications, and language of DSP.

First and foremost, DSP is exciting! The author attempts to give readers a sense of this excitement immediately by starting with a nonmathematical crash course in DSP. In the chapters that follow, examples frequently focus on real-life sounds—such as speech, whale songs, and seismic vibrations—and on real-life images—such as fingerprints, bacteria, and airport X-rays. The accompanying CD provides links between graphs and sounds, and allows readers to hear for themselves the "before" and "after" of processing. Once the necessary theory has been covered, in-depth applications of speech recognition, image processing, motor control, and encryption are studied, among others.

All key concepts of DSP are covered in this text, including details of how to perform transforms and design filters. This coverage is heavily supported by examples throughout. To make the ideas as accessible as possible, no calculus is used at all in the main text. While the mathematical techniques that are used are not trivial, they are always presented in as straightforward a manner as possible. Even students with strong mathematical backgrounds will appreciate the chance to focus on issues of DSP rather than mathematics. Essential mathematical topics that are prerequisite to understanding the material in the text are included in an appendix.

Chapter 1 contains the "Crash Course in Digital Signal Processing." It provides a surface treatment of all the major topics of the book, without going into details of the underlying mathematics. Chapter 2 explains how to obtain a digital signal from the analog signals that surround us, and Chapter 3 provides some experience with defining and handling digital signals. Chapters 4 through 8 contain the majority of the important underlying theory for DSP. Topics covered in these chapters include difference equations, digital convolution, z transforms, discrete time Fourier transforms, and discrete Fourier series. The essential concepts of filter, transfer function, frequency response, and spectrum are developed. Filter design is taken care of in Chapters 9 and 10, for both finite impulse response and infinite impulse response filters. Practical aspects are covered beginning with Chapter 11, which discusses discrete and fast Fourier transforms, followed by Chapters 12 and 13, which examine DSP hardware and programming issues. Applications of DSP for sounds and images are investigated in Chapters 14 and 15. Finally, Chapter 16 provides a description of wavelet theory and applications. Appendix A contains "The Math You Need," while the other appendices prove claims made in the text so that the text may stand on its own, without the need for outside references. End-of-chapter summaries and questions are provided for each chapter. The accompanying CD includes sample sounds, images, data, and video sequences for most chapters, as well as software for spectrograms and wavelets. The sun symbol in the margin of the text indicates that illuminating material is available on the CD. Matlab files and examples of how to use them are included on the CD to permit the reader to verify methods presented in the text. These files can also be used to solve many end-of-chapter problems with ease. Quick tests of basic chapter concepts are provided on the CD as well. The instructor's manual includes solutions to text problems, laboratories based on Matlab and the Analog Devices ADSP-2181 EZ-KIT Lite DSP development kit, and laboratory guides. It is accompanied by a CD containing PowerPoint® slides of key text graphics, as well as laboratory documents and solution set files. All comments, suggestions, and reports of errors in the text or software will be most appreciated. They may be sent to Joyce Mills (née Van de Vegte) at *millsj@camosun.bc.ca*.

Because the mathematical requirements of this text are moderate, the book can be used as early as the second year of a college engineering or technology program. Increasingly, DSP will be considered an essential technical skill. Perhaps this text can be of use as the pressure grows to teach DSP earlier in the curriculum.

I must thank my colleagues in the Computer and Electronics Engineering Technology Department of Camosun College for lending their expert advice, both solicited and unsolicited, on many topics addressed herein. Faculty in other departments, particularly Stewart Langton and Mile Erlic, were also most generous with their time, and Jon Jacox and other students were kind enough to proof many of the question and solution sets. Thanks go also to the Dean of Technology, Baldev Pooni, and the Vice President of the College, Bob Priebe, for their support of this initiative.

Several individuals at Prentice Hall helped to make the text a reality. I wish to thank my editor, Charles Stewart, for his contagious excitement about the project when it first began, and also editorial representative Carmen Batsford for her excellent advice and good sense of humor throughout. A special thank you must go to Delia Uherec, assistant editor, who promptly and expertly answered my questions and did everything possible to remove obstacles from my path. Delia's well-timed words of encouragement in the final phases of manuscript preparation were most appreciated.

As this text was being prepared, a number of reviewers provided constructive comments and suggestions that have certainly improved the finished product: Kefu Xue, Wright State University; Anthony Oxtoby, Purdue University; Mark Hihghum, Bay De Noc Community College; Charles A. Cipari, Arizona State University; and Charles J. Eckard, ITT Technical Institute. Also, my father, Dr. J. Van de Vegte, a textbook author himself, painstakingly edited not one but two complete draft manuscripts. He was surely my harshest critic, but the book is many times better as a result of his input, and my mother assures me that "rewrite this section" is just his way of saying "I love you."

In my home, my children whisper the word "textbook" reverently, as if it were one of the great and mysterious wonders of the world. I am humbled by how generously my family has accommodated my obsession. While I was ensconced at my computer, my dear husband Don juggled job, children, and housework, and only occasionally reminisced aloud about how life used to be. Indeed, the only downside to finishing the text is that I will have to start doing dishes again. My children—Stevin, Jesika, and Eric—will be joyful when my time is theirs again, as will I.

Jessy and Eric say this book will make you smarter. I hope they're right. Of course, their other idea was to use it to start the campfire: There's a lesson here somewhere.

CONTENTS

1 CRASH COURSE IN DIGITAL SIGNAL PROCESSING   1

1.1   Signals and Systems   1

1.2   Analog-to-Digital and Digital-to-Analog Conversions   6

1.3   Digital Signals and Their Spectra   11

1.4   Digital Filtering   14

1.5   Speech, Music, Images, and More   16

Chapter Summary   24

Review Questions   24

2 ANALOG-TO-DIGITAL AND DIGITAL-TO-ANALOG CONVERSION   29

2.1   A Simple DSP System   29

2.2   Sampling   30

2.2.1   Nyquist Sampling Theory   30

2.2.2   The Frequency View of Sampling   35

2.3   Quantization   44

2.4   Analog-to-Digital Conversion   51

2.5   Digital-to-Analog Conversion   52

Chapter Summary   55

Review Questions   58

3 DIGITAL SIGNALS   63

3.1   Pictures of Digital Signals   63

3.2   Notation for Digital Signals   65

3.3   Digital Functions   69

    3.3.1   Impulse Functions   69

    3.3.2   Step Functions   72

    3.3.3   Power and Exponential Functions   79

    3.3.4   Sine and Cosine Functions   82

3.4   Composite Functions   89

3.5   Two-Dimensional Digital Signals   93

    Chapter Summary   94

    Review Questions   95

4 DIFFERENCE EQUATIONS AND FILTERING   99

4.1   Filtering Basics   99

4.2   Analog Filters versus Digital Filters   104

4.3   Linear, Time-Invariant, Causal Systems   107

4.4   Difference Equation Structure   108

4.5   Superposition   111

4.6   Difference Equation Diagrams   115

    4.6.1   Nonrecursive Difference Equations   115

    4.6.2   Recursive Difference Equations   119

4.7   The Impulse Response   123

4.8   The Step Response   131

    Chapter Summary   134

    Review Questions   135

5 CONVOLUTION AND FILTERING   143

5.1   Convolution Basics   143

5.2   Difference Equations and Convolution   155

5.3   Moving Average Filters   157

5.4   Filtering Digital Images   161

    Chapter Summary   165

    Review Questions   166

6 *z* TRANSFORMS   170

6.1   *z* Transform Basics   170

6.2   Transfer Functions   176

    6.2.1   Transfer Functions and Difference Equations   176

    6.2.2   Transfer Functions and Impulse Responses   179

    6.2.3   Finding Filter Outputs   180

    6.2.4   Cascade and Parallel Combinations of Transfer Functions   181

6.3   Back to the Time Domain   184

    6.3.1   Standard Form   184

    6.3.2   Simple Inverse *z* Transforms   186

    6.3.3   Inverse *z* Transforms by Long Division   188

    6.3.4   Inverse *z* Transforms by Partial Fraction Expansion   190

6.4   Transfer Functions and Stability   198

    6.4.1   Poles and Zeros   198

    6.4.2   Stability   201

    6.4.3   First Order Systems   204

    6.4.4   Second Order Systems   209

          Chapter Summary   221

          Review Questions   222

7 FOURIER TRANSFORMS AND FILTER SHAPE   230

7.1   Fourier Transform Basics   230

7.2   Frequency Responses and Other Forms   234

    7.2.1   Frequency Responses and Difference Equations   234

    7.2.2   Frequency Responses and Transfer Functions   235

    7.2.3   Frequency Responses and Impulse Responses   236

7.3   Frequency Response and Filter Shape   236

    7.3.1   Filter Effects on Sine Wave Inputs   236

    7.3.2   Magnitude Response and Phase Response   240

    7.3.3   Analog Frequency f and Digital Frequency Ω 256

    7.3.4   Filter Shape from Poles and Zeros   261

    7.3.5   First Order Filters   267

    7.3.6   Second Order Filters   269

          Chapter Summary   273

          Review Questions   274

8 DIGITAL SIGNAL SPECTRA 281

8.1 The Meaning of the Spectrum 281

8.2 Nonperiodic Digital Signals 283

8.3 Periodic Digital Signals 291

Chapter Summary 307

Review Questions 307

9 FINITE IMPULSE RESPONSE FILTERS 314

9.1 Finite Impulse Response Filter Basics 315

9.2 Moving Average Filters Revisited 316

9.3 Phase Distortion 320

9.4 Approximating an Ideal Low Pass Filter 326

9.5 Windows 332

9.5.1 Rectangular Window 332

9.5.2 Hanning Window 335

9.5.3 Hamming Window 336

9.5.4 Blackman Window 337

9.5.5 Kaiser Window 338

9.6 Low Pass FIR Filter Design 342

9.6.1 Design Guidelines 342

9.6.2 Steps for Low Pass FIR Filter Design 344

9.7 Band Pass and High Pass FIR Filters 353

9.8 Band Stop FIR Filters 364

9.9 Equiripple FIR Filter Design 371

9.10 Hazards of Practical FIR Filters 373

Chapter Summary 375

Review Questions 376

10 INFINITE IMPULSE RESPONSE FILTERS 382

10.1 Infinite Impulse Response Filter Basics 382

10.2 Low Pass Analog Filters 384

10.3 Bilinear Transformation 386

10.4 Butterworth Filter Design 395

10.5 Chebyshev Type I Filter Design 404

10.6 Impulse Invariance IIR Filter Design   413

10.7 "Best Fit" Filter Design   417

10.8 Band Pass, High Pass, and Band Stop IIR Filters   417

10.9 Hazards of Practical IIR Filters   426

Chapter Summary   429

Review Questions   430

11 DFT AND FFT PROCESSING   434

11.1 DFT Basics   434

11.2 Relationship to Fourier Transform   458

11.3 Relationship to Fourier Series   463

11.4 DFT Window Effects   464

11.5 Spectrograms   470

11.6 FFT Basics   476

11.7 2D DFT/FFT   480

Chapter Summary   480

Review Questions   481

12 HARDWARE FOR DSPs   487

12.1 Digital Signal Processor Basics   487

12.2 DSP Architectures   488

12.3 Fixed Point and Floating Point Number Formats   492

12.4 DSP Hardware Units   497

12.4.1 Multiplier/Accumulators   497

12.4.2 Shifters   499

12.4.3 Address Generators   500

12.5 DSP Assembly Language   501

12.6 How to Choose a DSP   503

12.6.1 Fixed Point or Floating Point   503

12.6.2 Data Width   503

12.6.3 Hardware and Software Features   504

12.6.4 Speed   504

12.6.5 Memory   505

12.6.6 Power   505

12.6.7 Supporting Hardware 505

12.6.8 Convenience 505

12.6.9 Cost 506

12.6.10 Application 506

12.7 DSP Manufacturers 507

12.7.1 Analog Devices 507

12.7.2 Texas Instruments 509

12.7.3 Motorola 511

Chapter Summary 512

Review Questions 513

13 PROGRAMMING DSPs 516

13.1 ADSP-2181 Processor 516

13.2 EZ-KIT Lite Development Board 518

13.3 Number Formats and Scaling 519

13.4 Registers 528

13.5 Assembly Language Instructions 529

13.5.1 Instructions for ADSP-21529 Family 529

13.5.2 Assembly Language Examples 531

13.6 Set-up and Initialization of the EZ-KIT Lite Board 537

13.7 Running Programs on the EZ-KIT Lite 548

13.8 DSP Applications 549

13.8.1 Sine Wave Generator 549

13.8.2 FIR Filter 553

13.8.3 IIR Filter 556

Chapter Summary 557

Review Questions 558

14 SIGNAL PROCESSING 561

14.1 Digital Audio 561

14.1.1 Digital Audio Basics 561

14.1.2 Oversampling and Decimation 562

14.1.3 Zero Insertion and Interpolation 565

14.1.4 Dithering and Companding 568

14.1.5 Audio Processing 572

14.2   Speech Recognition   575

14.3   Voice and Music Synthesis   583

14.4   Geophysical Processing   592

14.5   Encryption   601

    14.5.1   Encryption Basics   601

    14.5.2   Pretty Good Privacy   603

    14.5.3   Data Encryption Standard   604

    14.5.4   DSP for Security   612

14.6   Motor Control   612

    Chapter Summary   617

15 IMAGE PROCESSING   619

15.1   Image Processing Basics   619

15.2   Histograms and Histogram Equalization   621

15.3   Combining Images   628

15.4   Warping and Morphing   632

15.5   Filtering Images   637

15.6   Pattern Recognition   643

    15.6.1   Identifying Features   643

    15.6.2   Object Classification   650

15.7   Image Spectra   654

    15.7.1   Image Spectra Basics   654

    15.7.2   Tomography   660

15.8   Image Compression   667

    Chapter Summary   673

    Review Questions   674

16 WAVELETS   682

16.1   Wavelet Basics   682

16.2   Families of Wavelets   692

16.3   Coding a Signal   698

16.4   Multiresolution Analysis   708

16.5   The Discrete Wavelet Transform   712

     16.5.1   Discrete Wavelet Transform Basics   712

     16.5.2   A Frequency View of Wavelet Analysis   716

     16.5.3   A Frequency View of Wavelet Synthesis   717

     16.5.4   Calculating the Discrete Wavelet Transform   718

     16.5.5   2D DWT   731

16.6   Tiling the Time-Scale Plane   735

16.7   Wavelet Compression   738

     16.7.1   Wavelet Compression Basics   738

     16.7.2   The FBI Fingerprint Image Compression Standard   745

     Chapter Summary   747

     Review Questions   748

REFERENCES   748

APPENDICES   760

A   THE MATH YOU NEED   760

A.1   Functions   760

A.2   Degrees and Radians   761

A.3   Rational Functions   761

A.4   Lowest Terms   761

A.5   Lowest Common Multiple   762

A.6   Reciprocals   762

A.7   Logarithms   762

A.8   Power and Exponential Functions   763

A.9   Sinusoidal Functions   763

A.10   Decibels   765

A.11   Decimal, Binary, and Hexadecimal Number Systems   766

A.12   Area and Perimeter   767

A.13   Complex Numbers   767

A.14   Absolute Value   774

A.15   Quadratic Formula   774

A.16 Σ Sums   775

B SIGNAL-TO-NOISE RATIO   777

C DIRECT FORM 2 REALIZATION OF RECURSIVE FILTERS   779

D CONVOLUTION IN THE TIME DOMAIN AND MULTIPLICATION IN THE FREQUENCY DOMAIN   781

E SCALING FACTOR IN DISCRETE FOURIER SERIES AND DISCRETE FOURIER TRANSFORM   783

F INVERSE DISCRETE TIME FOURIER TRANSFORM   785

G IMPULSE RESPONSE OF IDEAL LOW PASS FILTER   786

H SAMPLING PROPERTY   787

I SPECTRUM OF DIGITAL COSINE SIGNAL   788

J SPECTRUM OF IMPULSE TRAIN   790

K SPECTRAL EFFECTS OF SAMPLING   792

L BUTTERWORTH RECURSIVE FILTER ORDER   794

M CHEBYSHEV TYPE I RECURSIVE FILTER ORDER   796

N WAVELET RESULTS   798

1

CRASH COURSE IN DIGITAL SIGNAL PROCESSING

This chapter skates across the surface of the topics to be studied in greater depth in the remainder of the text. The chapter:

> ➤ distinguishes between analog and digital signals
> ➤ presents the basic steps in analog-to-digital conversion
> ➤ presents the basic steps in digital-to-analog conversion
> ➤ introduces the relationship between a signal and its spectrum
> ➤ explains the basic concepts of filtering
> ➤ discusses applications of digital signal processing

1.1 SIGNALS AND SYSTEMS

Computers operate on digital signals. As computers proliferate, the need for efficient handling of digital signals increases. Furthermore, high speed processing capabilities of modern computers attract applications that use digital signals, which further drives the development of digital signal techniques. Digital signal processing, or DSP, is essential to an enormous variety of old and new applications, a few of which are listed in Figure 1.1.

At the heart of DSP lie the signals to be processed. Signals are variations that carry information from one place to another. The outside world, for example, provides variations in pressure or light intensity that humans can perceive. Changes in pressure at the eardrum are heard as sounds. Variations in light intensity at the retina are seen as images.

• Touch-Tone™ telephones	• Speech synthesis
• Edge detection in images	• Echo cancellation
• Digital signal and image filtering	• Cochlear implants
• Seismic analysis	• Antilock brakes
• Text recognition	• Signal and image compression
• Speech recognition	• Noise reduction
• Magnetic resonance image (MRI) scans	• Companding
• Music synthesis	• High definition television (HDTV)
• Bar code readers	• Digital audio
• Sonar processing	• Encryption
• Satellite image analysis	• Motor control
• Digital mapping	• Remote medical monitoring
• Cellular telephones	• Smart appliances
• Digital cameras	• Home security
• Detection of narcotics and explosives	• High speed modems

FIGURE 1.1
Applications of DSP.

These signals are **analog signals**. They can take any value from a continuum of values and are defined at every instant of time. Sounds are one-dimensional analog signals: The size, or amplitude, of pressure variations changes with time. As another example, the voltage available from an electrical outlet in North America varies smoothly from its minimum to its maximum and back to its minimum again, 60 times per second. Figure 1.2 supplies a few examples of one-dimensional signals. Images are two-dimensional analog signals: Brightness varies along both the horizontal and vertical axes of the image. Figure 1.3 shows a sample of a black and white image, and Figure 1.4 shows four frames from a high speed digital video sequence.

In order to process signals, they must first be captured. Sound signals, for example, are acquired using a microphone, which converts acoustic signals into electrical signals. Images, on the other hand, are captured using an analog or digital camera. In analog cameras, light signals control chemical reactions on a photographic film. In digital cameras, light signals from objects in view create charge packets that are converted into electrical signals on a two-dimensional grid. These electrical signals are, like the light signals that produced them, analog in nature. Because they carry information at an infinite number of levels and points in time, analog signals are not suited to computer processing. They must be sampled and converted into digital form before they can be processed. **Digital signals** are perfectly suited for computer processing because they are defined at only a finite number of levels and points in time.

Both analog and digital signals are present in most digital processing systems. Analog signals at the input to the system are converted to digital form for processing. After processing, signals in digital form are converted back to analog form for output. It is in the pro-

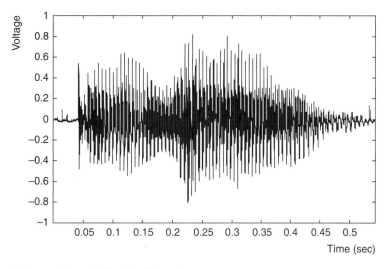

away.wav

(a) Speech Sample: The Word "away"

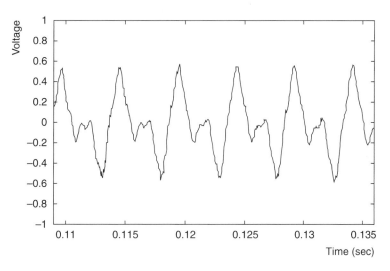

ooo.wav

(b) Speech Sample: The Vowel "ooo"

FIGURE 1.2
Examples of signals.

cessing stage, where only digital signals exist, that the flexibility and speed of DSP are realized.

Each of the applications listed in Figure 1.1 is a **system**. A system analyzes, combines, modifies, records, or plays back signals. An image compression system, for example, recodes digital images so they occupy less memory space and therefore take less time to transmit through the Internet. A speech recognition system is designed to understand

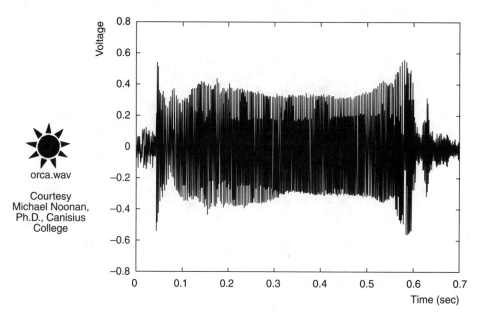

orca.wav

Courtesy
Michael Noonan,
Ph.D., Canisius
College

(c) Killer Whale Sounds

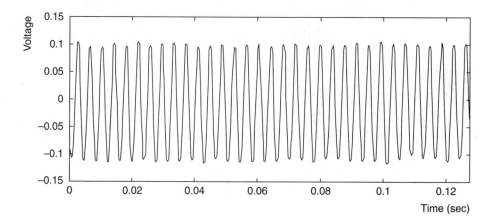

tf256.wav

(d) 256 Hz Tuning Fork Signal

FIGURE 1.2
Continued

human speech automatically. A digital filter, to take another example, allows certain signal frequencies to pass through the system while blocking others. For digital signals, the possibilities for processing are limitless. Any operation that is possible for numbers is also possible for digital signals, and with specialized DSP hardware and software, the operations can occur at lightning speeds.

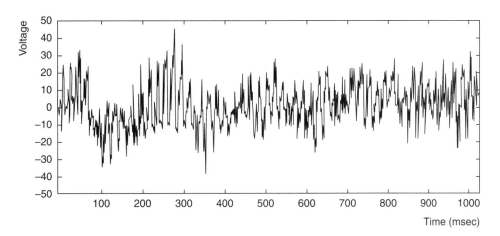

(e) Electroencephalogram (EEG)

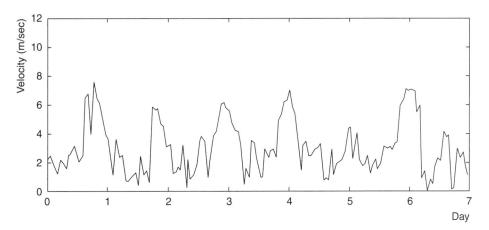

(f) Wind Velocities

FIGURE 1.2
Continued

Digital systems have many advantages over analog systems. Analog systems are circuits constructed from hardware components whose properties can vary quite widely within a manufacturer's tolerance. The properties can also change with temperature, altering the circuit's behavior. In contrast, digital systems behave in a predictable, repeatable way. Their behavior is almost completely unaffected by these problems, because it is mostly determined by software. For the same reason, digital systems are much less affected by noise than analog systems. In addition, digital systems are smaller and consume less power than their analog counterparts. Perhaps the greatest advantage of digital technology is the flexibility that allows a design to be modified by simply changing a few lines in a program. Most analog redesign involves building new circuits from scratch.

FIGURE 1.3
Black and white image.
© Snap-shot.com

nozzle.avi

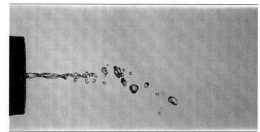
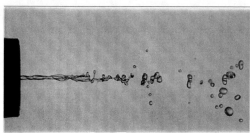
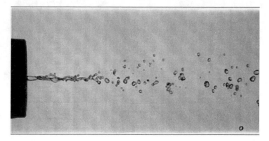

FIGURE 1.4
Four frames from high speed video sequence. © Vision Research, Inc., Wayne, N.J., USA

1.2 ANALOG-TO-DIGITAL AND DIGITAL-TO-ANALOG CONVERSIONS

Sights and sounds from the world around us are analog signals. In order to process these signals, sensors tuned to the signals' characteristics must be used. Sensors exist for a vast array of signals. Microphones are common sound sensors. Changes in light can be captured by semiconductor devices like photodiodes, phototransistors, or CCD (charge-coupled de-

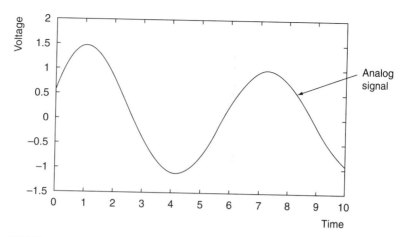

FIGURE 1.5
Analog voltage signal.

vice) chips, whose ability to carry current changes with incident light. Temperature can be measured by thermistors, semiconductor devices whose resistance changes with temperature, or by thermocouples, consisting of two dissimilar metals that react differently to temperature and develop a voltage difference. Accelerometers measure acceleration. Other sensors include strain sensors, pressure sensors, and flow sensors.

The most common output for all of these sensors is an electrical signal, voltage or current, proportional to the signal being measured. It is this analog electrical signal that must be converted into digital form. A voltage signal is shown in Figure 1.5. The values that the signal may take between its minimum and maximum values are not restricted, and the signal has values at every instant of time.

Analog-to-digital (A/D) conversion occurs in two steps. The first step is sampling. Sampling instants normally occur at regular intervals called sampling periods. At each sampling point, the analog signal is sampled, and the value of the signal is held steady until the next sampling point. This process is called **sample and hold**. Figure 1.6 shows the sample-and-hold signal for the analog signal in Figure 1.5. The vertical dashed lines mark the sampling points. Sampling must be fast enough to capture the most rapid changes in the signal being sampled. If sampling is too slow, important signal characteristics can be lost, a problem called **aliasing**. Note that the act of sampling takes a finite amount of time, called the **acquisition time** of the sampler. In Figure 1.6 the length of the acquisition time is exaggerated, so the sample-and-hold signal briefly appears to follow the signal being sampled. In subsequent illustrations, the acquisition time is assumed to be negligible.

The second step in the conversion between an analog signal and a digital signal, which can begin at the completion of sample acquisition, is to quantize and digitize the analog values. The hold interval normally gives ample time for this. As soon as possible after each sampling instant, the converter selects the **quantization level** that approximates the sample-and-hold value as closely as possible and then assigns a binary **digital code** that identifies the quantization level. This completes the analog-to-digital conversion process. Figure 1.7 supplies the digital signal that follows from Figure 1.6. The **digital signal** shows

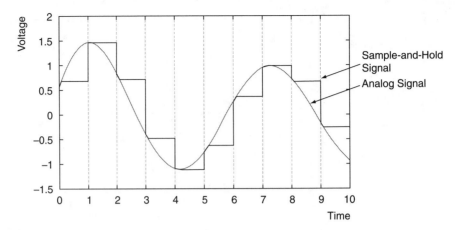

FIGURE 1.6
Sample-and-hold signal.

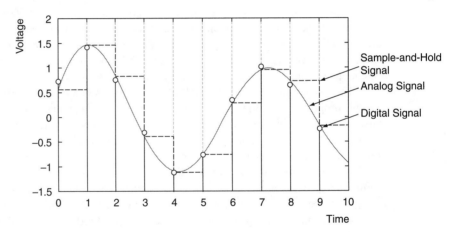

FIGURE 1.7
Quantization and digitization.

the quantization levels (rather than the digital codes) at each sample point. It is represented by vertical lines at each sampling point, topped with a small circle. The digital signal has values only at the discrete points in time marked by the sampling instants.

Notice that the digital signal values in Figure 1.7 do not, in general, coincide with the analog signal values at the sampling points. Perfect agreement is impossible because of the way digital numbers are stored in a computer. The digital values computers work with are stored in memory locations in binary form. Binary values are expressed exclusively in ones and zeros called **bits**, or binary digits. Memory locations are hardware constructions in a computer that store groups of bits. A single location might hold a group of 8, 16, or 32 bits. The number of bits per memory location limits the accuracy that is possible when an ana-

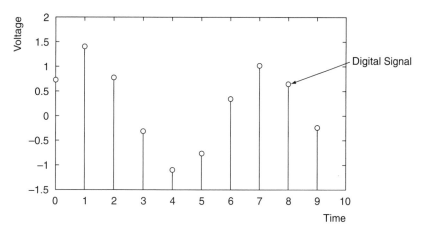

FIGURE 1.8
Digital signal.

log value is converted to digital. The conversion process is called **quantization**. Suppose, for example, analog voltages between -2.5 V and $+1.5$ V are converted to 2-bit digital values. In a 2-bit system, only four digital values are possible: 00, 01, 10, and 11. Together, these codes must be capable of representing any possible input voltage. For example, the voltages between -0.5 V and $+0.5$ V might map to the code 00, the voltages from $+0.5$ V to $+1.5$ V to the code 01, and so on. Since many possible voltages map to each code, most A/D conversions cause a **quantization error**. Four possible digital codes mean that only four quantization levels are defined. To minimize quantization errors, these levels are normally assigned to the center of the analog range corresponding to a single digital code. Any analog voltage between -0.5 V and $+0.5$ V, for example, converts to the digital code 00. The quantization level for this code would be in the center of the analog interval, at 0 V, meaning that errors with magnitudes up to 0.5 V occur in the conversion process. The larger the number of bits used by a computer, the smaller these errors will be, but they cannot be avoided completely.

The digital signal that results from the A/D conversion process has two important characteristics. First, the number of permitted signal values is limited by the number of bits used by the computer. Second, the digital signal has values only at the sampling instants: It is not defined between sampling points. After processing, a digital signal is normally converted back into analog form. Digital signals are not, for instance, suitable for driving speakers. To re-create a sound correctly, an analog signal is required. The first step in the **digital-to-analog (D/A) conversion** process is to convert each digital code into an analog voltage level that is proportional to the size of the digital number. Figure 1.8 shows a digital signal, where the height at each sample corresponds to the analog voltage obtained from the digital code. This analog voltage level is held steady for the duration of a sampling period, called **zero order hold**. Figure 1.9 shows the zero order hold (ZOH) signal superimposed on the digital signal. The ZOH signal is analog, but its staircase shape is not consistent with the analog signal that was originally sampled. The second step in D/A conversion

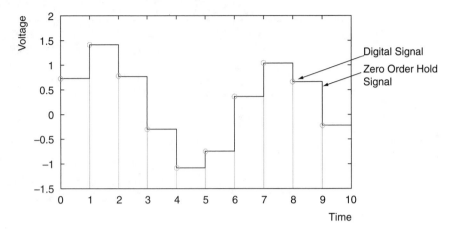

FIGURE 1.9
Zero order hold signal.

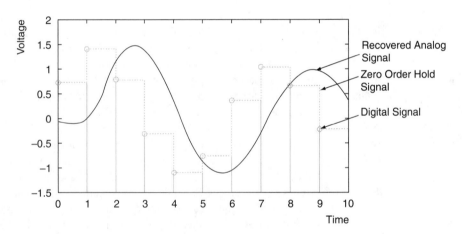

FIGURE 1.10
Recovered analog signal after smoothing.

is therefore to smooth the zero order hold signal. This smoothing step introduces a shift in time as a side effect. Figure 1.10 shows the final analog signal.

Figure 1.11 displays all steps in the A/D and the D/A conversions. Processing can occur once the signal is in digital form. If no digital signal processing occurs, the digital signal resulting from A/D conversion will be identical with the digital signal driving the D/A process. In this case, if sampling is fast enough and the number of bits used to digitize the analog signal is large enough, the input and output analog signals will be extremely close. Since the digital signal in Figure 1.10 is identical with the one in Figure 1.7, the analog signal in Figure 1.10 should be close to the one in Figure 1.7, apart from a time shift.

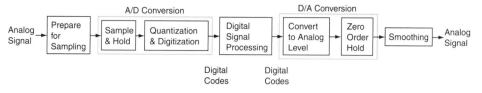

FIGURE 1.11
A/D and D/A conversion.

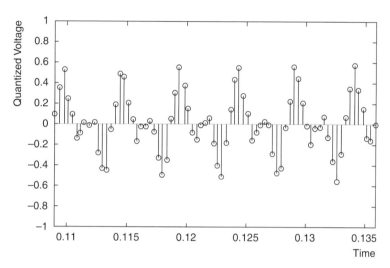

FIGURE 1.12
Digital speech signal.

1.3   DIGITAL SIGNALS AND THEIR SPECTRA

Any of the signals of Figure 1.2 can be converted into digital form. Figure 1.12, for example, is a digital version of the speech signal of Figure 1.2(b). The voltage of the speech signal is recorded and quantized at every point in time at which a sample occurs. Figure 1.12 is a **time domain** presentation of information about the digital signal. This means that the signal variations are displayed against time.

In DSP, it is frequently important to know what frequencies are present in a signal, in addition to its time-based behavior. It is common knowledge, for example, that female voices are generally higher pitched than male voices, but it is the precise frequency elements present in a speech sound that are important for automatic speech recognition. A **spectrum**[1] is a presentation of the frequency elements that are present in a signal. The importance of a frequency element in the signal is shown by the magnitude of the spectrum at

[1]   As will be discussed in Chapter 8, a spectrum consists of a magnitude spectrum and a phase spectrum. In this chapter and the next, "spectrum" refers to magnitude spectrum.

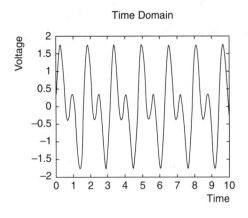

(a) Slowly changing signal

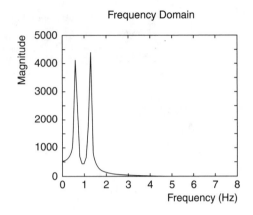

Spectrum of signal

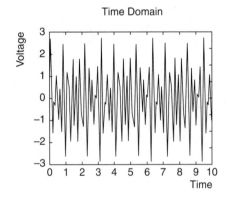

(b) Quickly changing signal

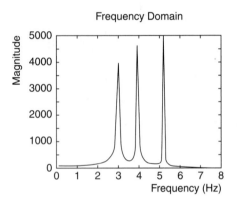

Spectrum of signal

FIGURE 1.13
Signals and spectra.

that frequency. The spectrum is plotted as magnitude against frequency and, as such, is a **frequency domain** presentation of information about the signal. It is most often calculated using an FFT (fast Fourier transform). Low frequency signals are constant or change slowly over time. They appear at the left end of the spectrum. High frequency signals, on the other hand, change rapidly with time. They appear at the right end of the spectrum.

Figure 1.13 shows both the time domain and the frequency domain information for two different signals. The spectrum for signal (a) can provide the useful information that the signal contains two distinct frequencies only, as marked by the two sharp spectral spikes at 0.64 and 1.27 Hz. The spikes in the spectrum for signal (b) indicate that this signal contains three separate frequencies: 2.86, 3.82, and 5.09 Hz. The frequencies present in the quickly changing signal are much higher than those in the slowly changing signal, as expected. For each signal, the spectrum gives clear information about the signal's characteristics that was not available in the time domain presentation.

C.wav

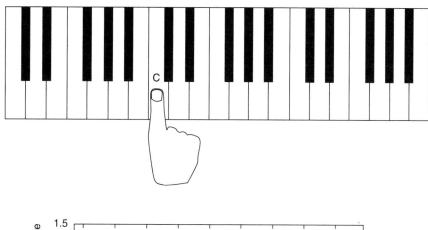

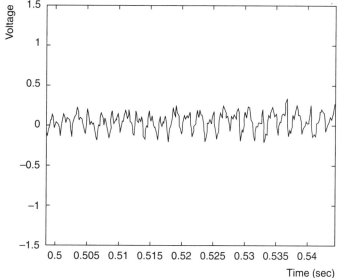

FIGURE 1.14
A single piano note (middle C) and its time trace.

The time trace for a portion of a single strike of a piano key is shown in Figure 1.14. Its spectrum is shown in Figure 1.15. Each piano note consists of many **harmonic frequencies**, all multiples of a **fundamental frequency**, which means that the full spectrum contains many spikes. In the portion of the spectrum shown, a single spike appears at 262 Hz, which corresponds to the fundamental frequency of the middle C note. Figure 1.16 shows a time trace for part of a CEG chord, and Figure 1.17 shows a portion of its spectrum. This spectrum features three spikes, one for each note present in the signal. The spike from middle C is still located at 262 Hz. The spikes due to E and G are located at higher frequencies, 330 Hz and 392 Hz, since these keys are located further to the right on the piano keyboard.

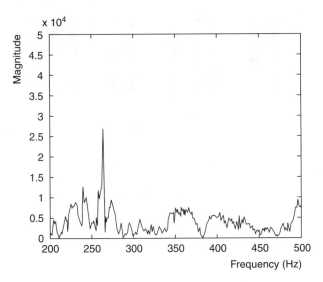

FIGURE 1.15
Spectrum of middle C (200–500 Hz).

The spectra of the signals in Figure 1.2 are shown in Figure 1.18. From each, some knowledge about the signal can be gleaned. The tuning fork spectrum, for example, confirms the fork's 256 Hz resonant frequency, while the spikes in the EEG spectrum provide indications of brain activity. The largest peak in the spectrum for wind speeds suggests a cycle that repeats once a day.

1.4 DIGITAL FILTERING

Digital filters provide a convenient means to change the nature of a signal. The most common filters change the frequency characteristics of a signal, letting some frequencies in the signal pass while blocking others. A low pass filter, for example, lets low frequencies through while blocking high frequencies. A high pass filter does just the opposite. A band pass filter allows a band of frequencies to pass, while a band stop filter allows all frequencies outside a band to pass.

If a low pass filter were applied to the sounds of a singing choir, the filter would tend to extract the bass voices and block the soprano voices. A high pass filter would pass the soprano voices while blocking, or attenuating, the bass voices. Figure 1.19 shows the normal range for bass, tenor, alto, and soprano voices. The notes included in the bass range correspond to frequencies between 65 and 330 Hz. The shape of the low pass filter that would best extract this range is shown in Figure 1.20, though elements of the other voices would be captured by this filter as well. The corner frequency of the filter, 330 Hz, is referred to as its **cut-off frequency**. Figure 1.21 shows a high pass filter that would best extract the soprano range, between 262 and 1319 Hz. This filter has a cut-off frequency of 262 Hz. The band pass filter shown in Figure 1.22, with cut-off frequencies 165 and 880 Hz, might be best suited to extracting the alto range.

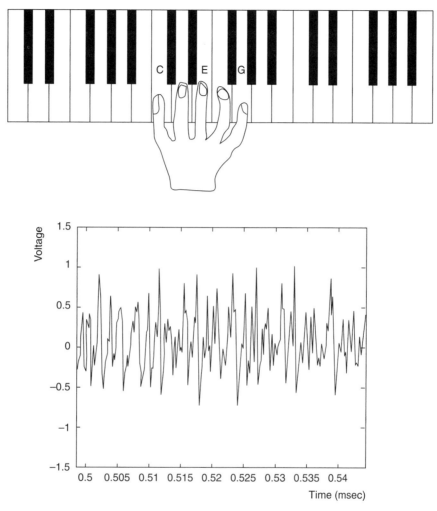

FIGURE 1.16
A piano chord (CEG) and its time trace.

As another example, the piano chord signal of Figure 1.16 can be filtered by several low pass filters with different results each time. Figure 1.23 shows the spectrum of the CEG chord from 250 to 2500 Hz. The three fundamental frequencies observed in Figure 1.17 are marked by C, E, and G. The locations of the other harmonics for each note are shown as well, marked with lowercase letters. Figure 1.24 shows the effect of low pass filtering the piano chord. When a low pass filter with a cut-off of 456 Hz is applied to the CEG chord, none of the fundamental components are removed but the rest of the harmonics are eradicated. This removal has a dramatic effect on the sound of the signal. The bright sound of the chord is lost, leaving a muffled remainder. Moving the low pass cut-off down to 361 Hz severely attenuates the G portion of the chord, and moving the cut-off down to 296 Hz removes the E and G components, leaving only a single fundamental for middle C.

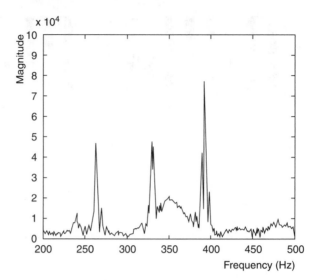

FIGURE 1.17
Spectrum of CEG chord (200–500 Hz).

Low pass and high pass filters for images exist as well. Low frequency parts of images are those where the colors change slowly, while the high frequency parts correspond to edges and other sharp changes in color. Low pass filters tend to blur images as the high frequency elements are removed, while high pass filters can be used to sharpen edges and locate the boundaries of an object in a digital image.

Digital filters are nothing more than equations defined by a list of filter coefficients. These equations form part of a digital filter program that accepts raw data as input and produces filtered data as output. The beauty of such programs is that filter redesign does not require a hardware change: Changing filter behavior is accomplished by simply changing the list of filter coefficients. Digital filter programs can be implemented on any processor, but they are implemented most efficiently on hardware that is specially designed to perform filtering and other DSP operations at enormous speeds. This DSP hardware enables extremely complex tasks to be performed in negligible amounts of time, which is the reason it is so often found at the core of computation-intensive applications.

1.5 SPEECH, MUSIC, IMAGES, AND MORE

DSP has many exciting applications in a wide range of fields, and the number of applications grows every year. What the applications have in common is that they are defined primarily by software, that is, by a list of program instructions that execute on a particular hardware platform. In other words, the very same DSP hardware may be the basis for a wide variety of applications. Among the most accessible are those applications that operate on familiar one-dimensional signals like speech or music, or two-dimensional signals like images.

Large vocabularies of continuous speech can be recognized using information from digital speech signals. Most often, speech recognition methods are based on a spectral

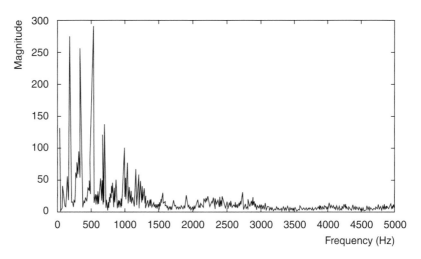

(a) Spectrum of Speech Sounds: The Word "away"

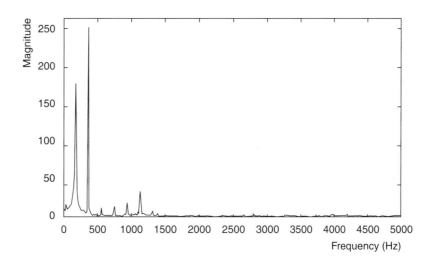

(b) Spectrum of Speech Sounds: The Vowel "ooo"

FIGURE 1.18

Examples of spectra.

analysis of speech sounds. Though this analysis can be time-consuming, fast DSP hardware makes real-time recognition a reality. The larger the population of users of a speech recognizer, the more variable the speech and the more difficult the task. Greater vocabulary size also makes discrimination between words more challenging. The AT&T telephone company has, for several years, used a word-spotting technique for its directory assistance system. Word spotting helps operators respond more quickly by searching for numbers and simple words embedded in continuous speech and then collating basic information about a

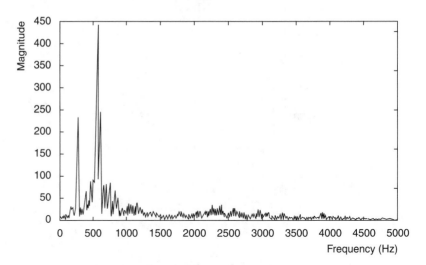

(c) Spectrum of Killer Whale Sounds

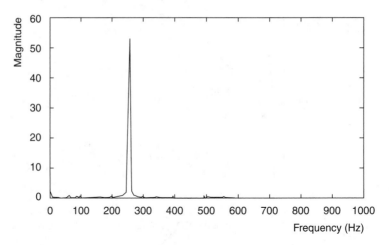

(d) Spectrum of 256 Hz Tuning Fork Signal

FIGURE 1.18
Continued

request. Automatic voice recognition for radio control makes hands-free radio use possible in jet cockpits, and similar technology allows hands-free use of cellular phones in automobiles. In complementary technology, advanced speech synthesis programs generate natural-sounding speech for automobile warnings ("The left door is ajar"), telephone information messages ("The number 555-1234 has been changed; the new number is 555-4321"), and telephone banking ("Your balance is $18.43"). Indeed, advanced models of human speech developed for speech synthesizers have been used to produce DSP-based hearing aids that are tuned to compensate precisely for an individual's specific hearing deficiencies.

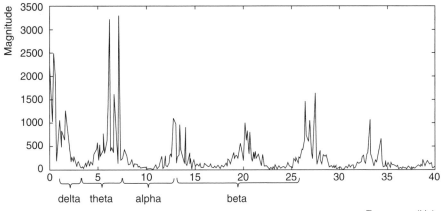

(e) Spectrum of Electroencephalogram (EEG)

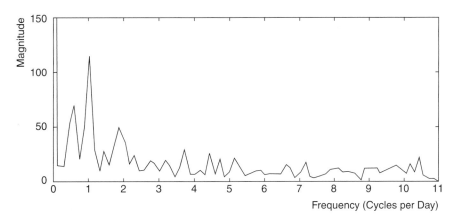

(f) Spectrum of Wind Speeds

FIGURE 1.18
Continued

DSP also makes contributions in the arena of music and other sounds. Old recordings of music and audio tracks can be cleaned up by removing clicking and buzzing sounds in the background. Also, the sounds of musical instruments can be closely copied for electronic synthesizers. Combinations of actual recordings and mathematical models produce high quality synthesis for many instruments, including piano, violin, and flute. In some unusual DSP synthesis research, the voice of a castrato was re-created for the soundtrack of a film (Depalle, Garcia, and Rodet, 1994). DSP can be applied to animal sounds as well. Whale songs have been analyzed to study whale communications patterns (Noonan, Chalupka, Viksjo, and Perri, 1999) and also to look for family signatures (Ford, 1992).

Fantastic effects are possible when DSP is applied to imagery. Images can be combined, as when Forrest Gump was digitally added to a film of President Nixon. Background speckling

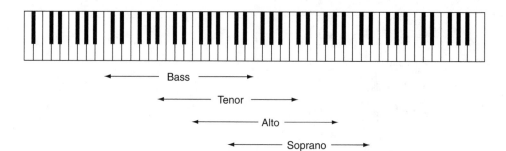

FIGURE 1.19
Bass, tenor, alto, and soprano singing ranges.

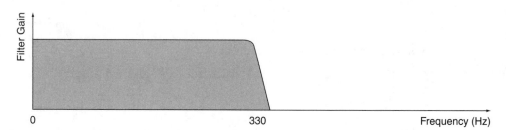

FIGURE 1.20
Low pass filter to extract bass voices.

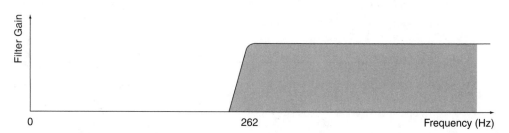

FIGURE 1.21
High pass filter to extract soprano voices.

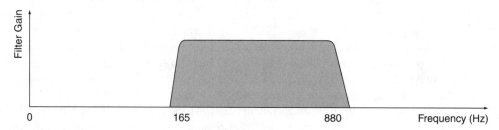

FIGURE 1.22
Band pass filter to extract alto voices.

CEGnl.wav

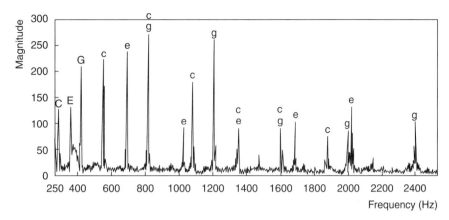

FIGURE 1.23
Spectrum for CEG chord (250–2500 Hz).

can be removed by averaging several pictures of the same scene. Movement can be detected by subtracting one image from another of the same scene, highlighting the places where the two images differ. Color maps can be altered, and contrast improved. Morphing can allow one image to change smoothly into another, a technique used by Star Trek's changeling, Odo. When several digital views of an object are available, they can be combined mathematically to create a three-dimensional view of the object, a technology that is exploited in medical images like MRIs (magnetic resonance images) and CT (computed tomography) scans. These images are normally assessed and interpreted manually by doctors. Visual inspection and robot vision systems, on the other hand, rely on automatic identification of objects. This is accomplished through detection of object edges coupled with sophisticated pattern recognition schemes.

In communications, DSP plays several important roles, particularly in support of cellular phones, digital modems, and audio and video transmission technologies. Digital cellular telephones use DSP for two main tasks: to code speech as compactly as possible while maintaining intelligibility, and to transmit coded speech reliably to a receiver across a wireless radio frequency link. Voice coding algorithms, called vocoders, match voice data to a model of speech that requires only a few parameters. These parameters are transmitted to the receiver instead of the speech signal, where the parameters are applied to the same model again to recover a reasonable copy of the original voice data. Use of vocoders decreases dramatically the size of the transmissions needed to support an individual call, providing a commensurate increase in the efficiency of the cellular network.

Since radio frequency links cannot support the transmission of digital data directly, they must employ a modulator. In one common scheme, quadrature amplitude modulation (QAM), several bits of digital data at a time determine the amplitude and phase of a sine wave that is transmitted in their place. At the receiver, the amplitude and phase of the received signal are detected by a demodulator before the original digital data may be identified. The two main obstacles to clean transmission of a call are signal fading and noise. Several strategies are used to mitigate their effects. For example, error codes that can be checked by the receiver are appended to some portions of the transmitted data. Also, parameters from different blocks of speech are interleaved with one another to reduce the

filtCEG.wav

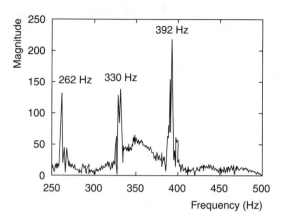

(a) Extracting CEG Fundamentals (Cut-off Frequency 456 Hz)

filtCE.wav

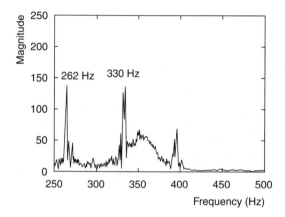

(b) Extracting CE Fundamentals (Cut-off Frequency 361 Hz)

filtC.wav

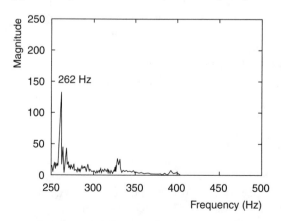

(c) Extracting C Fundamental (Cut-off Frequency 296 Hz)

FIGURE 1.24
Low pass filtering a piano chord.

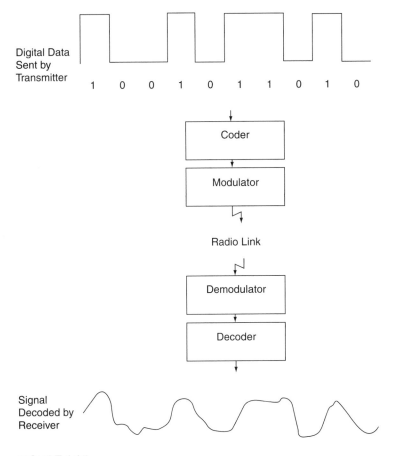

FIGURE 1.25
Wireless communication system.

chance that the integrity of an entire block will be destroyed by a single burst of noise. As a result of noise and fading, the received signal may be quite unlike the transmitted sequence, as Figure 1.25 suggests. To assist in signal recovery, a known pattern of bits is sometimes sent from the transmitter to the receiver to allow the receiver to estimate the characteristics of the link and therefore have the greatest chance of identifying the bits in the received signal correctly.

Modem technology permits computers to communicate using ordinary telephone lines or television cable. Many of the algorithms that permit data to be carried at very high rates on these media rely on complex calculations that must be completed quickly to keep up with incoming data. Digital data must first be coded for wired transmission in a manner that gives high throughput and reliable detection. Telephone line impairments like noise, fading, and crosstalk can, however, introduce errors. To save retransmissions, error codes can be added to the transmitted data that permit the receiver not only to detect but also to correct errors. As for cellular phones, digital modems sometimes exchange known signals to monitor the quality of the communications channel. Some modems can even adjust their transmission rates in accordance with the quality of the line, to help keep errors to a minimum. In addition to coding and error control, modems can apply compression to reduce the amount of data to be transmitted. In one compression scheme, for example, short codes are substituted for the most common bit patterns, codes that are replaced by the original patterns at the receiver. For secure transmissions, some modems provide encryption capability as well.

Whether transmission is wired or wireless, the growth of the Internet has occasioned an enormous increase in the movement of audio and video files. To save space and also transmission time, compression strategies have developed far beyond simple substitutional codes. Video compression schemes like MPEG (Moving Picture Experts Group), for example, rely on the fact that neighboring frames of a video sequence have very similar content. Instead of coding all pixels in all frames, the differences between frames are coded. MPEG also incorporates an audio compression standard, MP3, which exploits the characteristics of the human auditory system. For example, sounds that are masked by other, louder sounds are not coded at all, nor are quiet sounds that lie below the threshold of hearing. The compressed audio can still produce CD-quality sound.

The results of DSP can be admired and enjoyed without any knowledge of the underlying mathematics. Active analysis, filtering, or transformation of signals does, however, require some understanding of the math that drives DSP. The chapters that follow are designed to give competence in all the fundamental areas of digital signal processing, relying where necessary on noncalculus mathematics of real and complex numbers.

CHAPTER SUMMARY

1. An analog signal is defined at every point in time and may take any amplitude. A digital signal is defined only at sampling instants and may take only a finite number of amplitudes.
2. An analog signal is converted to a digital signal through sampling and quantization. A digital signal is converted to an analog signal by converting digital codes to analog levels and smoothing.
3. A digital signal is said to lie in the time domain. Its spectrum, which describes its frequency content, lies in the frequency domain.
4. Filtering modifies the spectrum of a signal by eliminating one or more frequency elements from it.
5. Digital signal processing has many applications, including speech recognition, music and voice synthesis, image processing, cellular phones, modems, and audio and video compression.

REVIEW QUESTIONS

1.1 Describe the differences between analog and digital signals.
1.2 Identify three reasons why digital systems are better than analog systems.
1.3 Name the quantity each of the following sensors measures:
   a. Microphone
   b. Photodiode
   c. Thermistor

  d. Phototransistor
  e. Accelerometer
  f. CCD chip
  g. Thermocouple

1.4 An analog voltage is sampled by a sample-and-hold circuit every 5 msec. The sampling points are marked in Figure 1.26. Draw the output from the circuit assuming 250 μ sec are needed to acquire each sample. (A dashed line showing the input signal is included for convenience.)

1.5 If an analog signal is sampled at a rate of 11.025 kHz, and the acquisition time of the sampler is 10 μsec, how much time is available for quantization and digitization?

1.6 How can the effects of quantization errors be minimized when an analog signal is converted to a digital signal?

1.7 An analog signal is converted to digital and then back to analog again, without intermediate DSP. In what ways will the analog signal at the output differ from the one at the input?

1.8 In Figure 1.27, a digital signal is applied to a zero order hold circuit. Draw the output from the circuit.

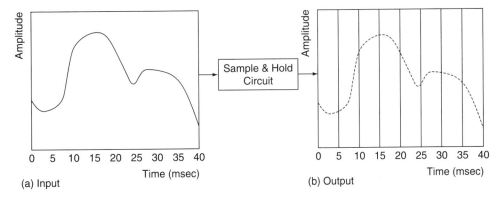

FIGURE 1.26
Sample and hold for Question 1.4.

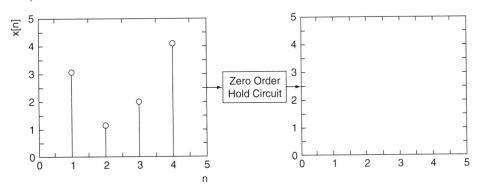

FIGURE 1.27
Applying zero order hold to signal for Question 1.8.

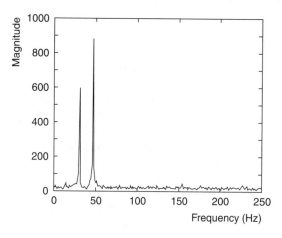

(a) Spectrum A

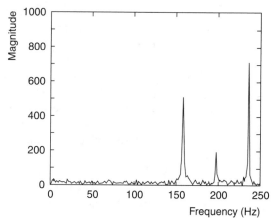

(b) Spectrum B

FIGURE 1.28
Spectra for Question 1.9.

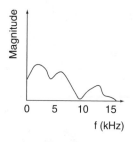

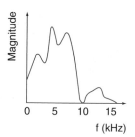

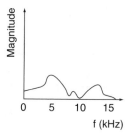

FIGURE 1.29
Magnitude spectra for Question 1.10.

1.9 Which of the spectra in Figure 1.28 belongs to a more rapidly changing signal?

1.10 An old recording of Edith Piaf is degraded by a high frequency hiss. Three 5-second magnitude spectra are collected, as shown in Figure 1.29. The high frequency parts of the spectra that remain constant are assumed to result from noise.

  a. Draw the shape of the filter that will recover the music but block the noise.

  b. What type of filter is this?

  c. What is the approximate cut-off frequency of the filter?

1.11 A low frequency motor control signal is disrupted by a 60 Hz buzz.

  a. What type of filter might best clean up the signal?

  b. Draw the filter's shape.

1.12 Each time a Touch-Tone™ telephone key is pressed, an audible two-tone signal is produced. The lower tone identifies the row the key is in, and the higher tone identifies the column. The spectra for the Touch-Tone™ telephone digits "1," "5," and "9" are shown in Figure 1.30(i), (ii), and (iii).

    a. Describe a filter capable of capturing all row frequencies. Include the filter's type (low pass, high pass, or band pass) and cut-off frequency.

    b. Describe a filter shape for a filter capable of capturing all column frequencies. Include the filter's type (low pass, high pass, or band pass) and cut-off frequency.

1.13 A choir is made up of bass, alto, and soprano voices. With only the low pass, high pass, and band pass filters shown in Figures 1.20, 1.21, and 1.22, suggest how to extract:

    a. The elements both the bass and alto voices are capable of producing.

    b. The elements either the alto or soprano voices can produce.

FIGURE 1.30

Three Touch-Tone™ digits.

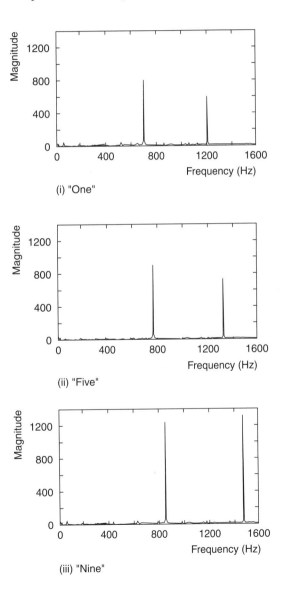

1.14 Devise a filter shape that will extract the signal produced by middle C from the spectrum in Figure 1.23. Include all harmonics of middle C, but remove all evidence of the E and G notes.

1.15 James Bond brings a photograph of a top secret list of operatives back to MI6. The photograph was taken in KGB headquarters using a tiny camera lodged in the second button of Bond's tuxedo. Will Q be more likely to apply a low pass or a high pass filter to the photograph?

2

ANALOG-TO-DIGITAL AND DIGITAL-TO-ANALOG CONVERSION

Most real-world signals are analog. The purpose of this chapter is to explore the steps needed to convert an analog signal into a digital signal that is more suitable for computer processing. The steps required to transform a processed digital signal back into an analog signal are also discussed. The chapter:

➤ presents the elements in a complete DSP system
➤ introduces the important element of sampling in the conversion of an analog signal to a digital signal
➤ defines the minimum sampling rate for a signal
➤ discusses the effects of sampling too slowly
➤ introduces the benefits of sampling more quickly than necessary
➤ explains the need for quantization in the conversion of an analog signal to a digital signal
➤ calculates errors caused by quantization
➤ illustrates the steps in an analog-to-digital conversion
➤ illustrates the steps in a digital-to-analog conversion

2.1 A SIMPLE DSP SYSTEM

The heart of digital signal processing is the manipulation of digital signals. Countable signals, like the number of days of rain in a year, can be represented as digital signals directly.

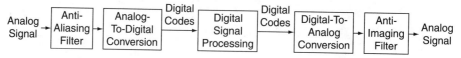

FIGURE 2.1

Typical digital signal processing system.

All the signals we perceive through our senses, though, whether speech or music or images, are analog signals. These analog signals must be converted into digital signals before processing can begin. Unfortunately the conversion process is never perfect, and the digital signals are not perfect representations of their analog counterparts. The differences between them are side effects of the conversion process, described in this chapter.

Once a digital signal that closely approximates the analog signal has been found, digital signal processing can proceed. For example, high frequency noise in speech can be removed, the bass frequencies in a piece of music can be emphasized, or the edges in an image can be highlighted. Because digital signals cannot exist in an analog world, the processed digital signals must be converted back into analog signals at the end of the process. Figure 2.1 shows the major elements in a digital signal processing system. The presence of the first and last blocks in this diagram is explained in the following sections.

2.2 SAMPLING

2.2.1 Nyquist Sampling Theory

Analog signals are defined at every point in time. To process analog signals, therefore, requires processing an infinite number of pieces of information. Signal processing relies heavily on the use of computers, which are not able to accommodate infinite data sets. Sampling solves this problem by reducing the number of points to be handled to a manageable level. Figure 2.2 shows an example of an analog signal. The signal is given the designation $x(t)$, where t is time, to indicate that it has a value for each instant of time. Figure 2.3 shows the sample-and-hold signal that corresponds to the analog signal, where single samples are taken from the analog signal at regular sampling intervals. In contrast with Figure 1.6, Figure 2.3 assumes that the time taken to acquire each sample is negligible. Assuming also that quantization errors are negligible, the digital signal obtained from the sample-and-hold signal is as shown in Figure 2.4. The digital signal is designated as $x[n]$, where n is the sampling instant, to show that it has a value only at each sampling point and not at times between sampling points. The **sampling interval**, or **sampling period**, is the time in seconds between samples. The **sampling frequency**, or **sampling rate**, is the number of samples per second, measured in Hertz. Therefore,

$$\text{Sampling frequency} = \frac{1}{\text{Sampling interval}}$$

or,

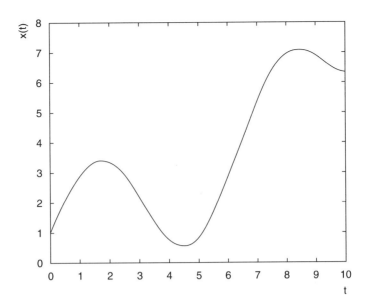

FIGURE 2.2
Analog signal.

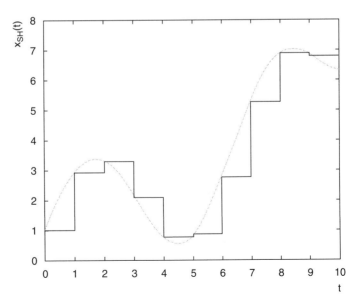

FIGURE 2.3
Sample-and-hold signal (shown with analog signal).

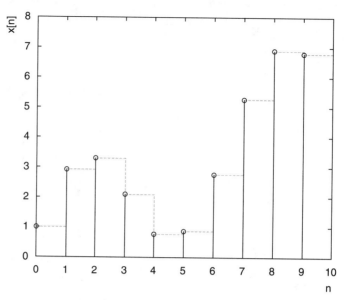

FIGURE 2.4

Digital signal (shown with sample-and-hold signal).

$$f_S = \frac{1}{T_S}$$

In Figure 2.4 the value of the digital signal at each sampling point is marked by a small circle at the top of a line. Both the original analog signal and its sampled digital counterpart give time information about the signal. Therefore, they are both time domain descriptions: They describe the variations of the signal as time proceeds.

It is not obvious from Figure 2.4 how a set of samples can uniquely represent the analog signal. Figure 2.5, for example, shows two possible analog signals that would produce the same set of sample values at the sampling points shown. As it turns out, when the sampling frequency is adequate, there is no ambiguity about which analog signal corresponds to a given set of samples. The **sampling theorem**, due to Nyquist, guarantees that an analog signal can be perfectly re-created from its sample values, provided the sampling interval is chosen correctly. The right sampling interval is determined from the characteristics of the signal being sampled. According to Nyquist theory, a signal with maximum frequency of W Hz must be sampled at least $2W$ times per second to make it possible to reconstruct the original signal from the samples. This minimum sampling frequency is called the **Nyquist sampling rate**. As an example, a signal containing frequencies up to 20 kHz must be sampled a minimum of 40,000 times per second: The Nyquist rate for the signal is 40 kHz. The **Nyquist frequency**, on the other hand, refers to a frequency that is half the sampling rate of a system. The range of frequencies between zero and the Nyquist frequency is called the **Nyquist range**.

The sampling rate used in Figure 2.4 was selected to be exactly twice the maximum frequency present in the analog signal of Figure 2.2. This maximum frequency controls the maximum steepness of the signal at any point in time. The signals shown in Figure 2.5

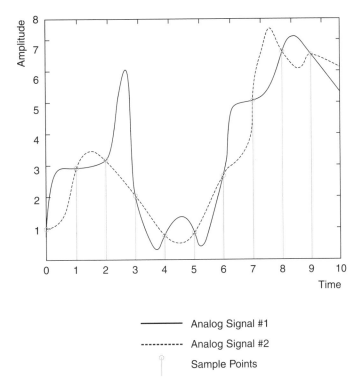

FIGURE 2.5
Undersampled analog signals.

clearly contain higher frequency elements than the signal in Figure 2.2, as the steep signal transitions show. The sampling rate used in Figure 2.4 is not adequate to sample either of the analog signals in Figure 2.5. The higher frequency signal content means a higher sampling frequency would have to be chosen to obtain a set of samples that would be adequate to reconstruct the original signals. Only when the sampling rate is too low is there ambiguity about the analog signal that produced the samples. Analog signal #1 and analog signal #2 are just two examples of signals that might produce the sample points in Figure 2.5. However, when the sampling rate is high enough, there is no ambiguity about the source signal: Only one signal can produce a given set of samples.

The time domain effects of undersampling are made clear in Figure 2.6. Here a sampling rate of 40 kHz is used to sample a group of signals, from 10 kHz to 80 kHz. The sample points, consistent for all signals, are represented as dashed vertical lines. According to Nyquist, only signals with frequencies up to 20 kHz can be perfectly reconstructed using a sampling rate of 40 kHz. Naturally a 30 kHz signal can be sampled with the same 40 kHz sampling rate, but the insufficient sample points trace out a signal that appears to have a frequency of 10 kHz. For the 40 kHz signal, the samples appear to lie on a horizontal line. The pattern continues for the higher frequencies: The apparent frequency never exceeds 20 kHz. This is an example of aliasing. Frequencies above the Nyquist frequency, half the sampling rate, are folded back and recovered as lower frequency signals.

FIGURE 2.6

Aliasing in the time domain with 40 kHz sampling (adapted from Pohlmann, 1994).

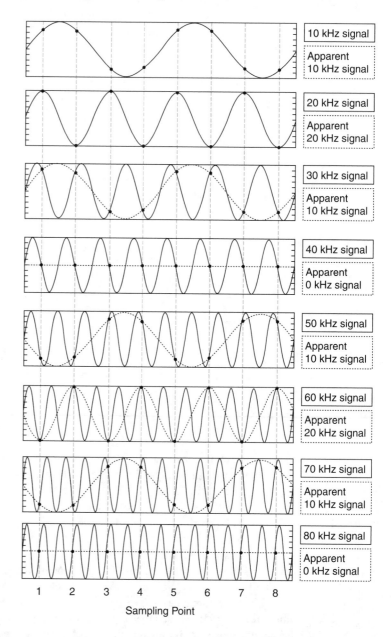

Once a sampling rate for a system has been selected, steps must be taken to ensure that signal elements with frequencies greater than the Nyquist frequency are excluded from the system. Many signals contain noisy or other nonessential high frequency elements that must be removed before sampling, as suggested by the spectrum in Figure 2.7(a). This is the job of the **antialiasing filter** introduced in Figure 2.1 and illustrated in Figure 2.7(b). This filter removes all signal elements above the Nyquist frequency from the signal to be sampled, and so ensures that Nyquist sampling will be sufficient to completely record the signal. At the

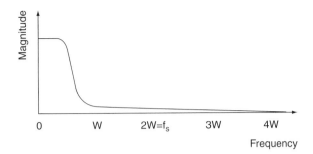

(a) Analog Signal Spectrum

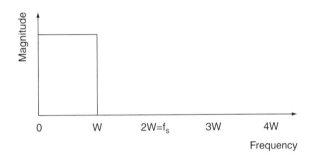

(b) Filter Shape for Analog Antialiasing Filter

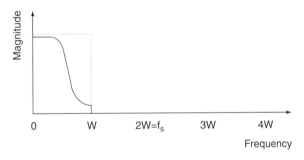

(c) Filtered Analog Signal Spectrum

FIGURE 2.7
Antialiasing Filter.

same time, all noise above the Nyquist frequency is eliminated, which prevents high frequency noise from interfering with the signal of interest. Figure 2.7(c) presents the spectrum of the signal after filtering, ready to be sampled at $2W$ samples per second.

2.2.2   The Frequency View of Sampling

An analog signal with an identifiable maximum frequency, by nature or through filtering, is called a **band-limited** signal. Such a signal and its spectrum are shown in Figure 2.8(a) and (b). The maximum frequency present in the signal is marked on the spectrum as W Hz.

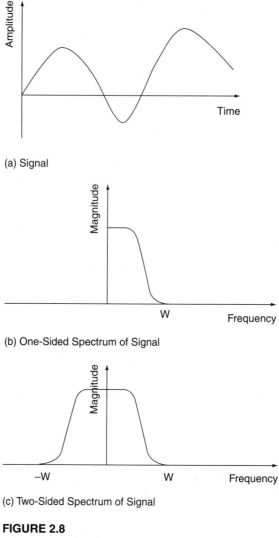

(a) Signal

(b) One-Sided Spectrum of Signal

(c) Two-Sided Spectrum of Signal

FIGURE 2.8
Signal and its spectrum.

Figure 2.8(c) shows the two-sided spectrum of the signal. It is created by placing a mirror image of the one-sided spectrum to the left of the 0 Hz axis. As a result, the spectrum is described between $-W$ and W Hz. Though negative frequencies do not have physical meaning, they are needed to explain the effects of sampling in the frequency domain.

The act of sampling an analog signal produces a series of sample values in the time domain, as suggested in the previous section. Sampling has a dramatic effect in the frequency domain as well. Figure 2.9(a) shows the two-sided spectrum of a signal, and Figure 2.9(b) shows the spectra of the same signal after sampling, for three different sampling rates f_S. As a result of sampling, copies of the original two-sided signal spectrum, called **images**, are placed at every multiple of the sampling frequency, that is, at 0, $\pm f_S$, $\pm 2f_S$, $\pm 3f_S$, ...

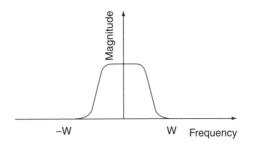

(a) Original Two-Sided Signal Spectrum

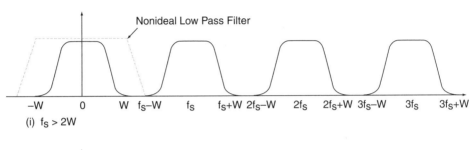

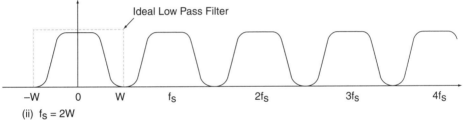

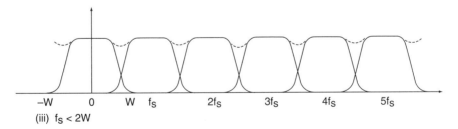

(b) Spectrum of Sampled Signal

FIGURE 2.9

Spectra of original and sampled signals.

A mathematical explanation of this result is provided in Appendix K. Note that the spectrum for the sampled signal is also two-sided, in the sense that it extends symmetrically for both negative and positive frequencies.

The three different spectra shown in Figure 2.9(b) differ only in the relationship between sampling rate, f_S, and the maximum frequency in the signal, W. The reasons behind the Nyquist sampling rate are evident. When $f_S > 2W$, copies of the original spectrum do not overlap; when $f_S < 2W$, copies do overlap. Where images overlap, spectral elements sum, as shown by the dashed lines in Figure 2.9(b)(iii). The condition $f_S = 2W$ marks the limiting case, the Nyquist sampling rate.

The issue of overlap is important when it is time to recover the original signal from the samples. This is done by a process called low pass filtering. In Figure 2.1, the low pass filter is identified as the **anti-imaging filter**. Low frequencies are passed by the filter, while higher frequency components are attenuated. Since the important elements of the signal lie below W Hz, the cut-off for the anti-imaging filter must occur no lower than W Hz. The objective is to pick out from all of the images in the frequency domain the one image that matches the original spectrum. The images that are removed give this filter its name. At the same time as the anti-imaging filter removes extraneous high frequency signals, it also removes out-of-band noise.

The dotted boxes in Figure 2.9(b) show the low pass filter shapes that might be used to accomplish this goal. In Figure 2.9(b)(i), where $f_S > 2W$, a filter with a relatively shallow roll-off slope can easily pick out the original spectrum. What this implies for the time domain is that the original signal can be perfectly reconstructed from the sample values. In Figure 2.9(b)(ii), only an ideal low pass filter with a cut-off at half the sampling frequency and an infinitely sharp roll-off would be capable of extracting the original spectrum, and such a filter does not exist. In Figure 2.9(b)(iii), however, there is no filter that can pick out the original spectrum: The overlap between spectral images, or aliasing, makes this impossible. When aliasing occurs, the signal that results after low pass filtering differs from the original. The frequency pictures make it easy to see why the sampling frequency should be more than twice the maximum frequency in the signal: If this condition cannot be met, there is no chance of reconstructing the signal from its samples.

For a sine wave, the Nyquist rate requires that two samples be collected every cycle. At this minimum sampling rate, though, the samples seem incapable of re-creating the sine wave, as illustrated in Figure 2.10. The pair of samples collected during each cycle seems to fit a square wave or triangle wave as easily as a sine wave. It is difficult to accept Nyquist's assertion that no ambiguity exists. In fact, the same low frequency component appears in the spectra for all square, triangle, and sine waves that repeat at the same rate. Square and triangle waves, however, have many high frequency elements as well. The anti-imaging filter removes these high frequency elements, leaving only the sine wave behind.

As Figure 2.9 showed, sampling causes images of a signal's spectrum to appear at every multiple of the sampling frequency. For a signal with frequency f, the sampled spectrum has frequency components at $kf_S \pm f$ Hz, where f_S is the sampling frequency and where k stands for all integers. Thus, the spectrum for a sampled signal has an infinite number of images, from which the anti-imaging filter must recover a signal. This calculation works whether or not the Nyquist sampling requirement is met. When it is not, the calculation provides the aliased frequency or frequencies. For example, Figure 2.11 shows the two-sided spectra of two sine waves sampled at 40 kHz. The first, with a frequency of 10 kHz, is below the Nyquist frequency for the chosen sampling rate. Some of the images lie at -40 ± 10, 0 ± 10, and 40 ± 10 kHz, or $-50, -30, -10, 10, 30$, and 50 kHz. Only one of these frequencies lies between zero and half the sampling rate. Thus, the signal's frequency is correctly identified as 10 kHz. The second sine wave in Figure 2.11(b), with a frequency of

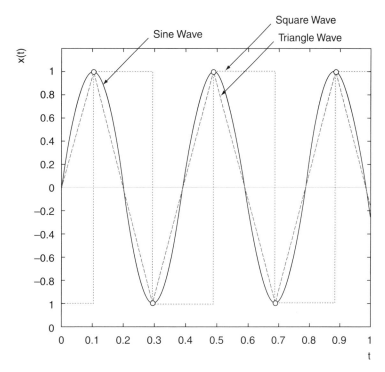

FIGURE 2.10
Sine wave sampled at Nyquist rate.

30 kHz, lies above the Nyquist frequency. Nevertheless, the image frequencies can still be computed. For example, with 40 kHz sampling, the 30 kHz signal produces, among others, components at -40 ± 30, 0 ± 30, and 40 ± 30 kHz. Only one of these components lies in the Nyquist range between 0 and 20 kHz, so 10 kHz is the aliased frequency of the recovered signal. Note that the 10 kHz and 30 kHz signals produce identical sampled spectra for a 40 kHz sampling rate, a finding that agrees with the observations of Figure 2.6.

The alias recovered from the 30 kHz signal is a **baseband** copy of the original frequency, which means that it lies between zero and the Nyquist limit. Such an alias can be distinguished from true signal by changing the sampling rate a little. In general, if the peak in the baseband moves, it is an alias; if it does not, it represents true signal. The calculation of image frequencies $kf_S \pm f$ Hz can be used for complex signals as well as for sine waves. The signal described by the spectrum in Figure 2.9(a) contains all the frequencies between 0 and W Hz. Thus, the first image in the sampled spectrum lies between $f_S - W$ and $f_S + W$ Hz, the second image lies between $2f_S - W$ and $2f_S + W$ Hz, and so on.

When signals are band-limited to a range such as $W_1 < f < W_2$ rather than $0 < f < W$, there is no need to sample at twice the highest frequency, or $2W_2$. Instead, minimum sampling limits depend on the bandwidth of the signal, $W_2 - W_1$, as well as the position of the bandwidth in the spectrum. The sampling rate must be at least twice the bandwidth, but may need

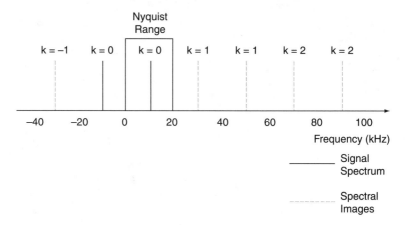

(a) 10 kHz sine wave sampled at 40 kHz

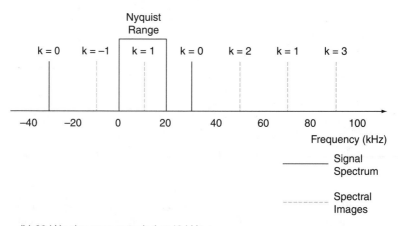

(b) 30 kHz sine wave sampled at 40 kHz

FIGURE 2.11
Spectra of sampled sine waves.

to be higher. The key is to ensure that no aliased copies of the signal overlap. Figure 2.12 shows two-sided spectra for a signal, band-limited between 120 and 160 kHz, sampled at three different rates. According to the usual Nyquist limits, 320 kHz would be the minimum sampling rate. As the figure shows, however, lower sampling rates can work as well. Of the three rates shown in the figure, only 120 kHz is acceptable for the 40 kHz bandwidth signal. The 100 kHz sampling rate gives no possibility of recovering an unaliased copy of the signal: As in Figure 2.9(b)(iii), overlapping spectral images sum, destroying any possibility of extracting an untainted copy of the original spectrum. The 80 kHz rate, on the other hand, demands an ideal anti-imaging filter. As in the example of Figure 2.11, the signals recovered from the Nyquist range in Figure 2.12 are baseband versions of the original. In the case of 80 kHz sampling, the signal spectrum is spectrally inverted in the baseband: The lowest fre-

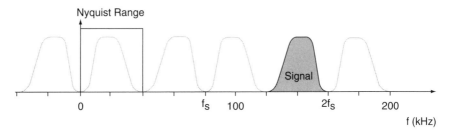

(a) f_s = 80 kHz

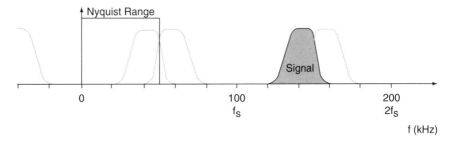

(b) f_s = 100 kHz

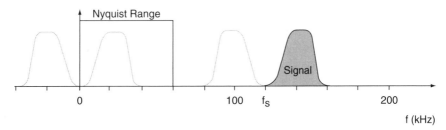

(c) f_s = 120 kHz

FIGURE 2.12
Sampling band-limited signals.

quencies in the signal alias to the highest frequencies in the baseband and vice versa. With 120 kHz sampling, no such spectral inversion occurs. Because Nyquist limits are not observed in sampling the band-limited signal in Figure 2.12, the approach is termed **undersampling**. One channel of a radio cell phone is an example of a high frequency, band-limited signal. Copying the spectrum to the baseband avoids impracticably high sampling rates. One cell phone occupies a bandwidth of 30 kHz in the 900 MHz range. Through undersampling, a sampling rate a little higher than 60 kHz, instead of 1.8 GHz, permits data to be reconstructed.

Though it is normally avoided, aliasing can, at times, be useful. In automatic target detection, the presence or absence of a signal of a particular frequency must be detected amongst signals of other frequencies. For targets that occur in the range 2 to 2.4 MHz, for example, Nyquist would require 4.8 MHz sampling, a very high rate. Instead, a band pass filter can be

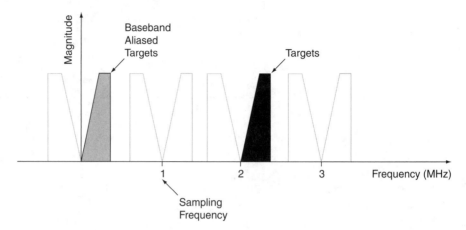

FIGURE 2.13
Undersampling in target detection.

used to filter out all but the signals in this range of frequencies. Then, according to the preceding discussion, the resulting signal can be sampled at 1 MHz, a little more than twice its $2.4 - 2 = 0.4$ MHz bandwidth. Since the targets are known to lie between 2 and 2.4 MHz, aliasing is definitely occurring, but is now exploited to deduce the correct frequencies of the targets. A 2 MHz target will appear, through aliasing, at 0 MHz; and a 2.4 MHz target will show up at 0.4 MHz, as Figure 2.13 illustrates. These aliased frequencies lie in the baseband, the range of frequencies between zero and the Nyquist frequency, half the sampling rate.

Many signals are not band-limited. If no maximum frequency can be designated, aliasing is impossible to avoid. To solve this problem, signals are low pass filtered prior to sampling, by the analog low pass filter identified as the antialiasing filter in Figure 2.1. This filter removes all frequencies above W Hz from the incoming signal. The procedure effectively band-limits the signal, which means that a sampling frequency that is high enough to avoid aliasing can be chosen. This feature, in fact, gives the filter its name. Naturally it is impossible to create an ideal antialiasing filter, just as it was impossible to create an ideal anti-imaging filter. When a nonideal low pass filter is used on a signal that is not band-limited, small amounts of signal at higher frequencies remain even after the antialiasing step. Thus the filtered signal is not truly band-limited, and small amounts of aliasing occur that cannot be completely eliminated. This effect will not have a large impact on the system provided the filter reduces signals above the Nyquist frequency to below quantization noise levels, discussed in the next section.

Aliasing occurs because ideal antialiasing filters are impossible to create. The steeper the roll-off of the filter, the smaller the amount of aliasing, but the greater the cost of the analog filter, since a steeper roll-off can be produced only by increasing the **order**, or complexity, of the filter. A trick that can be used to address this problem is to **oversample** at a frequency f_S chosen to accommodate the roll-off of a low order analog filter. For a signal with maximum frequency W Hz, the minimum sampling rate is $2W$ Hz. As suggested in Figure 2.9, sampling produces spectral images of the original signal spectrum at each integer multiple of the sampling frequency. The higher the sampling rate, the farther apart these images are, which means that the antialiasing filter can be permitted a shallower roll-off than that depicted in Figure 2.7. Figure 2.14(a) shows how a low order antialiasing filter can be

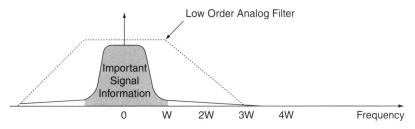

(a) Analog Filtering Before Sampling

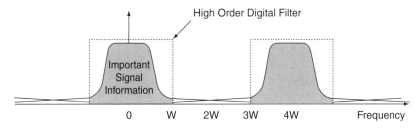

(b) Digital Filtering After Sampling

FIGURE 2.14
Oversampling.

used instead. When the output of this low order analog filter is sampled at $4W$ Hz, copies of its spectral content appear at every integer multiple of this frequency, as indicated in Figure 2.14(b). Significant aliasing occurs, for example, in the range between W and $3W$ Hz, but this aliasing does not affect the important signal information. As the figure shows, the aliased portions of the spectrum can be filtered out using a high order digital filter once sampling is complete. This trick saves the expense and complexity of a high order analog filter and can reduce phase distortion problems associated with analog filters. Oversampling is discussed in more detail in Section 14.1.2.

EXAMPLE 2.1

Most people have experienced the illusion of the wheels of a car appearing to turn backward in a movie or television show. This is a direct result of aliasing, meaning that the frames of film were not being recorded quickly enough to capture the correct rotation of the wheels. An average tire with a diameter of about 0.6 m has a circumference of 1.88 m. This is the distance traveled in one complete revolution of the tire. The car's speedometer records a speed in kilometers per hour. A speed of v km per hour is equivalent to $1000v/3600 = 0.278v$ m/sec. The number of cycles (revolutions) per second for the tire is given by:

$$\text{Frequency} = \frac{0.278v \, \dfrac{\text{m}}{\text{sec}}}{1.88 \, \dfrac{\text{m}}{\text{cycle}}} = 0.1479v \text{ Hz}$$

In order to satisfy Nyquist's rule, snapshots of the rotating tire must be taken with a frequency at least twice the frequency of rotation. That is,

$$\text{Minimum sampling frequency} = 2 \, (\text{Frequency}_{\text{max}}) = 0.2958v \text{ Hz}$$

Most commercial 16 mm cameras offer variable recording rates from 2 to 64 frames per second. A common choice is 16 frames per second. At this recording rate, the maximum speed that could be accommodated is found from

$$16 = 0.2958v_{\text{max}}$$

In other words, for speeds above $16/0.2958 = 54.1$ km/h, the film will not correctly record the rotating wheel.

2.3 QUANTIZATION

As mentioned in the previous section, analog signals have two characteristics that make them ill-suited for computer processing. First, analog signals are defined at every point in time. Sampling solves this problem by reducing the number of points to be used for processing to a finite level. Second, analog signals can take any amplitude value between their physical minimum and maximum levels. The output from an operational amplifier, for example, is a continuous voltage that can take any value between the limits of the power supplies that drive the circuit. Computers use groups of bits to represent numbers. The number of bits used limits the number of values that can be represented by the computer. For example, if 2 bits are used, only four digital codes—00, 01, 10, and 11—are possible, each with an associated quantization level. If the analog signal being coded lies between 0 and 2 V, for example, these digital codes might correspond to quantization levels 0.25 V, 0.75 V, 1.25 V, and 1.75 V. The analog signal can take the value 0.8 V, but the digital signal, defined by the legal quantization levels, cannot. An analog sample is coded by choosing the closest quantization level available, so errors will always exist when the number of bits is finite. When N bits are used, 2^N possible values can be represented by the computer. The larger the number of bits used, the more closely the digital signal is able to correspond to the analog signal, but the more time-consuming the calculations become.

When an analog signal with a certain range of values is coded using N bits, each sample must be coded to one of 2^N levels. The gap between levels is called a **quantization step**:

$$Q = \frac{R}{2^N} \tag{2.1}$$

where R is the full scale analog range and N is the number of bits. The quantization step size is sometimes referred to as the **resolution** of the quantizer. For a given range, the quantization step grows smaller as the number of bits increases. A simple quantization scheme divides the range into 2^N equal intervals and assigns each section a digital code. In Figure 2.15 analog values are mapped to one of the eight digital codes in a 3-bit system. The figure assumes **unipolar** analog inputs, which vary between zero and some positive maximum. Eight analog intervals are spaced evenly across the full scale range of the analog in-

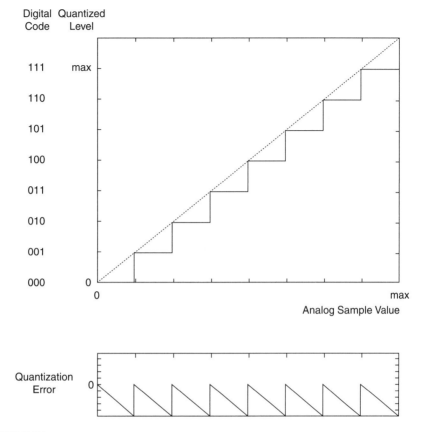

FIGURE 2.15
Quantization of unipolar data (maximum error = full step).

put. Each of these intervals maps to one of the available digital codes. Ideally all mappings would lie on the diagonal line where the digital value equals the analog value. Because a finite number of bits is used, this direct correspondence cannot be achieved. For every input value, the quantization error that occurs is the difference between the quantized value and the actual value of the sample.

$$\text{Quantization error} = \text{Quantized value} - \text{Actual value}$$

The quantization errors in the quantization scheme of Figure 2.15 are shown at the bottom of the diagram. Apparently, errors as large as one full quantization step can occur in this coding scheme. That is, the distance between the dashed line, which represents the ideal case, and the coding steps can be as large as a full step. The errors can be reduced if the quantization levels are shifted to lie symmetrically around the diagonal. This can be achieved by assigning 000 to the first half step rather than the first full step from the bottom of the analog range, as shown in Figure 2.16. All other codes follow as usual at full quantization step intervals. In Figure 2.16, the coding levels that would be obtained for the

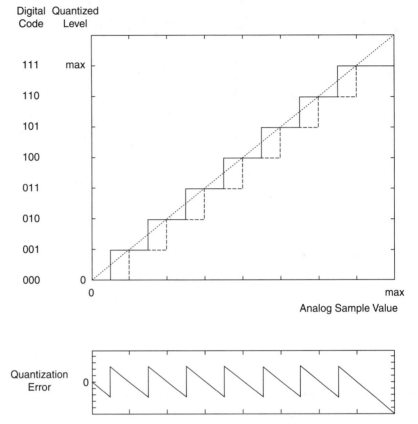

FIGURE 2.16
Quantization of unipolar data (maximum error = half step).

simple quantization scheme of Figure 2.15 are shown as a dashed line. Notice that the new
and improved quantization scheme guarantees that quantization errors are, on average, reduced by half. In fact, the only place where a full step error can occur is in the last interval,
as Example 2.2 will illustrate. This is no great penalty, especially when the number of bits,
and hence the number of coding levels, is large.

EXAMPLE 2.2

Analog pressures are recorded, using a pressure transducer, as voltages between 0 and 3 V.
The signal must be quantized using a 3-bit digital code. Indicate how the analog voltages
will be converted to digital values.

Since the range of the signal is 3 V, the quantization step size is

$$Q = \frac{3 \text{ V}}{2^3} = 0.375 \text{ V}$$

TABLE 2.1
Quantization Table for Example 2.2

Digital Code	Quantization Level (V)	Range of Analog Inputs Mapping to This Digital Code (V)
000	0.0	$0.0 \leq x < 0.1875$
001	0.375	$0.1875 \leq x < 0.5625$
010	0.75	$0.5625 \leq x < 0.9375$
011	1.125	$0.9375 \leq x < 1.3125$
100	1.5	$1.3125 \leq x < 1.6875$
101	1.875	$1.6875 \leq x < 2.0625$
110	2.25	$2.0625 \leq x < 2.4375$
111	2.625	$2.4375 \leq x \leq 3$

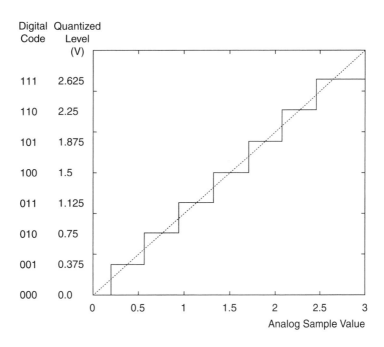

FIGURE 2.17
Quantization diagram for Example 2.2.

and half a quantization step is 0.1875 V wide. Table 2.1 shows all eight digital codes and their associated analog ranges, as is easily verified by inspecting Figure 2.17. Note that the first code covers only a half-step range, so the last code must cover a range equal to one and a half quantization steps. All other codes cover a range equal to a full quantization step. The quantization level corresponding to a digital code is given in the center column of the table. It matches the point where the diagonal ideal characteristic intersects the stepped quantization curve.

As mentioned, the only place where errors larger than half a quantization step can occur in this quantization scheme is for the largest inputs. The 111 code (with quantization level 2.625 V) must serve the range of inputs from 2.4375 to 3 V. Thus, an error of a full quantization step, or 0.375 V, will occur in coding an input sample of 3 V.

Bipolar analog data vary between a negative minimum and a positive maximum. Quantization errors affect bipolar data in the same way as they affect unipolar data. The best scheme for bipolar data begins with the unipolar quantization diagram of Figure 2.16 and simply extends the quantization steps in the negative direction. Symmetry around zero is maintained to keep errors small. Figure 2.18 shows the arrangement for bipolar data that will keep the largest average errors within one half of a quantization step. As in the unipolar case, the bottom range is half a step wide and the top range is one and a half steps wide. Note that the digital codes in Figure 2.18 use a two's complement representation (to be

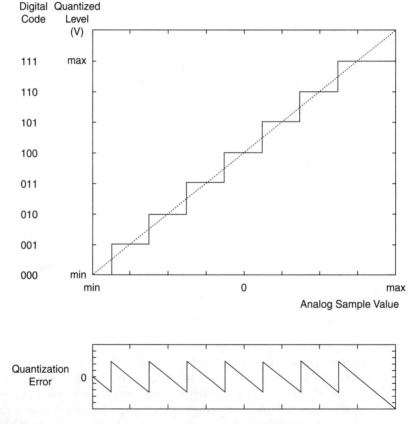

FIGURE 2.18
Quantization of bipolar data (maximum error = half step).

TABLE 2.2
Quantization Table for Example 2.3

Digital Code	Quantization Level (V)	Range of Analog Inputs Mapping to This Digital Code (V)
100	-5.0	$-5.0 \leq x < -4.375$
101	-3.75	$-4.375 \leq x < -3.125$
110	-2.5	$-3.125 \leq x < -1.875$
111	-1.25	$-1.875 \leq x < -0.625$
000	0.0	$-0.625 \leq x < 0.625$
001	1.25	$0.625 \leq x < 1.875$
010	2.5	$1.875 \leq x < 3.125$
011	3.75	$3.125 \leq x \leq 5.0$

described in Section 12.3), which requires the leading bit for all negative codes to be 1. Example 2.3 illustrates the quantization of bipolar data.

EXAMPLE 2.3

An analog voltage between -5 V and 5 V must be quantized using 3 bits. Quantize each of the following samples, and record the quantization error for each:

    a. -3.4 V
    b. 0.0 V
    c. 0.625 V

Using the quantization step size $10V/2^3 = 1.25$ V, the quantization table in Table 2.2 can be constructed.

    a. The analog sample -3.4 V produces the digital code 101, with a quantization error of $-3.75 - (-3.4) = -0.35$ V.
    b. The analog sample 0.0 V codes as 000, with zero quantization error.
    c. The analog sample 0.625 V generates the digital code 001, with a quantization error of $1.25 - 0.625 = 0.625$ V. For the midrange, this represents a worst-case quantization error, equal to half a quantization step.

Since the worst-case quantization errors are determined by the size of the quantization step, errors can be reduced by increasing the number of bits used to represent each sample. Unfortunately, they cannot be entirely eliminated, and their combined effect is sometimes called **quantization noise**. The **dynamic range** of the quantizer is the number of levels it can distinguish in noise. It is a function of the range of signal values and the range of error values, and is expressed in units of **decibels**, or **dB**, described in Appendix A.10. An analog signal takes values over a range R. Each quantized value lies somewhere between half a quantization step below and half a quantization step above the actual sample values, meaning that the errors lie between $-Q/2$ and $+Q/2$, where Q is the quantization step size. Thus, the errors take values over a range Q. The number of distinct levels that can be identified without error, then, is R/Q. From Equation (2.1), this equals 2^N, where N is the number of bits. Expressing the ratio in dB gives the dynamic range of the quantizer as

$$\text{Dynamic range} = 20 \log\left(\frac{R}{Q}\right) = 20 \log(2^N) = N(20 \log 2) = 6.02N \text{ dB}$$

The dynamic range improves as the number of bits N increases. The concept of dynamic range is connected to the concept of **signal-to-noise ratio**, which measures how easily a signal can be discerned from noise. The larger the signal-to-noise ratio, or SNR, the stronger the signal compared with the noise. SNR can be calculated as

$$\text{SNR} = 10 \log\left(\frac{\text{Signal power}}{\text{Noise power}}\right)$$

or,

$$\text{SNR} = 20 \log\left(\frac{\text{Signal amplitude}}{\text{Noise amplitude}}\right)$$

and is measured in dB. For the specific case of quantizing a sine wave, for example, the signal-to-noise ratio is $6.02N + 1.76$ dB, as calculated in Appendix B. When a maximum permissible quantization error level is set, Equation (2.1) can be used to determine how many quantization bits are needed. Solving this equation for N gives $2^N = R/Q$, or

$$N = \log_2\left(\frac{R}{Q}\right) = \frac{\log\left(\dfrac{R}{Q}\right)}{\log 2}$$

In performing this calculation, it is important to remember that Q is the size of the quantization step, and that the maximum midrange quantization error is half of this, or $Q/2$.

EXAMPLE 2.4

An analog signal whose range lies between 0 and 5 V must be quantized, with midrange quantization errors no bigger than 6×10^{-5} V. How many bits are required to meet this requirement?

If the maximum allowable quantization error is 6×10^{-5}, then the quantization step must be no greater than 12×10^{-5}. For an analog signal with a 5 V range, the number of quantization bits would then be

$$N = \log_2\left(\frac{R}{Q}\right) = \log_2\left(\frac{5}{12 \times 10^{-5}}\right) = 15.35$$

Thus, a 16-bit quantizer is adequate.

In practice, quantization is handled by analog-to-digital converters. A wide variety of converters is available commercially. They may be unipolar or bipolar, with many possible analog ranges, and between 8 and 24 bits. Quantization errors will always be smallest when

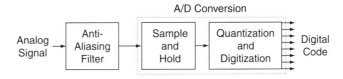

FIGURE 2.19
Analog-to-digital conversion.

the signal being quantized utilizes the maximum analog range of the converter. When it occupies only a small part of the converter's range, all errors are proportionally larger, and the effect of quantization noise on the signal grows, reducing the signal-to-noise ratio.

2.4   ANALOG-TO-DIGITAL CONVERSION

When data are inherently digital, digital signal processing can begin immediately. The number of newspapers sold each day, for example, requires no sampling, because the data is already sampled and needs no quantization,[1] because the data are integer-valued. Otherwise, an analog signal must first be converted into digital form. As shown in Figure 2.19, the analog-to-digital (A/D) conversion process consists of sample and hold, followed by quantization and digitization. Before it is sampled, an analog signal is first filtered by a low pass antialiasing filter to eliminate as much as possible the effects of aliasing. This is followed by sampling, accomplished by a sample-and-hold circuit. At each sampling point, this circuit acquires the current value of the analog signal as quickly as possible, and then holds it steady until the next sampling point. Thus, the sample-and-hold circuit defines the sampling instants and also freezes each sample value while it is quantized and converted to a digital code. If the signal were allowed to change during conversion, significant errors would be introduced. Figure 2.19 pictures 8-bit A/D conversion, since 8 bits are produced at the output. Usually digital codes generated by an A/D converter are presented in parallel to the DSP that will process them, as shown in the figure. For some low frequency converters, though, serial output may be used.

The output of the sample-and-hold circuit is still an analog signal. As long as the sampler takes a negligible amount of time to acquire each new sample, the sample-and-hold signal has the look of a staircase. When this acquisition time is significant, the sample-and-hold signal briefly follows the analog signal being sampled until a level hold state is reached, as shown in Figure 1.6. The sample-and-hold signal is quantized and converted to a digital code by an analog-to-digital converter. In fact, quantization and digitization occur at the same time. The quantization level closest to a sample-and-hold value is selected as the best equivalent, and the digital code for this quantization level is assigned to the sample. From this discussion, the nature of the signals in Figure 2.19 can be pictured, as shown in Figure 2.20. Figure 2.21 shows the relationships between the signals that result from each step in the A/D process. In

[1]   If the maximum number of newspapers sold exceeds the largest integer that can be represented by the processor, then quantization will be required and quantization errors will occur.

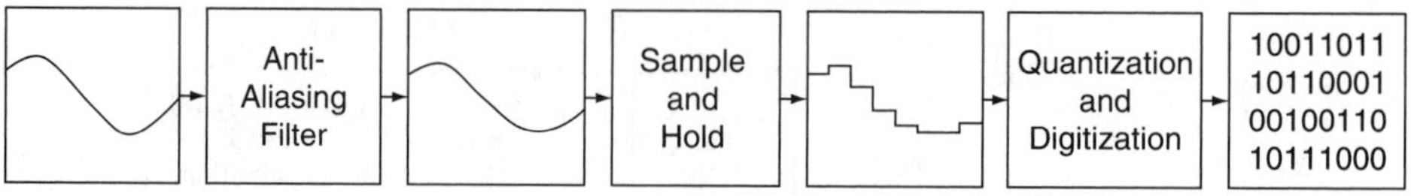

FIGURE 2.20
A/D.

TABLE 2.3
Quantization Table for Figure 2.21

Digital Code	Quantization Level (V)	Range of Analog Inputs Mapping to This Digital Code (V)
000	0.000	$0.000 \leq x < 0.0625$
001	0.125	$0.0625 \leq x < 0.1875$
010	0.250	$0.1875 \leq x < 0.3125$
011	0.375	$0.3125 \leq x < 0.4375$
100	0.500	$0.4375 \leq x < 0.5625$
101	0.625	$0.5625 \leq x < 0.6875$
110	0.750	$0.6875 \leq x < 0.8125$
111	0.875	$0.8125 \leq x \leq 1.000$

this illustrative example, the analog signal has a range from 0 to 1 V. It is sampled every 0.2 msec, or 5000 times per second, and quantized using the 3-bit quantization scheme shown in Table 2.3. The quantization errors and the digital codes that result from quantization and digitization are shown at the bottom of Figure 2.21, one for each sample point. These digital codes are strung together to form a serial bitstream in this example, as shown in Figure 2.22. In this 3-bit system, every group of 3 bits corresponds to a sample.

It is clear from Figure 2.21 that the digital signal provides an approximation to the analog signal, but that significant quantization errors do occur. These errors can be reduced by increasing the number of bits used for quantization. **Bit rate**, a measure of the rate at which bits are generated, is frequently quoted as a measure of A/D converter performance. It is defined as

$$\text{Bit rate} = Nf_S$$

where N is the number of bits per sample and f_S is the number of samples per second. For this example, the bit rate is 3 bits/sample $\times$ 5 ksamples/sec = 15 kbps, where bps stands for bits per second.

2.5 DIGITAL-TO-ANALOG CONVERSION

In some cases, the digital codes that result from processing can be used directly, to drive a device such as a stepper motor that does not require analog input. In most cases, though,

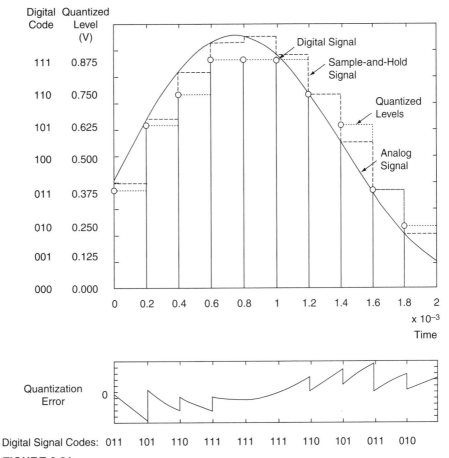

FIGURE 2.21

Three-bit A/D conversion.

Stream of Digital Codes: 0 1 1 1 0 1 1 1 0 1 1 1 1 1 1 1 1 1 1 0 1 0 1 0 1 1 0 1 0

FIGURE 2.22

Serial digital bitstream.

the digital code must be converted into an analog signal, in order, for example, to be seen or heard. In the digital-to-analog (D/A) conversion process presented in Figure 2.23, a circuit first maps 8-bit digital codes to analog levels proportional to the size of the digital number. These levels are then held steady for one full sample period through zero order hold, until a new digital code is presented at the beginning of the next cycle. The analog output of the D/A converter, therefore, resembles a staircase, similar to the sample-and-hold signal

D/A Conversion

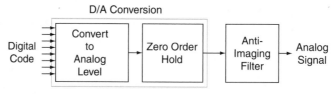

FIGURE 2.23

Digital-to-analog conversion.

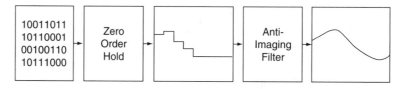

FIGURE 2.24

D/A.

that was produced during the analog-to-digital conversion process. In the last step, a low pass filter anti-imaging filter smoothes the steplike zero order hold signal. Figure 2.24 illustrates the types of signals found at each stage in the conversion process.

Figure 2.25 demonstrates D/A conversion for a 3-bit digital signal. The staircase signal that results from the zero order hold process is clearly unlike the smooth analog input to the DSP system, whose frequency content was limited to a maximum of W Hz by the antialiasing filter. The sharp edges of the staircase signal in Figure 2.25 contain much higher frequencies, which are introduced by the conversion process. They are removed by the final, anti-imaging filter. This filter has a cut-off frequency of W Hz and ensures that all unwanted frequencies are eliminated. Figure 2.9 showed why this filter is necessary from a frequency point of view: It effectively removes spectral images that arise from sampling. In the time domain, the effect of the anti-imaging filter is to smooth away the sharp edges of the staircase signal, as seen in Figure 2.25. If ideal filters were available and if Nyquist limits were observed, then applying an A/D conversion followed directly by a D/A conversion would reproduce the original analog signal, apart from quantization errors and a time shift due to filtering.

Perfect recovery is possible only if the anti-imaging filter can perfectly eliminate spurious frequency elements from the zero order hold signal. As discussed in Section 2.2.2, the impossibility of constructing an ideal filter necessitates additional spacing in the frequency domain between copies of the signal spectrum. Section 14.1.3 explains how such spacing may be obtained by effectively increasing the sampling rate. The greater the increase in the sampling rate, the simpler and cheaper the analog anti-imaging filter can be.

Hearing and seeing the effects of the sample and hold, quantization, zero order hold, and anti-imaging can help to clarify the relationships among these signals. Figure 2.26(a) shows an analog signal. This signal is passed to a sample-and-hold circuit, which produces the waveform in Figure 2.26(b). This staircase waveform is the basis for quantizing and digitizing, which ultimately produce a digital version of the original analog waveform. If no

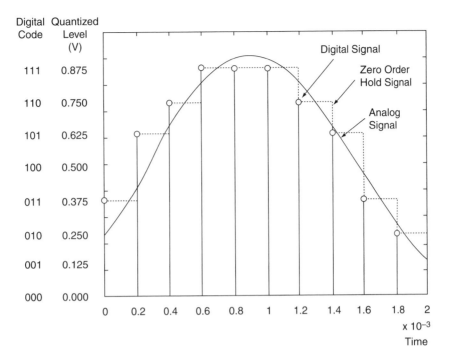

FIGURE 2.25
Three-Bit D/A conversion.

processing takes place, the digital signal is passed to a digital-to-analog converter, which produces a proportional analog level. Through zero order hold, this analog level is held steady for the duration of the sampling period. The analog zero order hold signal is shown in Figure 2.26(c). It differs from the sample-and-hold signal because of quantization errors. Finally, the zero order hold signal is filtered by the anti-imaging filter to produce a recon-struction of the original signal. This reconstruction is shown in Figure 2.26(d). It is a smoothed version of the zero order hold signal. As Figure 2.26(a) shows, the anti-imaging filter delays the reconstructed signal with respect to the original signal. Otherwise, the two signals differ mainly as a result of quantization.

CHAPTER SUMMARY

Matlab
Support

1. Sampling is the act of collecting values from an analog signal at regular intervals. Nyquist theory states that the sampling rate for a signal must be at least twice the max-imum frequency present in the signal.
2. The Nyquist sampling rate is twice the highest frequency in the signal being sampled. Oversampling refers to sampling at a rate greater than the Nyquist rate. It eases the de-sign of the antialiasing filter. Undersampling, on the other hand, refers to sampling at a

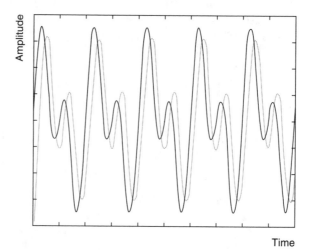

addaorig.wav

(a) Original Signal with Reconstructed Signal in Background

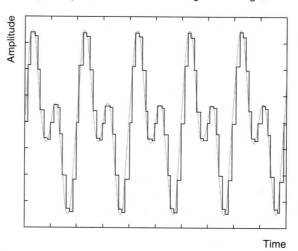

addash.wav

(b) Sample-and-Hold Signal with Original Signal in Background

FIGURE 2.26
Comparing signals in the A/D/D/A chain.

rate that is slower than the Nyquist rate. It results in aliasing, which alters the spectrum of the signal.

3. The first step in converting an analog signal to a digital signal is to apply a low pass antialiasing filter, which eliminates elements above the Nyquist frequency. The next step is sample and hold, which defines the sampling instants and freezes an analog value for conversion. The third step is quantization. For an N-bit A/D converter, 2^N possible

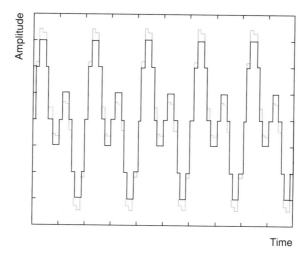

(c) Zero Order Hold Signal with Sample-and-Hold Signal
in Background

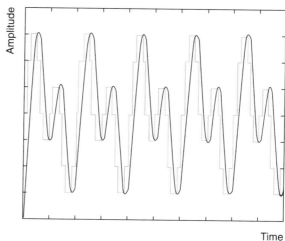

(d) Reconstructed Signal with Zero Order Hold Signal
in Background

FIGURE 2.26
Continued

addazoh.wav

addarecon.wav

quantization levels are available, and the one closest to the amplitude of the sample is
chosen. A digital code that matches the chosen quantization level is assigned to the sample, which completes A/D conversion.

4. After processing is complete, the digital signal must be converted to an analog signal.
The digital signal is first converted to a proportionate analog level, which is held steady

between sampling instants using zero order hold. The last step in D/A conversion is to smooth the signal using an anti-imaging filter. This filter removes the spectral images caused by sampling that are responsible for the staircase shape of the zero order hold signal.

5. Aliasing is one source of error for A/D conversion. Because a limited number of quantization levels are available, quantization is another source of error. The larger the number of bits used, the smaller the errors are, which translates to a larger dynamic range for the A/D converter.

REVIEW QUESTIONS

2.1 Humans can hear sounds at frequencies between 0 and 22.05 kHz. What minimum sampling rate should be chosen to permit perfect recovery from samples?

2.2 Determine the Nyquist sampling rate for each of the following analog signals:
   a. $x(t) = \cos(20t + 12°)$
   b. $x(t) = 2\sin(5000\pi t/3)$
   c. $x(t) = \sin(3000\pi t/7 + \pi/10)$

2.3 A voice signal is sampled at 8000 samples per second.
   a. What is the time between samples?
   b. What is the maximum frequency that will be recovered from the signal?

2.4 Five periods of an analog signal $x(t) = \cos(4000t)$ are oversampled at four times the Nyquist sampling rate. How many samples are collected?

2.5 Five periods of an analog signal $x(t) = 5\sin(2500\pi t + 10°)$ are undersampled at 7/8 of the Nyquist rate. How many samples are collected?

2.6 The spectrum for an analog signal is shown in Figure 2.27. The signal is sampled at 600 Hz. Draw the spectrum of the sampled signal between -200 and 1200 Hz.

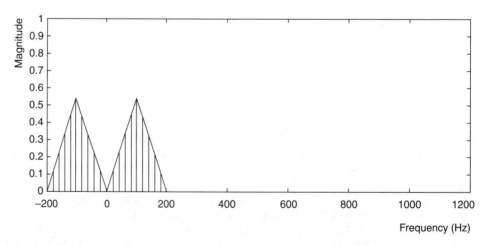

FIGURE 2.27
Spectrum for Question 2.6.

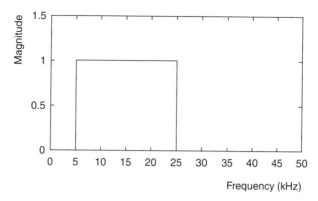

FIGURE 2.28
Magnitude spectrum for Question 2.7.

2.7 An analog signal's one-sided spectrum is shown in Figure 2.28. The signal is sampled. Draw the spectrum of the sampled signal from 0 to 150 kHz. If aliasing occurs, draw the true aliased spectrum by summing overlapping spectral elements. The sampling rate is:
  a. 60 kHz
  b. 40 kHz

2.8 A cell phone transmits voice signals on a carrier. The transmissions lie in the range 900 MHz to 900.03 MHz. What minimum sampling rate should be chosen to ensure that the transmissions will be recoverable from the digital samples?

2.9 Determine the locations of the peaks in the sampled spectrum for the following sine waves:
  a. Signal frequency 300 Hz, sampling frequency 1 kHz
  b. Signal frequency 600 Hz, sampling frequency 1 kHz
  c. Signal frequency 1.3 kHz, sampling frequency 1 kHz

2.10 Find the aliased frequencies for the following band-limited signals sampled at 1 kHz. For each, draw the spectrum for the sampled signal between -500 and 1500 Hz and indicate whether spectral inversion occurs in the baseband:
  a. Signal band-limited between 1100 and 1400 Hz, shown in Figure 2.29.
  b. Signal band-limited between 800 and 950 Hz, shown in Figure 2.30.

2.11 A 25 kHz signal is sampled at 8 kHz. Find the aliased frequency of the signal.

2.12 Show that $x_1(t) = \cos(120\pi t)$ and $x_2(t) = \cos(420\pi t)$ have the same samples when sampled at 150 Hz.

2.13 A radar signal ranging in frequency from 900 to 900.5 MHz is undersampled at 2 MHz. If a 200 kHz target appears in the baseband, what is the actual frequency of the target?

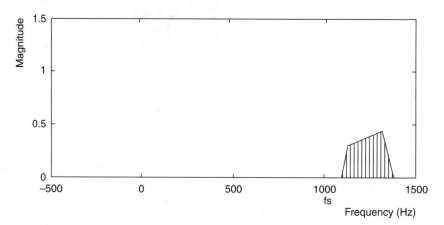

FIGURE 2.29
Spectrum for Question 2.10(a).

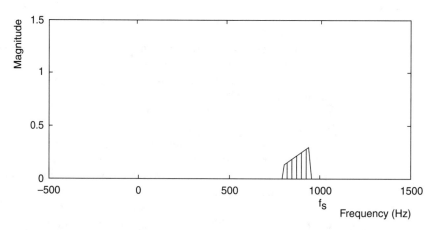

FIGURE 2.30
Spectrum for Question 2.10(b).

2.14 A bicycle travels down the road at 15 km/h. The tire of the bicycle has an outside diameter of 63.5 cm. How many snapshots of the bicycle must be taken every second if the snapshots are to be used to record correctly the true motion of the wheels?

2.15 A signal $x(t) = \sin(2\pi ft)$ with a frequency less than 1 kHz is sampled at 600 Hz and appears as a 150 Hz signal. The same signal is sampled at 550 Hz and appears as a 200 Hz signal. What is the frequency of the signal?

2.16 Voltages between -3 V and $+3$ V must be digitized. Find the quantization step size when samples are quantized using:

a. 4 bits

b. 8 bits

c. 16 bits

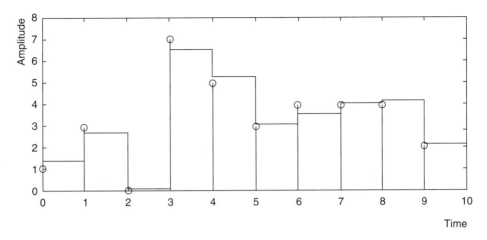

FIGURE 2.31
Signals for Question 2.19.

TABLE 2.4
Data for Question 2.19

n	0	1	2	3	4	5	6	7	8	9
Sample-and-hold level	1.4	2.7	0.1	6.6	5.3	3.1	3.6	4.1	4.2	2.2
Quantized level	1.0	3.0	0.0	7.0	5.0	3.0	4.0	4.0	4.0	2.0

2.17 How many different digital codes can be produced by an A/D converter using:
  a. 8 bits?
  b. 10 bits?
  c. 12 bits?

2.18 Explain why error-free quantization is impossible.

2.19 An analog sample-and-hold signal and its quantized digital counterpart are plotted in Figure 2.31 from Table 2.4. Compute the quantization errors for each of the 10 samples, assuming that the time for the sampler to acquire the signal at each sampling point is negligible.

2.20 Compile a quantization table and draw a quantization diagram for a 4-bit quantization of a bipolar analog voltage between −2 V and 2 V. Quantization errors should be no greater than half a quantization step in midrange.

2.21 A group of analog samples, listed in Table 2.5, is digitized using the quantization table in Table 2.6. Determine the digital codes, the quantized level, and the quantization error for each sample.

TABLE 2.5
Analog Samples for Question 2.21

n	0	1	2	3	4	5	6	7	8	9
Analog Sample (V)	0.5715	4.9575	0.625	3.6125	4.0500	0.9555	2.7825	1.5625	2.7500	2.8755

TABLE 2.6
Quantization Table for Question 2.21

Digital Code	Quantized Level (V)	Range of Analog Inputs Mapping to This Digital Code (V)
000	0.000	$0 \leq x < 0.3125$
001	0.625	$0.3125 \leq x < 0.9375$
010	1.250	$0.9375 \leq x < 1.5625$
011	1.875	$1.5625 \leq x < 2.1875$
100	2.500	$2.1875 \leq x < 2.8125$
101	3.125	$2.8125 \leq x < 3.4375$
110	3.750	$3.4375 \leq x < 4.0625$
111	4.375	$4.0625 \leq x < 5.0$

2.22 Find the dynamic range in dB of a quantizer using:
   a. 4 bits
   b. 8 bits
   c. 16 bits
2.23 The dynamic range of an A/D converter is 60.2 dB. How many data bits does it produce?
2.24 An analog signal ranging between 0 and 1 V must be quantized. The midrange quantization errors must be no greater than 0.1 V. How many quantization bits must be used to satisfy this requirement?
2.25 Compute the bit rate for an 8 kHz, 16-bit sampler.
2.26 The maximum midrange quantization error allowed is 1% of the full scale range of an analog signal. If the sampling frequency of an A/D converter is 16 kHz, what is the converter's minimum bit rate?
2.27 A 3-bit D/A converter produces a 0 V output for the digital code 000 and a 5 V output for the code 111, with other codes distributed evenly between 0 and 5 V. Draw the zero order hold output from the converter for an input: 111 101 011 101 000 001 011 010 100 110.
2.28 What effect does an anti-imaging filter have on a zero order hold signal?

3

DIGITAL SIGNALS

All digital signal processing is performed on digital signals. This chapter:

> ➤ identifies a graphical representation and a notation for digital signals
> ➤ explains shifting and scaling of digital signals
> ➤ introduces important elemental digital functions
> ➤ links analog frequencies and digital frequencies
> ➤ interprets composite digital signals
> ➤ introduces digital images

3.1 PICTURES OF DIGITAL SIGNALS

Digital signals provide the necessary input for all digital signal processing tasks. They originate, in most cases, from the analog-to-digital conversion of a measured analog signal sampled at regular intervals. Digital signals are represented graphically by circles at the top of vertical lines. Each line marks a sampling point that is labeled with an integer referring to the number of elapsed sampling periods. The value of the digital signal at a given sample point, marked by a circle, is the quantization level selected at the time of A/D conversion as being closest to the sampled analog value. It is more helpful to use quantization levels than digital codes to represent the digital signal, because quantization levels can give a clear visual sense of the increases or decreases in the signal that digital codes cannot. Notation for digital signals is presented in Section 3.2, and the most common elemental digital signals—impulse, step, exponential, and sinusoidal— are introduced in Section 3.3.

EXAMPLE 3.1

A portion of a speech signal takes values in the range 0 V to 1 V. The A/D process, using a 3-bit quantization scheme given in Table 3.1, produces the digital codes: 010 110 000 001 011 100 110 111 100 010. Plot the digital signal.

Figure 3.1 shows the graph of the digital signal. Note that the signal can take only a finite number of possible values, which are the quantization levels from Table 3.1 that correspond to the digital codes. Once quantization has occurred, no other values can be achieved.

TABLE 3.1

Three-Bit Quantization Table for Example 3.1

Digital Code	Quantization Level (V)	Range of Analog Inputs Mapping to This Digital Code (V)
000	0.000	$0.000 \leq x < 0.0625$
001	0.125	$0.0625 \leq x < 0.1875$
010	0.250	$0.1875 \leq x < 0.3125$
011	0.375	$0.3125 \leq x < 0.4375$
100	0.500	$0.4375 \leq x < 0.5625$
101	0.625	$0.5625 \leq x < 0.6875$
110	0.750	$0.6875 \leq x < 0.8125$
111	0.875	$0.8125 \leq x \leq 1.000$

FIGURE 3.1

Digital signal from codes for Example 3.1.

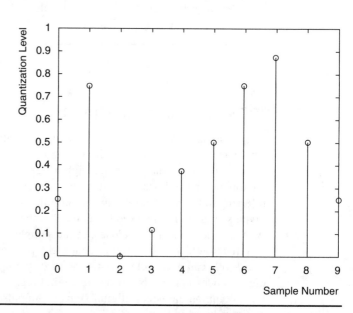

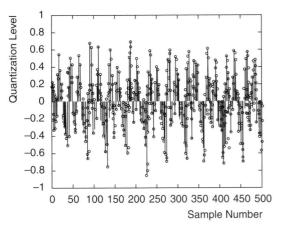

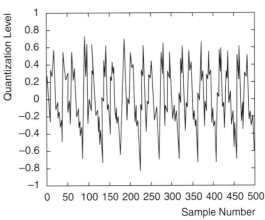

(a) As Individual Samples

(b) Samples Joined by Smooth Line

FIGURE 3.2
Plotting large numbers of digital samples.

In this text, digital signals are, as a rule, depicted using the "lollipop" style exemplified in Figure 3.1. When the sampling rate is very high, or when many samples must be presented, however, this style becomes too crowded. In these cases a smooth line is used to join the sample points. Figure 3.2 illustrates this point. Where this mechanism is used, a note will appear stating that, for clarity, the envelope rather than the individual samples of the digital signal is plotted.

3.2   NOTATION FOR DIGITAL SIGNALS

In an equation, a digital signal x is designated as $x[n]$, where the n is an integer referring to the number of the sample. Thus, $x[0]$ refers to the value of sample 0, $x[1]$ refers to the value of sample 1, and so on. The collection of all the samples together is the sequence $x[n]$. The signal $x[n-1]$, on the other hand, refers to the sequence of samples shifted to the right by one sample, while $x[n+1]$ refers to the sequence shifted to the left by one sample. In general, $x[n-N]$ refers to a time shift of the entire sequence by N samples to the right. Time shifting is required by many DSP operations, including difference equations and convolution. Scaled forms like $x[kn]$, where k is an integer, are used in decimation and wavelet transforms. They have the effect of selecting every k^{th} sample from the signal. Both shifting and scaling are illustrated in Example 3.2. Understanding manipulations of these kinds is necessary to comprehend DSP theory fully, as well as to use DSP software effectively.

EXAMPLE 3.2

For the signal in Figure 3.3, assume that all samples before $n = 0$ and after $n = 9$ have the value zero. The signal is plotted with trailing zeros.

Find: **a.** $x[0]$
      b. $x[5]$
      c. $x[n-1]$
      d. $x[n-2]$
      e. $x[2n]$
      f. $x[3n]$

a. The sample has the value $x[0] = 0.25$. While $x[n]$ refers to the entire signal, $x[0]$ refers to a single sample value in the signal.

b. The sample at $n = 5$ has the value $x[5] = 0.5$.

c. The shape of $x[n-1]$ is found by shifting the sample values for $x[n]$ by one time step to the right, as may be verified by using Table 3.2. The sample numbers and signal values $x[n]$ are listed in the first two columns. These columns are used to determine the contents of the last two. The highlighted row may be used as an example. When $n = 2, n - 1 = 1$. Therefore, $x[n] = x[2] = 0$, while $x[n-1] = x[1] = 0.75$. Thus, when $n = 2, x[n-1] = 0.75$. When the same analysis is applied to the other points in the table, the result is a shifting of all sample values to the right by one time step. This is also evident by comparing Figure 3.4, where the signal $x[n-1]$ is plotted against n, with Figure 3.3.

d. The signal $x[n-2]$ is shifted to the right by two time steps with respect to $x[n]$, as is seen by comparing Figure 3.5 and Figure 3.3.

e. The signal $x[2n]$ may be found using Table 3.3. The doubled value of n selects alternate samples of the signal, as shown in Figure 3.6.

f. The signal $x[3n]$ selects every third sample from the signal, as shown in Figure 3.7.

FIGURE 3.3
Digital signal $x[n]$ for Example 3.2.

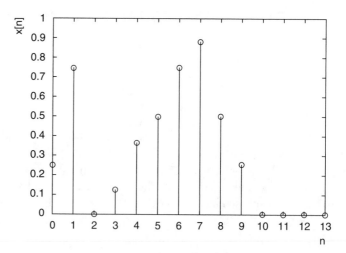

TABLE 3.2
Calculating $x[n-1]$ for Example 3.2

n	-2	-1	0	1	**2**	3	4	5	6	7	8	9	10	11	12	13
$x[n]$	0.0	0.0	0.25	0.75	**0.0**	0.125	0.375	0.5	0.75	0.875	0.5	0.25	0	0	0	0
$n-1$	-3	-2	-1	0	**1**	2	3	4	5	6	7	8	9	10	11	12
$x[n-1]$	0.0	0.0	0.0	0.25	**0.75**	0.0	0.125	0.375	0.5	0.75	0.875	0.5	0.25	0.0	0.0	0.0

FIGURE 3.4
Digital signal $x[n-1]$ for Example 3.2.

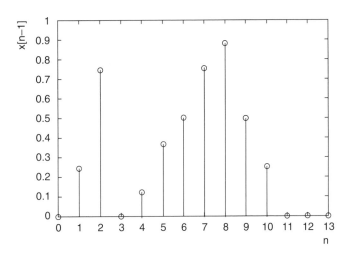

FIGURE 3.5
Digital signal $x[n-2]$ for Example 3.2.

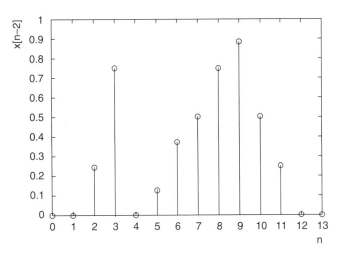

TABLE 3.3
Calculating x[2n] for Example 3.2

n	0	1	2	3	4	5	6
$x[n]$	0.25	0.75	0.0	0.125	0.375	0.5	0.75
$2n$	0	2	4	6	8	10	12
$x[2n]$	0.25	0.0	0.375	0.75	0.5	0.0	0.0

FIGURE 3.6
Digital signal x[2n] for Example 3.2.

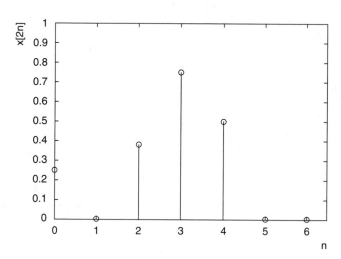

FIGURE 3.7
Digital signal x[3n] for Example 3.2.

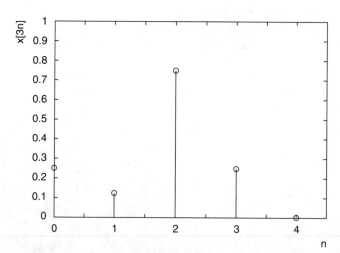

3.3 DIGITAL FUNCTIONS

3.3.1 Impulse Functions

Unit impulse functions are fundamental functions in the digital domain. In fact, all digital signals can be constructed from impulse functions alone. The **unit impulse function**, also called the delta function, has the value zero everywhere, except for a single spike at $n = 0$. It is defined as

$$\delta[n] = \begin{cases} 0 & n \neq 0 \\ 1 & n = 0 \end{cases} \tag{3.1}$$

and is depicted in Figure 3.8. The letter δ (delta) is reserved for the impulse function. The following examples exercise the definition of the impulse function.

FIGURE 3.8
Unit impulse function.

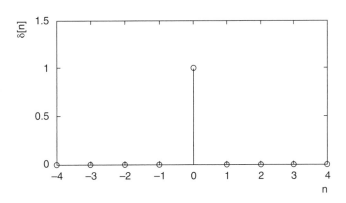

EXAMPLE 3.3
Determine the values $\delta[0]$, $\delta[3]$, and $\delta[-2]$.

The value of the impulse function $\delta[n]$ can be found at any point in time by using the definition in Equation (3.1). From this equation or the graph in Figure 3.8, $\delta[0] = 1$, $\delta[3] = 0$, and $\delta[-2] = 0$.

EXAMPLE 3.4
Draw the signals:
    a. $x[n] = 4\delta[n]$
    b. $x[n] = -2\delta[n]$
    c. $x[n] = \delta[n-3]$

    a. The signal is plotted in Figure 3.9.
    b. The signal is plotted in Figure 3.10.
    c. This signal shifts the basic impulse function three time steps to the right, as seen in Figure 3.11.

FIGURE 3.9
Signal for Example 3.4(a).

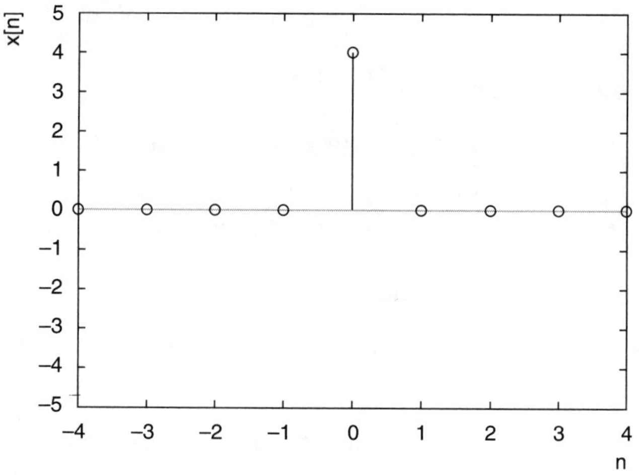

FIGURE 3.10
Signal for Example 3.4(b).

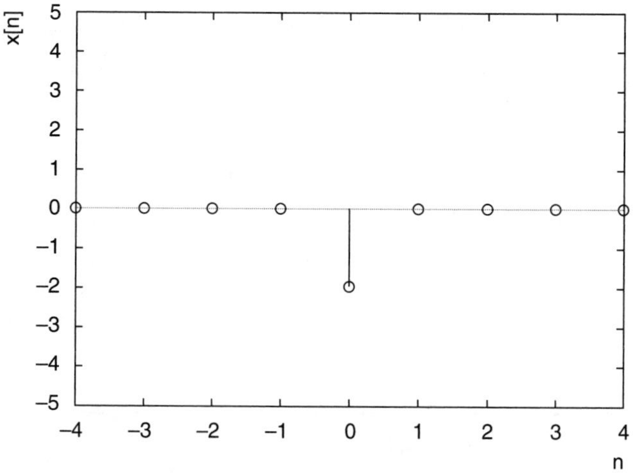

FIGURE 3.11
Signal for Example 3.4(c).

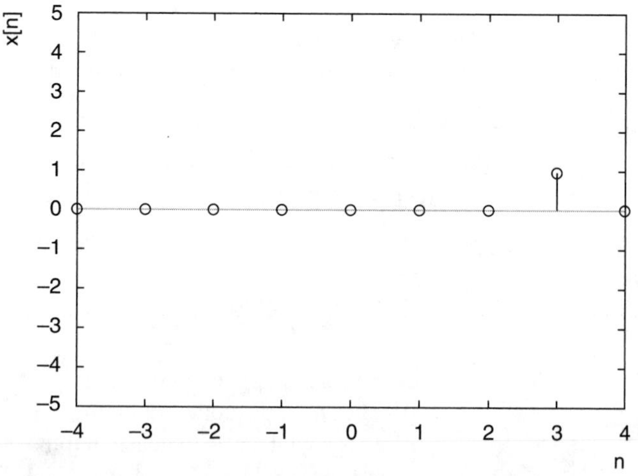

EXAMPLE 3.5

Draw the signals:

    a. $x[n] = \delta[n] + \delta[n-1] + \delta[n-2] + \delta[n-3]$
    b. $x[n] = 5\delta[n] + 4\delta[n-1] - \delta[n-3]$

    a. The signal is plotted in Figure 3.12.
    b. The figure is plotted in Figure 3.13.

FIGURE 3.12

Signal for Example 3.5(a).

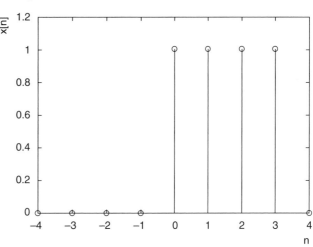

FIGURE 3.13

Signal for Example 3.5(b).

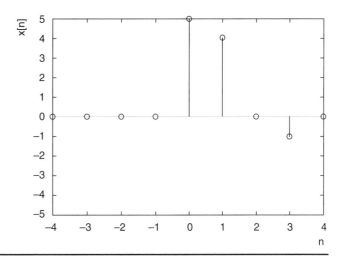

EXAMPLE 3.6

A digital function is defined as $x[n] = \delta[n] + 0.5\delta[n-1] + 0.2\delta[n-2]$. Write the function $x[n-1]$.

Substituting $(n-1)$ for each n gives $x[n-1] = \delta[n-1] + 0.5\delta[n-2] + 0.2\delta[n-3]$.

EXAMPLE 3.7

Write a function to describe the graph given in Figure 3.14, assuming that all sample values outside the window are zero.

Any digital signal may be written as the sum of impulse functions. Four nonzero samples constitute this function. Mathematically,

$$x[n] = 4\delta[n] - 2\delta[n-1] + 3\delta[n-2] - \delta[n-3]$$

FIGURE 3.14

Signal for Example 3.7.

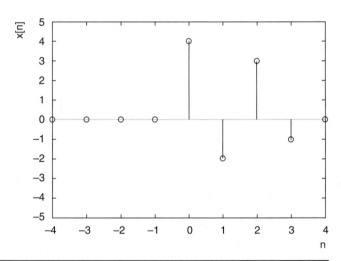

Example 3.7 showed how to write a digital function in terms of impulse functions. This is a general result. For any digital signal $x[n]$,

$$x[n] = x[0]\delta[n] + x[1]\delta[n-1] + x[2]\delta[n-2] + \ldots$$

where $x[0]$, $x[1]$, $x[2]$, ... refer to the signal sample values.

3.3.2 Step Functions

The unit step function is shown in Figure 3.15. It may be defined as:

$$u[n] = \begin{cases} 0 & n < 0 \\ 1 & n \geq 0 \end{cases} \qquad (3.2)$$

The letter u is reserved for the step function. The first few nonzero samples of $u[n]$ are drawn in Figure 3.15. The unit step function is frequently used to signal a "turn on" event. For example, the samples obtained as a 5 V DC power supply is switched on might be represented as $5u[n]$. The following illustrations make use of the definition of the step function.

FIGURE 3.15
Unit step function.

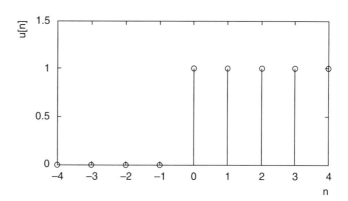

EXAMPLE 3.8

Determine the values of $u[-1]$, $u[0]$, and $u[1]$.

Using Equation (3.2) or Figure 3.15, $u[-1] = 0$, $u[0] = 1$, and $u[1] = 1$.

EXAMPLE 3.9

Draw the signal $x[n] = 3u[n]$.

The signal shown in Figure 3.16 is simply an amplification of the basic unit step function.

FIGURE 3.16
Signal for Example 3.9.

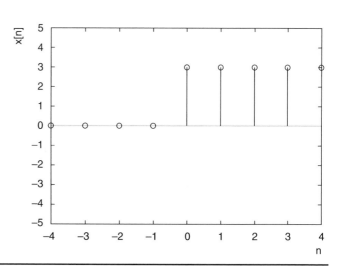

EXAMPLE 3.10

Draw $x[n] = u[-n]$.

For every positive value of n, the index $(-n)$ is negative; and for every negative index, the step function is zero. For every negative value of n, the index $(-n)$ becomes positive, and the step function for every positive index is one. The value at zero is unchanged. As Figure 3.17 shows, the function $u[-n]$ is a reflection of $u[n]$ across the $n = 0$ axis.

FIGURE 3.17

Signal for Example 3.10.

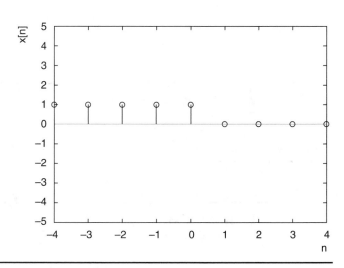

EXAMPLE 3.11

Draw the digital signals:
    a. $x[n] = u[n-3]$
    b. $x[n] = u[3-n]$

    a. The function $u[n-3]$ is identical with the function $u[n]$ but shifted three positions to the right, as shown in Figure 3.18. All samples of $x[n]$ for $n \geq 3$ are equal to one. Any point may be verified by direct evaluation. For example, $x[4] = u[4-3] = u[1] = 1$, while $x[2] = u[2-3] = u[-1] = 0$.

    b. The signal $u[3-n]$ may be written as $u[-(n-3)]$. It is a copy of $u[-n]$ shifted three positions to the right, as indicated in Figure 3.19. The change from one to zero now occurs where $3 - n = 0$, or $n = 3$.

FIGURE 3.18
Signal for Example 3.11(a).

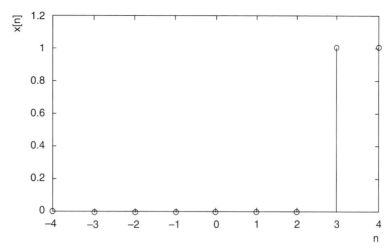

FIGURE 3.19
Signal for
Example
3.11(b).

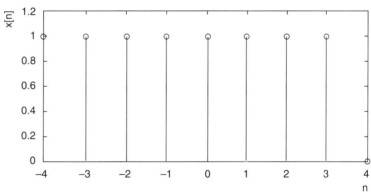

EXAMPLE 3.12
Draw $x[n] = u[n] + 2u[n-2]$.

This signal is formed as the point-by-point sum of $u[n]$ and $2u[n-2]$, as detailed in Figure 3.20.

FIGURE 3.20
Sum of signals for
Example 3.12.

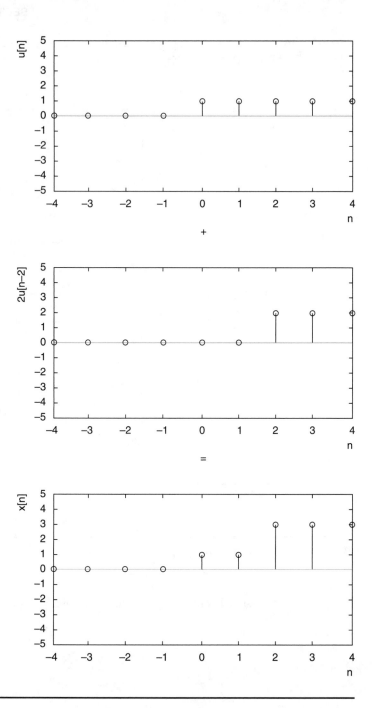

EXAMPLE 3.13

Draw $x[n] = u[n] - u[n-3]$.

This signal is produced by subtracting the signal $u[n-3]$ from the signal $u[n]$, as detailed in Figure 3.21. The result shows a sampled DC level turned on and then off again.

FIGURE 3.21
Difference of signals for
Example 3.13.

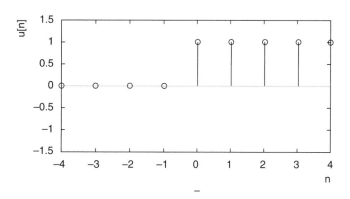

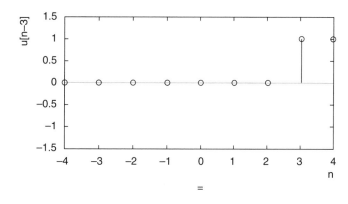

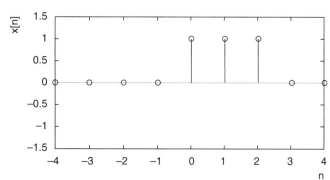

EXAMPLE 3.14

Write a function to describe the signal shown in Figure 3.22, assuming that the signal value is zero for $n < -4$ and remains at -4 for $n > 4$.

The signal may be described using impulse functions or step functions, but step functions are most efficient. Using impulse functions,

$$x[n] = 2\delta[n] + 2\delta[n-1] - 4\delta[n-2] - 4\delta[n-3] - 4\delta[n-4] - \ldots$$

$$= 2(\delta[n] + \delta[n-1]) - 4\sum_{k=2}^{\infty}\delta[n-k]$$

Using step functions,

$$x[n] = 2u[n] - 6u[n-2]$$

FIGURE 3.22
Signal for
Example 3.14.

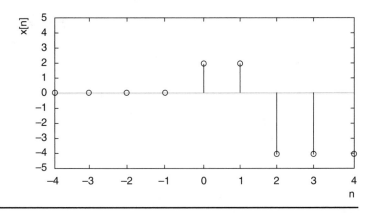

EXAMPLE 3.15

A digital signal is described as $x[n] = 4(u[n] - u[n-1])$. Write the function that describes $x[n-3]$.

Substituting $(n - 3)$ for n gives $x[n-3] = 4(u[n-3] - u[n-4])$.

Useful connections can be made between the step and impulse functions. The step function can be written as the sum of impulse functions, and the impulse function can be written as a difference of step functions:

$$u[n] = \delta[n] + \delta[n - 1] + \delta[n - 2] + \delta[n - 3] + \ldots = \sum_{m=0}^{\infty}\delta[n - m]$$

$$\delta[n] = u[n] - u[n-1]$$

3.3.3   Power and Exponential Functions

Digital power functions are defined as:

$$x[n] = A\alpha^{\beta n}$$

and the special case of exponential functions as:

$$x[n] = Ae^{\beta n}$$

where α has been set to the irrational number $e = 2.71828.\ldots$ When $\beta > 0$, the values of these functions grow in size; when $\beta < 0$, they shrink. The value of A determines the value of the function when $n = 0$.

EXAMPLE 3.16

Draw the signal $x[n] = e^{-0.5n}$.

The values of the signal may be calculated for any value of n. Table 3.4 shows the values for $-2 \le n \le 4$, which are also plotted in Figure 3.23.

TABLE 3.4
Values of $x[n] = e^{-0.5n}$ for Example 3.16

n	$x[n]$
-2	2.7183
-1	1.6487
0	1.0000
1	0.6065
2	0.3679
3	0.2231
4	0.1353

FIGURE 3.23
Signal for
Example 3.16.

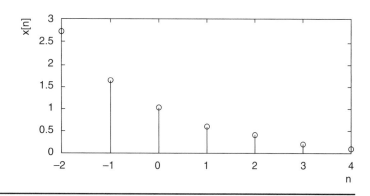

EXAMPLE 3.17

A signal is defined by $x[n] = (-0.6)^n$.

    a. Draw the signal.

    b. Write a function describing the shifted signal $x[n-1]$.

    a. Table 3.5 shows the signal values, and the signal is plotted in Figure 3.24.

    b. The function is $x[n-1] = (-0.6)^{n-1}$.

TABLE 3.5
Values of $x[n] = (-0.6)^n$ for Example 3.17

n	$x[n]$
-2	2.7778
-1	-1.6667
0	1.0000
1	-0.6000
2	0.3600
3	-0.2160
4	0.1296

FIGURE 3.24
Signal for Example 3.17.

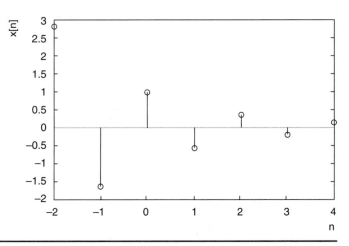

The complex exponential

$$x[n] = e^{j\beta n}$$

has an important place in DSP. It appears in the definitions of the discrete time Fourier transform (DTFT), the discrete Fourier series (DFS), and the discrete Fourier transform

(DFT). Euler's identity, given in Equation (A.4), states that $e^{j\theta} = \cos\theta + j\sin\theta$. This identity permits the complex exponential to be decomposed as

$$x[n] = \cos\beta n + j\,\sin\beta n$$

This signal has a complex value for every value of n. Moreover, as noted in Appendix A.13.2.5, the complex number $e^{j\beta n} = \cos\beta n + j\,\sin\beta n$ always lies a distance of one from the origin of the complex plane. For functions of the form $e^{-j\beta n}$, the alternative form of Euler's identity, $e^{-j\theta} = \cos\theta - j\sin\theta$, from Equation (A.5), permits the decomposition $e^{-j\beta n} = \cos\beta n - j\,\sin\beta n$.

EXAMPLE 3.18

Plot the first eight samples of the digital signal $x[n] = e^{-j\pi n/6}$.

The signal may be evaluated using the alternative form of Euler's identity:

$$x[n] = \cos\left(\frac{\pi n}{6}\right) - j\,\sin\left(\frac{\pi n}{6}\right)$$

Eight samples of $x[n]$ are computed in Table 3.6 and plotted in Figure 3.25. Since the values of the signal are complex, they cannot be plotted in the usual format. Instead, the location of each sample is marked on the complex plane. As expected, all points lie a distance of one from the origin, forming a circle of unit radius. Note that larger values of n will produce additional points on the same circle, and for this particular function, the samples in $x[n]$ will begin to repeat when n reaches 12.

TABLE 3.6
Values of $x[n] = e^{-j\pi n/6}$ for Example 3.18

n	$x[n]$
0	1.0000
1	$0.8660 - j0.5000$
2	$0.5000 - j0.8660$
3	$-j1.0000$
4	$-0.5000 - j0.8660$
5	$-0.8660 - j0.5000$
6	-1.0000
7	$-0.8660 + j0.5000$

FIGURE 3.25
Plot of $x[n] = e^{-j\pi n/6}$ for Example 3.18.

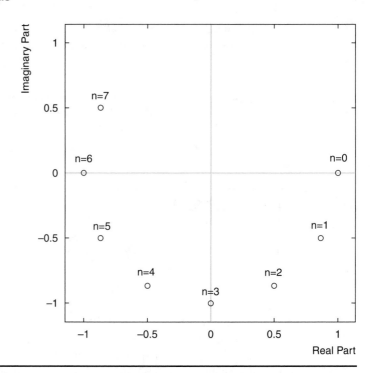

3.3.4 Sine and Cosine Functions

Digital sinusoids are especially useful in testing the response of a system to an input of a single frequency. Digital sine and cosine functions have the form

$$x[n] = A \sin(n\Omega) \tag{3.3}$$

or,

$$x[n] = A \cos(n\Omega)$$

where A is the amplitude and Ω determines the frequency of repetition of the digital sequence. Figure 3.26, for example, plots the signal $x[n] = 3\sin(n\pi/8)$, with $A = 3$ and $\Omega = \pi/8$. The situation, however, is not as straightforward as it appears. Unlike their analog counterparts, digital sine and cosine functions are not necessarily periodic, and Ω is not equal to the frequency of the analog signal being sampled.

An analog sine wave can be described by the function

$$x(t) = A \sin(\omega t)$$

where ω is the frequency of the sine wave in radians per second. The signal $x(t)$ is defined at every value of time t. When this sine wave is sampled, samples are collected every T_S

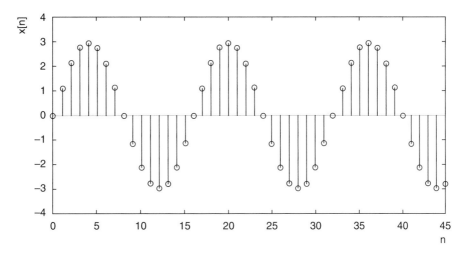

FIGURE 3.26

Plot of $x[n] = 3\sin\left(n\dfrac{\pi}{8}\right)$.

seconds, or once every sampling period. The sample times can be described by $t = nT_S$, so, in the analog domain, the sample values are described by $x(nT_S)$. The same sample values are designated as $x[n]$ in the digital domain, that is,

$$x(nT_S) = x[n] \tag{3.4}$$

when no quantization errors occur. Equation (3.4) allows a relationship between analog frequency and digital frequency to be drawn. At the sample times $t = nT_S$, the analog signal produces the values

$$x(nT_S) = A\sin(\omega nT_S)$$

where n is the sample number. The frequency ω in radians per second may be converted to a frequency f in Hz through the equation $\omega = 2\pi f$, defined in Appendix A.9; and the sampling interval T_S may be replaced by $1/f_S$, where f_S is the sampling frequency. With these substitutions,

$$x(nT_S) = A\sin\left(n2\pi\frac{f}{f_S}\right) \tag{3.5}$$

Using Equation (3.4) to equate Equations (3.3) and (3.5) gives:

$$A\sin\left(n2\pi\frac{f}{f_S}\right) = A\sin(n\Omega)$$

This equality requires that

$$\Omega = 2\pi\frac{f}{f_S} \tag{3.6}$$

Equation (3.6) relates analog frequency f and digital frequency Ω. It will be used often in the chapters that follow. Ω is called the **digital frequency** of a digital sinusoid because its position in the function $\sin(\Omega n)$ is analogous to the position of the analog frequency ω in $\sin(\omega t)$. While analog frequency is measured in radians per second, however, digital frequency is measured in radians. This is a consequence of the fact that, in the digital world, time elapsed is measured in numbers of samples, not seconds. The products Ωn and ωt both have units of radians.

Figure 3.27 shows plots of two digital sinusoids, $x_1[n] = \sin(n4\pi/7)$ and $x_2[n] = \sin(n13/7)$. In the first graph, the digital signal sequence begins to repeat after seven samples. In the second graph, the digital sequence does not repeat. For a sequence to repeat, N sampling intervals T_S must fit exactly into M periods T of the analog waveform being sampled, for some pair of integers N and M. That is,

$$NT_S = MT$$

or,

$$\frac{N}{M} = \frac{T}{T_S} = \frac{f_s}{f}$$

where $f = 1/T$ is the frequency of the analog sinusoid and $f_S = 1/T_S$ is the sampling frequency. Since $\Omega = 2\pi f/f_S$, $f_S/f = 2\pi/\Omega$, so

$$\frac{N}{M} = \frac{2\pi}{\Omega} \tag{3.7}$$

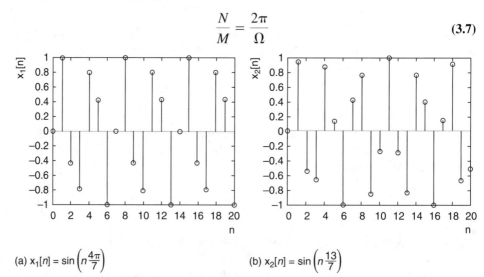

(a) $x_1[n] = \sin\left(n\frac{4\pi}{7}\right)$

(b) $x_2[n] = \sin\left(n\frac{13}{7}\right)$

FIGURE 3.27
Repeating and nonrepeating digital sinusoids.

where N is the number of samples it takes for the digital sequence to repeat, and M is the number of analog cycles that elapse while the N samples are collected. To find N and M, the fraction $2\pi/\Omega$ must be reduced to its lowest terms. N is referred to as the **digital period**, or the period of the digital signal. It must not be confused with the period of the analog waveform.

EXAMPLE 3.19

A digital signal is defined as $x[n] = \cos(2n)$.

  a. Is this digital sequence periodic?
  b. Find the first eight elements in this sequence.

  a. Since $x[n] = A\cos(n\Omega)$, $\Omega = 2$ rads, and $2\pi/\Omega = 2\pi/2 = \pi$. This number is irrational, meaning that it cannot be expressed as the ratio of integers. Therefore, the digital sequence is not periodic.
  b. As shown in Table 3.7 and Figure 3.28, no repetition is found within the first eight samples. Even when more samples are taken, the sequence does not repeat.

TABLE 3.7
Signal Values for Example 3.19

n	$x[n]$
0	1.0000
1	−0.4161
2	−0.6536
3	0.9602
4	−0.1455
5	−0.8391
6	0.8439
7	0.1367

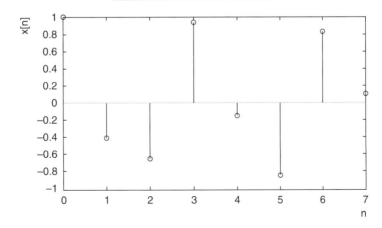

FIGURE 3.28
Signal for Example 3.19.

EXAMPLE 3.20

A digital signal is defined as $x[n] = \cos(n4\pi/5)$.

    a. Is this digital sequence periodic?
    b. Find the first eight elements in this sequence.

    a. The digital frequency is $\Omega = 4\pi/5$, so $2\pi/\Omega = 5/2$. The result, $N = 5$ and $M = 2$, means that the sequence repeats every five samples, and that these five samples are collected over two complete cycles of the analog signal being sampled.
    b. Note that the sequence calculated in Table 3.8 and shown in Figure 3.29 repeats every five samples.

TABLE 3.8
Signal Values for Example 3.20

n	$x[n]$
0	1.0000
1	−0.8090
2	0.3090
3	0.3090
4	−0.8090
5	1.0000
6	−0.8090
7	0.3090

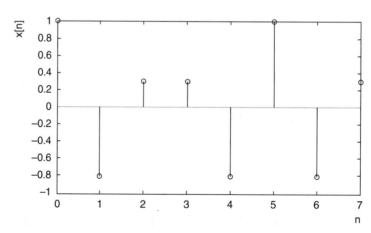

FIGURE 3.29
Signal for Example 3.20.

EXAMPLE 3.21

Figure 3.30 plots the signals:

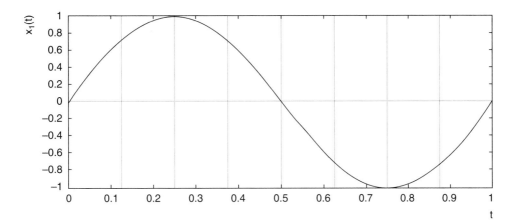

(a) 1 Hz Signal

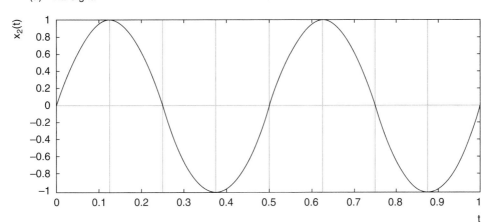

(b) 2 Hz Signal

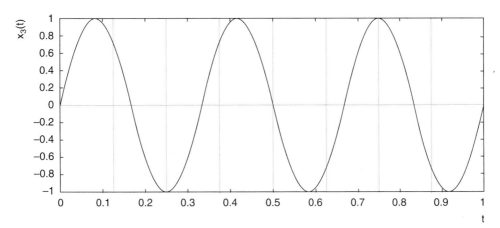

(c) 3 Hz Signal

FIGURE 3.30

Common sampling of three sine waves for Example 3.21.

a. $x_1(t) = \sin(2\pi t)$

b. $x_2(t) = \sin(4\pi t)$

c. $x_3(t) = \sin(6\pi t)$

All of the signals are sampled at the rate of $f_S = 8$ samples/second. The sample points are marked by solid vertical lines. In each case, check the periodicity of the sampled signal.

a. The frequency of the first signal is $f = \omega/2\pi = 1$ Hz. For this signal, $\Omega = 2\pi f/f_S = 2\pi/8$. Thus, $2\pi/\Omega = 8/1$, which means that the digital sequence is periodic and that the same 8 digital samples will be collected during each period of the analog sine wave. This is confirmed in Figure 3.30(a).

b. The frequency of the second signal is 2 Hz. For this signal, $2\pi/\Omega = 4/1$, so the periodic digital sequence repeats every 4 samples, over one cycle of the analog sine wave. Figure 3.30(b) shows two full cycles of the 2 Hz analog signal. During each of these cycles, exactly 4 digital samples are collected.

c. The frequency of the third signal is 3 Hz, so $2\pi/\Omega = 8/3$. This means that the digital sequence is periodic, and that 8 repeating samples cover 3 periods of the analog sine wave being sampled. In part (c) of Figure 3.30, 3 cycles of the analog signal are plotted. The digital sample values do not begin to repeat until all 3 analog cycles complete.

The most general form of a digital sinusoid is

$$x[n] = A\sin(n\Omega + \theta)$$

or

$$x[n] = A\cos(n\Omega + \theta)$$

where θ is a phase shift, measured in radians or degrees. When $x[n]$ is evaluated, θ must be expressed in radians, to agree with the units of $n\Omega$. A phase shift θ produces a translation of a digital signal in the time domain, as Example 3.22 illustrates. The size of the translation can be determined by solving the equation $n\Omega + \theta = 0$ for the sample number n. When the result is positive, the digital sequence shifts to the right; when it is negative, the sequence shifts to the left.

EXAMPLE 3.22

Plot $x_1[n] = \sin(n2\pi/9)$ and $x_2[n] = \sin(n2\pi/9 - 3\pi/5)$.

The samples of the signals can be computed directly for any value of n. The plots are shown in Figure 3.31. Both sequences repeat every 9 samples, but one is shifted from the other. The shift experienced by $x_2[n]$, computed from $n2\pi/9 - 3\pi/5 = 0$, equals $n = (3\pi/5)(9/2\pi) = 2.7$. Because this result is positive, $x_2[n]$ is shifted to the right from $x_1[n]$. Since the result is not an integer, however, the repeating sequence of values seen in $x_2[n]$ does not match the sequence observed in $x_1[n]$.

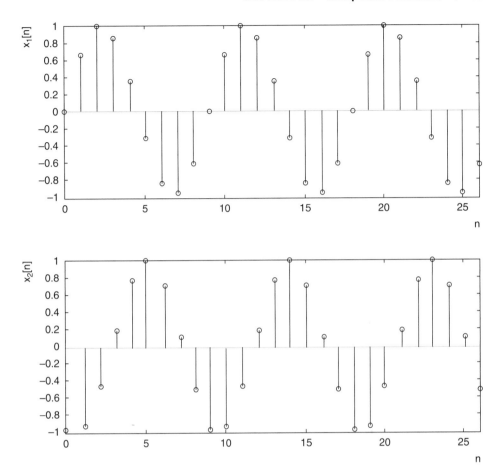

FIGURE 3.31
Phase shifting a digital sinusoid for Example 3.22.

3.4   COMPOSITE FUNCTIONS

Composite functions are combinations of functions. Such combinations give flexibility in defining digital signals that goes beyond the definitions of the elemental signals defined in the previous section. To evaluate a composite function, each basic function is constructed first. Then, the basic signals are multiplied, added, or subtracted, as required. Four composite functions are presented in the following examples.

EXAMPLE 3.23

Draw the signal $x[n] = u[n]u[3-n]$.

This signal is the product of the signals $u[n]$ and $u[3-n]$. As shown in Figure 3.32, the two signals are multiplied point by point to produce $x[n]$. Naturally, the same signal $x[n]$ could alternatively have been expressed as a sum of impulse functions:

$$x[n] = \delta[n] + \delta[n-1] + \delta[n-2] + \delta[n-3]$$

FIGURE 3.32
Product of step functions for
Example 3.23.

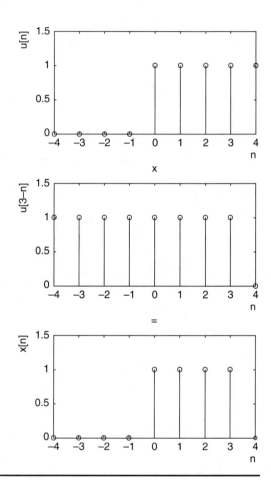

EXAMPLE 3.24

A digital signal is described as $x[n] = e^{-2n}u[n]$.

    a. Draw the signal.

    b. Write the function $x[n-2]$.

    a. As in the previous example, the simplest way to construct this signal is to find each component part and multiply. The $u[n]$ term has the effect of turning on the function e^{-2n} at $n = 0$. Because $u[n]$ is zero for $n < 0$, $x[n]$ is also zero for this range of n. For $n \geq 0$, $u[n]$ has the value one, so the samples of $x[n]$ for this range are the same as samples of e^{-2n} alone, as shown in Figure 3.33.

    b. The shifted signal is $x[n-2] = e^{-2(n-2)}u[n-2]$.

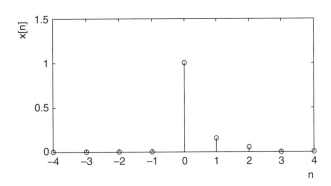

FIGURE 3.33
"Turn on" function for Example 3.24.

EXAMPLE 3.25
Draw the signal $x[n] = 3\sin(n\pi/5 - 1)u[n]$.

The sine wave is multiplied by a step function to produce the signal in Figure 3.34.

FIGURE 3.34
Signal for Example 3.25.

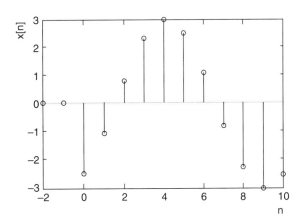

EXAMPLE 3.26
Plot the digital signal $x[n] = 0.5e^{-0.2n}\sin(n\pi/9)u[n]$.

The signal $x[n]$, plotted in Figure 3.35(d), is formed as the product of the three signals shown in (a), (b), and (c). Because $x[n]$ is the product of a decaying exponential and a sinusoid, it is called a **damped sinusoid**.

FIGURE 3.35
Product of functions for Example 3.26.

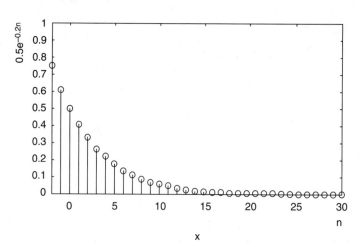

(a)

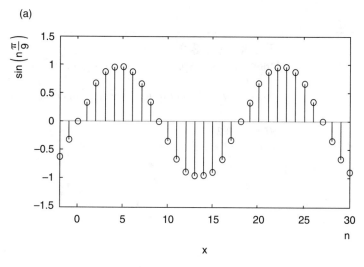

(b)

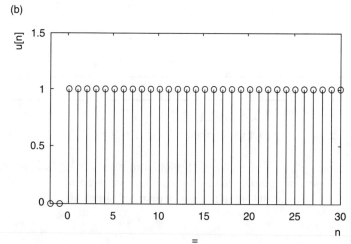

=

(c)

FIGURE 3.35
Continued

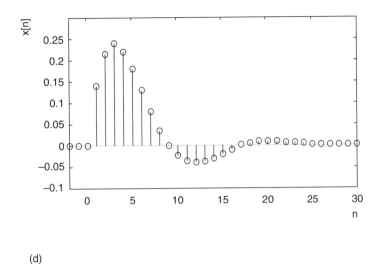

(d)

3.5 TWO-DIMENSIONAL DIGITAL SIGNALS

The digital signals described so far have been one-dimensional. Such digital signals are suitable for describing speech, music, or voltages, but not images. Describing images requires the use of two-dimensional digital signals. A two-dimensional digital signal is a **matrix**, or grid, of numbers. Each number in the matrix corresponds to one **pixel** of the digital image, which records the color of the image at the pixel's location. In a gray scale image x, $x[m, n]$ records a **gray scale level** for the pixel in row m and column n. For an 8-bit gray scale image, $2^8 = 256$ levels are possible, so each gray scale level can take a value between 0 (black) and 255 (white). The pattern of grays produces the shapes in the image, as exemplified by Figure 3.36. With 16 rows and columns of pixels, and 8 bits assigned to each pixel, the image is described as being $16 \times 16 \times 8$-bit, or $16 \times 16 \times 256$. The matrix in Figure 3.37 displays the gray scale levels for each pixel, with lower numbers cor-

FIGURE 3.36
$16 \times 16 \times 256$ digital gray scale image.

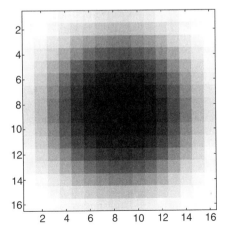

Column

223	208	194	182	172	164	159	158	159	164	172	182	194	208	223	240
208	191	176	163	151	143	137	135	137	143	151	163	176	191	208	225
194	176	159	144	131	121	115	113	115	121	131	144	159	176	194	213
182	163	144	128	113	101	93	90	93	101	113	128	144	163	182	202
172	151	131	113	96	81	71	68	71	81	96	113	131	151	172	193
164	143	121	101	81	64	50	45	50	64	81	101	121	143	164	186
159	137	115	93	71	50	32	23	32	50	71	93	115	137	159	182
158	135	113	90	68	45	23	0	23	45	68	90	113	135	158	180
159	137	115	93	71	50	32	23	32	50	71	93	115	137	159	182
164	143	121	101	81	64	50	45	50	64	81	101	121	143	164	186
172	151	131	113	96	81	71	68	71	81	96	113	131	151	172	193
182	163	144	128	113	101	93	90	93	101	113	128	144	163	182	202
194	176	159	144	131	121	115	113	115	121	131	144	159	176	194	213
208	191	176	163	151	143	137	135	137	143	151	163	176	191	208	225
223	208	194	182	172	164	159	158	159	164	172	182	194	208	223	240
240	225	213	202	193	186	182	180	182	186	193	202	213	225	240	255

Row

FIGURE 3.37

Gray scale values for digital image in Figure 3.36.

responding to the dark heart of the image. For color images, each pixel would have three numbers associated with it, one for each of the red, green, and blue components. Figure 3.38 shows a digital image with a larger number of pixels. This image contains 829 rows and 1173 columns, and each pixel in the image has an 8-bit gray scale value. Thus, the image is 829 × 1173 × 256.

CHAPTER SUMMARY

Matlab
Support

1. Digital signals $x[n]$ are represented by their quantization levels, marked at every sampling instant n by a vertical line with a circle at the top.

2. The function $x[n-m]$ refers to a copy of $x[n]$ shifted m steps to the right. The function $x[kn]$ selects every k^{th} sample from the function $x[n]$.

3. The impulse function $\delta[n]$ has all samples equal to zero, except for the sample at $n = 0$, which has the value one. All digital signals can be expressed as the sum of shifted impulse functions.

4. The step function $u[n]$ has the value zero for all samples $n < 0$, and the value one for all samples $n \geq 0$.

5. Exponential functions $Ae^{\beta n}$ grow when $\beta > 0$ and decay when $\beta < 0$. Complex exponentials $e^{j\beta n}$ may be expressed using Euler's identity as $\cos\beta n + j\sin\beta n$.

6. Digital sine and cosine functions $A\sin(n\Omega)$ and $A\cos(n\Omega)$ are periodic only if $2\pi/\Omega = N/M$, a ratio of integers, where N is the digital period of the sequence. With phase shifts, the sinusoids become $A\sin(n\Omega + \theta)$ and $A\cos(n\Omega + \theta)$.

7. Digital frequency Ω is linked to analog frequency f through the equation $\Omega = 2\pi f/f_S$.

8. Composite digital signals can be produced by adding, subtracting, or multiplying digital signals.

9. Digital images are two-dimensional digital signals. Each pixel in a gray scale image is assigned a gray scale level that records its color.

FIGURE 3.38
Eight-bit digital image.

REVIEW QUESTIONS

3.1 A signal $x[n]$ is depicted in Figure 3.39.
    a. Find:
        (i) $x[0]$
        (ii) $x[3]$
        (iii) $x[-1]$
    b. Sketch:
        (i) $x[n-2]$
        (ii) $x[n+1]$

FIGURE 3.39
Signal for Question 3.1.

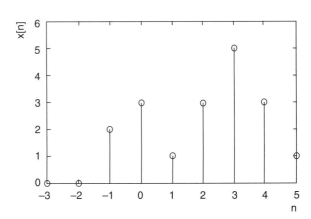

3.2 Evaluate:
  a. $\delta[-4]$
  b. $\delta[0]$
  c. $\delta[2]$

3.3 Draw:
  a. $x[n] = \delta[-n]$
  b. $x[n] = 2\delta[n-2]$
  c. $x[n] = 5\delta[n] - 3\delta[n-1]$

3.4 Evaluate:
  a. $u[-3]$
  b. $u[0]$
  c. $u[2]$

3.5 Draw:
  a. $x[n] = 4u[n-1]$
  b. $x[n] = -2u[n]$
  c. $x[n] = 2u[-n]$
  d. $x[n] = u[n-2]$
  e. $x[n] = u[2-n]$

3.6 Draw:
  a. $x[n] = u[n] + u[n-2]$
  b. $x[n] = u[n] - u[n-2]$

3.7 **a.** Draw the first 10 samples of $x[n] = \displaystyle\sum_{k=1}^{\infty} 0.1u[n-k]$.

  b. Draw the first 10 samples of $x[n] = \displaystyle\sum_{k=0}^{\infty} 0.1k\delta[n-k]$.

3.8 Draw the signal $x[n] = \displaystyle\sum_{k=0}^{2}(u[n-5k] - u[n-5k-2])$.

3.9 Write a function to describe the digital signal in Figure 3.40. Assume that the signal is zero outside the window.

FIGURE 3.40
Signal for Question 3.9.

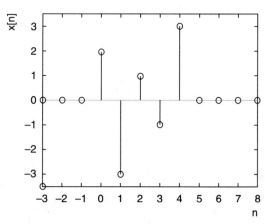

3.10 Write a function to describe the signal in Figure 3.41 using:
a. Impulse functions
b. Step functions

FIGURE 3.41
Signal for Question 3.10.

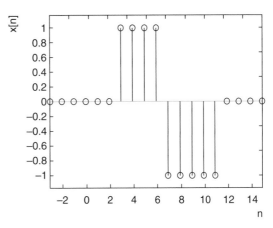

3.11 Write a function that uses step functions to describe the digital signal in Figure 3.42, assuming the signal maintains the value 3 for $n > 8$.

FIGURE 3.42
Signal for Question 3.11.

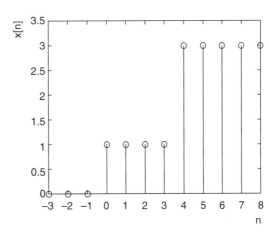

3.12 Plot the first 10 nonzero samples of $x[n] = 3e^{-0.2n} u[n]$.
3.13 Write a function to describe the digital signal in Figure 3.43. Assume that the pattern of the signal continues for $n > 5$.
3.14 **a.** Rewrite $x[n] = e^{-j\pi n/4}$ using Euler's identity.
b. Use the result in (a) to predict the digital period of $x[n]$.
3.15 **a.** Plot $x[n] = e^{-j\pi n/3}$ for $0 \le n \le 5$.
b. Demonstrate that $|e^{-j\pi n/3}|$ is equal to one for each point in (a).
3.16 An analog signal $x(t) = 5\sin 200t$ is sampled every 25 msec. Write the function that describes the digital signal $x[n] = A\sin(n\Omega)$, assuming no quantization errors occur.
3.17 Determine whether the following digital sinusoids are periodic. For those that are periodic, state how many digital samples occur before the sequence repeats.
a. $x[n] = \cos(n4/5)$

FIGURE 3.43

Signal for Question 3.13.

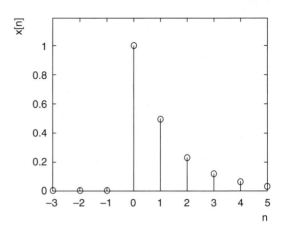

b. $x[n] = \sin(n6\pi/7)$

c. $x[n] = 4\cos(n2\pi/3)$

3.18 Seven samples are collected during every three cycles of the analog signal $x(t) = \cos(1000\pi t)$.

a. What is the sampling rate?

b. Is the sampling rate adequate to avoid aliasing?

3.19 The digital signal $x[n] = 2\sin(n4\pi/5)$ is obtained by sampling an analog signal.

a. Draw a graph of the digital signal for $0 \le n \le 12$.

b. Using Equation (3.7), show that the signal is periodic. How many samples N must be collected before the sequence begins to repeat? How many analog periods elapse during the collection of each group of N samples?

c. Use information from (b) to add a line to the graph in (a) to show how the analog signal fits the digital samples.

3.20 An analog signal $x(t) = \cos(2500\pi t)$. The signal obtained by sampling is $x[n] = \cos(n\pi/3)$. What sampling rates are possible?

3.21 The signal $x_1[n] \sin(n2\pi/9 + 2\pi/5)$ is a phase-shifted version of $x_2[n] = \sin(n2\pi/9)$.

a. Find the size and the direction of the shift.

b. Will the samples in one digital period of $x_1[n]$ be the same as the samples in one digital period of $x_2[n]$?

c. Evaluate $x_1[0]$ and $x_1[1]$.

d. Write a function $x_3[n]$ that phase-shifts $x_2[n]$ by exactly 2 samples to the right. Will the samples in one digital period of $x_3[n]$ match those of $x_2[n]$?

3.22 Draw the following functions for $-2 \le n \le 12$:

a. $x[n] = \sin(n2\pi/5)(u[n] - u[n-9])$

b. $x[n] = e^{0.5n}u[4-n]$

c. $x[n] = e^{-0.1n}\sin(n2\pi/7)u[n-2]$

d. $x[n] = 2\cos(n\pi/4 + \pi/5)u[n]$

3.23 a. A digital sinusoid that is produced through 4 kHz sampling of an analog sinusoid repeats every 10 samples. What possible frequencies can the analog signal have?

b. How is the answer in (a) connected with aliasing?

3.24 How many gray scale values are possible in a 16-bit image?

3.25 Describe the gray scale values for the 128×128 pixel, 8-bit image of a black and white checkerboard. (A checkerboard contains eight rows and eight columns.)

4

DIFFERENCE EQUATIONS AND FILTERING

This chapter introduces filtering, one of the essential elements of digital signal processing. Digital filters are implemented using difference equations. This chapter:

> ➢ introduces basic filtering concepts, including gain and bandwidth
> ➢ compares analog and digital filters
> ➢ defines what is meant by a linear, time-invariant, causal system
> ➢ implements filters using difference equations, both recursive and nonrecursive
> ➢ explains superposition, the combination of multiple inputs to a filter
> ➢ indicates how difference equations may be expressed in diagram form, including implementation considerations
> ➢ introduces the impulse response
> ➢ introduces the step response

4.1 FILTERING BASICS

Digital signal processing systems can record, reproduce, or transform digital signals. Filters are systems that change a signal by altering its frequency characteristics in a specific way. Common categories of filters include high pass, low pass, band pass, and band stop. A filter might be used, for example, to remove the background hiss from an old music recording. For this purpose a low pass filter would be needed, to retain the low and medium

FIGURE 4.1
DTMF tones.

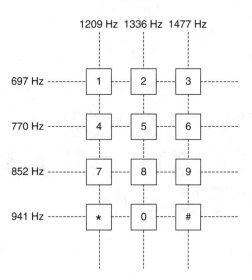

frequencies that contain the major elements of the musical content, but to reduce the high frequency elements that include the hiss. For sonar applications, target identification can be made easier if low frequency ship and sea noises are removed from the sonar signal. To do this job, a high pass filter is needed. A high pass filter will retain the target signatures while eliminating unwanted low frequency noise. Band pass filters can be used for the decoding of DTMF (dual tone multifrequency) signals in digital telephone systems. As Figure 4.1 shows, each key press is characterized by a pair of frequencies from a keypad grid. At the receiver, a bank of band pass filters can distinguish one key press from another by detecting which two of the seven frequencies are present in the signal. Band stop filters allow all frequencies except those in a specified band to pass. This type of filter is useful for masking out a single tone in a signal or for suppressing narrow band noise, which occupies a narrow range of frequencies only.

The easiest way to describe what a filter does is to present its shape in the frequency domain. Figure 4.2 shows the shapes of the most common filters. Ideal filters have perfect rectangular shapes. These filters are not rectangular because they are nonideal. The higher the order of a filter, the steeper its **roll-off**, and the more closely it approximates ideal behavior. A filter's **gain**[1] at a certain frequency determines the amplification factor that the filter applies to an input at this frequency. A gain may have any value. A filter is said to pass signals in the range of frequencies where filter gains are high, called the **pass band** of the filter, and attenuate, or block, signals in the range of frequencies where filter gains are low, called the filter's **stop band**. The cut-off frequency of the filter occurs where the gain is $1/\sqrt{2} \approx 0.707$ of its maximum value. This point is often also considered the edge of the

[1]   Unless otherwise specified, the term "gain" in this text refers to voltage gain.

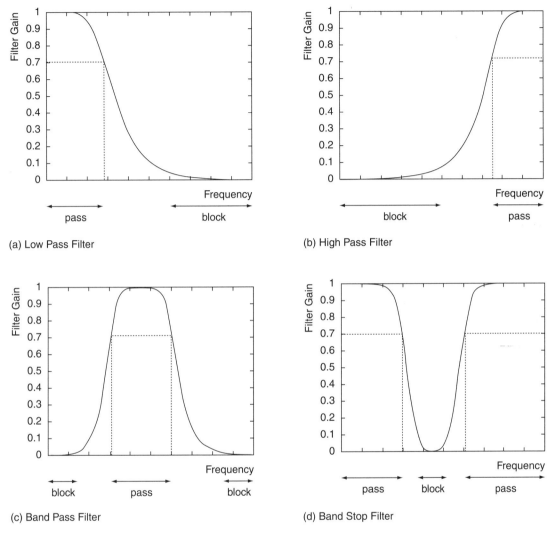

FIGURE 4.2
Common filter types.

pass band. Gains are frequently expressed in decibels, or dB. A gain in decibels is calculated as

$$\text{Gain (dB)} = 20 \log (\text{Gain})$$

Thus, a gain of 0.707 corresponds to a gain in decibels of -3 dB. For this reason, the cutoff frequencies are also referred to as -3 dB frequencies. These points define the **bandwidth** of a filter, the range of frequencies for which signals are passed.

EXAMPLE 4.1

Find the bandwidth of the band pass filter shown in Figure 4.3.

FIGURE 4.3

Band pass filter for Example 4.1.

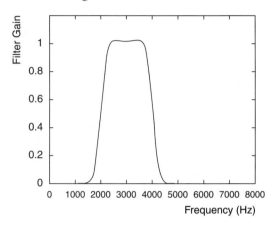

The edges of the pass band occur where the gain equals 0.707. As shown in Figure 4.4, the bandwidth of the filter is just under 2000 Hz.

FIGURE 4.4

Bandwidth calculation for Example 4.1.

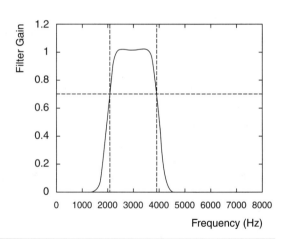

Each filter type has a unique effect on an input signal. Low pass filters tend to smooth signals by averaging out sudden changes. High pass filters, on the other hand, tend to emphasize sharp transitions. Figure 4.5 illustrates this point using an analog square wave signal. Note how the low and high pass results combined tend to reconstitute the original square wave. Figure 4.6 demonstrates how filtering a speech signal in a variety of ways can give useful information about its frequency content. The word "two" has been sampled at 22.05 kHz and filtered using a group of tenth order Butterworth filters (to be introduced in Section 10.4). Low pass, high pass, and band pass filters are used. Their shapes are plotted as gains against frequency, where frequency is measured in Hz. The original signal, shown

FIGURE 4.5

Effects of low and high pass filters.

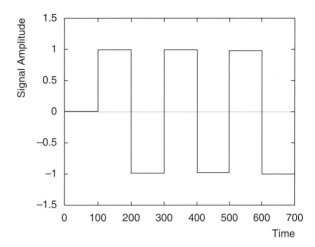

(a) Square Wave

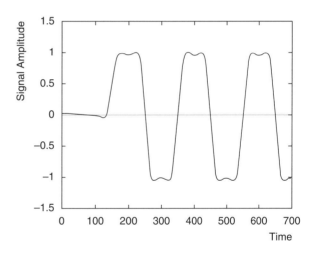

(b) Low Pass Filtered Square Wave

in Figure 4.6(a), consists of two main parts: a noisy high frequency burst from the "t" sound, and a lower frequency periodic portion due to the "oo" sound, as evidenced by the train of regularly spaced spikes between samples 3200 and 5000. The low pass filter picks out the low frequency part while attenuating the high frequency content, as shown in Figure 4.6(b). As Figure 4.6(c) shows, the original signal does not have much frequency content in the range picked up by the high pass filter. Most of the initial burst due to the "t" sound consists of the midrange frequencies picked up by the band pass filter, as seen in Figure 4.6(d). Figure 4.6(e) is the sum of low pass, high pass, and band pass filtered signals. It closely approximates the original signal, differing slightly only because the filters used are not ideal.

FIGURE 4.5
Continued

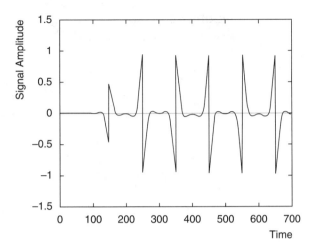

(c) High Pass Filtered Square Wave

FIGURE 4.6
Filtering a speech signal.

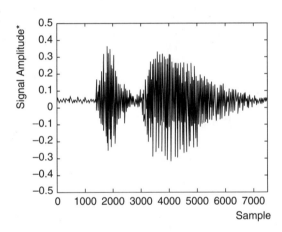

(a) The Original Speech Sample ("TWO")

4.2 ANALOG FILTERS VERSUS DIGITAL FILTERS

Signal filtering as described in the previous section can be accomplished using either analog filters or digital filters. Digital filters, however, offer advantages that cannot be duplicated by analog filters and that make them the preferred choice for many applications. Analog filters are circuits, built from components like resistors, capacitors, and inductors. As such, their operation is quite sensitive to the values of the components used, and the behavior of some of these components can change significantly with temperature. Digital filters, on the other hand, are defined in software and rely very little on hardware. Filtering software is nothing more than a list of program instructions. It does run on a hardware platform, frequently a single integrated circuit chip, but the hardware does not determine filter

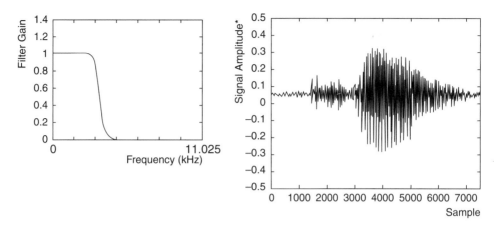

(b) The Low Pass Filter and the Speech Signal After Low Pass Filtering

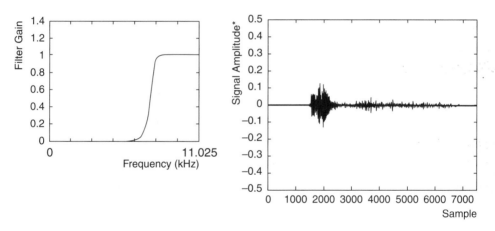

(c) The High Pass Filter and the Speech Signal After High Pass Filtering

FIGURE 4.6

Continued

behavior: This is defined by a list of numerical coefficients. Redesigning a digital filter, then, is a simple matter of redefining the coefficients provided to the filter program. In fact, digital filter coefficients can be changed "on the fly" to change the filter's behavior while it is operating. Redesigning an analog filter, on the other hand, requires completely re-designing and rebuilding a circuit. As the order of the filter increases, more and more com-ponents are required, multiplying the difficulty of dealing with component tolerances. By contrast, a higher order digital filter is easy to produce: The list of coefficients used by the filter is just longer than before. This feature makes digital filters incredibly flexible and easy to use.

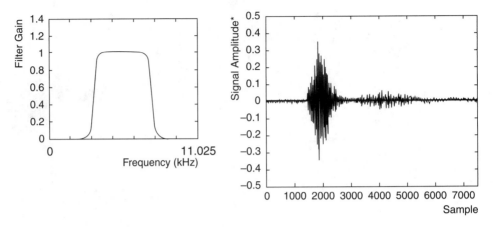

(d) The Band Pass Filter and the Speech Signal After Band Pass Filtering

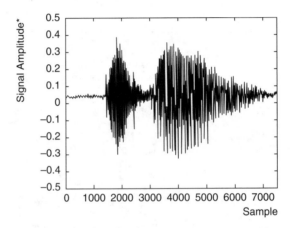

(e) The Sum of the Low Pass, High Pass, and Band Pass
Filtered Signals

*For clarity, the envelopes rather than the individual
samples of the signals are plotted.

FIGURE 4.6
Continued

Digital circuits, in which signals may have values only at a defined set of quantization levels, are inherently less sensitive to electrical noise than analog circuits, in which signals may take any one of an infinitely fine gradation of values. While it is an advantage in this sense, quantization is, of course, also a source of noise, affecting digital, but not analog, signals. Aliasing, caused by sampling, is another source of error that affects only digital signals. These unavoidable errors are the reason that the most demanding audiophiles still insist that the best results for high-end audio systems are obtained with analog circuitry. Unless a great deal of care is taken to reduce noise in analog circuits, though, the effects of noise are, in most cases, greater than the combined effects of aliasing and quantization.

FIGURE 4.7
Linear systems.

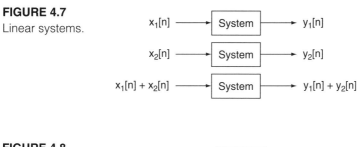

FIGURE 4.8
Time invariance.

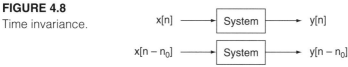

Digital filter programs implement filtering in two main ways. The first uses the filter's difference equation, introduced in the next section, to calculate filter outputs. The second uses a process called convolution, discussed in Chapter 5, to compute outputs. The latter method requires the filter's impulse response, introduced in Section 4.7.

4.3   LINEAR, TIME-INVARIANT, CAUSAL SYSTEMS

A **system** is an entity that has some effect on a signal. While a filter is really only a particular example of a system, the terms are sometimes used interchangeably in the DSP literature. This text deals with linear, time-invariant digital systems only. **Linear systems** are those that obey **superposition**, which means that when input x_1 produces output y_1 and input x_2 produces output y_2, then an input that is the sum of input x_1 and input x_2 will produce an output that is the sum of output y_1 and output y_2. Figure 4.7 illustrates this idea. Linear systems have the form

$$\text{output} = a\,(\text{input}_1) + b\,(\text{input}_2) + c\,(\text{input}_3) + \dots$$

where a, b, and c are weighting factors. It is easy to show that a nonlinear system does not obey this property. Take, for example, a system that squares the input, that is,

$$\text{output} = (\text{input})^2$$

Input x_1 will give output x_1^2, input x_2 will give output x_2^2, but the input $(x_1 + x_2)$ gives the output $(x_1 + x_2)^2$, which is not the same as the sum of x_1^2 and x_2^2. Systems that are not linear can often be approximated as linear systems over narrow ranges of inputs in order to make use of linear digital techniques. Nonlinear methods are much more complex.

Time-invariant systems give the same output for an input no matter when that input is applied. In other words, if the input is delayed, then the output is delayed by the same amount. Figure 4.8 demonstrates this idea.

One further property, causality, is possessed by all practical systems. **Causal systems** are those whose outputs depend on present and previous data about the system, but never on future data. Causality is a normal attribute of physical systems.

4.4 DIFFERENCE EQUATION STRUCTURE

Difference equations can be used to describe how a linear, time-invariant, causal digital filter works. In general, an output from the filter can depend on present and previous inputs and also on past outputs. The letter x is usually used to refer to the filter inputs and the letter y to refer to the filter outputs. If the present input value is $x[n]$, then the preceding input is $x[n-1]$, and the one before that is $x[n-2]$. There is a time delay of one sample period between each of these values. Similarly, the past outputs are described as $y[n-1]$, $y[n-2]$, and so on. The most general expression of the difference equation is

$$a_0y[n] + a_1y[n-1] + a_2y[n-2] + \ldots + a_Ny[n-N]$$

$$= b_0x[n] + b_1x[n-1] + b_2x[n-2] + \ldots + b_Mx[n-M] \quad \textbf{(4.1)}$$

with the outputs on the left and the inputs on the right. The a_k and b_k values are weightings that determine the contribution of each input and output sample. These weightings are called **filter coefficients**. The general difference equation shown in Equation (4.1) has $N+1$ a_k coefficients and $M+1$ b_k coefficients. The equation can be presented more compactly as

$$\sum_{k=0}^{N} a_k y[n-k] = \sum_{k=0}^{M} b_k x[n-k] \quad \textbf{(4.2)}$$

where N is the number of past outputs required, normally referred to as the order of the filter, and M is the number of past inputs required.

There is no reason why a_0 cannot be set to 1: If necessary, all the other coefficients can be adjusted by dividing by a_0. For example,

$$2y[n] + y[n-1] = 3\,x[n]$$

becomes

$$y[n] + 0.5\,y[n-1] = 1.5\,x[n]$$

after dividing by two throughout. Once a_0 is equal to one, Equation (4.2) can be rearranged to obtain a new general expression for $y[n]$.

$$y[n] = -\sum_{k=1}^{N} a_k y[n-k] + \sum_{k=0}^{M} b_k x[n-k]$$

$$= -a_1y[n-1] - a_2y[n-2] - \ldots - a_Ny[n-N]$$

$$+ b_0x[n] + b_1x[n-1] + b_2x[n-2] + \ldots + b_Mx[n-M] \quad \textbf{(4.3)}$$

This form shows how each new output from the digital filter can be calculated using past outputs, present input, and past inputs. When a digital system relies on both inputs and past outputs, it is referred to as a **recursive filter**. Equation (4.3) gives the general form for this kind of filter. Recursive filters are studied in Chapter 10. When the digital filter relies only on inputs, and not on past outputs, it is referred to as a **nonrecursive filter.** Equation (4.4) gives the general form for this kind of filter:

$$y[n] = \sum_{k=0}^{M} b_k x[n - k]$$

$$= b_0 x[n] + b_1 x[n-1] + b_2 x[n-2] + \dots + b_M x[n-M] \qquad \textbf{(4.4)}$$

Nonrecursive filters are studied in Chapter 9. The following examples illustrate how difference equations are used.

EXAMPLE 4.2

A filter has the difference equation

$$y[n] = 0.5\, y[n-1] + x[n] \qquad \textbf{(4.5)}$$

a. Identify all coefficients a_k and b_k.
b. Is this a recursive or nonrecursive difference equation?
c. If the input $x[n]$ is as given in Figure 4.9, find the first 12 samples of the output, starting with $n = 0$.

FIGURE 4.9
Filter input for
Example 4.2.

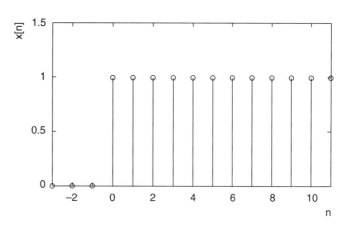

a. The values of the coefficients are detected most easily when the difference equation is rewritten with outputs on the left and inputs on the right:

$$y[n] - 0.5\, y[n-1] = x[n]$$

Using Equation (4.1), the filter coefficients can be identified as: $a_0 = 1.0$, $a_1 = -0.5$, and $b_0 = 1.0$. All coefficients other than a_0, a_1, and b_0 are zero.

b. Since the output $y[n]$ depends on a past output $y[n-1]$, the digital filter is recursive.

c. The outputs may be found by evaluating Equation (4.5) over and over, beginning with $n = 0$. The only difficulty arises for this first case, since the calculation requires the output $y[-1]$. In this text, all digital filters are assumed to be causal. This means that the output cannot begin to change until the input first becomes nonzero, at $n = 0$ in this case. Therefore, all outputs before $y[0]$ may be assumed to be zero. Once $y[0]$ has been calculated, it can be used for the calculation of $y[1]$. The first 12 output samples are:

$$y[0] = 0.5\, y[-1] + x[0] = 0.5(0.0000) + 1.0 = 1.0000$$
$$y[1] = 0.5\, y[0] + x[1] = 0.5(1.0000) + 1.0 = 1.5000$$
$$y[2] = 0.5\, y[1] + x[2] = 0.5(1.5000) + 1.0 = 1.7500$$
$$y[3] = 0.5\, y[2] + x[3] = 0.5(1.7500) + 1.0 = 1.8750$$
$$y[4] = 0.5\, y[3] + x[4] = 0.5(1.8750) + 1.0 = 1.9375$$
$$y[5] = 0.5\, y[4] + x[5] = 0.5(1.9375) + 1.0 = 1.9688$$
$$y[6] = 0.5\, y[5] + x[6] = 0.5(1.9688) + 1.0 = 1.9844$$
$$y[7] = 0.5\, y[6] + x[7] = 0.5(1.9844) + 1.0 = 1.9922$$
$$y[8] = 0.5\, y[7] + x[8] = 0.5(1.9922) + 1.0 = 1.9961$$
$$y[9] = 0.5\, y[8] + x[9] = 0.5(1.9961) + 1.0 = 1.9980$$
$$y[10] = 0.5\, y[9] + x[10] = 0.5(1.9980) + 1.0 = 1.9990$$
$$y[11] = 0.5\, y[10] + x[11] = 0.5(1.9990) + 1.0 = 1.9995$$

The output is plotted in Figure 4.10. Since the input has a constant value, the output ultimately also approaches a constant value.

FIGURE 4.10
Output for Example 4.2

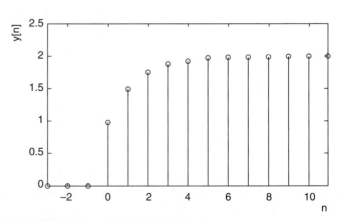

EXAMPLE 4.3

$$y[n] = 0.5\, x[n] - 0.3\, x[n-1]$$

a. Identify all coefficients a_k and b_k.
b. Is this a recursive or nonrecursive difference equation?
c. For input $x[n] = \sin(n2\pi/9)u[n]$, find the first 20 samples of the output.

a. $a_0 = 1.0$ $b_0 = 0.5$ $b_1 = -0.3$
b. The output does not depend on past outputs, so the digital filter is nonrecursive.
c. Because of the $u[n]$ factor in the input, the values of the input before $n = 0$ are zero. Table 4.1 shows the first 20 values of the inputs and the outputs. Figure 4.11 plots both the input and output samples. Notice that, while the amplitudes and phases of the two signals differ, they both have a sinusoidal nature and an identical digital period.

TABLE 4.1
Input and Output Values for Example 4.3

n	-1	0	1	2	3	4	5
$x[n]$	0.000	0.000	0.643	0.985	0.866	0.342	-0.342
$y[n]$	0.000	0.000	0.321	0.300	0.138	-0.089	-0.274

n	6	7	8	9	10	11	12
$x[n]$	-0.866	-0.985	-0.643	0.000	0.643	0.985	0.866
$y[n]$	-0.330	-0.233	-0.026	0.193	0.321	0.300	0.138

n	13	14	15	16	17	18	19
$x[n]$	0.342	-0.342	-0.866	-0.985	-0.643	0.000	0.643
$y[n]$	-0.089	-0.274	-0.330	-0.233	-0.026	0.193	0.321

FIGURE 4.11
Input and output signals for Example 4.3.

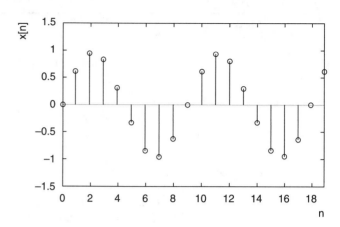

(a) Input

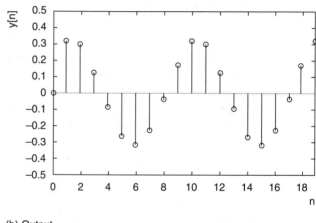

(b) Output

4.5   SUPERPOSITION

In some instances, several inputs may be applied to a filter at the same time. When this happens, the filter responds to the sum of these inputs through superposition. Fortunately, when the filter is linear, multiple inputs can be handled easily. As Figure 4.7 suggests, the overall output can be found in one of two possible ways. The first method applies each input separately, computes a distinct output for each, and sums these outputs to produce a final, overall output. The second method sums the inputs together immediately and applies the combined input to the filter to obtain the output. Both methods produce the same answer, as illustrated in Example 4.4.

EXAMPLE 4.4

A filter is described by the difference equation

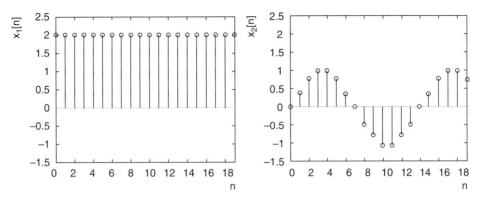

FIGURE 4.12
Input signals for Example 4.4.

$$y[n] = x[n] + 0.5x[n-1]$$

Two inputs, plotted in Figure 4.12, are summed and applied to the filter. They are:

$$x_1[n] = 2u[n]$$

$$x_2[n] = \sin\left(\frac{n\pi}{7}\right)u[n]$$

Find and plot the first 20 samples of the output resulting from the combined effect of the two inputs.

The input appears in two parts: $x_1[n] = 2u[n]$ and $x[n] = \sin(n\pi/7)u[n]$. The outputs $y_1[n]$ and $y_2[n]$ due to each input can be found independently, and the overall output found by summing the two output components. Note that inputs prior to $n = 0$ are zero. The first 20 output values are calculated in Table 4.2 and plotted in Figure 4.13. The row of calculations for $n = 3$ is highlighted in Table 4.2 for illustration. The part of the output due to $x_1[n]$, $y_1[n]$, is given by

$$y_1[n] = x_1[n] + 0.5x_1[n-1]$$

Thus, $y_1[3] = x_1[3] + 0.5x_1[2] = 2 + 0.5(2) = 3$. The part of the output due to $x_2[n]$, $y_2[n]$ is given by

$$y_2[n] = x_2[n] + 0.5x_2[n-1]$$

Thus, $y_2[3] = x_2[3] + 0.5x_2[2] = 0.975 + 0.5(0.782) = 1.37$. The total output due to both inputs is found as the sum of $y_1[n] + y_2[n]$. For $n = 3$, $y_1[3] + y_2[3] = 3 + 1.37 = 4.37$. This same result can be obtained by combining both portions of the input together. In the highlighted row, $y[3] = x[3] + 0.5x[2] = 2.975 + 0.5(2.782) = 4.37$. Indeed, the final output column $y_1[n] + y_2[n]$ can be computed directly from the summed inputs column $x_1[n] + x_2[n]$ without completing the intermediate columns for $y_1[n]$ and $y_2[n]$.

TABLE 4.2
Calculation of Output Samples for Example 4.4

n	$x_1[n]$	$x_2[n]$	$x[n] = x_1[n]+x_2[n]$	$y_1[n]$	$y_2[n]$	$y[n] = y_1[n]+y_2[n]$
0	2	0.00	2.00	2	0.00	2.00
1	2	0.434	2.434	3	0.43	3.43
2	2	0.782	2.782	3	1.00	4.00
3	2	0.975	2.975	3	1.37	4.37
4	2	0.975	2.975	3	1.46	4.46
5	2	0.782	2.782	3	1.27	4.27
6	2	0.434	2.434	3	0.82	3.82
7	2	0.000	2.000	3	0.22	3.22
8	2	−0.434	1.566	3	−0.43	2.57
9	2	−0.782	1.218	3	−1.00	2.00
10	2	−0.975	1.025	3	−1.37	1.63
11	2	−0.975	1.025	3	−1.46	1.54
12	2	−0.782	1.218	3	−1.27	1.73
13	2	−0.434	1.566	3	−0.82	2.18
14	2	0.000	2.000	3	−0.22	2.78
15	2	0.434	2.434	3	0.43	3.43
16	2	0.782	2.782	3	1.00	4.00
17	2	0.975	2.975	3	1.37	4.37
18	2	0.975	2.975	3	1.46	4.46
19	2	0.782	2.782	3	1.27	4.27

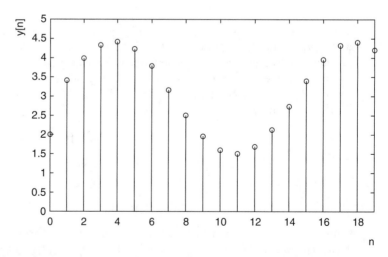

FIGURE 4.13
Output signal for Example 4.4.

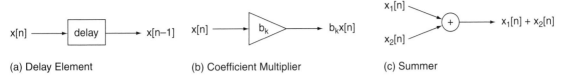

(a) Delay Element   (b) Coefficient Multiplier   (c) Summer

FIGURE 4.14
Difference equation diagram elements.

FIGURE 4.15
Nonrecursive difference equation diagram.

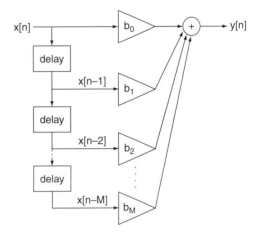

4.6   DIFFERENCE EQUATION DIAGRAMS

4.6.1   Nonrecursive Difference Equations

A nonrecursive difference equation requires only past and present inputs for the calculation of each new output. No past outputs are needed. All of the required inputs can be presented in diagram form by using delay elements like the one defined in Figure 4.14(a). The coefficients are provided as multipliers in triangles, as shown in Figure 4.14(b). Finally, all elements are added together using a summer, shown in Figure 4.14(c).

A general nonrecursive difference equation, as described in Equation (4.4), can be presented schematically as shown in Figure 4.15. Diagrams like this are often used as models for implementing digital filters in software. All such implementations require that the coefficients defining the filter be quantized. The quantization errors that result contribute to inaccuracies that accumulate as calculations proceed. All effects that are produced because the number of bits available to a processor is limited are called **finite word length effects**. In choosing a suitable implementation for a filter, steps are always taken to minimize these effects. One effective strategy is to break a high order filter into second order chunks, with two delay blocks each, cascaded together as suggested in Figure 4.16. When a high order filter is factored in this way, the coefficients in each second order section are,

FIGURE 4.16

Cascaded second order nonrecursive filter sections.

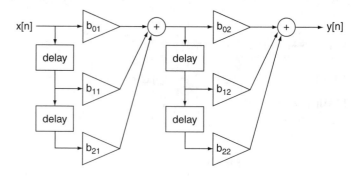

on average, larger than the coefficients of the original filter, and therefore less sensitive to quantization errors. When the order of the filter is odd, a single first order section is added to the group of second order sections. The methods for decomposing a difference equation into second order sections are normally handled by signal processing software and will not be discussed in this text, but an illustration will help to suggest the power of the technique. For example, the fourth order difference equation

$$y[n] = x[n] - 0.1x[n-1] + 0.1x[n-2] - 0.06x[n-3] + 0.02x[n-4]$$

can be implemented in two stages described by the pair of difference equations:

$$y_1[n] = x_1[n] - 0.5x_1[n-1] + 0.1x_1[n-2]$$
$$y_2[n] = x_2[n] + 0.4x_2[n-1] + 0.2x_2[n-2]$$

To check that this pair of equations has the same effect as the original difference equation, the output $y_1[n]$ from the first difference equation is applied as the input $x_2[n]$ to the second difference equation, that is, $x_2[n] = y_1[n]$. With this,

$$y_2[n] = x_2[n] + 0.4x_2[n-1] + 0.2x_2[n-2] = y_1[n] + 0.4y_1[n-1] + 0.2y_1[n-2]$$

$$= (x_1[n] - 0.5x_1[n-1] + 0.1x_1[n-2]) + 0.4(x_1[n-1] - 0.5x_1[n-2] + 0.1x_1[n-3])$$

$$+ 0.2(x_1[n-2] - 0.5x_1[n-3] + 0.1x_1[n-4])$$

$$= x_1[n] - 0.1x_1[n-1] + 0.1x_1[n-2] - 0.06x_1[n-3] + 0.02x_1[n-4]$$

which reproduces the original difference equation. Combining second order sections is easier to accomplish using z transforms, as will be seen in Section 6.2.4. Examples 4.5, 4.6, and 4.7 provide practice in producing and interpreting difference equation diagrams.

EXAMPLE 4.5

Draw a diagram for the difference equation

$$y[n] = 0.5\, x[n] + 0.4\, x[n-1] - 0.2\, x[n-2]$$

The diagram for this difference equation is shown in Figure 4.17.

FIGURE 4.17
Difference equation diagram for
Example 4.5.

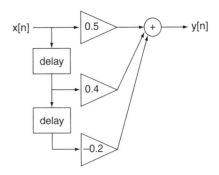

EXAMPLE 4.6
Write the difference equation that corresponds to the diagram in Figure 4.18.

FIGURE 4.18
Difference equation diagram for
Example 4.6.

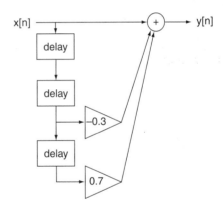

The multiplier for $x[n]$ is unity and the $x[n-1]$ term is absent. The difference equation is

$$y[n] = x[n] - 0.3\, x[n-2] + 0.7\, x[n-3]$$

EXAMPLE 4.7
Find the difference equation that corresponds to the cascaded sections shown in Figure
4.19.

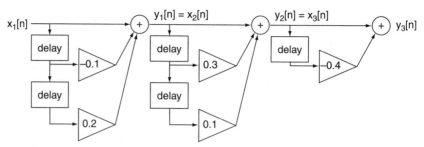

FIGURE 4.19
Difference equation diagram for Example 4.7.

The first stage produces the difference equation

$$y_1[n] = x_1[n] - 0.1x_1[n-1] + 0.2x_1[n-2]$$

The second stage produces the difference equation

$$y_2[n] = x_2[n] + 0.3x_2[n-1] + 0.1x_2[n-2]$$

The third stage produces the difference equation

$$y_3[n] = x_3[n] - 0.4x_3[n-1]$$

Setting the output of the first stage equal to the input of the second stage, or $x_2[n] = y_1[n]$, and setting the output of the second stage equal to the input of the third stage, or $x_3[n] = y_2[n]$, gives the overall output of the cascaded system $y_3[n]$. Beginning with the difference equation for the third stage and substituting the difference equation for the second stage gives:

$$y_3[n] = x_3[n] - 0.4x_3[n-1] = y_2[n] - 0.4y_2[n-1]$$

$$= (x_2[n] + 0.3x_2[n-1] + 0.1x_2[n-2]) - 0.4(x_2[n-1] + 0.3x_2[n-2] + 0.1x_2[n-3])$$

$$= x_2[n] - 0.1x_2[n-1] - 0.02x_2[n-2] - 0.04x_2[n-3]$$

Substituting the difference equation for the first stage into this expression gives:

$$y_3[n] = x_2[n] - 0.1x_2[n-1] - 0.02x_2[n-2] - 0.04x_2[n-3]$$

$$= y_1[n] - 0.1y_1[n-1] - 0.02y_1[n-2] - 0.04y_1[n-3]$$

$$= (x_1[n] - 0.1x_1[n-1] + 0.2x_1[n-2])$$

$$- 0.1(x_1[n-1] - 0.1x_1[n-2] + 0.2x_1[n-3])$$

$$- 0.02(x_1[n-2] - 0.1x_1[n-3] + 0.2x_1[n-4])$$

$$- 0.04(x_1[n-3] - 0.1x_1[n-4] + 0.2x_1[n-5])$$

$$= x_1[n] - 0.2x_1[n-1] + 0.19x_1[n-2] - 0.058x_1[n-3] - 0.008x_1[n-5]$$

which is the difference equation for the filter as a whole. This difference equation is much easier to obtain using z transforms, as shown in Example 6.14.

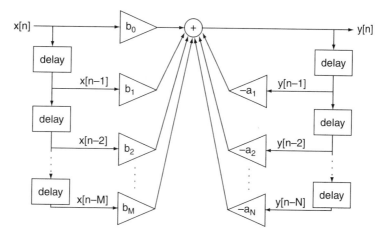

FIGURE 4.20
Recursive difference equation diagram, direct form 1 realization.

4.6.2 Recursive Difference Equations

4.6.2.1 Direct Form 1 Realization Using the diagram elements introduced in Section 4.6.1, recursive difference equations can also be represented in a general way in diagram form. The diagram in Figure 4.20 follows the form of Equation (4.3), where a_0 is assumed to be unity. This type of diagram is called a direct form 1 realization of a recursive difference equation. Calculations for $y[n]$ using this form require $M + 1$ input states, N output states, $M + N + 1$ coefficient multiplications, and $M + N$ additions. When more than two or three delays are needed, this particular implementation is very sensitive to the finite word length effects mentioned in the last section. Again, a solution is to cascade second order blocks together, using a first order block as well if the total number of output delays, N, is odd. Other realizations of recursive filters can give even more implementation advantages, as the next section will show.

EXAMPLE 4.8
Draw a direct form 1 difference equation diagram to describe the recursive digital filter

$$y[n] + 0.5y[n-2] = 0.8x[n] + 0.1x[n-1] - 0.3x[n-2]$$

From the equation, $a_0 = 1.0$, $a_1 = 0.5$, $b_0 = 0.8$, $b_1 = 0.1$, and $b_2 = -0.3$. However, the diagram is easiest to build if the difference equation is rearranged as

$$y[n] = -0.5y[n-2] + 0.8x[n] + 0.1x[n-1] - 0.3x[n-2]$$

This yields the diagram in Figure 4.21.

FIGURE 4.21
Difference equation
diagram for Example 4.8.

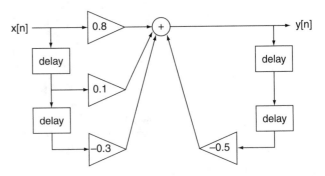

EXAMPLE 4.9

Write the difference equation that corresponds to the diagram in Figure 4.22.

FIGURE 4.22
Difference equation
diagram for Example 4.9.

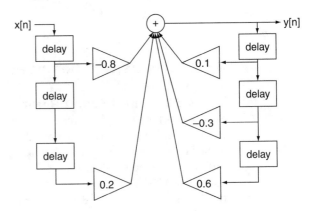

The difference equation is

$$y[n] = 0.1\, y[n-1] - 0.3\, y[n-2] + 0.6\, y[n-3] - 0.8\, x[n-1] + 0.2\, x[n-3]$$

4.6.2.2 Direct Form 2 Realization The direct form 1 realization described in Section 4.6.2.1 is not the most efficient way to implement a recursive difference equation. Much less storage is needed if a direct form 2 realization, derived in Appendix C, is used instead. This realization requires the use of an intermediate signal $w[n]$ that records salient information about the history of the filter in place of past inputs and past outputs. Instead of N past outputs and M past inputs, only $\max(N, M)$ samples of the intermediate signal need to be remembered to produce new filter outputs. The two equations that define the direct form 2 realization are:

FIGURE 4.23

Recursive difference
equation diagram, direct
form 2 realization.

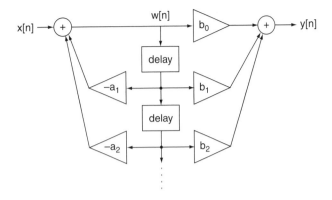

$$w[n] = x[n] - \sum_{k=1}^{N} a_k w[n - k] \tag{4.6}$$

$$y[n] = \sum_{k=0}^{N} b_k w[n - k] \tag{4.7}$$

where, again, a_0 is assumed to be unity. Equation (4.6) computes a new value for the intermediate signal $w[n]$, while Equation (4.7) generates a new filter output $y[n]$. The two equations together can be used to construct the diagram for the direct form 2 realization shown
in Figure 4.23. Calculations for $y[n]$ using this implementation require $\max(N, M) + 1$ intermediate states, $M + N + 1$ coefficient multiplications, and $M + N$ additions. Thus, the
savings in using direct form 2 over direct form 1 come from a reduced need to store old input and output states. Because it requires two summers, though, the direct form 2 realization runs a greater risk of causing arithmetic overflows when implemented in DSP hardware. Nevertheless, storage efficiencies make direct form 2, also called canonical form, a
popular filter realization choice.

The transpose of the direct form 2 realization, illustrated in Figure 4.24, provides another common implementation model, obtained by reversing the flows of information from
Figure 4.23. For a second order transposed model with two delay elements, the output of
the bottom summer in the figure is $b_2 x[n] - a_2 y[n]$. These terms are delayed by one step on
the way up to give $b_2 x[n-1] - a_2 y[n-1]$. At the middle summer, two additional terms are
added, to give an output of $b_2 x[n-1] - a_2 y[n-1] + b_1 x[n] - a_1 y[n]$. These terms are in
turn delayed to give $b_2 x[n-2] - a_2 y[n-2] + b_1 x[n-1] - a_1 y[n-1]$, which is added to one
final term, $b_0 x[n]$, at the top summer. The result is

$$y[n] = b_2 x[n-2] - a_2 y[n-2] + b_1 x[n-1] - a_1 y[n-1] + b_0 x[n]$$

which defines a general second order recursive difference equation with $a_0 = 1$. To minimize
finite word length effects, second order blocks of the direct form 2 realization or its transpose
are frequently connected together in cascade or parallel combinations to produce high order filter implementations. For example, a seventh order filter can be constructed from three second

FIGURE 4.24
Transpose of direct
form 2 realization.

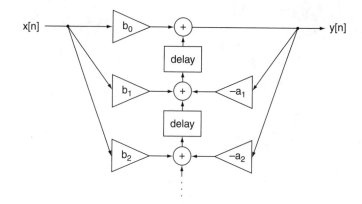

order sections plus a first order section. As a rule, the coefficients in second order sections tend
to be larger than the original coefficients, lessening the effects of quantization.

EXAMPLE 4.10

Find the difference equation for the filter whose difference equation diagram appears in
Figure 4.25.

FIGURE 4.25
Difference equation
diagram for Example
4.10.

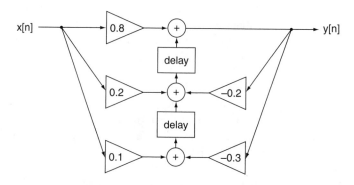

The output from the bottom summer is $0.1x[n] - 0.3y[n]$. This output is delayed by one step
to give $0.1x[n-1] - 0.3y[n-1]$. The output of the middle summer in the first stage is, then,
$0.1x[n-1] - 0.3y[n-1] + 0.2x[n] - 0.2y[n]$. This output is delayed by one more step and
added to $0.8x[n]$ to produce:

$$y[n] = 0.1x[n-2] - 0.3y[n-2] + 0.2x[n-1] - 0.2y[n-1] + 0.8x[n]$$

This is the difference equation for the filter. Expressed in a more standard order:

$$y[n] = -0.2y[n-1] - 0.3y[n-2] + 0.8x[n] + 0.2x[n-1] + 0.1x[n-2]$$

Several other realization structures exist for recursive filters (Ifeachor and Jervis,
1993; Oppenheim and Schafer, 1999). The goal is to choose a realization that minimizes the

FIGURE 4.26

Impulse response.

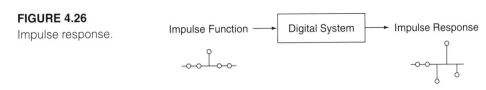

piano.wav

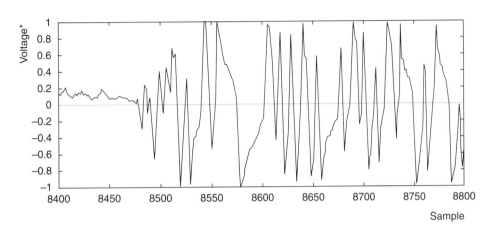

(a) Piano

FIGURE 4.27

Impulse responses for musical instruments.

need for storage of old data, that reduces the number of multiplications and additions required, and that is as insensitive as possible to the finite word length effects that are inevitable in computer calculations.

4.7   THE IMPULSE RESPONSE

The **impulse response** for a filter is, as the name suggests, the response of the filter to an impulse. In other words, when the input to a filter is a unit impulse function, the output from the filter is the unit impulse response, as illustrated in Figure 4.26. A physical example of an impulse response might be the striking of a tuning fork or of a key on a piano. The impulse-like input to the filter creates a sustained output. Figure 4.27 shows the impulse responses of a piano, a pipe organ, and a harpsichord, sampled at 22.05 kHz. Eventually each of these impulse responses will die away to zero, but because these musical instruments are designed to sustain notes well, their impulse responses take some time to disappear. The effect on a car's suspension of driving over a speed bump is another good example of an impulse response. If the suspension is well-designed, the impulse response should die away rapidly, to give a smooth ride.

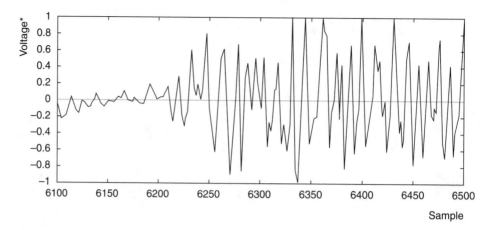

organ.wav

(b) Pipe Organ

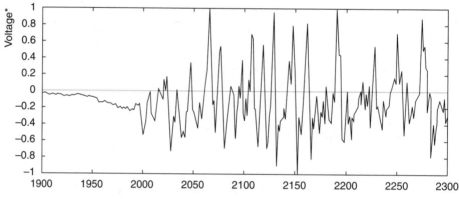

harpsi.wav

(c) Harpsichord

*For clarity, the envelopes rather than the individual samples of these digital signals are plotted.

FIGURE 4.27

Continued

The difference equation for a digital filter can be used to calculate the impulse response for the filter, as illustrated in Example 4.11. Since the input must be an impulse function, the input $x[n]$ in the difference equation can be replaced by $\delta[n]$. The impulse response for a filter is normally designated as $h[n]$. Thus, the output $y[n]$ in the difference equation should be replaced by $h[n]$. Once these substitutions have been made, samples of $h[n]$ can be calculated step by step in the usual way. The impulse response is an essential characteristic of a filter. Since all digital signals can be constructed from impulse functions, as shown in Section 3.3.1, the impulse response can be used to predict outputs for inputs of all kinds. This idea is explored further with Example 4.14.

EXAMPLE 4.11

Find the first six samples of the impulse response for the difference equation

$$y[n] - 0.4y[n-1] = x[n] - x[n-1]$$

First, replace $x[n]$ with $\delta[n]$, and $y[n]$ with $h[n]$ to give:

$$h[n] - 0.4h[n-1] = \delta[n] - \delta[n-1]$$

or

$$h[n] = 0.4h[n-1] + \delta[n] - \delta[n-1]$$

Starting with $n = 0$:

$$h[0] = 0.4h[-1] + \delta[0] - \delta[-1]$$

The values for the impulse function $\delta[n]$ are known: At $n = 0$, it has the value one, and at all other values of n it has the value zero. The filter can be assumed to be causal, which means that the impulse response is zero before $n = 0$. Therefore,

$$h[0] = 0.4(0.0) + 1.0 - 0.0 = 1.0$$

Notice that $\delta[-1] = 0$ because zero is the value of the function $\delta[n]$ when $n = -1$, not a consequence of causality.

The subsequent impulse response samples are:

$$h[1] = 0.4h[0] + \delta[1] - \delta[0] = 0.4(1.0) + 0.0 - 1.0 = -0.6$$

$$h[2] = 0.4h[1] + \delta[2] - \delta[1] = 0.4(-0.6) + 0.0 - 0.0 = -0.24$$

$$h[3] = 0.4h[2] + \delta[3] - \delta[2] = 0.4(-0.24) + 0.0 - 0.0 = -0.096$$

$$h[4] = 0.4h[3] + \delta[4] - \delta[3] = 0.4(-0.096) + 0.0 - 0.0 = -0.0384$$

$$h[5] = 0.4h[4] + \delta[5] - \delta[4] = 0.4(-0.0384) + 0.0 - 0.0 = -0.01536$$

The impulse function and impulse response are shown in Figure 4.28(a) and (b).

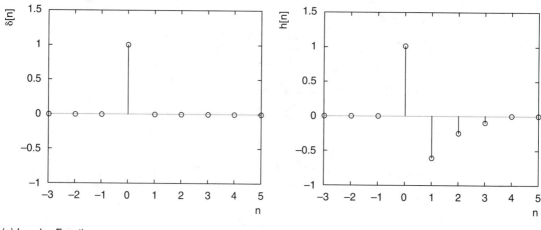

(a) Impulse Function

(b) Impulse Response

FIGURE 4.28

Impulse function and impulse response for Example 4.11.

Notice in Figure 4.28(b) that, although the impulse response samples get smaller and smaller, they never settle to zero, even though the input activity ceased after $n = 0$. This characteristic belongs to recursive filters in general. Because new outputs depend on old outputs, the impulse response never dies away. This response is called an **infinite impulse response (IIR)** and is typical for recursive difference equations. According to the general recursive difference equation in Equation (4.3), outputs up to N steps ago and inputs up to M steps ago can affect the calculation of each new output. Since N past outputs are used in the calculation of each new output, nonzero outputs can be obtained even when the input contribution is already zero.

EXAMPLE 4.12

Find the first six samples in the impulse response for the digital filter

$$y[n] = 0.25(x[n] + x[n-1] + x[n-2] + x[n-3]) \tag{4.8}$$

Substituting $\delta[n]$ for $x[n]$ and $h[n]$ for $y[n]$,

$$h[n] = 0.25(\delta[n] + \delta[n-1] + \delta[n-2] + \delta[n-3]) \tag{4.9}$$

so that

$$h[0] = 0.25(\delta[0] + \delta[-1] + \delta[-2] + \delta[-3]) = 0.25(1.0 + 0.0 + 0.0 + 0.0) = 0.25$$

$$h[1] = 0.25(\delta[1] + \delta[0] + \delta[-1] + \delta[-2]) = 0.25(0.0 + 1.0 + 0.0 + 0.0) = 0.25$$

$$h[2] = 0.25(\delta[2] + \delta[1] + \delta[0] + \delta[-1]) = 0.25(0.0 + 0.0 + 1.0 + 0.0) = 0.25$$

$$h[3] = 0.25(\delta[3] + \delta[2] + \delta[1] + \delta[0]) = 0.25(0.0 + 0.0 + 0.0 + 1.0) = 0.25$$

$$h[4] = 0.25(\delta[4] + \delta[3] + \delta[2] + \delta[1]) = 0.25(0.0 + 0.0 + 0.0 + 0.0) = 0.0$$

$$h[5] = 0.25(\delta[5] + \delta[4] + \delta[3] + \delta[2]) = 0.25(0.0 + 0.0 + 0.0 + 0.0) = 0.0$$

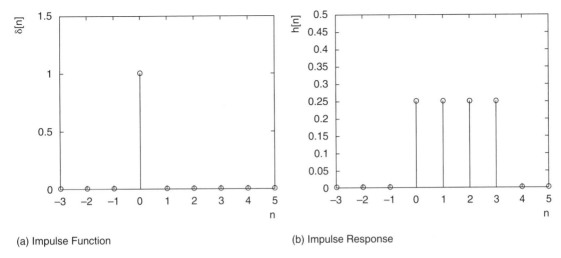

(a) Impulse Function

(b) Impulse Response

FIGURE 4.29
Impulse function and impulse response for Example 4.12.

It is easy to see that all the samples in the impulse response will be zero for $n \geq 4$. Figure 4.29 shows the impulse function input and the impulse response.

In Example 4.12, the impulse response drops to zero after a finite number of nonzero samples. This type of **finite impulse response (FIR)** is characteristic of nonrecursive filters, defined by Equation (4.4), which use inputs up to M steps in the past, but no past outputs, to calculate each new output. The number of samples it takes for the impulse response to become zero depends on the number of past inputs used in the calculation. When the input is an impulse function, nonrecursive filter outputs can be affected by the impulse input only until the single spike in the input reaches a point M steps in the past. After that the filter output goes to zero. The filter in Example 4.12 has one more notable feature: Its impulse response contains 4 elements, each with amplitude 1/4. An impulse response of this form has the effect of averaging every 4 samples in the input signal, as may be seen by inspecting the difference equation for the filter. For this reason, a filter with this type of impulse response is called a **moving average filter**.

A comparison of the difference equation in Equation (4.8) and the impulse response function in Equation (4.9) suggests a relationship that is true in general: For a nonrecursive filter, the impulse response samples give the coefficients for the difference equation. An impulse response $h[n]$ with M nonzero samples may be expressed as a sum of impulse functions according to

$$h[n] = h[0]\delta[n] + h[1]\delta[n-1] + \ldots + h[M]\delta[n-M] \qquad \textbf{(4.10)}$$

On the other hand, the nonrecursive difference equation

$$y[n] = b_0 x[n] + b_1 x[n-1] + \ldots + b_M x[n-M] \qquad \textbf{(4.11)}$$

gives rise to the equation for the impulse response

$$h[n] = b_0\delta[n] + b_1\delta[n-1] + \ldots + b_M\delta[n-M] \qquad (4.12)$$

Since Equations (4.10) and (4.12) must be equal, it follows that $b_k = h[k]$ for all values of k. Using this conclusion, it is also possible to rewrite Equation (4.11) as:

$$y[n] = h[0]x[n] + h[1]x[n-1] + \ldots + h[M]x[n-M] \qquad (4.13)$$

This relationship is illustrated in Example 4.13.

EXAMPLE 4.13

Write the difference equation for the filter whose impulse response is shown in Figure 4.30.

FIGURE 4.30

Impulse response for Example 4.13.

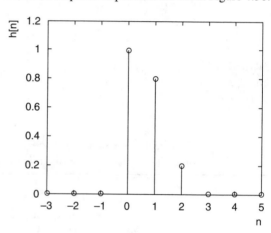

The impulse response can be written as a sum of impulse functions.

$$h[n] = \delta[n] + 0.8\delta[n-1] + 0.2\delta[n-2]$$

so the difference equation has the parallel structure

$$y[n] = x[n] + 0.8x[n-1] + 0.2x[n-2]$$

Since the number of nonzero samples in the impulse response is finite, the difference equation has a finite impulse response (FIR) characteristic.

If the impulse response for a linear filter is known, then the output for any other input can easily be calculated. Since every digital input can be expressed as the sum of impulse functions, the output will be the sum of impulse responses. Example 4.14 develops this connection, one that makes the impulse response a powerful tool for understanding how a filter behaves. In fact, because all outputs that a system is capable of producing are built from copies of the system's impulse response, the impulse response can be used to provide a complete specification for the system.

EXAMPLE 4.14

The signal $x[n]$ shown in Figure 4.31 is applied at the input of a linear filter whose impulse response $h[n]$ is given in Figure 4.32. Find the output $y[n]$ of the filter by breaking the input signal into impulse functions and finding the response to each.

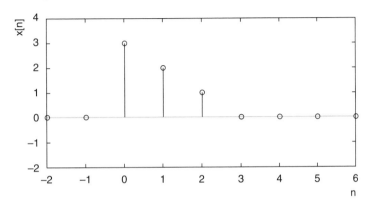

FIGURE 4.31
Input signal for Example 4.14.

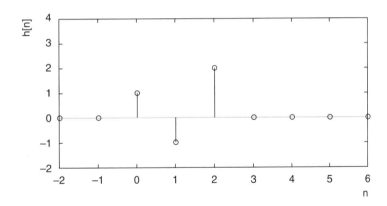

FIGURE 4.32
Impulse response for Example 4.14.

The input signal in Figure 4.31 may be described by the equation

$$x[n] = 3\delta[n] + 2\delta[n-1] + \delta[n-2]$$

As the equation and the graph show, the signal consists of three impulse functions of different amplitudes, $3\delta[n]$, $2\delta[n-1]$, and $\delta[n-2]$. Since the response to an impulse $\delta[n]$ is the impulse response $h[n]$, the response to $3\delta[n]$ will be $3h[n]$. Similarly, the response to $2\delta[n-1]$ will be $2h[n-1]$, and the response to $\delta[n-2]$ will be $h[n-2]$. All of these functions are graphed in Figure 4.33. As expected, summing the three impulse functions on the left

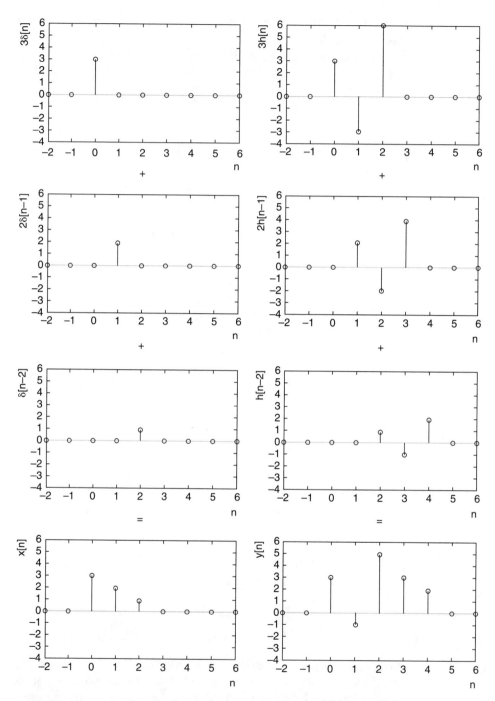

FIGURE 4.33
Output signal for Example 4.14.

reconstitutes the input signal at the bottom left. Since the filter is linear, the sum of the three impulse functions at the input will produce as output the sum of the three impulse responses. This sum appears at the bottom right of the diagram. In other terms,

$$y[n] = 3h[n] + 2h[n-1] + h[n-2]$$

4.8 THE STEP RESPONSE

The **step response** for a filter is the filter's response to a unit step function. It provides the response of the system to a change in level at the input. The step function is designated as $u[n]$ and the step response as $s[n]$. There are two simple ways to find the step response for a filter. One way is to use the difference equation directly, with $u[n]$ as input. This approach is illustrated in Example 4.15. Another way to determine the step response for a filter is to find a cumulative sum of the impulse response samples, given the fact that a step function is an infinite sum of shifted impulse functions. Each of these individual impulse functions applied to the filter will produce an impulse response. For a linear filter, applying the sum of the impulse functions as input should produce the sum of the impulse responses as output, as suggested by Example 4.14. Example 4.16 demonstrates the cumulative sum approach in graphical form and gives the same results as Example 4.15. When a filter's impulse response is known, finding a cumulative sum is the easiest way to compute the step response.

EXAMPLE 4.15
Find the step response for the filter

$$y[n] - 0.2y[n-2] = 0.5x[n] + 0.3x[n-1]$$

For the step function input $x[n] = u[n]$ shown in Figure 4.34(a), the output $y[n] = s[n]$ is

$$s[n] = 0.2s[n-2] + 0.5u[n] + 0.3u[n-1]$$

so that

$$s[0] = 0.2s[-2] + 0.5u[0] + 0.3u[-1] = 0.2(0) + 0.5(1) + 0.3(0) = 0.5$$

$$s[1] = 0.2s[-1] + 0.5u[1] + 0.3u[0] = 0.2(0) + 0.5(1) + 0.3(1) = 0.8$$

$$s[2] = 0.2s[0] + 0.5u[2] + 0.3u[1] = 0.2(0.5) + 0.5(1) + 0.3(1) = 0.9$$

$$s[3] = 0.2s[1] + 0.5u[3] + 0.3u[2] = 0.2(0.8) + 0.5(1) + 0.3(1) = 0.96$$

$$s[4] = 0.2s[2] + 0.5u[4] + 0.3u[3] = 0.2(0.9) + 0.5(1) + 0.3(1) = 0.98$$

$$s[5] = 0.2s[3] + 0.5u[5] + 0.3u[4] = 0.2(0.96) + 0.5(1) + 0.3(1) = 0.992$$

$$s[6] = 0.2s[4] + 0.5u[6] + 0.3u[5] = 0.2(0.98) + 0.5(1) + 0.3(1) = 0.996$$

$$s[7] = 0.2s[5] + 0.5u[7] + 0.3u[6] = 0.2(0.992) + 0.5(1) + 0.3(1) = 0.9984$$

The step response for the filter is plotted in Figure 4.34(b).

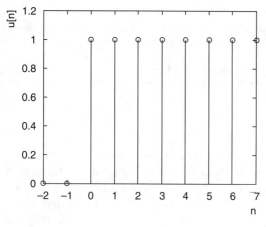

(a) Step Function

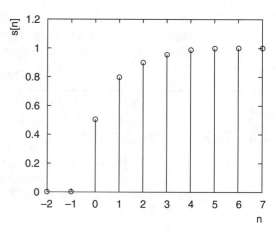

(b) Step Response

FIGURE 4.34
Step function and step response for Example 4.15.

EXAMPLE 4.16

The impulse response for the filter

$$y[n] - 0.2y[n-2] = 0.5x[n] + 0.3x[n-1]$$

from Example 4.15 can also be found using

$$h[n] - 0.2h[n-2] = 0.5\delta[n] + 0.3\delta[n-1]$$

The values for $h[n]$ are listed in the second column in Table 4.3. Step response values can easily be obtained from impulse response values by forming a cumulative sum, as illustrated in Figure 4.35. That is, each step response value is the sum of all impulse response values up to that value of n. For example, $s[2]$ is found as the sum $h[0] + h[1] + h[2] = 0.5 + 0.3 + 0.1 = 0.9$.

TABLE 4.3
Impulse Response and Step Response Values for Example 4.16

n	$h[n]$	$s[n]$
0	0.5	0.5
1	0.3	0.8
2	**0.1**	**0.9**
3	0.06	0.96
4	0.02	0.98
5	0.012	0.992
6	0.004	0.996
7	0.0024	0.9984

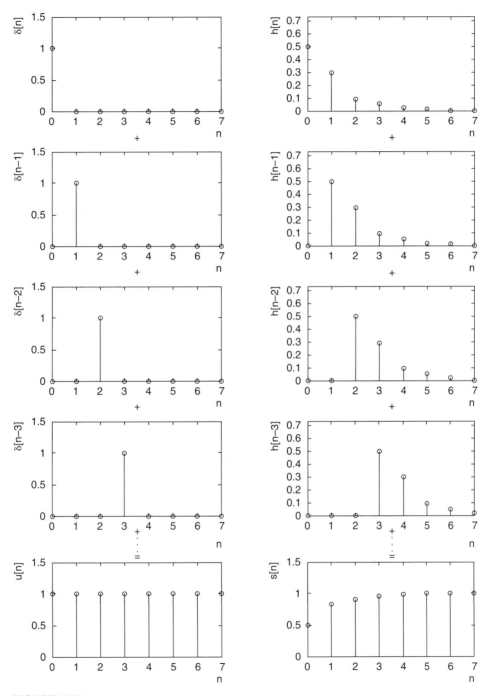

FIGURE 4.35

Calculation of step response given impulse response for Example 4.16.

As stated in Section 4.7, knowing the impulse response for a filter means knowing everything there is to know about the filter. The same can be said of the step response. Since $\delta[n] = u[n] - u[n-1]$ and the filters are linear, knowing the step response permits the impulse response to be calculated as $h[n] = s[n] - s[n-1]$. Knowing the impulse response means that the output for an arbitrary input can be calculated.

CHAPTER SUMMARY

Matlab
Support

1. A filter removes frequency elements from a signal. Its gain at a particular frequency determines the amplification factor applied to an incoming signal at that frequency. The filter passes signals at frequencies for which the filter gain is high, greater than $1/\sqrt{2}$ of its maximum value, and attenuates signals at frequencies where the filter gain is low. If the maximum filter gain is unity, the cut-off frequency for the filter is marked by a gain of -3 dB, also the point that delimits the bandwidth of the filter.

2. Low pass filters pass low frequency signals, and high pass filters pass high frequency signals. Band pass filters pass a band of midrange frequencies, which band stop filters block.

3. Digital filters have many advantages over analog filters, including noise insensitivity and ease of redesign. Also, very high order digital filters are no more difficult to create than low order ones, which is not the case for analog filters.

4. A filter is an example of a system. Practical filters are linear, time-invariant, and causal.

5. A difference equation is used to calculate filter outputs. A nonrecursive filter has the form

$$y[n] = b_0x[n] + b_1x[n-1] + \ldots + b_Mx[n-M]$$

It relies on inputs only to compute a new filter output $y[n]$. A recursive filter has the form

$$y[n] = -a_1y[n-1] - a_2y[n-2] - \ldots - a_Ny[n-N]$$
$$+ b_0x[n] + b_1x[n-1] + \ldots + b_Mx[n-M]$$

It relies on inputs as well as past outputs to compute a new filter output $y[n]$.

6. For linear systems, superposition can be used to apply multiple inputs at the same time.

7. Implementations of difference equations are often presented as difference equation diagrams. Those that minimize storage requirements and finite word length effects are best. Cascade or parallel combinations of low order filters are used to implement high order filters.

8. An impulse response is the response of a filter to an impulse function input. A filter's output to an arbitrary input can always be expressed as the sum of shifted impulse responses.

9. A step response is the response of a filter to a step function input.

REVIEW QUESTIONS

4.1 **a.** For the filter shape in Figure 4.36, determine which frequencies are passed by the filter.
b. What kind of filter is this?
c. What is the bandwidth of this filter?

FIGURE 4.36

Filter shape for Question 4.1.

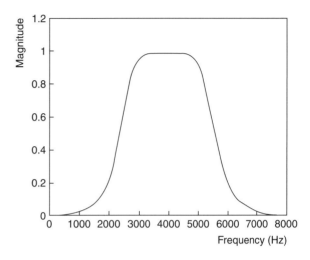

4.2 If the bandwidth of a low pass filter is 2 kHz, what is its cut-off frequency?
4.3 Find the bandwidth of the high pass filter shown in Figure 4.37. The sampling rate is 4 kHz.

FIGURE 4.37

Filter shape for Question 4.3.

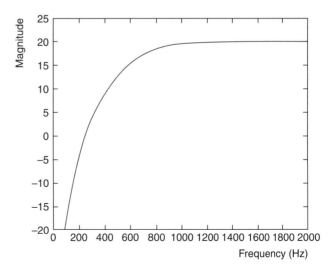

4.4 Three separate tones must be extracted from a noisy signal. The frequencies of the tones are 100 Hz, 300 Hz, and 450 Hz. The sampling rate is 1 kHz. Describe a low pass, a band pass, and a high pass filter that could be used to detect the individual tones. The cut-off frequency (or frequencies) for each filter should be 50 Hz from the tone it must detect. Identify the filter types, the cut-off frequencies, and the bandwidths.

4.5 The recording of the vowel "ahhh" is filtered using a low pass and a high pass filter whose shapes are given in Figure 4.38.

ah.wav

FIGURE 4.38

Filter shapes for Question 4.5.

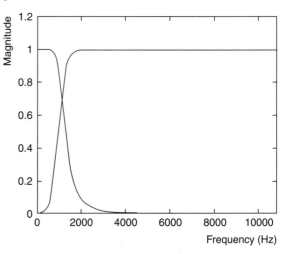

a. In Figure 4.39, identify which filter produces each output signal shown.

b. Since the two filters in Figure 4.38 cover all frequencies of interest, the two signals in Figure 4.39 carry most of the information in the original signal. Sketch an approximation to the original vowel signal using the signals in Figure 4.39.

FIGURE 4.39

Filter outputs for Question 4.5.

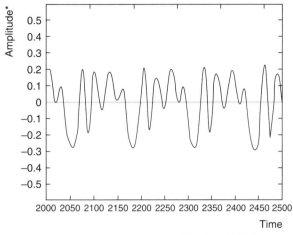

_____Pass Filter Output

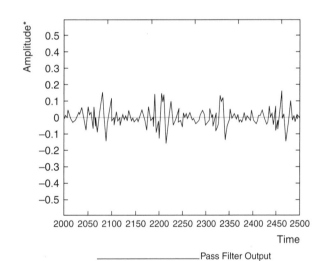

_____Pass Filter Output

*For clarity, the envelopes rather than the individual samples of the
signals are plotted.

4.6 Identify which systems are linear and which are nonlinear:

a. $y[n] = 2x_1[n] + 3x_2[n] - 5x_3[n]$

b. $y[n] = x_1^2[n] + x_2^2[n]$

c. $y[n] = x_1[n] - \sin(x_2[n])$

d. $y[n] = 0.5x[n] - 0.2x[n-1] - 0.4x[n-2]$

4.7 The input and the corresponding output of a filter are shown in Figure 4.40. If the filter is time-invariant, sketch the output for the new input shown in Figure 4.41.

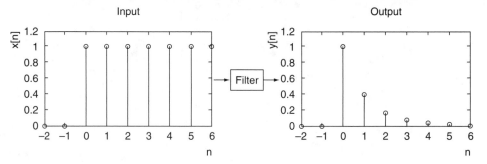

FIGURE 4.40
Input and output for Question 4.7.

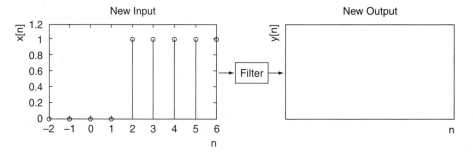

FIGURE 4.41
New input for Question 4.7.

4.8 Rewrite the following difference equations with $a_0 = 1$ and solve for $y[n]$:
   a. $4y[n] + y[n-1] = 3x[n] - x[n-1]$
   b. $-2y[n] + 2y[n-1] - 2x[n] - x[n-1] = 0$

4.9 Identify which of the following filters are recursive and which are nonrecursive. For each filter, identify the coefficients a_k and b_k.
   a. $y[n] = (x[n] + x[n-1] + x[n-2])/3$
   b. $y[n] - 0.2y[n-1] = x[n]$
   c. $y[n] = -0.5y[n-1] + x[n] - 0.4x[n-1]$

4.10 An input $x[n]$ whose sample values are listed in Table 4.4 is shown in Figure 4.42. Find and plot the first ten samples of the output that results when this input is applied to the causal filter:
   a. $y[n] = x[n] - 0.1x[n-1]$
   b. $y[n] = 0.6\,y[n-1] + x[n]$
   c. $y[n] = -0.9y[n-1] + x[n] - 0.7x[n-1]$
   d. $y[n] = (x[n] + x[n-1])/2$

TABLE 4.4
Sample Values for Question 4.10

n	0	1	2	3	4	5	6	7	8	9
$x[n]$	1.0	0.0	3.0	2.0	1.0	0.0	0.0	0.0	0.0	0.0

FIGURE 4.42
Input signal for Question 4.10.

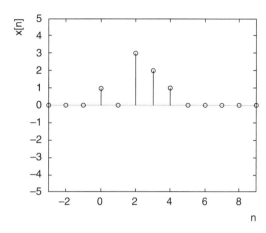

4.11 The difference equation for a recursive filter is given by $y[n] + 0.25y[n-1] = x[n]$. Find 10 samples of the output from the filter for the input $x[n] = u[n] - u[n-5]$.

4.12 Find and plot the first 10 output samples from the filter $y[n] = 0.3x[n] - 0.25x[n-1] + 0.1x[n-2]$ for input $x[n] = 2\delta[n] - \delta[n-2]$.

4.13 Two inputs, $x_1[n] = 0.1nu[n]$ and $x_2[n] = \sin(n\pi/4)(u[n] - u[n-4])$, are applied to a digital filter as shown in Figure 4.43. The difference equation of the filter is $y[n] = x[n] - 0.45x[n-1]$. Find the first 10 samples of the output $y[n]$.

FIGURE 4.43
System for Question 4.13.

$$x_1[n] + x_2[n] \longrightarrow \boxed{\text{Filter}} \longrightarrow y[n]$$

4.14 Compute the output of the filter $y[n] = -0.8y[n-1] + x[n]$ for the input $x[n] = (1 - e^{-0.5n})u[n]$.

4.15 Draw a diagram to represent the difference equation

$$y[n] = x[n] - 0.8x[n-1] + 0.5x[n-3]$$

4.16 Write the difference equation for the diagram in Figure 4.44.

FIGURE 4.44
Difference equation diagram for Question 4.16.

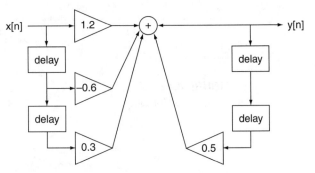

4.17 Find the difference equation that corresponds to the diagram in Figure 4.45.

FIGURE 4.45
Difference equation diagram for Question 4.17.

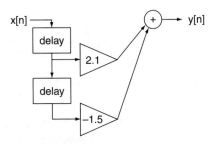

4.18 Write the difference equation for the diagram in Figure 4.46.

FIGURE 4.46
Difference equation diagram for Question 4.18.

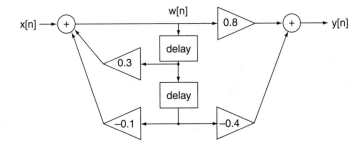

4.19 Find the difference equation for the cascaded filter in Figure 4.47.

FIGURE 4.47
Difference equation diagram for Question 4.19.

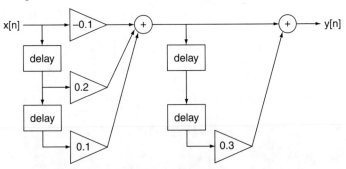

4.20 Draw a transposed direct form 2 realization for the difference equation

$$y[n] + 0.2y[n-1] - 0.3y[n-2] = x[n] - 0.5x[n-1]$$

4.21 Write equations for the direct form 2 realization of the difference equation

$$y[n] - 1.2y[n-1] + 0.5y[n-2] = x[n] - 0.2x[n-1]$$

4.22 A filter has the difference equation

$$y[n] + 0.14y[n-1] + 0.38y[n-2] = x[n]$$

a. Draw a direct form 1 diagram for this filter.
b. Draw a direct form 2 diagram for this filter.

4.23 Find the first 10 samples of the impulse response for

$$y[n] = x[n] - 1.2x[n-1] - 0.8x[n-2] + 0.4x[n-3]$$

4.24 A digital filter has the impulse response shown in Figure 4.48. Find the difference equation for the filter.

FIGURE 4.48

Impulse response for Question 4.24.

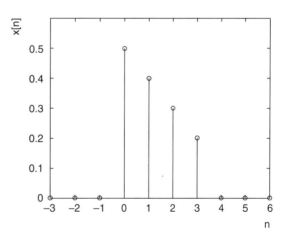

4.25 The impulse response for a nonrecursive filter is described by the function $h[n] = (0.3)^n(u[n] - u[n-4])$. Find the difference equation for the filter.

4.26 Find the first 10 samples of the impulse response for

$$y[n] - 0.8y[n-1] = x[n] - 0.5x[n-1]$$

4.27 A filter is described by the difference equation

$$y[n] = x[n] - 0.8x[n-1] - 0.5y[n-1]$$

Find the first 10 samples of the impulse response and the step response for the filter.

4.28 Find and plot 10 samples of the impulse response for a five-term moving average filter.

4.29 An impulse response is shown in Figure 4.49. Is this a recursive or nonrecursive filter? Why?

FIGURE 4.49

Impulse response for
Question 4.29.

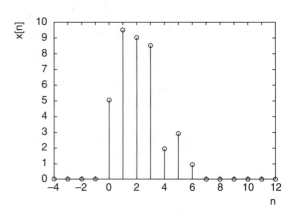

4.30 An impulse response for a linear digital filter is shown in Figure 4.50. Find the response of the filter to:
a. An impulse function
b. $x[n] = 0.8\delta[n] + 0.5\delta[n-1]$

FIGURE 4.50

Impulse response for
Question 4.30.

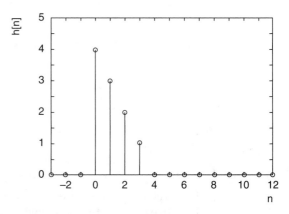

4.31 Compute the first 10 samples of the step response for the nonrecursive difference equation

$$y[n] = x[n] - 0.5x[n-1] - 0.7x[n-2]$$

4.32 A filter has the difference equation

$$y[n] + 0.75y[n-1] = x[n] - 0.5x[n-1]$$

Compute 10 samples of:
a. The impulse response
b. The step response

5

CONVOLUTION AND FILTERING

This chapter continues the theme of filtering with a second implementation method, convolution. The chapter:

- ➤ defines and explains filtering by convolution
- ➤ explains boundary effects
- ➤ identifies transient and steady state parts of an output response
- ➤ makes a connection between difference equations and convolution
- ➤ investigates moving average filters
- ➤ explains how to filter digital images

5.1 CONVOLUTION BASICS

The impulse response was introduced in Section 4.7 as the response of a system to an impulse function input. Every filter has an impulse response. In digital convolution, this response is used to calculate the output for a general input. Convolution is an essential tool for understanding digital signal processing. For one, it provides an alternative to difference equations for filter implementation. In addition, it accounts for the creation of spectral copies by sampling and for the aliasing errors that sampling can introduce. To see how convolution works, start by expressing the input signal as a sum of impulse functions:

$$x[n] = \sum_{k=-\infty}^{\infty} x[k]\delta[n - k] \qquad (5.1)$$

Since $\delta[n-k]$ equals 1 for $k = n$ and 0 for $k \neq n$, it follows that, for each value of n, only one of the infinite series of terms is nonzero, namely, the term for which $k = n$, and that $x[n]$ is equal to the corresponding sample value $x[k]$ at this point. This formula is merely a generalization of the relation

$$x[n] = x[0]\delta[n] + x[1]\delta[n-1] + x[2]\delta[n-2] + \ldots$$

introduced in Section 3.3.1.

EXAMPLE 5.1

A signal $x[n]$ is depicted in Figure 5.1. Find an expression for $x[n]$.

FIGURE 5.1
Signal for Example 5.1.

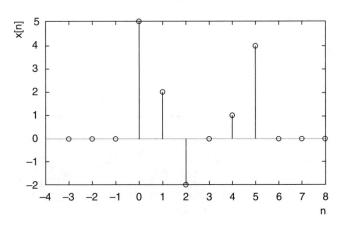

Following the method of Section 3.3.1, the function $x[n]$ may be written in terms of impulse functions as:

$$x[n] = x[0]\delta[n] + x[1]\delta[n-1] + x[2]\delta[n-2] + x[3]\delta[n-3] + x[4]\delta[n-4] + x[5]\delta[n-5]$$

$$= 5\delta[n] + 2\delta[n-1] - 2\delta[n-2] + \delta[n-4] + 4\delta[n-5]$$

Another way to write the same equation is as an infinite sum:

$$x[n] = \sum_{k=-\infty}^{\infty} x[k]\delta[n-k]$$

where $x[k]$ is defined in the following table:

k	-2	-1	0	1	2	3	4	5	6	7
$x[k]$	0	0	5	2	-2	0	1	4	0	0

For $n = 2$, for example, only the term for $k = 2$ in the series is nonzero, and $x[n]$ equals the corresponding value $x[k]$, or $x[2]$.

For each impulse function $\delta[n-k]$ that forms part of the input $x[n] = \sum_{k=-\infty}^{\infty} x[k]\delta[n-k]$ to a digital filter, the output is an impulse response $h[n-k]$. That is, every $\delta[n-k]$ in Equation (5.1) becomes $h[n-k]$ as it passes through the filter. The sample values $x[k]$ provide the weightings for each impulse function. For example, an input sample $x[5] = 4.0$ corresponds to a system input $4\delta[n-5]$ and gives a filter output $4h[n-5]$. Thus, the total output $y[n]$ due to the input $x[n]$ is the sum of all of these weighted impulse responses, or

$$y[n] = x[n]*h[n] = \sum_{k=-\infty}^{\infty} x[k]h[n-k] \tag{5.2}$$

Equation (5.2) defines **digital convolution**, where * is the convolution operator. The output $y[n]$ depends on the input $x[n]$ and the system impulse response $h[n]$. In fact, the idea of digital convolution is just a generalization of the approach used in Example 4.14. Equation (5.2) can be written in expanded form as:

$$y[n] = \dots + x[-2]h[n+2] + x[-1]h[n+1] + x[0]h[n]$$
$$+ x[1]h[n-1] + x[2]h[n-2] + \dots$$

An alternative, but equivalent, form to Equation (5.2) is

$$y[n] = h[n]*x[n] = \sum_{k=-\infty}^{\infty} h[k]x[n-k] \tag{5.3}$$

with the expanded form:

$$y[n] = \dots + h[-2]x[n+2] + h[-1]x[n+1]$$
$$+ h[0]x[n] + h[1]x[n-1] + h[2]x[n-2] + \dots$$

The relationship between Equations (5.2) and (5.3) is developed in Appendix D.

Figure 5.2 shows the dependence of the output signal $y[n]$ on the input $x[n]$ and the impulse response $h[n]$. As the above equations show, the output is obtained by convolving an input to a digital system with the system's impulse response. When the digital system is a digital filter, the output $y[n]$ is said to be a filtered version of $x[n]$.

It is helpful to consider what the convolution sum looks like for the first few values of n. For $n = 0, 1,$ and 2:

$$y[0] = \sum_{k=-\infty}^{\infty} x[k]h[-k]$$

$$y[1] = \sum_{k=-\infty}^{\infty} x[k]h[1-k]$$

$$y[2] = \sum_{k=-\infty}^{\infty} x[k]h[2-k]$$

The sequence $h[-k]$ is needed for the first calculation. This sequence is a mirror image of $h[k]$, reflected across the vertical axis. This mirror image is time-shifted to the right by one

FIGURE 5.2

Digital convolution.

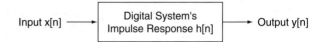

Input x[n] ⟶ Digital System's Impulse Response h[n] ⟶ Output y[n]

sample period to give $h[1-k]$, by two samples to give $h[2-k]$, and so on. Thus, the convolution calculations require that the input values $x[k]$ be multiplied by reflected and shifted copies of $h[k]$. Example 5.2 reviews how to reflect and shift a digital signal, and Example 5.3 introduces a straightforward graphical way to perform convolution calculations.

EXAMPLE 5.2

An impulse response is given by $h[n] = 3\delta[n] - 2\delta[n-1] + \delta[n-3] + \delta[n-4]$. Draw:

    a. $h[n]$
    b. $h[-n]$
    c. $h[1-n]$
    d. $h[7-n]$

    a. The samples of the impulse response $h[n]$ are listed in Table 5.1 and graphed in Figure 5.3.

TABLE 5.1

Impulse Response Values $h[n]$ for Example 5.2(a)

n	-4	-3	-2	-1	0	1	2	3	4	5	6	7
$h[n]$	0	0	0	0	3	-2	0	1	1	0	0	0

FIGURE 5.3

Plot of $h[n]$ for Example 5.2(a).

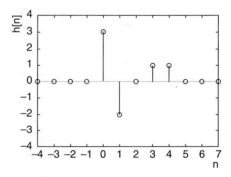

    b. The values for $h[-n]$ in Table 5.2 are readily found from the table of values for $h[n]$ in part (a). The signal is plotted in Figure 5.4. As expected, $h[-n]$ is a mirror image of $h[n]$ across the vertical axis.

TABLE 5.2
Shifted Impulse Response Values for Example 5.2(b)

n	-4	-3	-2	-1	0	1	2	3	4	5	6	7
$-n$	4	3	2	1	0	-1	-2	-3	-4	-5	-6	-7
$h[-n]$	1	1	0	-2	3	0	0	0	0	0	0	0

FIGURE 5.4
Plot of $h[-n]$ for Example 5.2(b).

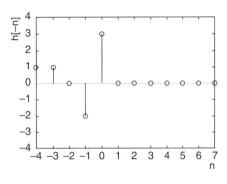

c. The table and plot for $h[1-n]$, shown in Table 5.3 and Figure 5.5, readily follow from those in (b). Notice that $h[1-n]$ is simply a copy of $h[-n]$ shifted by one sample position to the right.

TABLE 5.3
Shifted Impulse Response Values for Example 5.2(c)

n	-4	-3	-2	-1	0	1	2	3	4	5	6	7
$1-n$	5	4	3	2	1	0	-1	-2	-3	-4	-5	-6
$h[1-n]$	0	1	1	0	-2	3	0	0	0	0	0	0

FIGURE 5.5
Plot of $h[1-n]$ for Example 5.2(c).

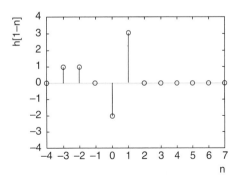

d. The same approach as in (c) gives the signal $h[7-n]$ described in Table 5.4 and Figure 5.6. The function $h[7-n]$ is a copy of $h[-n]$ shifted seven sample positions to the right.

TABLE 5.4
Shifted Impulse Response Values for Example 5.2(d)

n	-4	-3	-2	-1	0	1	2	3	4	5	6	7
$7-n$	11	10	9	8	7	6	5	4	3	2	1	0
$h[7-n]$	0	0	0	0	0	0	0	1	1	0	-2	3

FIGURE 5.6
Plot of $h[7-n]$ for Example 5.2(d).

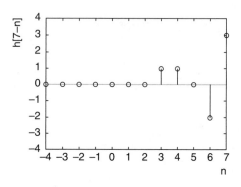

EXAMPLE 5.3

The input $x[n]$ to a digital system, and the system's impulse response $h[n]$, are shown in Figure 5.7. Find the output $y[n]$ from the system.

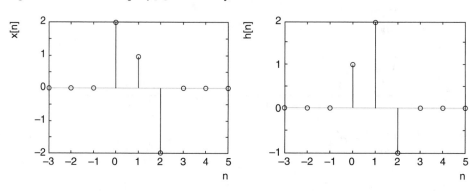

(a) Input   (b) Impulse Response

FIGURE 5.7
System input and impulse response for Example 5.3.

To find the output, it is necessary to compute the convolution of $x[n]$ and $h[n]$. The definition of convolution in Equation (5.2) suggests that the graphs of $x[k]$ versus k and $h[n-k]$ versus k can be used to perform the digital convolution graphically, as shown in Figure 5.8. Each $y[n]$ value is found by multiplying $x[k]$ and $h[n-k]$ point by point and summing the results. Output values for $n < 0$ and $n > 5$ are zero. Figure 5.9 plots the output signal values computed in Figure 5.8.

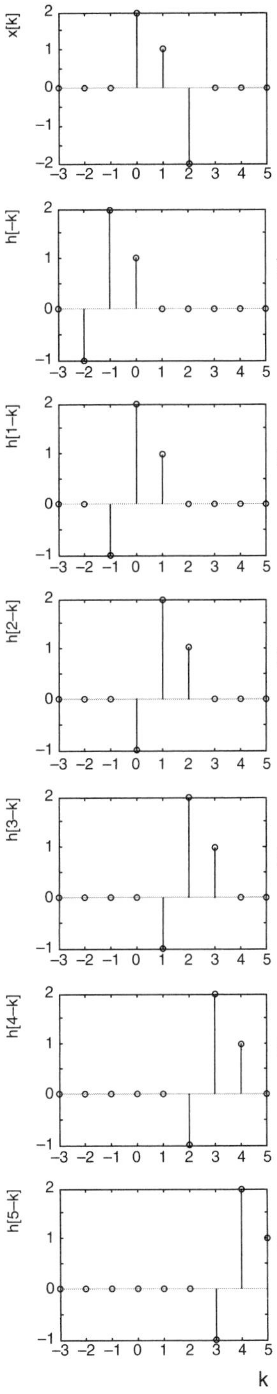

$$y[0] = \sum_{k=-\infty}^{\infty} x[k]h[-k]$$
$$= (0)(-1) + (0)(2) + (2)(1) + (1)(0) + (-2)(0) + (0)(0) + (0)(0) + (0)(0)$$
$$= 2$$

$$y[1] = \sum_{k=-\infty}^{\infty} x[k]h[1-k]$$
$$= (0)(0) + (0)(-1) + (2)(2) + (1)(1) + (-2)(0) + (0)(0) + (0)(0) + (0)(0)$$
$$= 5$$

$$y[2] = \sum_{k=-\infty}^{\infty} x[k]h[2-k]$$
$$= (0)(0) + (0)(0) + (2)(-1) + (1)(2) + (-2)(1) + (0)(0) + (0)(0) + (0)(0)$$
$$= -2$$

$$y[3] = \sum_{k=-\infty}^{\infty} x[k]h[3-k]$$
$$= (0)(0) + (0)(0) + (2)(0) + (1)(-1) + (-2)(2) + (0)(1) + (0)(0) + (0)(0)$$
$$= -5$$

$$y[4] = \sum_{k=-\infty}^{\infty} x[k]h[4-k]$$
$$= (0)(0) + (0)(0) + (2)(0) + (1)(0) + (-2)(-1) + (0)(2) + (0)(1) + (0)(0)$$
$$= 2$$

$$y[5] = \sum_{k=-\infty}^{\infty} x[k]h[5-k]$$
$$= (0)(0) + (0)(0) + (2)(0) + (1)(0) + (-2)(0) + (0)(-1) + (0)(2) + (0)(1)$$
$$= 0$$

FIGURE 5.8

Graphical digital convolution for Example 5.3.

FIGURE 5.9

Output for Example 5.3.

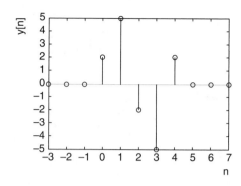

TABLE 5.5

Steps in Digital Convolution

1. List the input sample values. Spaces left blank are zeros.
2. List the impulse response values in reverse order. Line up the $n = 0$ sample of the impulse response with the $n = 0$ sample of the input.
3. Multiply and sum wherever the two sequences are both nonzero to find an output value.
4. Shift the impulse response values to the right by one position.
5. Go back to Step 3 (as long as some nonzero overlap between the sequences remains).
6. All other output samples are zero.

TABLE 5.6

Tabular Digital Convolution

$x[k]$:				2	1	-2				
$h[-k]$:	-1	2	1							$y[0] = 2$
$h[1-k]$:		-1	2	1						$y[1] = 5$
$h[2-k]$:			-1	2	1					$y[2] = -2$
$h[3-k]$:				-1	2	1				$y[3] = -5$
$h[4-k]$:					-1	2	1			$y[4] = 2$
$h[5-k]$:						-1	2	1		$y[5] = 0$

To avoid the need for graphs, the process in Example 5.3 can be modified by using a table. Table 5.5 summarizes the steps needed to perform tabular digital convolution, and Table 5.6 shows the results for Example 5.3. Each output is calculated by summing the products across a row, giving the same answers as before.

It is important to interpret filter outputs in a sensible way. If the inputs are known to be preceded with and followed by zeros, as in the last example, then the computed outputs make physical sense. In most cases, however, nothing is known about the input prior to the beginning of sampling. When the impulse response sequence overlaps with unknown input samples, the computed outputs cannot be considered reliable, because the actual outputs

TABLE 5.7
Boundary Effects

$x[k]$:				1	−2	3	1	5	2	0	1	2	4				
$h[-k]$:	−3	2	−1	4													?$y[0] = 4$
$h[1-k]$:		−3	2	−1	4												?$y[1] = -9$
$h[2-k]$:			−3	2	−1	4											?$y[2] = 16$
$h[3-k]$:				−3	2	−1	4										$y[3] = -6$
. . .																	. . .
$h[9-k]$:									−3	2	−1	4					$y[9] = 16$
$h[10-k]$:										−3	2	−1	4				?$y[10] = -3$
$h[11-k]$:											−3	2	−1	4			?$y[11] = 2$
$h[12-k]$:												−3	2	−1	4		?$y[12] = -12$

? = output samples affected by boundary effects.

may be affected by the unknown inputs that were applied before sampling began. This happens both at the beginning and at the end of the calculations. Only when the input sequence completely overlaps the impulse response can the calculations be considered meaningful. Table 5.7 illustrates how these difficulties, called **boundary effects**, arise for finite input sequences. In this example, an input of length 10 is filtered by an impulse response of length 4. Whenever the impulse response samples lie partly outside the range of input samples, the outputs are in doubt. In this case, six of the output samples (numbers 0, 1, 2, 10, 11, and 12) are influenced by boundary effects. Sample three is the first sample for which the input samples completely overlap the impulse response samples. This can be achieved when the number of samples in the impulse response is finite, as is the case for all FIR filters. For FIR filter output, the intervals at the beginning and end that are influenced by boundary effects each last for the length of the impulse response, less one. When the number of nonzero impulse response samples is infinite, as is the case for IIR filters, complete overlap between input and impulse response cannot in general be achieved. Boundary effects, however, diminish as the impulse response samples become small. There are a number of ways of handling the boundary effect problem, including extending the input with zeros, repeating edge values, and periodically repeating the entire input sequence (Ambardar, 1999). Fortunately, for large numbers of input samples, typical in most applications, boundary effects affect only a small percentage of the results.

Another way to interpret boundary effects relates to a system's transient and steady state output behavior. The **transient** part of the output signal is its relatively short-term behavior, while the **steady state** part is its long-term behavior. When a system starts up, or when conditions change, the output takes some time to settle down to a new steady state. The length of this initial transient period depends on the system: It may be a few microseconds, several minutes, or many hours. The steady state behavior of the output will mimic that of the input. When the input is constant, the output will approach a constant value; when the input is a repeating signal, the output will ultimately repeat in a similar way. In digital filtering, the transient effects are, in fact, boundary effects. Steady state behavior begins where boundary effects either terminate (for FIR filters) or become small (for IIR filters). The following examples illustrate.

EXAMPLE 5.4

The input to a system is the unit step. The impulse response of the system is given by $h[n] = 0.4\delta[n] - \delta[n-1] + 0.7\delta[n-2]$. Find the output of the system, $y[n]$, and identify the transient and steady state portions of the output.

TABLE 5.8
Calculations for Example 5.4

$x[k]$:			1	1	1	1	1	1	1	
$h[-k]$:	0.7	−1	0.4							$y[0] = 0.4$
$h[1-k]$:		0.7	−1	0.4						$y[1] = -0.6$
$h[2-k]$:			0.7	−1	0.4					$y[2] = 0.1$
$h[3-k]$:				0.7	−1	0.4				$y[3] = 0.1$
$h[4-k]$:					0.7	−1	0.4			$y[4] = 0.1$
$h[5-k]$:						0.7	−1	0.4		$y[5] = 0.1$
$h[6-k]$:							0.7	−1	0.4	$y[6] = 0.1$

The unit step was defined in Section 3.3.2 as a signal that is zero before $n = 0$ and one thereafter. This signal has been entered into Table 5.8 as $x[n]$. The rest of the table shows the reflected and shifted impulse response values, as well as the output from the system.

FIGURE 5.10
Output for Example 5.4

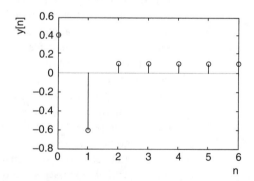

The output, shown graphically in Figure 5.10, reaches a steady state at 0.1 units after just two samples. These two sample periods are enough to complete the transient part of the signal; they contain the boundary effects of the filter. Note that, because the input step stays at a constant level, the output ultimately reaches a constant value also. This steady state behavior begins with the sample at $n = 2$.

EXAMPLE 5.5

A digital filter's input is $x[n] = \cos(n\pi/4)u[n]$. The filter's impulse response is $h[n] = 2\delta[n] - \delta[n-1] + 3\delta[n-2]$. Find the output of the filter, and identify the transient and steady state portions of the response.

The output may be found by multiplying and summing using Table 5.9. Only the first 11 samples of the input and output signals are shown. Figure 5.11 plots the first 20 samples of each. Note that the steady state behavior of the output is sinusoidal with the same frequency as the input: Both signals repeat every eight samples. The first two output samples form the output transient; the rest follow the steady state pattern.

TABLE 5.9
Calculations for Example 5.5

				1	$\frac{1}{\sqrt{2}}$	0	$-\frac{1}{\sqrt{2}}$	-1	$-\frac{1}{\sqrt{2}}$	0	$\frac{1}{\sqrt{2}}$	1	$\frac{1}{\sqrt{2}}$	0	
$x[n]$:															
$h[-k]$:	3	−1	2												$y[0] = 2.00$
$h[1-k]$:		3	−1	2											$y[1] = 0.41$
$h[2-k]$:			3	−1	2										$y[2] = 2.29$
$h[3-k]$:				3	−1	2									$y[3] = 0.71$
$h[4-k]$:					3	−1	2								$y[4] = -1.29$
$h[5-k]$:						3	−1	2							$y[5] = -2.54$
$h[6-k]$:							3	−1	2						$y[6] = -2.29$
$h[7-k]$:								3	−1	2					$y[7] = -0.71$
$h[8-k]$:									3	−1	2				$y[8] = 1.29$
$h[9-k]$:										3	−1	2			$y[9] = 2.54$
$h[10-k]$:											3	−1	2		$y[10] = 2.29$

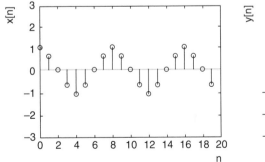
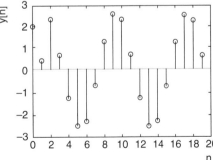

FIGURE 5.11
Input and output signals for Example 5.5.

EXAMPLE 5.6

Use convolution to find the step response for the system whose impulse response is $h[n] = (-0.55)^n u[n]$.

The first few calculations appear in Table 5.10, and the step response is shown in Figure 5.12. In this case, the impulse response is infinite, so complete overlap by the input signal cannot be achieved. Although true steady state cannot be reached for this reason, the figure shows that the system does settle to a nearly constant value within a dozen samples. In the case of IIR filters, an approximate steady state is often identified as the point where the output level changes by less than 1% from the previous sample. For this example, "steady state" is reached at $n = 10$. In the case of sinusoidal inputs to IIR filters, the same point is reached when the samples in one cycle change by less than 1% from the samples in the previous cycle.

TABLE 5.10
Calculations for Example 5.6

$x[k]$:			1	1	1	1	1	1	1	
$h[-k]$:	0.30	−0.55	1.00							$y[0] = 1.0$
$h[1-k]$:	−0.17	0.30	−0.55	1.00						$y[1] = 0.45$
$h[2-k]$:	0.09	−0.17	0.30	−0.55	1.00					$y[2] = 0.75$
$h[3-k]$:	−0.05	0.09	−0.17	0.30	−0.55	1.00				$y[3] = 0.59$
$h[4-k]$:	0.03	−0.05	0.09	−0.17	0.30	−0.55	1.00			$y[4] = 0.68$
$h[5-k]$:	−0.02	0.03	−0.05	0.09	−0.17	0.30	−0.55	1.00		$y[5] = 0.63$
$h[6-k]$:	0.01	−0.02	0.03	−0.05	0.09	−0.17	0.30	−0.55	1.00	$y[6] = 0.66$

FIGURE 5.12

Step response for Example 5.6.

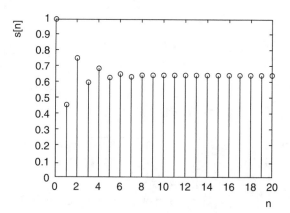

5.2   DIFFERENCE EQUATIONS AND CONVOLUTION

In Section 4.4, the general formula for the difference equation of a filter was given as:

$$\sum_{k=0}^{N} a_k y[n-k] = \sum_{k=0}^{M} b_k x[n-k] \tag{5.4}$$

Section 5.1 provided the formulas for digital convolution as:

$$y[n] = \sum_{k=-\infty}^{\infty} x[k]h[n-k] \tag{5.5a}$$

or,

$$y[n] = \sum_{k=-\infty}^{\infty} h[k]x[n-k] \tag{5.5b}$$

Note that Equation (5.5) describes a nonrecursive relationship, since the calculation of $y[n]$ does not depend on past outputs. It turns out that Equations (5.4) and (5.5) can be used to describe the same system. In the case of Equation (5.4), the coefficients a_k and b_k, called the filter coefficients, are needed to specify the system. In the case of Equation (5.5), the impulse response values are needed. For an infinite impulse response (IIR), an infinite number of $h[n]$ values are necessary, as seen in Example 5.7. For finite impulse response (FIR) systems, the number of $h[n]$ values is finite, as shown in Example 5.8. In the FIR case, Equation (5.5b) provides a practical means of constructing a difference equation from a filter's impulse response, a generalization of the observations made in Section 4.7 and embodied by Equation (4.13).

EXAMPLE 5.7
Express the difference equation

$$y[n] = 0.25y[n-1] + x[n]$$

in the form of Equation (5.5).

The impulse response for this system can be calculated from:

$$h[n] = 0.25h[n-1] + \delta[n]$$

The first seven samples are shown in Table 5.11. As expected for this recursive difference equation, the impulse response is infinite. That is, it has an infinite number of nonzero samples.

TABLE 5.11
Impulse Response for Example 5.7

n	0	1	2	3	4	5	6
$h[n]$	1.0000	0.2500	0.0625	0.0156	0.0039	0.0010	0.0002

Since the impulse response values are available, it is easiest to use Equation (5.5b). The difference equation can be expressed as:

$$y[n] = \ldots + h[-2]x[n+2] + h[-1]x[n+1] + h[0]x[n]$$

$$+ h[1]x[n-1] + h[2]x[n-2] + \ldots$$

$$= x[n] + 0.25x[n-1] + 0.0625x[n-2]$$

$$+ 0.0156x[n-3] + 0.0039x[n-4] + \ldots$$

$$= \sum_{k=0}^{\infty} (0.25)^k x[n-k]$$

The final equation has the form of Equation (5.5), as required.

EXAMPLE 5.8

Express the difference equation

$$y[n] = 0.1x[n] + 0.2x[n-1] + 0.3x[n-2]$$

in the form of Equation (5.5). The filter coefficients are $a_0 = 1$, $b_0 = 0.1$, $b_1 = 0.2$, $b_2 = 0.3$. Since this is a nonrecursive difference equation, only the b_k coefficients are of interest. Replacing $y[n]$ by $h[n]$ and $x[n]$ by $\delta[n]$ in the difference equation gives the impulse response

$$h[n] = 0.1\delta[n] + 0.2\delta[n-1] + 0.3\delta[n-2]$$

Since the impulse response (and all other digital signals) can be written as the sum of impulse functions, this expression can also be interpreted in the form

$$h[n] = h[0]\delta[n] + h[1]\delta[n-1] + h[2]\delta[n-2]$$

which gives $h[0] = 0.1$, $h[1] = 0.2$, and $h[2] = 0.3$. From Equation (5.5b),

$$y[n] = \ldots + h[-2]x[n+2] + h[-1]x[n+1] + h[0]x[n]$$

$$+ h[1]x[n-1] + h[2]x[n-2] + \ldots$$

$$= 0.1x[n] + 0.2x[n-1] + 0.3x[n-2]$$

$$= \sum_{k=0}^{2} h[k]x[n-k]$$

where $h[0] = 0.1$, $h[1] = 0.2$, $h[2] = 0.3$. In other words, though the expressions may look different on the surface, the difference equation and digital convolution calculations are identical for nonrecursive filters.

As discussed in this section and in the last chapter, convolution and difference equations are both capable of generating filter outputs. Because impulse responses for FIR filters consist of, by definition, a finite number of nonzero samples, convolution and difference

FIGURE 5.13

Impulse response for five-term moving average filter.

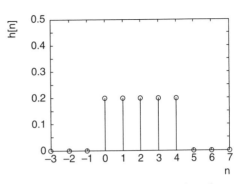

equations are equally useful choices for implementation. For IIR filters, whose impulse responses contain an infinite number of nonzero terms, difference equations are by far a more practical implementation option.

5.3   MOVING AVERAGE FILTERS

Moving average filters are a class of nonrecursive filters, introduced in Example 4.12, that tend to smooth an incoming signal. They do this by acting as low pass filters to remove the higher frequency elements. The outputs from a moving average filter such as

$$y[n] = \frac{1}{5}(x[n] + x[n-1] + x[n-2] + x[n-3] + x[n-4])$$

are the mathematical averages over several input values. This difference equation corresponds to a five-term moving average filter, for which the five most recent input values are added together, and the result divided by five, to produce each output. It is easy to find the impulse response $h[n]$ for the difference equation by replacing $y[n]$ with $h[n]$, and $x[n]$ with the impulse function $\delta[n]$:

$$h[n] = \frac{1}{5}(\delta[n] + \delta[n-1] + \delta[n-2] + \delta[n-3] + \delta[n-4])$$

The impulse response is plotted in Figure 5.13. It has the rectangular shape characteristic of moving average filters.

Using a five-term moving average filter on a rapidly changing input signal produces an output signal with more gradual transitions. Additional smoothing can be obtained by using a larger number of input values. A seven-term moving average filter, for example, averages over the last seven inputs according to the following equation:

$$y[n] = \frac{1}{7}(x[n] + x[n-1] + x[n-2] + x[n-3]$$

$$+ x[n-4] + x[n-5] + x[n-6] + x[n-7])$$

$$= \frac{1}{7}\sum_{k=0}^{7} x[n-k]$$

There are two ways to find filter outputs for the moving average filter: One way is to use the difference equation directly; the other is to use the impulse response and digital convolution. Example 5.9 demonstrates both approaches.

EXAMPLE 5.9

A five-term moving average filter has the difference equation

$$y[n] = \frac{1}{5}(x[n] + x[n-1] + x[n-2] + x[n-3] + x[n-4])$$

Find the output from this filter for the input shown in Figure 5.14.

FIGURE 5.14

Input for Example 5.9.

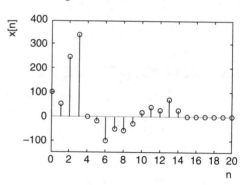

Table 5.12 shows the output values obtained by using the difference equation. Note that both input and output are zero prior to $n = 0$, the input is zero beyond $n = 14$, and the output falls to zero beyond $n = 18$. Also note that the first and the last four samples are influenced by boundary effects, since the length of the impulse response is five.

The output of the moving average filter is plotted in Figure 5.15. Notice that the output signal is smoother than the input signal shown in Figure 5.14. The removal of the sharp transitions in the input is evidence of the filter's low pass behavior. The same output can also be calculated by using the impulse response. The impulse response for the filter is

$$h[n] = \frac{1}{5}(\delta[n] + \delta[n-1] + \delta[n-2] + \delta[n-3] + \delta[n-4])$$

Using a convolution table like the one in Table 5.6 produces exactly the same output values as obtained previously. The first few calculations are shown in Table 5.13.

TABLE 5.12
Input and Output for Example 5.9

n	$x[n]$	$y[n]$
0	100	20
1	50	30
2	250	80
3	350	150
4	0	150
5	−20	126
6	−100	96
7	−50	36
8	−70	−48
9	−30	−54
10	20	−46
11	40	−18
12	30	−2
13	70	26
14	20	36
15	0	32
16	0	24
17	0	18
18	0	4
19	0	0
20	0	0

FIGURE 5.15
Output for Example 5.9.

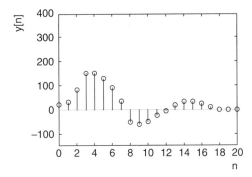

TABLE 5.13
Output Calculation by Digital Convolution for Example 5.9

$x[k]$:					100	50	250	350	0	−20	−100	−50	−70	
$h[k]$:	0.2	0.2	0.2	0.2	0.2									$y[0] = 20$
$h[1-k]$:		0.2	0.2	0.2	0.2	0.2								$y[1] = 30$
$h[2-k]$:			0.2	0.2	0.2	0.2	0.2							$y[2] = 80$
$h[3-k]$:				0.2	0.2	0.2	0.2	0.2						$y[3] = 150$
$h[4-k]$:					0.2	0.2	0.2	0.2	0.2					$y[4] = 150$
$h[5-k]$:						0.2	0.2	0.2	0.2	0.2				$y[5] = 126$

EXAMPLE 5.10

To further illustrate the effect of a five-term moving average filter, consider the input signal shown in Figure 5.16, which consists of many samples with magnitude one, with occasional samples significantly higher or lower. The filter output may be obtained using either difference equations or convolution, and is shown in Figure 5.17. The leading and trailing edges are influenced by boundary effects, produced while the impulse response only partially overlaps the input signal at the beginning and end. In the center portion, the output values are close to one throughout. The large departures from one evident in the input signal are gone because the moving average filter has smoothed them away.

FIGURE 5.16

Input to moving average filter for Example 5.10.

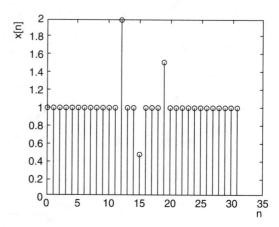

FIGURE 5.17

Output from moving average filter for Example 5.10.

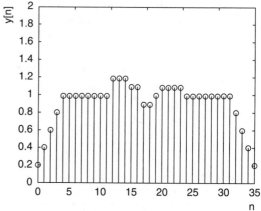

EXAMPLE 5.11

Moving average filters can be useful in finding trends in rapidly changing data such as stock market prices. Figure 5.18 shows a graph of simulated daily stock prices. The prices show many sharp changes during one year. Smoothing the data with a moving average filter gives a clearer view of how the prices are changing over time. Averaging the stock prices over every three days, shown in Figure 5.19, provides some degree of smoothing. Averaging over every nine days, shown in Figure 5.20, provides additional smoothing, to give a good estimate of the stock's general trends during the year.

FIGURE 5.18
Original stock prices for one year for Example 5.11.

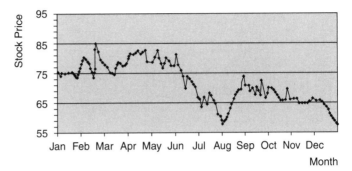

FIGURE 5.19
Stock prices smoothed over three days for Example 5.11.

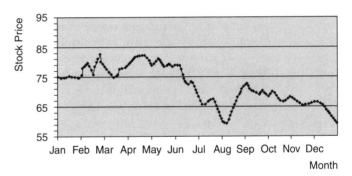

FIGURE 5.20
Stock prices smoothed over nine days for Example 5.11.

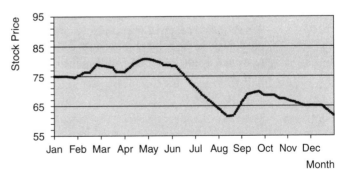

5.4 FILTERING DIGITAL IMAGES

Digital convolution can be used to filter two-dimensional signals like digital images as well as one-dimensional signals like speech or music. The process is very similar to the

FIGURE 5.21

Gray scale values for digital image $x[m, n]$ to be filtered.

$$\begin{bmatrix} 0 & 0 & 0 & 0 & 0 & 0 & 0 & 0 \\ 0 & 0 & 0 & 0 & 0 & 0 & 0 & 0 \\ 255 & 255 & 255 & 255 & 255 & 255 & 255 & 255 \\ 255 & 255 & 255 & 255 & 255 & 255 & 255 & 255 \\ 255 & 255 & 255 & 255 & 255 & 255 & 255 & 255 \\ 255 & 255 & 255 & 255 & 255 & 255 & 255 & 255 \\ 0 & 0 & 0 & 0 & 0 & 0 & 0 & 0 \\ 0 & 0 & 0 & 0 & 0 & 0 & 0 & 0 \end{bmatrix}$$

FIGURE 5.22

3×3 convolution kernel $h[m, n]$ for smoothing 2D filter.

$$\frac{1}{9}\begin{bmatrix} 1 & 1 & 1 \\ 1 & 1 & 1 \\ 1 & 1 & 1 \end{bmatrix} = \begin{bmatrix} \frac{1}{9} & \frac{1}{9} & \frac{1}{9} \\ \frac{1}{9} & \frac{1}{9} & \frac{1}{9} \\ \frac{1}{9} & \frac{1}{9} & \frac{1}{9} \end{bmatrix}$$

one-dimensional case described in Section 5.1. The main difference is that the impulse response used to filter images must be two-dimensional. This two-dimensional impulse response is usually referred to as a **convolution kernel**.

To perform 2D convolution, the convolution kernel is rotated by 180°. This 180° rotation is the 2D analog of finding the mirror image of the impulse response in the one-dimensional case. The kernel is then passed over the image, multiplying and summing at each point to find the filtered output. The 3×3 convolution kernel of a 2D smoothing filter is

$$\frac{1}{9}\begin{bmatrix} 1 & 1 & 1 \\ 1 & 1 & 1 \\ 1 & 1 & 1 \end{bmatrix}$$

It is the 2D equivalent of a three-term moving average filter for one-dimensional data. The 3×3 filter is centered over a pixel in the image. It averages the gray scale values over the nine pixels in the neighborhood of the center pixel. A 5×5 version of the same kind of filter would average over 25 pixels in the neighborhood of a central pixel using the convolution kernel:

$$\frac{1}{25}\begin{bmatrix} 1 & 1 & 1 & 1 & 1 \\ 1 & 1 & 1 & 1 & 1 \\ 1 & 1 & 1 & 1 & 1 \\ 1 & 1 & 1 & 1 & 1 \\ 1 & 1 & 1 & 1 & 1 \end{bmatrix}$$

To illustrate the process of 2D filtering, an image containing black and white stripes is filtered by a 2D low pass filter. The original image is shown in Figure 5.26(a), while its gray scale values are presented in Figure 5.21. The convolution kernel for the smoothing filter is shown in Figure 5.22.

FIGURE 5.23

Two-dimensional digital convolution.

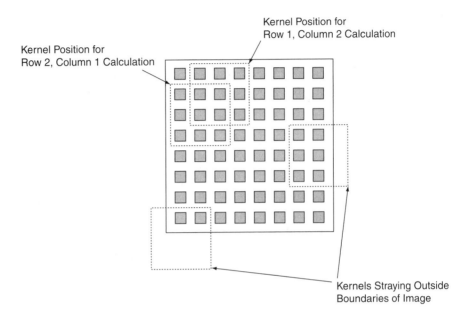

FIGURE 5.24

Position of convolution kernel in digital image.

Consider the highlighted pixel with gray scale value 0 in the second row of the image $x[m,n]$ in Figure 5.21. Because the convolution kernel is symmetrical, rotating by 180° does not change the matrix. The next step is to center the kernel over the pixel of interest. Figure 5.23 shows the required part of the input image with the convolution kernel superimposed. Multiplying the values in each position and summing gives

$$(0)(1/9) + (0)(1/9) + (0)(1/9) + (0)(1/9) + (0)(1/9)$$

$$+ (0)(1/9) + (255)(1/9) + (255)(1/9) + (255)(1/9) = 85$$

The value 85 is placed in the output image at [1, 1], the same position as the 0 in the input image. Then the convolution kernel is shifted one position to the right to calculate the output for [1, 2]. The position of the convolution kernel for this calculation is shown in Figure 5.24. When this row is complete, the kernel is positioned at the beginning of the next row.

FIGURE 5.25
Smoothed digital image.

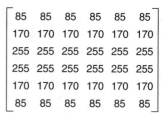

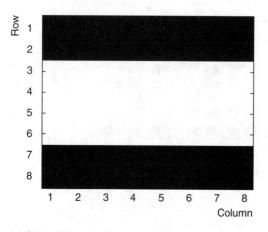

(a) Original Image

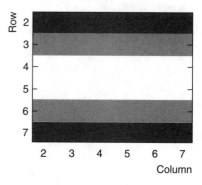

(b) Low Pass Filtered Image

FIGURE 5.26
Low pass filtering an image.

Clearly there will be a problem near the edges of the image, where the convolution kernel overlaps with areas outside the image. Two examples are shown in Figure 5.24. This is the two-dimensional version of the boundary effects introduced in Section 5.1. There are various ways to handle the problem. The simplest is to accept only those outputs obtained when the kernel lies completely within the image. This means that the output image will be smaller than the input image by an amount that depends on the size of the kernel. When a 3×3 filter is used, a row is lost at the top and bottom of the image, and also at the left and right of the image. Thus, an $N \times N$ input image will produce an $(N-2) \times (N-2)$ filtered image. Since most digital images contain many rows and columns of pixels, this loss is probably acceptably small. If this method is used for an 8×8 image, the final output will be a 6×6 matrix. The highlighted 0 used in Figure 5.23 is, in fact, the first pixel in the image beginning from [0, 0] for which the kernel does not stray outside the boundaries of the image.

The 6×6 convolution results for the gray scale levels in Figure 5.21 are presented in Figure 5.25. The low pass filtered image with these gray scale levels is depicted in Figure 5.26(b). As expected, it is smaller by two rows and two columns than the original. Not only the size but also the content of the image changes as a result of filtering. In the original image of Figure 5.26(a), there is a clear boundary between the top and bottom black stripes and the middle white stripe. In the filtered image shown in Figure 5.26(b), the boundary is no longer as sharp, as evidenced by the medium and light gray bars replacing the sharp black to white transitions. The smoothing filter tends to remove high frequency elements like edges from the image, reducing the overall contrast. The interior edges of the black objects become dark gray, and the pixels immediately outside the objects become light gray.

EXAMPLE 5.12

This example illustrates the effect of the low pass filter on a 152×217 pixel digital image containing the words "Digital Signal Processing." Figure 5.27(a) shows the original image. Figure 5.27(b) shows the same image after low pass filtering.

Digital
Signal
Processing

(a) Original Image

Digital
Signal
Processing

(b) Low Pass Filtered Image

FIGURE 5.27
Low pass filtering text for Example 5.12.

Additional digital image filters are investigated in Chapter 15.

CHAPTER SUMMARY

Matlab
Support

1. If a filter's impulse response is known, filter outputs can be calculated using convolution.
2. Convolution is an operation that combines a filter input $x[n]$ with the filter's impulse response $h[n]$ to produce an output $y[n]$. It is described by the equation

$$y[n] = x[n]*h[n] = \sum_{k=-\infty}^{\infty} x[k]h[n-k]$$

but is computed most easily using a graphical or tabular method.
3. In convolution, an impulse response sequence inches past an input sequence, and an output is calculated at each step. Boundary effects occur whenever the input sequence fails to encompass the impulse response sequence completely. For FIR filters, boundary

effects have a well-defined end. For IIR filters, boundary effects never truly disappear, but do become small.

4. The initial output samples, which are significantly influenced by boundary effects, form the transient part of an output. The steady state portion of the output begins where the transient part ceases. For a constant input, the steady state output will be constant. For a sinusoidal input, the steady state output will be sinusoidal.

5. A difference equation can be reexpressed as a convolution, and vice versa.

6. A moving average filter smoothes a signal by producing a running average of its sample values.

7. The two-dimensional version of an impulse response is a convolution kernel. The convolution kernel is convolved with a digital image to produce a filtered image.

8. A low pass convolution kernel blurs an image by averaging gray scale levels over neighborhoods of pixels.

REVIEW QUESTIONS

5.1 a. Write the function $x[n] = e^{-0.5n}(u[n] - u[n-3])$ as a sum of impulse functions.

b. If $x[n]$ must be written in the form $\sum\limits_{k=-\infty}^{\infty} x[k]\delta[n - k]$, what values of k give nonzero components?

5.2 The impulse response, $h[n]$, of a system is shown in Figure 5.28. This is the response of the system to an impulse input, $x[n] = \delta[n]$. Find and plot the system's output for $x[n] = \delta[n] + \delta[n-1]$.

FIGURE 5.28

Impulse response for Question 5.2.

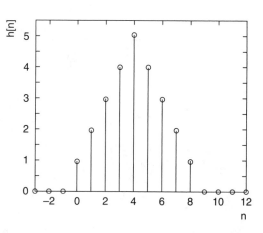

5.3 The impulse response of a system is $h[n] = \delta[n] - 0.5\delta[n-1] + 0.25\delta[n-2]$. Draw:
  a. $h[n-2]$
  b. $h[n+1]$
  c. $h[-n]$
  d. $h[3-n]$

5.4 The impulse response for a filter is given by $h[n] = (1 + n)(u[n] - u[n-4])$. The input to the filter is $x[n] = -n(u[n] - u[n-3])$. Find the output $y[n]$ from the filter using digital convolution.

5.5 The input to a filter is $x[n] = (n + 1)u[n]$, and the filter's impulse response is $h[n] = \delta[n-3]$.
  a. Find the output of the filter.
  b. What effect does a filter with the impulse response $\delta[n-k]$ have on a signal?

5.6 The impulse response of a filter is given by $h[n] = \delta[n] - 0.4\delta[n-1] - 0.2\delta[n-2]$. The input to the filter is $x[n] = u[n] - u[n-7]$.
  a. Find the output.
  b. In (a), identify those output samples influenced by boundary effects.

5.7 A filter has the impulse response $h[n] = (0.85)^n u[n]$. After how many samples does the step response of the filter reach an approximate steady state?

5.8 The impulse response for a filter is $h[n] = (-0.6)^n u[n]$, and the input to the filter is $x[n] = \sin(n\pi/2)u[n]$.
  a. Find the first 20 samples of the output.
  b. How many samples does it take for the output to reach an approximate steady state?

5.9 The input $x[n] = 5u[n]$ is applied to the filter $h[n] = (0.5)^n u[n]$.
  a. Plot the first 10 samples of the output.
  b. What is the approximate steady state output level for the filter, and how many samples are required to reach it?

5.10 The impulse response for a digital filter is $h[n] = (n + 2)(u[n] - u[n-3])$. The input $x[n] = \sin(n2\pi/5)u[n]$ is applied to the filter.
  a. What is the digital period of the input?
  b. Find 10 samples of the output from this system.
  c. Plot the output signal obtained in (b). Mark the transient and steady state parts of the signal.
  d. What is the digital period of the steady state output?

5.11 The impulse response of a filter is $h[n] = u[n-3] - u[n-8]$.
  a. Use convolution to find the filter's step response.
  b. Identify the transient and the steady state parts of the step response.

5.12 Two filters are cascaded together as shown in Figure 5.29. The impulse responses for the filters are $h_1[n] = \delta[n] - 0.1\delta[n-2] + 0.2\delta[n-4]$ and $h_2[n] = e^{-0.5n}(u[n] - u[n-4])$.
  a. Find the impulse response of the cascaded system.
  b. Find the step response of the cascaded system.

FIGURE 5.29
Cascaded system for Question 5.12.

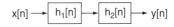

$x[n] \rightarrow \boxed{h_1[n]} \rightarrow \boxed{h_2[n]} \rightarrow y[n]$

5.13 The impulse response of a system is given by $h[n] = e^{-n}(u[n] - u[n-3])$. Find the step response of the system using convolution.

5.14 Express the difference equation $y[n] + 0.8\,y[n-1] = x[n]$ in the form of the convolution equation $y[n] = \displaystyle\sum_{k=-\infty}^{\infty} h[k]x[n - k]$.

5.15 Identify the difference equation that matches the impulse response

$$h[n] = e^{-0.3n}(u[n] - u[n-4])$$

Show that the difference equation and convolution give the same step response.

5.16 Find the step response of a nine-term moving average filter using convolution, and identify the transient portion of the response.

5.17 A 2048-point signal is filtered by an 11-point moving average filter. In the filtered signal, how many points in all are influenced by boundary effects?

5.18 The signal shown in Figure 5.30 is filtered by a five-term moving average filter. Plot 16 samples of the filtered signal. What effect does the filter have on the signal?

FIGURE 5.30

Input signal for Question 5.18.

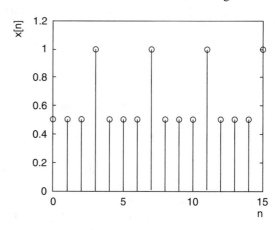

5.19 a. Write the difference equation and the impulse response for a three-term moving average filter.

b. The input $x[n] = \sin(n\pi/6)u[n]$ is passed through a three-term moving average filter. Draw the first 15 samples of the input and the output.

c. Identify the transient and the steady state parts of the output.

5.20 The input to an FIR filter is shown in Figure 5.31(a) and the output in Figure 5.31(b). The signals are zero for all samples not shown.
    a. Deduce the impulse response for the filter.
    b. Find the output of the filter for the input $x[n] = 2\delta[n] - \delta[n-2] + 0.5\delta[n-4]$.

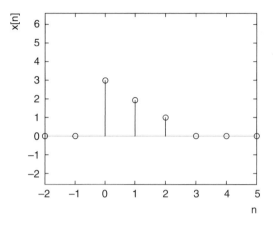

(a) Input Signal

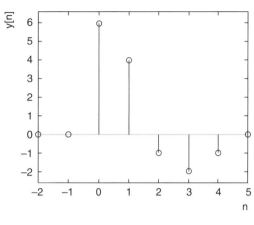

(b) Output Signal

FIGURE 5.31
Signals for Question 5.20.

5.21 Each square in the image in Figure 5.32 contains twelve rows and twelve columns. All pixels in the white squares have the gray scale value 255. All pixels in the black squares have the gray scale value 0. Filter this image with a 3×3 low pass smoothing filter, and describe the appearance of the filtered image.

FIGURE 5.32
Image for Question 5.21.

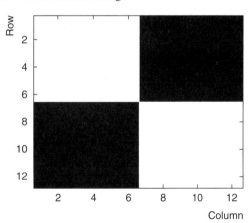

5.22 If a 7×7 low pass filter is applied to a 1024×512 image, determine the size of the filtered image if pixels tainted by boundary effects are removed.

6

z TRANSFORMS

The z transform is an important tool for digital signal processing. This chapter:

- ➤ defines the z transform
- ➤ calculates simple z transforms
- ➤ explains the region of convergence of a z transform
- ➤ provides a table of transforms
- ➤ defines what is meant by a transfer function
- ➤ makes the connections between transfer functions, difference equations, and impulse responses
- ➤ uses z transforms to calculate filter outputs
- ➤ explains how to combine transfer functions in cascade or parallel
- ➤ surveys methods of finding inverse z transforms, including tables, long division, and partial fraction expansion
- ➤ explains how to identify the poles and zeros of a filter
- ➤ defines filter stability
- ➤ investigate first and second order system examples
- ➤ examines the effects of pole and zero positions on impulse and step response characteristics

6.1   z TRANSFORM BASICS

In the foregoing chapters, digital signals have been expressed in the time domain, with one signal value at each sample point. Digital filters have also been expressed in the time domain, in the form of difference equations or convolution. This chapter introduces a technique called the z transform. The purpose of the z transform is to make descriptions of digital signals and systems more compact, and to make calculations with digital signals easier.

The definition of the z **transform** for a digital signal $x[n]$ is:

$$X(z) = \sum_{n=0}^{\infty} x[n] z^{-n} \tag{6.1}$$

The z transform of $x[n]$, called $X(z)$, is said to lie in the z **domain**, which is a frequency domain containing complex numbers. The z transform may not be defined for all values of z in the z domain. The values of z for which it is defined form its **region of convergence**. The capital X is used to designate the z transform of a signal x in the time domain. The z transform of a signal $y[n]$, for example, would be $Y(z)$. For short, this would be written as

$$\mathbf{Z}\{y[n]\} = Y(z)$$

where $\mathbf{Z}\{\ \}$ indicates a z transform is taken. Computing $y[n]$ from $Y(z)$ requires the **inverse z transform,** denoted

$$y[n] = \mathbf{Z}^{-1}\{Y(z)\}$$

Notice that the z transform summation begins with $n = 0$, not $-\infty$. This is not a limitation because the initial sample of any signal can always be identified as the $n = 0$ sample.

EXAMPLE 6.1
Find the z transform $X(z)$ of the signal $x[n] = \delta[n]$.

This signal is nonzero at only a single place, $n = 0$. Thus,

$$\mathbf{Z}\{x[n]\} = X(z) = \sum_{n=0}^{\infty} \delta[n] z^{-n} = \delta[0] = 1$$

This z transform is defined for all values of z, so its region of convergence includes all z.

EXAMPLE 6.2
Find the z transform of $x[n] = \delta[n-1]$.

The signal is nonzero only at $n = 1$. Therefore,

$$\mathbf{Z}\{x[n]\} = X(z) = \sum_{n=0}^{\infty} \delta[n-1] z^{-n} = \delta[0] z^{-1} = z^{-1}$$

which is defined as long as $z \neq 0$, so its region of convergence is all z except $z = 0$.

EXAMPLE 6.3
Find $X(z)$ if $x[n] = u[n]$.

$$X(z) = \sum_{n=0}^{\infty} x[n]z^{-n} = \sum_{n=0}^{\infty} u[n]z^{-n} = \sum_{n=0}^{\infty} z^{-n} = 1 + z^{-1} + z^{-2} + z^{-3} + z^{-4} + z^{-5} + \ldots$$

This is a geometric series of the form $a + ar + ar^2 + \ldots$ with initial term a equal to one and multiplier r equal to z^{-1}. As shown in Appendix A.16, the sum of an infinite geometric series is given by

$$S_{\infty} = \frac{a}{1 - r}$$

as long as $|r| < 1$. Therefore,

$$X(z) = \frac{1}{1 - z^{-1}} = \frac{z}{z - 1}$$

provided $|z^{-1}| < 1$. That is, the region of convergence for this z transform is $|z| > 1$.

EXAMPLE 6.4

A signal $x[n]$ is depicted in Figure 6.1. Find the z transform of the signal.

The signal may be described as

$$x[n] = 2\delta[n] + \delta[n-1] + 0.5\delta[n-2]$$

It has only three nonzero elements, so the z transform contains the same number of terms. The z transform is

$$X(z) = \sum_{n=0}^{\infty} x[n]z^{-n} = x[0] + x[1]z^{-1} + x[2]z^{-2} = 2 + z^{-1} + 0.5z^{-2}$$

which is defined as long as $z \neq 0$.

FIGURE 6.1
Signal for Example 6.4.

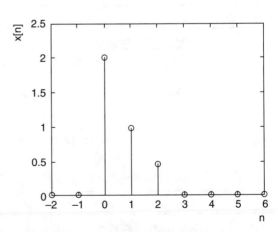

EXAMPLE 6.5

Find the *z* transform of the signal $x[n] = (-0.5)^n u[n]$.

Since $u[n] = 1$ for $n \geq 0$,

$$X(z) = \sum_{n=0}^{\infty} x[n]z^{-n} = \sum_{n=0}^{\infty} (-0.5)^n z^{-n} = \sum_{n=0}^{\infty} (-0.5z^{-1})^n$$

$$= 1 - 0.5z^{-1} + 0.25z^{-2} - 0.125z^{-3} + \dots$$

As in Example 6.3, this is an infinite geometric series. In this series, $a = 1$ and $r = -0.5z^{-1}$, so its sum is

$$X(z) = \frac{1}{1 + 0.5z^{-1}} = \frac{z}{z + 0.5}$$

The region of convergence for this *z* transform is $\left| -0.5z^{-1} \right| < 1$, or $|z| > 0.5$.

A selection of basic *z* transforms is listed in Table 6.1. The third transform, for example, is the general form for the type of signal studied in Example 6.5. This table may be used to perform *z* transforms and simple inverse *z* transforms. The time domain signals obtained from the *z* transforms include steps, decaying exponentials, sinusoids, and damped sinusoids, according to the types of roots in the denominator of the *z* transform. Section 6.3 introduces methods for handling both simple inverse *z* transforms and more complex inverse *z* transforms.

For any signal with a finite number of samples, the region of convergence is $z \neq 0$, that is, all *z* except $z = 0$. Table 6.1 includes the regions of convergence for a number of signals with an infinite number of nonzero samples. It is important to consider the region of convergence for a *z* transform because a signal $x_L[n]$ defined for $n < 0$ may have the same *z* transform as a signal $x_R[n]$ defined for $n \geq 0$, or a signal $x_{LR}[n]$ defined for all n, but all three signals have a different region of convergence. In this text, only "right-sided" signals like $x_R[n]$ are considered, so no confusion can occur. Regions of convergence are therefore not specified with the *z* transforms in the remainder of the chapter.

One important property of the *z* transform that must be established is the time-shifting property. Suppose that $X(z)$ is the *z* transform of a signal $x[n]$ whose value is zero prior to $n = 0$. The *z* transform of a time-shifted version of this signal, $x[n-1]$, is, by definition,

$$Y(z) = \sum_{n=0}^{\infty} x[n-1]z^{-n}$$

Let $m = n - 1$, so that $n = m + 1$. Then

$$Y(z) = \sum_{m=-1}^{\infty} x[m]z^{-m-1}$$

	z Transform $X(z)$	Region of Convergence
	1	all z
$u[n]$	$\dfrac{z}{z-1}$	$\lvert z\rvert > 1$
$\beta^n u[n]$	$\dfrac{z}{z-\beta}$	$\lvert z\rvert > \lvert\beta\rvert$
$nu[n]$	$\dfrac{z}{(z-1)^2}$	$\lvert z\rvert > 1$
$\cos(n\Omega)u[n]$	$\dfrac{z^2 - z\cos\Omega}{z^2 - 2z\cos\Omega + 1}$	$\lvert z\rvert > 1$
$\sin(n\Omega)u[n]$	$\dfrac{z\sin\Omega}{z^2 - 2z\cos\Omega + 1}$	$\lvert z\rvert > 1$
$\beta^n\cos(n\Omega)u[n]$	$\dfrac{z^2 - \beta z\cos\Omega}{z^2 - 2\beta z\cos\Omega + \beta^2}$	$\lvert z\rvert > \lvert\beta\rvert$
$\beta^n\sin(n\Omega)u[n]$	$\dfrac{\beta z\sin\Omega}{z^2 - 2\beta z\cos\Omega + \beta^2}$	$\lvert z\rvert > \lvert\beta\rvert$

Since $x[-1] = 0$,

$$Y(z) = \sum_{m=0}^{\infty} x[m]z^{-m}z^{-1} = z^{-1}\left(\sum_{m=0}^{\infty} x[m]z^{-m}\right) = z^{-1}X(z)$$

Hence, if the z transform of $x[n]$ is $X(z)$, then the z transform of $x[n-1]$ is $z^{-1}X(z)$. For instance, since the z transform of $\delta[n]$ is 1, the z transform of $\delta[n-1]$ is $(z^{-1})(1) = z^{-1}$. This confirms the finding of Example 6.2. Thus, a factor of z^{-1} in the z domain corresponds to a one-sample delay in the time domain, or

$$\mathbf{Z}\{x[n-1]\} = z^{-1}X(z)$$

where the one-step delay term, z^{-1}, must not be confused with the symbol for the inverse z transform, $\mathbf{Z}^{-1}$. The idea can be extended to longer delays. The signal $x[n-k]$, with a time domain delay of k samples compared to $x[n]$, has the z transform $z^{-k}X(z)$, that is,

$$\mathbf{Z}\{x[n-k]\} = z^{-k}X(z)$$

Note also that the z transform of $x[n+1]$ would be $zX(z)$, so a factor of z corresponds to a one-sample advance in the time domain.

The time-shifting property of the z transform suggests a notation change for difference equation diagrams. The delay blocks of Chapter 4 can be replaced by z^{-1} blocks, as

FIGURE 6.2

Delays in *z* domain.

shown in Figure 6.2. Even though this convention mixes the time and *z* domain notations, it is used in much of the DSP literature.

EXAMPLE 6.6

Find the *z* transform of the signal $x[n] = 2u[n-2]$.

Since $\mathbf{Z}\{u[n]\} = \dfrac{z}{z-1}$,

$$\mathbf{Z}\{u[n-2]\} = z^{-2}\frac{z}{z-1} = \frac{1}{z(z-1)}$$

Therefore,

$$X(z) = \frac{2}{z(z-1)}$$

EXAMPLE 6.7

Reexpress the nonrecursive difference equation diagram of Figure 4.15 using the z^{-1} notation.

The general form for a nonrecursive difference equation is

$$y[n] = b_0x[n] + b_1x[n-1] + b_2x[n-2] + \ldots + b_Mx[n-M]$$

The corresponding diagram is shown in Figure 6.3.

FIGURE 6.3

Difference equation diagram using z^{-1} notation for Example 6.7.

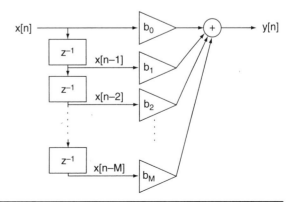

6.2 TRANSFER FUNCTIONS

6.2.1 Transfer Functions and Difference Equations

To find the z transform of a difference equation, the z transform of each term in the equation must be taken. This means taking the z transform of terms like $y[n-2]$ and $x[n-1]$. Remember that if the z transform of $y[n]$ is $Y(z)$, then the z transform of $y[n-2]$ is $z^{-2}Y(z)$. Similarly, if the z transform of $x[n]$ is $X(z)$, then the z transform of $x[n-1]$ is $z^{-1}X(z)$. After taking the z transform of each term in a difference equation, the ratio of output to input in the z domain may be found as

$$H(z) = \frac{\text{output}}{\text{input}} = \frac{Y(z)}{X(z)}$$

This ratio $H(z)$, a z transform in its own right, is given the special name of **transfer function**. The transfer function is a way of summarizing all information about a digital system's behavior. It is an extremely useful tool for evaluating digital filters and for calculating filter outputs. Term-by-term transformation of a general difference equation

$$\sum_{k=0}^{N} a_k y[n-k] = \sum_{k=0}^{M} b_k x[n-k]$$

or,

$$a_0 y[n] + a_1 y[n-1] + \ldots + a_N y[n-N]$$
$$= b_0 x[n] + b_1 x[n-1] + \ldots + b_M x[n-M]$$

yields

$$a_0 Y(z) + a_1 z^{-1} Y(z) + \ldots + a_N z^{-N} Y(z) = b_0 X(z) + b_1 z^{-1} X(z) + \ldots + b_M z^{-M} X(z)$$

or,

$$(a_0 + a_1 z^{-1} + \ldots + a_N z^{-N}) Y(z) = (b_0 + b_1 z^{-1} + \ldots + b_M z^{-M}) X(z)$$

Solving this equation for $\dfrac{Y(z)}{X(z)}$ gives the transfer function:

$$H(z) = \frac{Y(z)}{X(z)} = \frac{b_0 + b_1 z^{-1} + \ldots + b_M z^{-M}}{a_0 + a_1 z^{-1} + \ldots + a_N z^{-N}} = \frac{\displaystyle\sum_{k=0}^{M} b_k z^{-k}}{\displaystyle\sum_{k=0}^{N} a_k z^{-k}} \qquad \textbf{(6.2)}$$

The conversion of a difference equation into a transfer function by taking the z transform of each term is illustrated by the following examples.

EXAMPLE 6.8

Find the transfer function of the system described by the difference equation.

$$2y[n] + y[n-1] + 0.9y[n-2] = x[n-1] + x[n-4]$$

Taking z transforms term by term:

$$2Y(z) + z^{-1}Y(z) + 0.9z^{-2}Y(z) = z^{-1}X(z) + z^{-4}X(z)$$

where $Y(z)$ is the z transform of the filter output $y[n]$, and $X(z)$ is the z transform of the filter input $x[n]$. Factoring out $Y(z)$ on the left and $X(z)$ on the right:

$$(2 + z^{-1} + 0.9z^{-2})Y(z) = (z^{-1} + z^{-4})X(z)$$

Solving this for $\dfrac{Y(z)}{X(z)}$ gives the system transfer function

$$H(z) = \frac{\text{output}}{\text{input}} = \frac{Y(z)}{X(z)} = \frac{z^{-1} + z^{-4}}{2 + z^{-1} + 0.9z^{-2}}$$

EXAMPLE 6.9

Find the transfer function for the system with the difference equation

$$y[n] - 0.2y[n-1] = x[n] + 0.8x[n-1]$$

Virtually by inspection, the transfer function is

$$H(z) = \frac{1 + 0.8z^{-1}}{1 - 0.2z^{-1}}$$

EXAMPLE 6.10

Find the transfer function for the difference equation

$$y[n] = 0.75x[n] - 0.3x[n-2] - 0.01x[n-3]$$

The transfer function for this nonrecursive difference equation is

$$H(z) = 0.75 - 0.3z^{-2} - 0.01z^{-3}$$

Just as difference equations can be transformed into transfer functions, transfer functions can be converted to difference equations. To do this, express the transfer function $H(z)$ as $Y(z)/X(z)$ and cross-multiply with the system polynomials, as the following examples show.

EXAMPLE 6.11

Find the difference equation that corresponds to the transfer function

$$H(z) = \frac{1 + 0.5z^{-1}}{1 - 0.5z^{-1}}$$

Since the transfer function is the ratio of $Y(z)$ to $X(z)$,

$$\frac{Y(z)}{X(z)} = \frac{1 + 0.5z^{-1}}{1 - 0.5z^{-1}}$$

Cross-multiplication gives

$$Y(z)(1 - 0.5z^{-1}) = X(z)(1 + 0.5z^{-1})$$

or,

$$Y(z) - 0.5z^{-1}Y(z) = X(z) + 0.5z^{-1}X(z)$$

Inversely transforming term by term yields the difference equation

$$y[n] - 0.5y[n-1] = x[n] + 0.5x[n-1]$$

EXAMPLE 6.12

Find the difference equation that matches the transfer function

$$H(z) = \frac{z}{(2z - 1)(4z - 1)}$$

Multiplying out gives

$$H(z) = \frac{Y(z)}{X(z)} = \frac{z}{8z^2 - 6z + 1}$$

Cross-multiplying then gives

$$Y(z)(8z^2 - 6z + 1) = X(z)(z)$$

and inverse z transformation gives

$$(8z2 - 6z + 1)Y(z) = zX(z)$$

This difference equation looks unfamiliar because the freshest output is $y[n+2]$ rather than $y[n]$. The difference equation, however, simply expresses connections between data values that have different relative positions in time. As long as all parts are shifted equally, there is no change to the equation. After shifting all parts by two backward steps, the difference equation becomes

$$8y[n] - 6y[n-1] + y[n-2] = x[n-1]$$

or,

$$y[n] - 0.75y[n-1] + 0.125y[n-2] = 0.125x[n-1]$$

6.2.2 Transfer Functions and Impulse Responses

Figure 6.4 presents the three means of describing systems introduced so far: the difference equation, the impulse response, and the transfer function. Section 6.2.1 established a relationship between the difference equation and the transfer function. A connection between the impulse response and the transfer function can be established as well, by taking the z transform of the convolution Equation (5.3):

$$y[n] = \sum_{k=-\infty}^{\infty} h[k]x[n-k] = h[n] * x[n]$$

Appendix D shows that a step-by-step z transformation of this equation for a causal filter gives

$$Y(z) = H(z)X(z)$$

where $Y(z)$ is the z transform of $y[n]$, $X(z)$ is the z transform of $x[n]$, and, most importantly, $H(z)$ is the z transform of $h[n]$. In other words, the transfer function of a filter is the z transform of the filter's impulse response. Using the definition of the z transform,

$$H(z) = \mathbf{Z}\{h[n]\}$$

$$= \sum_{n=0}^{\infty} h[n]z^{-n} \qquad (6.3)$$

FIGURE 6.4
Finding filter outputs in time and z domains.

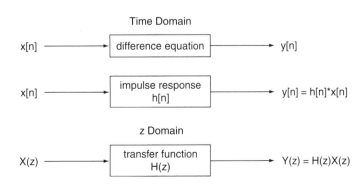

As noted in Section 6.1, starting the *z* transform at $n = 0$ is not a problem for digital signals because they can generally be defined to begin at or after $n = 0$. It is not a problem when the *z* transform is used to transform the impulse response $h[n]$ into the transfer function $H(z)$ either: All practical filters are causal, so the impulse response is zero prior to $n = 0$. Naturally, it follows from Equation (6.3) that the impulse response is the inverse *z* transform of the transfer function:

$$h[n] = \mathbf{Z}^{-1}\{H(z)\}$$

Examples of this calculation will be provided in Section 6.3.

EXAMPLE 6.13

The impulse response for a digital filter is

$$h[n] = \delta[n] + 0.4\delta[n-1] + 0.2\delta[n-2] + 0.05\delta[n-3]$$

Find the transfer function of the filter.

The transfer function for the filter is nothing more than the *z* transform of the impulse response:

$$H(z) = 1 + 0.4z^{-1} + 0.2z^{-2} + 0.05z^{-3}$$

Note that this transfer function leads to the difference equation

$$y[n] = x[n] + 0.4x[n-1] + 0.2x[n-2] + 0.05x[n-3]$$

6.2.3 Finding Filter Outputs

In Chapter 4, outputs from digital filters were computed directly from the difference equation. In Chapter 5, digital convolution was used to compute filter outputs. Both of these methods operate in the time domain. A filter's transfer function, $H(z)$, can be used to calculate system outputs using the *z* domain. Since $H(z) = Y(z)/X(z)$,

$$Y(z) = H(z)X(z) \tag{6.4}$$

as shown in Figure 6.4. In other words, the *z* transform of the output is the product of the transfer function in the *z* domain and the *z* transform of the input. The inverse *z* transform of $Y(z)$ then produces the output signal $y[n]$, that is,

$$y[n] = \mathbf{Z}^{-1}\{Y(z)\}$$

According to Appendix D, Equation (6.4) is the *z* transform of the convolution $y[n] = h[n] * x[n]$. This demonstrates the most significant advantage of calculating outputs in the *z* domain, namely, that multiplication is generally simpler than convolution. The biggest disad-

FIGURE 6.5

Cascade and parallel combinations of filters.

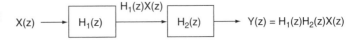

(a) Cascade Combination of Filters

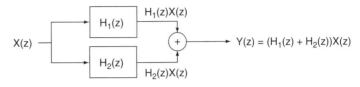

(b) Parallel Combination of Filters

vantage is that z transforms and inverse z transforms must be computed. Examples will be provided in Section 6.3.4.

6.2.4   Cascade and Parallel Combinations of Transfer Functions

The last section showed how to compute the output from a filter with transfer function $H(z)$ as $Y(z) = H(z)X(z)$. This idea can be extended for more complex systems, as suggested in Figure 6.5. Part (a) of the figure shows a cascade combination of filters. The output of the first filter feeds the second, so the overall filter output $Y(z)$ is the product of the transfer functions $H_1(z)$ and $H_2(z)$ with the input $X(z)$. Thus, the transfer function of the cascade combination is the product $H_1(z)H_2(z)$. Figure 6.5(b) shows a parallel combination of filters. The transfer function of the parallel combination is the sum of the transfer functions, $H_1(z) + H_2(z)$. Z transforms permit very complicated combinations of filters to be analyzed, much more conveniently than with the difference equations used in Section 4.6.

EXAMPLE 6.14

Find the difference equation that corresponds to the cascaded sections presented in Figure 6.6, repeated here from Figure 4.19.

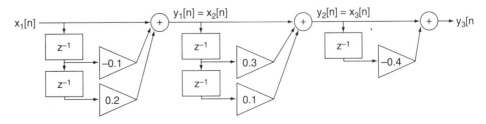

FIGURE 6.6

Difference equation diagram for Example 6.14.

In Example 4.7, the same cascaded system was investigated. The difference equations for each stage were identified as:

$$y_1[n] = x_1[n] - 0.1x_1[n-1] + 0.2x_1[n-2]$$
$$y_2[n] = x_2[n] + 0.3x_2[n-1] + 0.1x_2[n-2]$$
$$y_3[n] = x_3[n] - 0.4x_3[n-1]$$

These equations were combined to produce the difference equation for the cascaded filter. In this example, the same result will be achieved more easily with transfer functions. The transfer functions for the three filters are:

$$H_1(z) = 1 - 0.1z^{-1} + 0.2z^{-2}$$
$$H_2(z) = 1 + 0.3z^{-1} + 0.1z^{-2}$$
$$H_3(z) = 1 - 0.4z^{-1}$$

The overall transfer function is the product

$$H(z) = H_1(z)H_2(z)H_3(z)$$
$$= 1 - 0.2z^{-1} + 0.19z^{-2} - 0.058z^{-3} - 0.008z^{-5}$$

As in Example 4.7, the difference equation is

$$y_3[n] = x_1[n] - 0.2x_1[n-1] + 0.19x_1[n-2] - 0.058x_1[n-3] - 0.008x_1[n-5]$$

EXAMPLE 6.15

Find the transfer function of the transposed direct form 2 realization of the filter depicted in Figure 6.7.

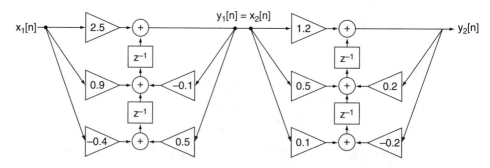

FIGURE 6.7
Difference equation diagram for Example 6.15.

The figure shows a cascade combination of two second order filters. Using the methods of Section 4.6.2.2, the difference equations for the two filters are:

$$y_1[n] = -0.1y_1[n-1] + 0.5y_1[n-2] + 2.5x_1[n] + 0.9x_1[n-1] - 0.4x_1[n-2]$$

$$y_2[n] = 0.2y_2[n-1] - 0.2y_2[n-2] + 1.2x_2[n] + 0.5x_2[n-1] + 0.1x_2[n-2]$$

The difference equation for the filter is most easily found by multiplying the transfer functions for each section:

$$H(z) = H_1(z)H_2(z)$$

$$= \left(\frac{2.5 + 0.9z^{-1} - 0.4z^{-2}}{1 + 0.1z^{-1} - 0.5z^{-2}}\right)\left(\frac{1.2 + 0.5z^{-1} + 0.1z^{-2}}{1 - 0.2z^{-1} + 0.2z^{-2}}\right)$$

$$= \frac{3 + 2.33z^{-1} + 0.22z^{-2} - 0.11z^{-3} - 0.04z^{-4}}{1 - 0.1z^{-1} + 0.32z^{-2} + 0.12z^{-3} - 0.1z^{-4}}$$

EXAMPLE 6.16

Find the transfer function for the parallel combination in Figure 6.8.

FIGURE 6.8
Difference equation diagram for Example 6.16.

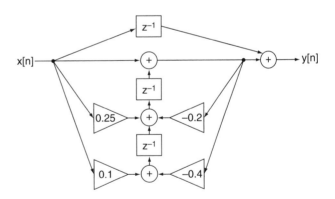

The transfer function for the top section is $H_1(z) = z^{-1}$. The transfer function for the bottom transposed direct form 2 section is

$$H_2(z) = \frac{1 + 0.25z^{-1} + 0.1z^{-2}}{1 + 0.2z^{-1} + 0.4z^{-2}}$$

The overall transfer function is

$$H(z) = \frac{1 + 0.25z^{-1} + 0.1z^{-2}}{1 + 0.2z^{-1} + 0.4z^{-2}} + z^{-1}$$

$$= \frac{z^2 + 0.25z + 0.1}{z^2 + 0.2z + 0.4} + \frac{1}{z}$$

$$= \frac{1 + 1.25z^{-1} + 0.3z^{-2} + 0.4z^{-3}}{1 + 0.2z^{-1} + 0.4z^{-2}}$$

6.3   BACK TO THE TIME DOMAIN

6.3.1   Standard Form

When z transforms are used, connection to the time domain must be made using the inverse z transform. For example, an inverse z transform is used to determine an impulse response $h[n]$ from a transfer function $H(z)$, and also to find a time domain signal $y[n]$ from its z transform $Y(z)$. Inverse transforms can be found in a number of ways, as the following sections will illustrate. For all methods, the inverse transform is found most easily when the z domain function is first expressed in **standard form**. This form requires that all exponents of z in the z transform be positive, and that the coefficient of the highest power term in both the numerator and the denominator be one. Assuming $N > M$, the transfer function in Equation (6.2) becomes

$$H(z) = \frac{b_0\left(1 + \frac{b_1}{b_0}z^{-1} + \dots + \frac{b_M}{b_0}z^{-M}\right)}{a_0\left(1 + \frac{a_1}{a_0}z^{-1} + \dots + \frac{a_N}{a_0}z^{-N}\right)}$$

$$= \frac{K\left(z^N + \frac{b_1}{b_0}z^{N-1} + \dots + \frac{b_M}{b_0}z^{N-M}\right)}{\left(z^N + \frac{a_1}{a_0}z^{N-1} + \dots + \frac{a_N}{a_0}\right)} \tag{6.5}$$

where $K = b_0/a_0$. Converting to standard form does not mathematically alter the transfer function. When factoring is required, the standard form permits the numerator and the denominator of a z transform to be expressed in terms of simple factors of the form $(z-\alpha)$. Example 6.17 illustrates how standard form may be obtained.

EXAMPLE 6.17

Express the following transfer function in standard form:

$$H(z) = \frac{z^{-1}}{4 - 2.5z^{-1} + z^{-2}}$$

The first step in the conversion to standard form is to make the exponents of all delay terms positive. If the term with the most negative exponent is z^{-N}, then every term in the transfer function should be multiplied by z^N to make all exponents positive. In this example, the largest delay is represented by the z^{-2} term in the denominator. Once each term has been multiplied by z^2, the transfer function becomes

$$H(z) = \frac{z}{4z^2 - 2.5z + 1}$$

The second step in the conversion is to ensure that the highest power denominator coefficient is one. For this reason, each term in the transfer function is divided by four to give the final standard form:

$$H(z) = \frac{0.25z}{z^2 - 0.625z + 0.25}$$

A transfer function expressed in standard form is a rational function consisting of a numerator polynomial divided by a denominator polynomial. The highest power in a polynomial is called its **degree**. In a **proper rational function**, the degree of the numerator is less than or equal to the degree of the denominator. In a **strictly proper rational function**, the degree of the numerator is less than the degree of the denominator. The degree of the numerator of an **improper rational function** is greater than the degree of its denominator. Some inverse transform methods require that a transfer function be strictly proper. A proper or improper rational function can be converted to a strictly proper rational function by using long division, as Example 6.18 suggests.

EXAMPLE 6.18

Convert the proper transfer function

$$H(z) = \frac{z^2 + 0.1z + 0.3}{z^2 - 0.5z + 0.9}$$

into a strictly proper expression.

Using long division:

$$\begin{array}{r} 1 \\ z^2 - 0.5z + 0.9 \overline{)\, z^2 + 0.1z + 0.3} \\ \underline{z^2 - 0.5z + 0.9} \\ 0.6z - 0.6 \end{array}$$

Thus,

$$H(z) = 1 + \frac{0.6z - 0.6}{z^2 - 0.5z + 0.9}$$

The rational part of the transfer function is strictly proper.

6.3.2 Simple Inverse *z* Transforms

Simple inverse transforms can be obtained using Table 6.1. As mentioned in Section 6.1, all of the *z* transforms studied in this text are assumed to belong to right-sided signals like those in the table, so the regions of convergence need not be specified. The goal in finding an inverse *z* transform using the table is to find a recognizable *z* transform from the list, in order to identify the digital signal to which it belongs. Example 6.19 provides a simple illustration. Use of the time-shifting property described in Section 6.1 can broaden the range of inverse *z* transforms that can be handled in this way. This latter point is illustrated in Examples 6.21, 6.22, and 6.23.

EXAMPLE 6.19

Find the signal $x[n]$ that corresponds to the *z* transform $X(z) = \dfrac{z}{z - 0.8}$.

Table 6.1 provides the inverse transform:

$$x[n] = \mathbf{Z}^{-1}\{X(z)\} = (0.8)^n u[n]$$

EXAMPLE 6.20

Find the inverse *z* transform of the function

$$X(z) = \frac{z^2 - 0.9z}{z^2 - 1.8z + 1}$$

A *z* transform in Table 6.1 that closely matches $X(z)$ is

$$X(z) = \frac{z^2 - z\cos\Omega}{z^2 - 2z\cos\Omega + 1}$$

where $\cos\Omega = 0.9$. Since $\Omega = \cos^{-1}(0.9) = 0.451$, the inverse transform is $x[n] = \cos(n\Omega)u[n] = \cos(0.451n)u[n]$.

EXAMPLE 6.21

A system's transfer function is

$$H(z) = \frac{z^{-2}}{1 + 0.25z^{-1}}$$

a. Find the difference equation for this system.
b. Find the impulse response for this system.

a. The difference equation for the system is $y[n] + 0.25\, y[n-1] = x[n-2]$.

b. The impulse response for the system is the inverse z transform of its transfer function. The transfer function is manipulated to isolate a transform found in Table 6.1.

$$H(z) = \frac{z^{-2}}{1 + 0.25z^{-1}} = z^{-2}\frac{1}{1 + 0.25z^{-1}} = z^{-2}\frac{z}{z + 0.25}$$

This expression shows that the inverse z transform of $H(z)$ will be delayed by two steps from the inverse transform of the function $z/(z + 0.25)$. The inverse z transform of this function, from Table 6.1, is $(-0.25)^n u[n]$, so the impulse response $h[n]$, after imposing the two-step delay, must be

$$h[n] = (-0.25)^{n-2}u[n-2] \tag{6.6}$$

EXAMPLE 6.22

The input to a digital filter is $x[n] = u[n]$. The output from the filter is $y[n] = (0.89)^n u[n]$.

a. Find the transfer function of the filter.

b. Find the impulse response of the filter.

a. From Table 6.1, $X(z) = \dfrac{z}{z - 1}$ and $Y(z) = \dfrac{z}{z - 0.89}$, so the transfer function is

$$H(z) = \frac{Y(z)}{X(z)} = \frac{z}{z - 0.89}\frac{z - 1}{z} = \frac{z - 1}{z - 0.89}$$

b. Taking an inverse z transform of the transfer function is the only easy way to find the impulse response from the information given. First $H(z)$ is converted to strictly proper form using long division, and then it is rewritten to isolate a transform found in Table 6.1:

$$H(z) = \frac{z - 1}{z - 0.89} = 1 - \frac{0.11}{z - 0.89} = 1 - z^{-1}\frac{0.11z}{z - 0.89}$$

From the table, the inverse transform of 1 is $\delta[n]$ and the inverse transform of $\dfrac{0.11z}{z - 0.89}$ is $0.11(0.89)^n u[n]$. The factor z^{-1} causes a one-step delay, so the impulse response is

$$h[n] = \mathbf{Z}^{-1}\{H(z)\} = \delta[n] - 0.11(0.89)^{n-1}u[n-1]$$

EXAMPLE 6.23

Find the time domain signal $x[n]$ that corresponds to the z transform

$$X(z) = \frac{5}{z^2 + 0.2z}$$

This z transform is in standard form. One solution is to factor $X(z)$ and isolate a transform from Table 6.1:

$$X(z) = \frac{5}{z(z + 0.2)} = z^{-2}\left(\frac{5z}{z + 0.2}\right)$$

The factor $5z/(z + 0.2)$ gives an inverse transform $5(-0.2)^n u[n]$. After the shift by two steps due to the factor z^{-2} term, the inverse transform is

$$x[n] = 5(-0.2)^{n-2}u[n-2] \tag{6.7}$$

6.3.3   Inverse z Transforms by Long Division

Another way of computing inverse z transforms is to divide a transfer function's numerator by its denominator and then take the inverse transform of each term. The advantage of this method is that it is relatively straightforward and can be applied to any rational function. The disadvantage is that in general a closed-form solution like those found in the preceding examples cannot be found. The following examples illustrate this method.

EXAMPLE 6.24

Find the inverse z transform of

$$H(z) = \frac{z^2 - 0.1z}{z^2 + 0.4z + 0.8}$$

Using long division,

$$
\begin{array}{r}
1 - 0.5z^{-1} - 0.6z^{-2} + 0.64z^{-3} - \cdots \\
z^2 + 0.4z + 0.8 \overline{)\, z^2 - 0.1z } \\
z^2 + 0.4z + 0.8 \\
\hline
-0.5z - 0.8 \\
-0.5z - 0.2 - 0.4z^{-1} \\
\hline
-0.6 + 0.4z^{-1} \\
-0.6 - 0.24z^{-1} - 0.48z^{-2} \\
\hline
0.64z^{-1} + 0.48z^{-2} \\
0.64z^{-1} + 0.256z^{-2} + 0.512z^{-3} \\
\hline
0.224z^{-2} - 0.512z^{-3} \\
\cdots
\end{array}
$$

So,

$$H(z) = 1 - 0.5z^{-1} - 0.6z^{-2} + 0.64z^{-3} - \dots$$

Inverse transforming term by term gives the impulse response

$$h[n] = \delta[n] - 0.5\delta[n-1] - 0.6\delta[n-2] + 0.64\delta[n-3] - \dots$$

EXAMPLE 6.25

Using long division, find the time domain signal $x[n]$ that corresponds to the z transform of Example 6.23, that is,

$$X(z) = \frac{5}{z^2 + 0.2z}$$

Long division gives:

$$
z^2 + 0.2z \ \overline{\smash{\big)}\ \begin{aligned}
&\ 5z^{-2} - z^{-3} + 0.2z^{-4} - 0.04z^{-5} + \dots \\
&\ 5 \\
&\ \underline{5 + z^{-1}} \\
&\ {-z^{-1}} \\
&\ \underline{-z^{-1} - 0.2z^{-2}} \\
&\ 0.2z^{-2} \\
&\ \underline{0.2z^{-2} + 0.04z^{-3}} \\
&\ {-0.04z^{-3}} \\
&\ \underline{-0.04z^{-3} - 0.008z^{-4}} \\
&\ 0.008z^{-4} \\
&\ \dots
\end{aligned}}
$$

This means that $X(z) = 5z^{-2} - z^{-3} + 0.2z^{-4} - 0.04z^{-5} + \dots$ It is now easy to find $x[n]$ by taking an inverse z transform of each term:

$$x[n] = 5\delta[n-2] - \delta[n-3] + 0.2\delta[n-4] - 0.04\delta[n-5] + \dots$$

In this particular case, it so happens that the pattern that defines $x[n]$ can be identified, so a solution in closed form can be found:

$$x[n] = \sum_{k=0}^{\infty} 5(-0.2)^k \delta[n-2-k] \tag{6.8}$$

The signal $x[n]$ is plotted in Figure 6.9. Though the description of $x[n]$ provided by Equation (6.8) may appear to be very different from that given by Equation (6.7), the digital signals are, in fact, identical.

FIGURE 6.9
Signal for Example 6.25.

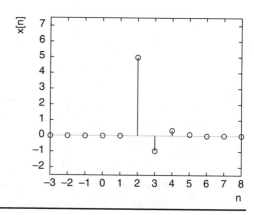

6.3.4 Inverse z Transforms by Partial Fraction Expansion

Partial fraction expansion is a third method of finding an inverse z transform for a strictly proper rational function in standard form. To illustrate this method, suppose that the input to a filter is $x[n] = u[n-1]$ and its impulse response $h[n] = (-0.25)^n u[n]$. The first step is to calculate the z transforms of both the input and the impulse response. Since

$$\mathbf{Z}\{u[n]\} = \frac{z}{z-1},$$

$$X(z) = \mathbf{Z}\{u[n-1]\} = z^{-1} \frac{z}{z-1} = \frac{1}{z-1}$$

The z transform of the impulse response is the transfer function of the system. From Table 6.1, the transfer function is

$$H(z) = \frac{z}{z+0.25}$$

Then, according to Equation (6.4),

$$Y(z) = H(z)X(z) = \frac{z}{(z+0.25)(z-1)} \tag{6.9}$$

This z transform is strictly proper, but is not one of the basic transforms from Table 6.1. The purpose of partial fraction expansion is to decompose it into terms that are in Table 6.1, one for each factor in the denominator. The partial fraction expansion has the form

$$Y(z) = \frac{z}{(z+0.25)(z-1)} = \frac{A}{z+0.25} + \frac{B}{z-1}$$

The coefficients A and B may be found by what is sometimes called the **cover-up method**.[1] The method requires the roots of the denominator, -0.25 and 1 in this case. The value for

[1]   This term was coined by Norm Eyres of the mathematics department of Camosun College.

A may be found by "covering up" the $(z + 0.25)$ factor in Equation (6.9) and evaluating $Y(z)$ for $z = -0.25$. The value for B is found by "covering up" the $(z - 1)$ term and evaluating $Y(z)$ for $z = 1$. For B, for example, this amounts to multiplying both sides of the partial fraction expansion equation by the denominator $(z - 1)$ and then setting $z = 1$. This leaves only B on the right side and cancels, or covers up, the $(z - 1)$ factor in the denominator of $Y(z)$ on the left. Using this method,

$$A = \frac{-0.25}{-0.25 - 1} = 0.2$$

and

$$B = \frac{1}{1 + 0.25} = 0.8$$

Therefore,

$$Y(z) = \frac{0.2}{z + 0.25} + \frac{0.8}{z - 1} \qquad (6.10)$$

As a check, the terms of $Y(z)$ may be recombined with a common denominator to verify that the transfer function matches Equation (6.9). To use the basic transforms in Table 6.1, $Y(z)$ can be rewritten as

$$Y(z) = z^{-1}\left(\frac{0.2z}{z + 0.25} + \frac{0.8z}{z - 1}\right)$$

This equation is mathematically identical with Equation (6.10). The portion within the brackets has the inverse transform $0.2(-0.25)^n u[n] + 0.8u[n]$. The z^{-1} term outside the brackets simply indicates a time shift by one step, producing the final inverse transform

$$y[n] = 0.2(-0.25)^{n-1}u[n-1] + 0.8u[n-1]$$

This expression can be evaluated for all values of n, remembering that $u[n-1]$ has the value one for $n \geq 1$ and the value zero for $n < 1$. The result is plotted in Figure 6.10. The output settles to 0.8 in the steady state.

In general, three kinds of roots are possible for a polynomial function of z: distinct real roots, repeated real roots, or complex conjugate roots. In the foregoing, the partial fraction expansion method was used to handle the first type of roots, in which the roots of the denominator are real and different from one another. In such cases, the denominator can be expressed as a product of simple terms like $(z - \alpha)(z - \beta)(z - \gamma)$, where α, β, and γ are all real numbers. The expansion has the form

$$\frac{A}{z - \alpha} + \frac{B}{z - \beta} + \frac{C}{z - \gamma}$$

FIGURE 6.10
Output signal obtained by cover-up method.

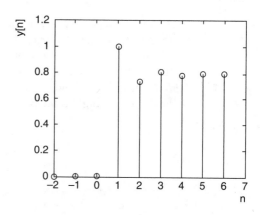

where A, B, and C can be found using the cover-up method. To avoid unnecessary complexity, the methods for the cases of repeated real or complex conjugate roots (Ambardar, 1999) will not be studied in this text. Fortunately, these cases can most often be handled satisfactorily using the long division method of the last section.

EXAMPLE 6.26

Find the inverse z transform of

$$Y(z) = \frac{0.5}{z(z - 1)(z - 0.6)}$$

The denominator is already factored into simple factors. The partial fraction expansion of $Y(z)$ has three terms, one for each of the roots in the denominator:

$$Y(z) = \frac{A}{z} + \frac{B}{z - 1} + \frac{C}{z - 0.6}$$

Covering up the z term in the denominator and evaluating $Y(z)$ at $z = 0$,

$$A = \frac{0.5}{(0 - 1)(0 - 0.6)} = \frac{5}{6}$$

Covering up the $(z - 1)$ term in the denominator and evaluating at $z = 1$,

$$B = \frac{0.5}{(1)(1 - 0.6)} = \frac{5}{4}$$

Covering up the $(z - 0.6)$ term and evaluating at $z = 0.6$,

$$C = \frac{0.5}{(0.6)(0.6 - 1)} = -\frac{25}{12}$$

Hence,

$$Y(z) = \frac{\frac{5}{6}}{z} + \frac{\frac{5}{4}}{z-1} + \frac{-\frac{25}{12}}{z-0.6} = z^{-1}\left(\frac{5}{6} + \frac{\frac{5}{4}z}{z-1} + \frac{-\frac{25}{12}z}{z-0.6}\right)$$

The inverse z transform is, using Table 6.1,

$$y[n] = \frac{5}{6}\delta[n-1] + \frac{5}{4}u[n-1] - \frac{25}{12}(0.6)^{n-1}u[n-1]$$

EXAMPLE 6.27

Find the impulse response for the system

$$H(z) = \frac{z^{-2}}{1 + 0.25z^{-1}}$$

The impulse response for this transfer function was computed in Example 6.21. In this example, partial fraction expansion is used to accomplish the same goal. Changing to standard form, the transfer function becomes:

$$H(z) = \frac{z^{-2}}{1 + 0.25z^{-1}} = \frac{1}{z^2 + 0.25z}$$

Its partial fraction expansion is

$$H(z) = \frac{1}{z(z+0.25)} = \frac{A}{z} + \frac{B}{z+0.25} = \frac{4}{z} + \frac{-4}{z+0.25} = z^{-1}\left(4 - \frac{4z}{z+0.25}\right)$$

The portion within the brackets gives the inverse transform $4\delta[n] - 4(-0.25)^n u[n]$, so the final inverse transform is

$$h[n] = 4\delta[n-1] - 4(-0.25)^{n-1}u[n-1] \qquad (6.11)$$

While Equations (6.6) and (6.11) appear very different, they are identical functions for all values of n.

EXAMPLE 6.28

Find the time domain signal $x[n]$ that corresponds to the z transform

$$X(z) = \frac{5}{z^2 + 0.2z}$$

This example is a repetition of Examples 6.23 and 6.25. This time, partial fraction expansion will be used to solve the problem. The denominator of $X(z)$ can be factored to give

$$X(z) = \frac{5}{z(z + 0.2)} \tag{6.12}$$

The partial fraction expansion is

$$X(z) = \frac{A}{z} + \frac{B}{z + 0.2} = \frac{25}{z} + \frac{-25}{z + 0.2} = z^{-1}\left(25 - 25\frac{z}{z + 0.2}\right)$$

Thus, the final inverse transform is

$$x[n] = 25\delta[n-1] - 25(-0.2)^{n-1}u[n-1] \tag{6.13}$$

Though this function looks completely different from Equations (6.7) and (6.8), it produces exactly the same values as those shown in Figure 6.9.

EXAMPLE 6.29

The difference equation for a digital filter is

$$y[n] + 0.1y[n-1] - 0.3y[n-2] = x[n] \tag{6.14}$$

a. Use z transforms to find the filter's response to the input $x[n] = \delta[n]$.
b. Use z transforms to find the filter's response to the input $x[n] = u[n]$.

a. The system's response to $\delta[n]$ is, of course, its impulse response. The input transform is

$$X(z) = 1$$

and the system transfer function is

$$H(z) = \frac{1}{1 + 0.1z^{-1} - 0.3z^{-2}}$$

so that the output transform is

$$Y(z) = H(z)X(z) = \frac{1}{1 + 0.1z^{-1} - 0.3z^{-2}}$$

In this case $Y(z)$ can be expressed in standard form by multiplying numerator and denominator by z^2:

$$Y(z) = \frac{z^2}{z^2 + 0.1z - 0.3}$$

Partial fraction expansion requires that the z domain function be strictly proper, that is, that the degree of the numerator be less than the degree of the denominator. After one step of long division,

$$
z^2 + 0.1z - 0.3 \overline{\smash{\big)}\ \begin{array}{l} \phantom{z^2 + 0.1z -{}} 1 \\ z^2 \\ \underline{z^2 + 0.1z - 0.3} \\ \phantom{z^2 +{}} -0.1z + 0.3 \end{array}}
$$

a strictly proper expression is produced:

$$
Y(z) = 1 + \frac{-0.1z + 0.3}{z^2 + 0.1z - 0.3}
$$

The first term can be identified as one of the basic transforms in Table 6.1. The second term can be handled by partial fraction expansion, which decomposes the rational expression into parts, one for each root of the denominator. The roots can be found using the quadratic formula:

$$
z = \frac{-0.1 \pm \sqrt{0.1^2 - 4(1)(-0.3)}}{2(1)} = \frac{-0.1 \pm 1.1}{2} = 0.5 \text{ and } -0.6
$$

Thus,

$$
Y(z) = 1 + \frac{-0.1z + 0.3}{(z - 0.5)(z + 0.6)} = 1 + \frac{A}{z - 0.5} + \frac{B}{z + 0.6} \qquad \textbf{(6.15)}
$$

The cover-up method provides the values for A and B. With the $(z - 0.5)$ term covered, the rational part evaluates to $5/22$. With the $(z + 0.6)$ term covered, the rational part evaluates to $-18/55$. $Y(z)$ now becomes

$$
Y(z) = 1 + \frac{\frac{5}{22}}{z - 0.5} + \frac{-\frac{18}{55}}{z + 0.6} = 1 + z^{-1}\left(\frac{\frac{5}{22}z}{z - 0.5} + \frac{-\frac{18}{55}z}{z + 0.6}\right)
$$

Table 6.1 then provides the inverse transform, which is the impulse response:

$$
y[n] = h[n] = \delta[n] + \frac{5}{22}(0.5)^{n-1}u[n-1] - \frac{18}{55}(-0.6)^{n-1}u[n-1] \quad \textbf{(6.16)}
$$

The impulse response is graphed in Figure 6.11(a). An advantage of using z transforms instead of a difference equation to find the impulse response is that a closed-form solution like Equation (6.16) is obtained. That is, $h[n]$ can be computed directly for any value of n, without prior calculations for lower values of n.

The problem presented in this example can also be solved by using partial fraction expansion with an alternate form of $Y(z)$:

$$Y(z) = z^2\left(\frac{1}{z^2 + 0.1z - 0.3}\right) = z^2\left(\frac{1}{(z - 0.5)(z + 0.6)}\right) = z^2\left(\frac{A}{z - 0.5} + \frac{B}{z + 0.6}\right)$$

$$= z^2\left(\frac{\frac{1}{1.1}}{z - 0.5} + \frac{-\frac{1}{1.1}}{z + 0.6}\right) = z\left(\frac{\frac{1}{1.1}z}{z - 0.5} + \frac{-\frac{1}{1.1}z}{z + 0.6}\right)$$

Therefore,

$$y[n] = h[n] = \frac{10}{11}(0.5)^{n+1}u[n+1] - \frac{10}{11}(-0.6)^{n+1}u[n+1]$$

The sample values are identical with those for Equation (6.16).

b. The step response can be found using the same method as in (a). The only difference is that the z transform of the input in this case is

$$X(z) = \frac{z}{z - 1}$$

The output transform is the product of the transfer function and the input transform, as indicated in Equation (6.4). Steps must be taken to obtain a strictly proper rational function before partial fraction expansion can be used. Multiplying $H(z)$ by $X(z)$ to obtain $Y(z)$ gives:

$$Y(z) = H(z)X(z) = \frac{z^2}{z^2 + 0.1z - 0.3}\frac{z}{z - 1} = \frac{z^3}{z^3 - 0.9z^2 - 0.4z + 0.3}$$

One step of long division gives

$$Y(z) = 1 + \frac{0.9z^2 + 0.4z - 0.3}{z^3 - 0.9z^2 - 0.4z + 0.3}$$

which has a strictly proper rational term. A partial fraction expansion of this term gives

$$Y(z) = 1 + \frac{0.9z^2 + 0.4z - 0.3}{(z - 0.5)(z + 0.6)(z - 1)}$$

$$= 1 + \frac{A}{z - 0.5} + \frac{B}{z + 0.6} + \frac{C}{z - 1}$$

$$= 1 + \frac{-\frac{5}{22}}{z - 0.5} + \frac{-\frac{27}{220}}{z + 0.6} + \frac{\frac{5}{4}}{z - 1}$$

$$= 1 + z^{-1}\left(\frac{-\frac{5}{22}z}{z - 0.5} + \frac{-\frac{27}{220}z}{z + 0.6} + \frac{\frac{5}{4}z}{z - 1}\right)$$

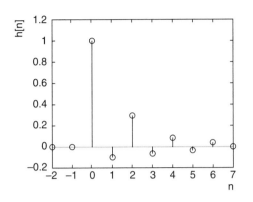

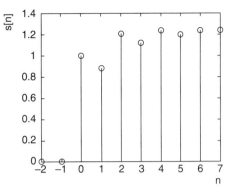

(a) Impulse Response

(b) Step Response

FIGURE 6.11
Impulse response and step response for Example 6.29.

The step response is given by the inverse z transform of this expression, which is

$$y[n] = s[n] = \delta[n] - \frac{5}{22}(0.5)^{n-1}u[n-1] - \frac{27}{220}(-0.6)^{n-1}u[n-1] + \frac{5}{4}u[n-1]$$

This signal is plotted in Figure 6.11(b). The only sustained component of the step response is the last term, as all other terms have a transient nature and decay to zero. Thus, the response settles to a steady state value of $5/4 = 1.25$.

Alternatively, the step response can be found by reformulating $Y(z)$ as

$$Y(z) = z^3\left(\frac{1}{(z^2 + 0.1z - 0.3)(z-1)}\right) = z^3\left(\frac{1}{(z-0.5)(z+0.6)(z-1)}\right)$$

$$= z^3\left(\frac{-\frac{20}{11}}{z-0.5} + \frac{\frac{25}{44}}{z+0.6} + \frac{\frac{5}{4}}{z-1}\right)$$

$$= z^2\left(-\frac{20}{11}\frac{z}{z-0.5} + \frac{25}{44}\frac{z}{z+0.6} + \frac{5}{4}\frac{z}{z-1}\right)$$

After the two-step advance required by the z^2 term is implemented, the inverse transform gives:

$$y[n] = s[n] = -\frac{20}{11}(0.5)^{n+2}u[n+2] + \frac{25}{44}(-0.6)^{n+2}u[n+2] + \frac{5}{4}u[n+2]$$

This function also produces the sample values plotted in Figure 6.11(b).

6.4   TRANSFER FUNCTIONS AND STABILITY

6.4.1   Poles and Zeros

Poles are the values of z that make the denominator of a transfer function zero. **Zeros** are the values of z that make the numerator of a transfer function zero. Of the two, poles have the biggest effect on the behavior of a digital filter. Zeros tend to modulate, to a greater or lesser degree depending on their positions relative to the poles, the behavior due to the poles. The poles and zeros for a digital filter can be found if its transfer function is known. From Equation (6.2), the general form for a transfer function is

$$H(z) = \frac{Y(z)}{X(z)} = \frac{b_0 + b_1 z^{-1} + \ldots + b_M z^{-M}}{a_0 + a_1 z^{-1} + \ldots + a_N z^{-N}}$$

The negative exponents, however, make finding poles and zeros difficult. The calculations are easiest if the transfer function is expressed in the standard form of Equation (6.5). While standard form is not strictly necessary for finding poles and zeros, it does make the process more straightforward, as the following example demonstrates.

EXAMPLE 6.30

Find the poles and zeros for the digital filter whose transfer function is

$$H(z) = \frac{4z^{-1}}{4 - 9z^{-1} + 2z^{-2}}$$

Converting to standard form:

$$H(z) = \frac{4z}{4z^2 - 9z + 2} = \frac{z}{z^2 - 2.25z + 0.5}$$

The denominator polynomial can be factored using the quadratic formula:

$$z = \frac{2.25 \pm \sqrt{2.25^2 - 4(1)(0.5)}}{2(1)} = 0.25 \text{ or } 2.00$$

Hence,

$$H(z) = \frac{z}{(z - 0.25)(z - 2)}$$

The factored form of the transfer function makes it easy to identify the poles and zeros. For this filter, there is a single zero, at $z = 0$, and two poles, one at $z = 0.25$ and one at $z = 2$.

In general, the numerator and denominator polynomials of a transfer function in standard form can always be factored. Once the expression in Equation (6.5) is factored, it has the form

$$H(z) = \frac{K(z - z_1)(z - z_2)...(z - z_M)}{(z - p_1)(z - p_2)...(z - p_N)}$$

where the z_j are the zeros of the filter, the p_j are the poles of the filter, and K is called the gain. A very powerful tool for digital filter analysis and design is a complex plane called the **z plane**, on which the poles and zeros of the transfer function are plotted. It is customary to identify the poles on this plane by Xs and the zeros by Os. The positions of the poles and zeros on the z plane can give clues about the way a filter will behave. These relationships are investigated in the remainder of this chapter.

EXAMPLE 6.31

Find and plot the poles and zeros for the transfer function

$$H(z) = \frac{1}{0.8 + 0.23z^{-1} + 0.15z^{-2}}$$

To find the poles and zeros, convert $H(z)$ to standard form:

$$H(z) = \frac{z^2}{0.8z^2 + 0.23z + 0.15} = \frac{1.25z^2}{z^2 + 0.2875z + 0.1875}$$

The zeros are located where $z^2 = 0$. In other words, there are two zeros, both located at $z = 0$. The pole locations can be found using the quadratic formula:

$$z = \frac{-0.2875 \pm \sqrt{0.2875^2 - 4(1)(0.1875)}}{2(1)}$$

$$= \frac{-0.2875 \pm j0.8169}{2} = -0.1438 \pm j0.4085$$

In Figure 6.12, the two zeros are plotted as Os, and the pair of complex poles are plotted as Xs.

FIGURE 6.12
Poles and zeros for
Example 6.31.

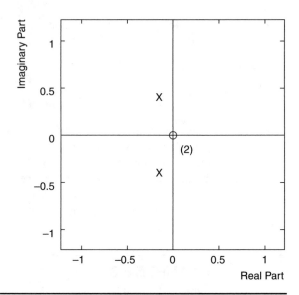

EXAMPLE 6.32

A digital filter has zeros at -0.2 and 0.4, poles at $-0.7 \pm j0.6$, and a gain of 0.5.

  a. Draw a pole-zero plot for the filter.
  b. Find the transfer function of the filter.

  a. The pole-zero plot is shown in Figure 6.13.
  b. Each zero produces a factor in the numerator of the transfer function, and each pole produces a factor in the denominator. The transfer function is

FIGURE 6.13
Pole-zero plot for Example 6.32.

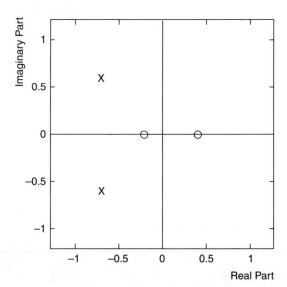

$$H(z) = \frac{K(z - z_1)(z - z_2)\ldots(z - z_M)}{(z - p_1)(z - p_2)\ldots(z - p_N)}$$

$$= \frac{0.5(z - (-0.2))(z - 0.4)}{(z - (-0.7 + j0.6))(z - (-0.7 - j0.6))}$$

Simplifying, the transfer function becomes

$$H(z) = \frac{0.5(z + 0.2)(z - 0.4)}{(z + 0.7)^2 + 0.6^2} = \frac{0.5(z^2 - 0.2z - 0.08)}{z^2 + 1.4z + 0.85}$$

$$= \frac{0.5 - 0.1z^{-1} - 0.04z^{-2}}{1 + 1.4z^{-1} + 0.85z^{-2}}$$

6.4.2 Stability

As Example 6.32 showed, specifying the gain of a filter along with its poles and zeros is enough to completely specify the system. Of these system characteristics, the poles of the filter have the greatest influence on overall filter behavior. They determine not only whether or not a filter is stable but also the general type of output response the filter will produce.

As long as the inputs are bounded, meaning finite in size, the outputs from a stable filter will always settle to some regular behavior because the transients are guaranteed to die away. When the input stays at a constant level, the output will settle to a constant level. When the input is a sinusoid, a stable filter output will eventually become sinusoidal. When a filter is unstable, however, outputs grow without bound. The output from an unstable filter can change dramatically even when the input changes by only the smallest amount. All useful filters are stable, and one important aspect of filter design is to guarantee stability.

FIGURE 6.14

Stable region of the z plane.

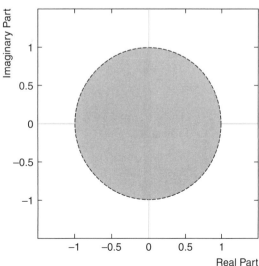

Imaginary Part

Real Part

The **unit circle** is a circle with radius one centered at the origin of the z plane. A filter is **stable** if all its poles are inside the unit circle. A filter with poles on the unit circle is said to be **marginally stable**. A filter with poles outside the unit circle is **unstable**. The shaded area of Figure 6.14 shows the region of the z plane that encloses stable poles. Mathematically, the region of stability can be described as

$$|z| < 1$$

where the double bars refer to the magnitude of the complex number. Recall that the magnitude of a complex number $z = \text{Re } z + j \text{ Im } z$ is calculated from the real and imaginary parts of the number by the equation

$$|z| = \sqrt{(\text{Re } z)^2 + (\text{Im } z)^2}$$

If the magnitude of each pole is less than one, the poles are less than one unit's distance from the center of the unit circle, and the filter is stable.

Every transfer function $H(z)$ has a region of convergence, as indicated in Table 6.1. The region of convergence for a stable transfer function must include the unit circle. Thus the transfer function $H(z) = z/(z - 1)$ cannot be stable because its region of convergence is $|z| > 1$, while the transfer function $H(z) = z/(z - \beta)$ is stable as long as $|\beta| < 1$, since its region of convergence is $|z| > |\beta|$. In other words, insisting on a region of convergence that includes the unit circle is identical with requiring that all poles lie inside the unit circle.

To demonstrate the effect pole position has on system behavior, Figure 6.15 displays impulse responses for a variety of simple systems with poles both inside and outside the unit circle. As the figure shows, all of the responses with poles inside the circle tend to die away to zero. Transients disappear like this only for stable systems. Poles outside the unit circle produce impulse responses that grow over time and never settle to zero, a characteristic of unstable systems.

EXAMPLE 6.33

The transfer function for a digital filter is

$$H(z) = \frac{1 - z^{-2}}{1 + 0.7z^{-1} + 0.9z^{-2}}$$

Is the filter stable?

In standard form,

$$H(z) = \frac{z^2 - 1}{z^2 + 0.7z + 0.9}$$

The zeros are located where $z^2 - 1 = 0$, or $z = \pm 1$. The poles are located at

$$z = \frac{-0.7 \pm \sqrt{0.7^2 - 4(1)(0.9)}}{2(1)} = \frac{-0.7 \pm j\sqrt{3.11}}{2} = -0.35 \pm j0.8818$$

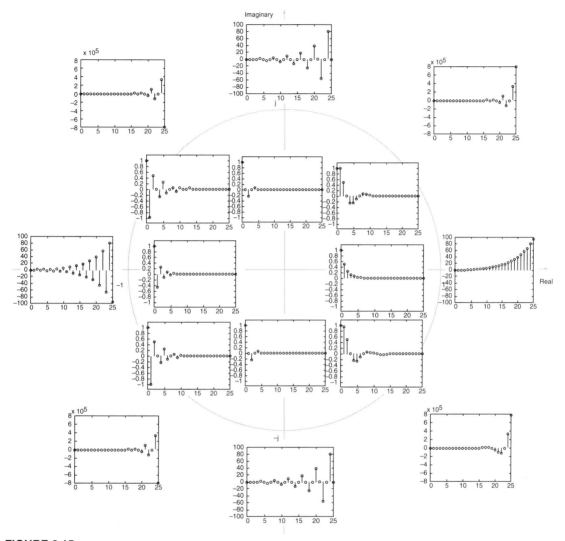

FIGURE 6.15
Stable and unstable impulse responses on the *z* plane.

For these poles, the distance from the center of the unit circle is

$$|z| = \sqrt{(-0.35)^2 + (0.8818)^2} = 0.9487$$

Because the distance is less than one, as illustrated in Figure 6.16, both poles are within the unit circle and the system is stable.

FIGURE 6.16
Pole-zero plot for Example 6.33.

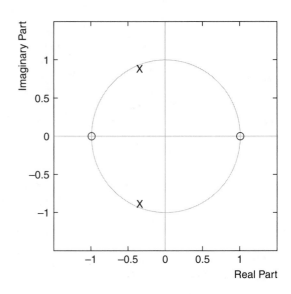

EXAMPLE 6.34

The difference equation for a filter is

$$y[n] + 0.8y[n-1] - 0.9y[n-2] = x[n-2]$$

Is the filter stable?

Poles are found most easily from the transfer function,

$$H(z) = \frac{z^{-2}}{1 + 0.8z^{-1} - 0.9z^{-2}} = \frac{1}{z^2 + 0.8z - 0.9}$$

The quadratic formula gives the pole locations as

$$z = \frac{-0.8 \pm \sqrt{0.8^2 - 4(1)(-0.9)}}{2(1)} = \frac{-0.8 \pm 2.059}{2} = 0.630 \text{ and } -1.430$$

The poles in this case are purely real, without any imaginary component. Clearly the pole at $z = -1.430$ lies outside the unit circle, so the system is unstable.

6.4.3 First Order Systems

A first order system, by definition, has just one pole. The transfer function for a simple first order system is

$$H(z) = \frac{1}{1 + \alpha z^{-1}} = \frac{z}{z + \alpha}$$

where α may be positive or negative. Since there is only one pole, at $z = -\alpha$, the requirement for stability is that $|\alpha| < 1$.

From Table 6.1, the impulse response is

$$h[n] = (-\alpha)^n u[n]$$

This is a good illustration of what stability means. When $|\alpha| > 1$, the impulse response grows without bound as n increases. As long as $|\alpha| < 1$, the impulse response settles down to zero.

Finally, the difference equation for this first order filter is

$$y[n] + \alpha y[n-1] = x[n]$$

Figure 6.17 shows the pole-zero plots, impulse responses, and step responses for two stable systems, one with $\alpha > 0$ and the other with $\alpha < 0$. The step responses settle to a constant value y_{ss} in steady state. Since the step function input has the constant value one, the final value can be predicted: At steady state, the difference equation becomes

$$y_{ss} + \alpha y_{ss} = 1$$

which gives $y_{ss} = \dfrac{1}{1 + \alpha}$.

EXAMPLE 6.35

A filter has the transfer function $H(z) = \dfrac{2}{1 + 0.4z^{-1}}$.

a. Find the pole-zero plot for the filter. Is the filter stable?
b. Find the impulse response for the filter.
c. Find the step response for the filter.

a. To find the poles and zeros, $H(z)$ is first expressed in standard form:

$$H(z) = \frac{2}{1 + 0.4z^{-1}} = \frac{2z}{z + 0.4}$$

There is a single zero at $z = 0$ and a single pole at $z = -0.4$. These points are plotted in Figure 6.18. Since the pole is within the unit circle, the filter is stable.

b. The impulse input has a z transform of $X(z) = 1$. The output $Y(z)$ is

$$Y(z) = H(z)X(z) = \frac{2z}{z + 0.4}$$

The inverse z transform, from Table 6.1, is

$$y[n] = h[n] = 2(-0.4)^n u[n]$$

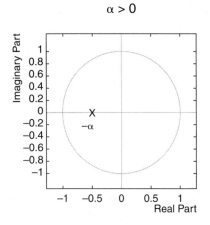
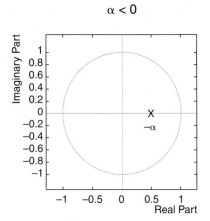

(a) Pole-Zero Plots

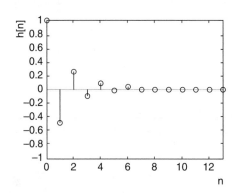
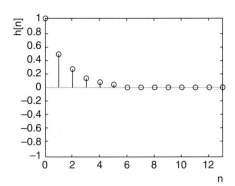

(b) Impulse Responses

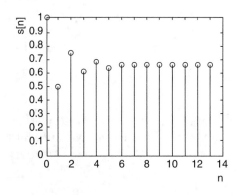
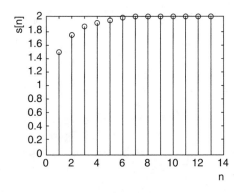

(c) Step Responses

FIGURE 6.17

First order filters.

FIGURE 6.18

Pole-zero plot for Example 6.35.

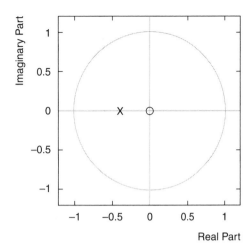

This impulse response is plotted in Figure 6.19.

c. The step input has a z transform of $X(z) = z/(z - 1)$. The output is

$$Y(z) = H(z)X(z) = \frac{2z}{z + 0.4} \frac{z}{z - 1}$$

$$= 2z^2 \left(\frac{1}{(z + 0.4)(z - 1)} \right) = 2z^2 \left(\frac{A}{z + 0.4} + \frac{B}{z - 1} \right)$$

$$= 2z^2 \left(\frac{-\frac{5}{7}}{z + 0.4} + \frac{\frac{5}{7}}{z - 1} \right) = z \left(\frac{-\frac{10}{7}z}{z + 0.4} + \frac{\frac{10}{7}z}{z - 1} \right)$$

Thus, the step response is

$$y[n] = s[n] = -\frac{10}{7}(-0.4)^{n+1}u[n+1] + \frac{10}{7}u[n+1]$$

FIGURE 6.19

Impulse response
for Example 6.35.

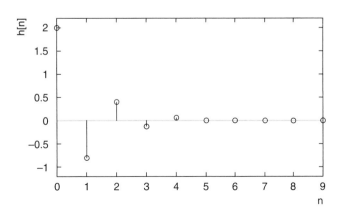

FIGURE 6.20

Step response for Example 6.35.

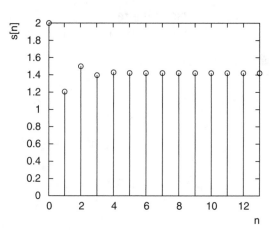

This response is plotted in Figure 6.20. The difference equation for this filter is $y[n] + 0.4y[n-1] = 2x[n]$. Since the input step function $x[n]$ equals one, the final value of the output in steady state may be found from $y_{ss} + 0.4y_{ss} = 2$, which gives a steady state value of $2/(1 + 0.4) = 1.428$, evident in Figure 6.20.

EXAMPLE 6.36

The transfer function of a filter (similar to the one in Example 6.35) is:

$$H(z) = \frac{2}{1 - 0.4z^{-1}}$$

a. Determine the difference equation of the filter.
b. Find the pole-zero plot and evaluate stability.
c. Find and plot the impulse response.

a. The difference equation is $y[n] - 0.4y[n - 1] = 2x[n]$.
b. The poles and zeros are found from

$$H(z) = \frac{2}{1 - 0.4z^{-1}} = \frac{2z}{z - 0.4}$$

There is a single zero at $z = 0$ and a single pole at $z = 0.4$, plotted in Figure 6.21. The pole is within the unit circle, so the filter is stable.

c. The impulse response, $h[n] = 2(0.4)^n u[n]$, is the inverse z transform of the transfer function $H(z)$ and is plotted in Figure 6.22.

FIGURE 6.21

Pole-zero plot for Example 6.36.

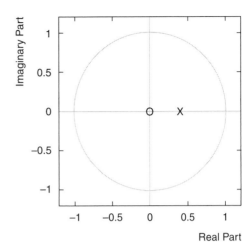

FIGURE 6.22

Impulse response for Example 6.36.

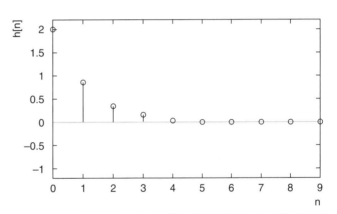

6.4.4   Second Order Systems

The transfer function for a simple second order system is

$$H(z) = \frac{1}{1 + \alpha z^{-1} + \beta z^{-2}} = \frac{z^2}{z^2 + \alpha z + \beta} = \frac{z^2}{(z - p_1)(z - p_2)}$$

where p_1 and p_2 are the poles of the transfer function. This particular second order system has two zeros at $z = 0$ as well. Second order systems are defined by the fact that they have two poles. The system is stable as long as both poles are inside the unit circle on the z plane. This requires $|p_1| < 1$ and $|p_2| < 1$. This second order system has the difference equation

$$y[n] + \alpha y[n-1] + \beta y[n-2] = x[n]$$

EXAMPLE 6.37

A second order system has poles at $z = 0.7 \pm j0.8$, no zeros, and a gain of one.

    a. Is the system stable?
    b. What is the transfer function of the system?

    a. The magnitude, or distance from the origin, for these poles is

$$\sqrt{0.7^2 + 0.8^2} = 1.06 > 1$$

The poles fall outside the unit circle, and the system is unstable.
    b. The transfer function is

$$H(z) = \frac{1}{(z - (0.7 + j0.8))(z - (0.7 - j0.8))} = \frac{1}{(z - 0.7)^2 + (0.8)^2}$$

$$= \frac{1}{z^2 - 1.4z + 1.13} = \frac{z^{-2}}{1 - 1.4z^{-1} + 1.13z^{-2}}$$

EXAMPLE 6.38

A second order transfer function is given by

$$H(z) = \frac{z^{-2}}{1 + 0.1z^{-1} + 0.05z^{-2}}$$

    a. Find the poles and zeros of the transfer function, and find the magnitude of each pole.
    b. Plot the impulse and step responses for this system.

    a. The transfer function is written in standard form as

$$H(z) = \frac{1}{z^2 + 0.1z + 0.05}$$

It has no zeros, and two poles at the locations

$$z = \frac{-0.1 \pm \sqrt{(0.1)^2 - 4(0.05)}}{2} = -0.05 \pm j0.2179$$

The magnitude of the first pole is $\sqrt{(-0.05)^2 + (0.2179)^2} = 0.2236$, and the magnitude of the second pole is $\sqrt{(-0.05)^2 + (-0.2179)^2} = 0.2236$ as well. The system is stable.

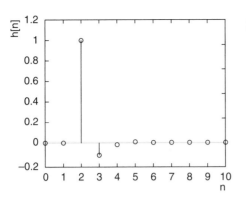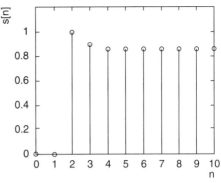

FIGURE 6.23
Impulse and step response for Example 6.38.

b. This particular system has complex poles. Since partial fraction expansion for complex poles has not been studied in the chapter, impulse and step response samples can be found using the difference equation for the filter,

$$y[n] = -0.1y[n-1] - 0.05y[n-2] + x[n-2]$$

Following the methods of Sections 4.7 and 4.8, the impulse response and the step response can be calculated from the equations

$$h[n] = -0.1h[n-1] - 0.05h[n-2] + \delta[n-2]$$

$$s[n] = -0.1s[n-1] - 0.05s[n-2] + u[n-2]$$

They are plotted in Figure 6.23.

EXAMPLE 6.39

Find the poles and zeros for the transfer function

$$H(z) = \frac{z^{-2}}{1 + 0.6z^{-1} + 0.05z^{-2}}$$

and determine the magnitudes of the poles.

The transfer function has no zeros. The quadratic formula

$$z = \frac{-0.6 \pm \sqrt{(0.6)^2 - 4(0.05)}}{2} = -0.3 \pm 0.2$$

gives poles at $z = -0.1$ and $z = -0.5$. The magnitude of the first pole is $\sqrt{(-0.1)^2 + (0)^2} = 0.1$, and the magnitude of the second pole is $\sqrt{(-0.5)^2 + (0)^2} = 0.5$, so the system is stable.

The easiest way to see the effect of pole location on system behavior is to study a sample group of second order systems. Eight transfer functions will be investigated:

a. $H(z) = \dfrac{z^{-2}}{1 + 0.2z^{-1} + 0.01z^{-2}}$

b. $H(z) = \dfrac{z^{-2}}{1 + 0.6z^{-1} + 0.13z^{-2}}$

c. $H(z) = \dfrac{z^{-2}}{1 - 1.2z^{-1} + 0.36z^{-2}}$

d. $H(z) = \dfrac{z^{-2}}{1 - z^{-1} + 0.5z^{-2}}$

e. $H(z) = \dfrac{z^{-2}}{1 - 1.15z^{-1} + 0.28z^{-2}}$

f. $H(z) = \dfrac{z^{-2}}{1 + 1.7z^{-1} + 0.7625z^{-2}}$

g. $H(z) = \dfrac{z^{-2}}{1 + 1.8z^{-1} + 0.81z^{-2}}$

h. $H(z) = \dfrac{z^{-2}}{1 - 1.6z^{-1} + 0.9425z^{-2}}$

Each of the transfer functions has the same form. For each, the poles, the impulse response, and the step response are presented in Figure 6.24. To determine the poles, the transfer functions can be written in standard form as

$$H(z) = \frac{z^{-2}}{1 + \alpha z^{-1} + \beta z^{-2}} = \frac{1}{z^2 + \alpha z + \beta} \tag{6.17}$$

and the impulse and step responses may be found using a difference equation of the form

$$y[n] = -\alpha y[n-1] - \beta y[n-2] + x[n-2]$$

Each response shows an initial two-step delay due to the $x[n-2]$ term.

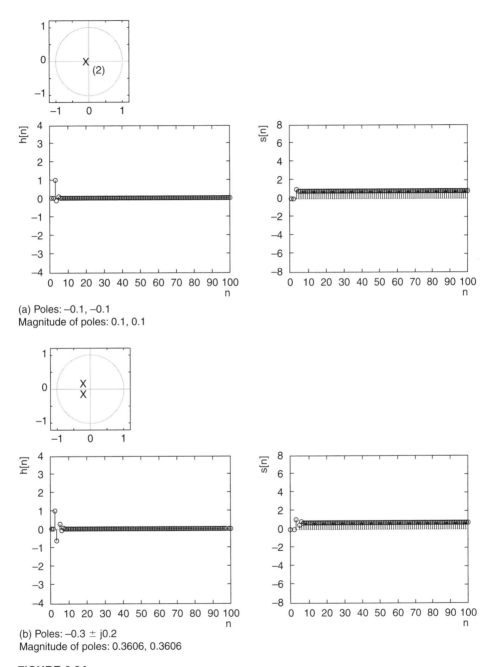

(a) Poles: −0.1, −0.1
Magnitude of poles: 0.1, 0.1

(b) Poles: −0.3 ± j0.2
Magnitude of poles: 0.3606, 0.3606

FIGURE 6.24
Collection of second order impulse and step responses.

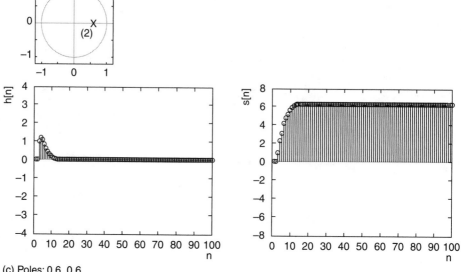

(c) Poles: 0.6, 0.6
Magnitude of poles: 0.6, 0.6

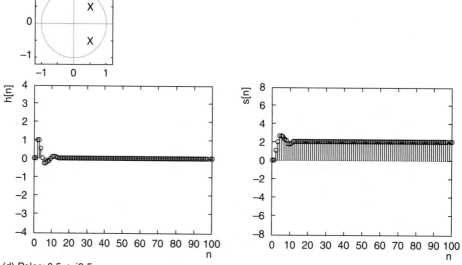

(d) Poles: 0.5 ± j0.5
Magnitude of poles: 0.7071, 0.7071

FIGURE 6.24

Continued

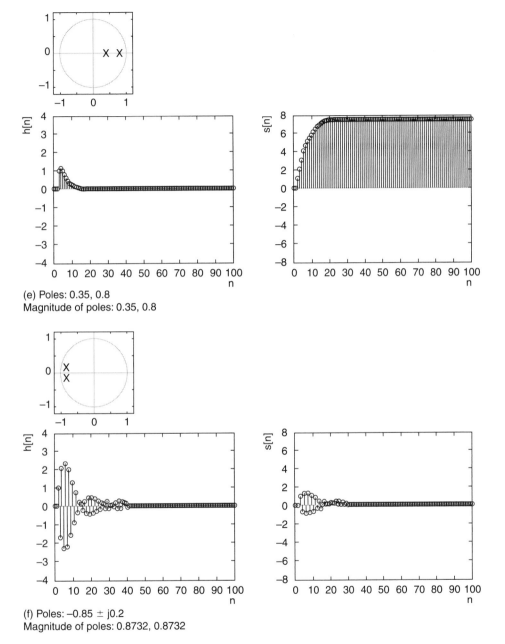

(e) Poles: 0.35, 0.8
Magnitude of poles: 0.35, 0.8

(f) Poles: $-0.85 \pm j0.2$
Magnitude of poles: 0.8732, 0.8732

FIGURE 6.24
Continued

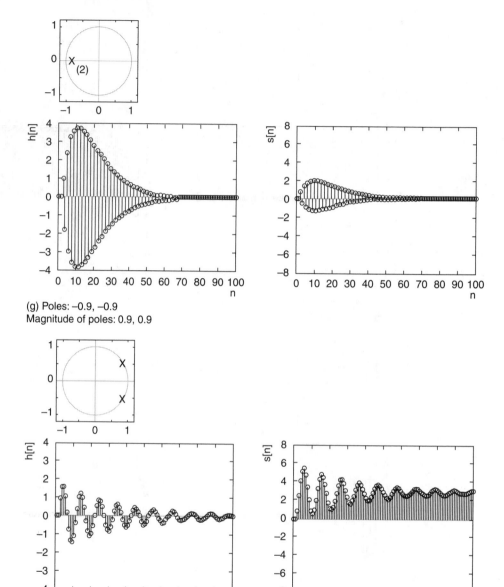

(g) Poles: −0.9, −0.9
Magnitude of poles: 0.9, 0.9

(h) Poles: 0.8 ± j0.55
Magnitude of poles: 0.9708, 0.9708

FIGURE 6.24
Continued

Figure 6.24 provides a good illustration of the effect of pole position on system response. When the poles have a negative real part, impulse response samples alternate between positive and negative values. When poles have a positive real part, impulse response samples do not show this alternation. Since the systems investigated in Figure 6.24 are all stable, the impulse responses all settle to zero and the step responses all settle to a constant value. For a stable transfer function in the form of Equation (6.17), the final output value can easily be found. As indicated earlier, the difference equation is

$$y[n] + \alpha y[n-1] + \beta y[n-2] = x[n-2]$$

For a unit step input, $x[n-2] = 1$ at steady state. For large $n, y[n] \approx y[n-1] \approx y[n-2] = y_{ss}$, so that

$$y_{ss}(1 + \alpha + \beta) = 1$$

Hence the steady state output value for a second order transfer function like Equation (6.17) is

$$y_{ss} = \frac{1}{1 + \alpha + \beta}$$

Thus, for a unit step input, the output of a stable system is guaranteed to approach a constant value at steady state. Similar reasoning shows that the output will always settle to zero when the input is an impulse function. In Example 6.40, the steady state output values are computed for some of the unit step responses in Figure 6.24, while Example 6.41 examines a new transfer function.

EXAMPLE 6.40

Compute the steady state outputs for the step responses (a), (b), and (c) in Figure 6.24. The transfer functions for the three systems are:

a. $H(z) = \dfrac{z^{-2}}{1 + 0.2z^{-1} + 0.01z^{-2}}$

b. $H(z) = \dfrac{z^{-2}}{1 + 0.6z^{-1} + 0.13z^{-2}}$

c. $H(z) = \dfrac{z^{-2}}{1 - 1.2z^{-1} + 0.36z^{-2}}$

The steady state outputs are:

a. $y_{ss} = \dfrac{1}{1 + \alpha + \beta} = \dfrac{1}{1 + 0.2 + 0.01} = 0.826$

b. $y_{ss} = \dfrac{1}{1 + \alpha + \beta} = \dfrac{1}{1 + 0.6 + 0.13} = 0.578$

c. $y_{ss} = \dfrac{1}{1 + \alpha + \beta} = \dfrac{1}{1 - 1.2 + 0.36} = 6.25$

The steady state outputs for the other step responses in Figure 6.24 may be confirmed in the same manner.

EXAMPLE 6.41

Find the steady state output value for the step response of the filter whose transfer function is

$$H(z) = \frac{0.8z^{-1}}{1 + 0.5z^{-1} - 0.3z^{-2}}$$

The poles of the filter can be found from $z^2 + 0.5z - 0.3 = 0$. The poles are $z = -0.852$ and $z = 0.352$, so the filter is stable. The difference equation for this filter is

$$y[n] + 0.5y[n-1] - 0.3y[n-2] = 0.8x[n-1]$$

Since the filter is stable, the output will settle to a constant value y_{ss}. The step input is one, so the difference equation becomes

$$y_{ss} + 0.5y_{ss} - 0.3y_{ss} = 0.8$$

This gives $y_{ss} = 0.8/(1 + 0.5 - 0.3) = 2/3$. This result is confirmed in Figure 6.25, which plots the step response for the filter.

FIGURE 6.25

Step response for Example 6.41.

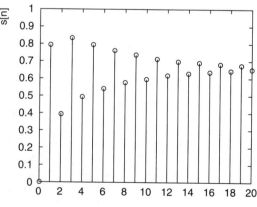

The magnitudes of the poles have a large impact on the time it takes the system to settle to its final value. In Figure 6.24, the systems are presented in order of increasing pole magnitude. To facilitate comparison, 100 samples are presented for each impulse and step response. The closer a pole is to the edge of the unit circle, the longer it takes for the output to settle. The closer a pole is to the center of the unit circle, the faster the output settles. In part (a), for example, the output takes less than three samples to settle. The number of samples required for settling increases with pole magnitude and exceeds 100 samples for part (f). Part (e) is the only system in Figure 6.24 with two poles whose magnitudes differ. The response due to the pole with the smaller magnitude of 0.35 disappears relatively quickly; the transient behavior is mostly due to the pole with the larger magnitude. That is, the pole at $z = 0.8$ dominates the output behavior. As the results of part (b) suggest, a pole with magnitude 0.35 should enable settling within just a few samples. The settling time in part (e) is about 25 samples, larger by a suitable margin than the settling time in part (d) where the poles have magnitude 0.7071. The greater the difference in magnitude of the two poles, the more the output behavior can be attributed to the dominant pole alone.

All of the systems in Figure 6.24 have two poles and no zeros. While poles have the greatest influence on output behavior, zeros can modify this behavior fairly dramatically, as Example 6.42 shows. As a general rule, the closer the zeros are to the poles, the greater their effect on system behavior. When a zero is far from the poles, it has a negligible impact.

EXAMPLE 6.42

Figure 6.26 presents the impulse responses and pole-zero plots for the systems:

a. $H(z) = \dfrac{1}{1 - 1.6z^{-1} + 0.9425z^{-2}}$

b. $H(z) = \dfrac{1 - 0.3z^{-1}}{1 - 1.6z^{-1} + 0.9425z^{-2}}$

c. $H(z) = \dfrac{1 - 0.8z^{-1}}{1 - 1.6z^{-1} + 0.9425z^{-2}}$

d. $H(z) = \dfrac{1 - 1.6z^{-1} + 0.8z^{-2}}{1 - 1.6z^{-1} + 0.9425z^{-2}}$

These four systems have identical poles at $z = 0.8 \pm j0.55$, the same as in part (h) of Figure 6.24, and zeros that come progressively closer to the pole positions. As seen in Figure 6.26, the impulse response decreases in amplitude as the zeros come closer to coinciding with the poles.

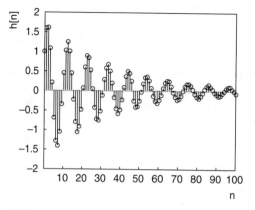
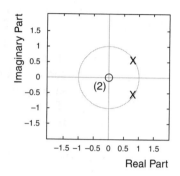

(a) Zeros: 0, 0

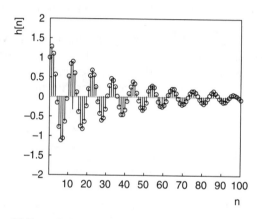
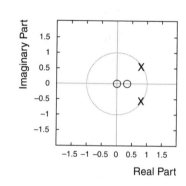

(b) Zeros: 0, 0.3

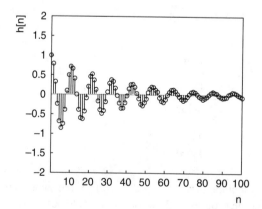
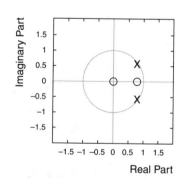

(c) Zeros: 0, 0.8

FIGURE 6.26

Effect of zero position on impulse response for Example 6.42.

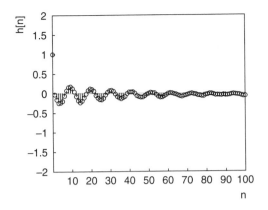
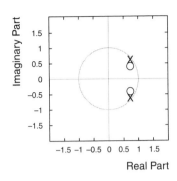

(d) Zeros: $0.8 \pm j0.4$

FIGURE 6.26
Continued

CHAPTER SUMMARY

Matlab
Support

1. The z transform of a signal $x[n]$ is calculated from

$$X(z) = \sum_{n=0}^{\infty} x[n] z^{-n}$$

2. Common z transforms are available in tables.
3. Each z transform $X(z)$ has a region of convergence that designates the values of z for which $X(z)$ exists. The region of convergence distinguishes between signals that have the same z transform but are defined for different intervals, $n < 0$, $n \geq 0$, or all n.
4. The factor z^{-k} in the z domain indicates a delay of k steps in the time domain. Thus, the z transform of $x[n - k]$ is $z^{-k}X(z)$.
5. A transfer function $H(z)$ is the ratio of the z transform of the output $Y(z)$ to the z transform of the input $X(z)$, that is,

$$H(z) = \frac{Y(z)}{X(z)}$$

6. A transfer function can be converted to a difference equation, and vice versa.
7. A transfer function $H(z)$ is the z transform of the impulse response $h[n]$. The impulse response is the inverse z transform of the transfer function.
8. The z transform of the output of a filter can be computed from $Y(z) = H(z)X(z)$, where $H(z)$ is the filter's transfer function and $X(z)$ is the z transform of the input. The output signal $y[n]$ is found by taking the inverse z transform of $Y(z)$.
9. Inverse z transforms may be found by using tables, long division, or partial fraction expansion.

10. Poles are the values of z that make the denominator of a transfer function $H(z)$ zero. Zeros are the values of z that make the numerator of the transfer function zero.

11. A filter is stable if all of its poles lie inside the unit circle on the complex plane.

12. The impulse response for a stable system will always settle to zero. The step response for a stable filter always settles to a constant value.

13. The closer the poles of a stable filter are to the unit circle, the longer it takes for the impulse response of the filter to settle to zero, and the longer it takes for the step response to settle to its final value.

REVIEW QUESTIONS

6.1 Find the z transforms of the following functions:
  a. $x[n] = 2\delta[n] - 3\delta[n-1] + \delta[n-3]$
  b. $x[n] = u[n] - u[n-2]$
  c. $x[n] = \delta[n-1] + 2u[n-3]$

6.2 Find the z transform of each of the following functions:
  a. $x[n] = (0.2)^n u[n]$
  b. $x[n] = 3 \sin(n\pi/2)u[n]$
  c. $x[n] = 0.5nu[n]$
  d. $x[n] = 4e^{-0.1n}\cos(n\pi/7)u[n]$
  e. $x[n] = -(0.8)^n \sin(n3\pi/4)u[n]$

6.3 Find the region of convergence for each of the z transforms in Question 6.1.

6.4 Find the region of convergence for each of the z transforms in Question 6.2.

6.5 a. Write the difference equation for the filter that corresponds to the diagram in Figure 6.27.
  b. Find the transfer function for the filter in (a).

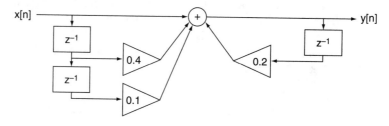

FIGURE 6.27
Difference equation diagram for Question 6.5.

6.6 Find the transfer functions for the difference equations:
  a. $y[n] - 0.7y[n-1] = x[n] + 0.2x[n-1]$
  b. $y[n] = 0.25y[n-2] + x[n] - 0.5x[n-1] + 0.25x[n-2]$
  c. $y[n] = 0.25(x[n] + x[n-1] + x[n-2] + x[n-3])$
  d. $y[n] = x[n]$

6.7 Convert the following transfer functions to difference equations:

a. $H(z) = \dfrac{1 - 0.4z^{-1}}{1 + 0.5z^{-1} + 0.75z^{-2}}$

b. $H(z) = 1 - 0.4z^{-1} + 0.1z^{-2}$

c. $H(z) = \dfrac{4z}{z^2 + 1.4z + 0.9}$

d. $H(z) = \dfrac{z}{(2z - 1)(z + 0.6)}$

6.8 Find the transfer function and the difference equation for the filter with the impulse response $h[n] = 2\delta[n] - 1.5\delta[n-1] + \delta[n-2] + 0.5\delta[n-3]$.

6.9 The impulse response of a filter is $h[n] = (-0.75)^{2n}(u[n-1] - u[n-4])$. What is the transfer function of the filter?

6.10 The difference equations for two filters are:

$$y_1[n] + 0.2y_1[n-1] = x_1[n] - 0.2x_1[n-1]$$

$$y_2[n] - 0.1y_2[n-1] + 0.4y_2[n-2] = 0.5x_2[n]$$

a. The filters are cascaded. What is the overall transfer function?

b. The filters are connected in parallel. What is the overall transfer function?

6.11 Two filters are cascaded together. The impulse responses for the filters are $h_1[n] = (0.2)^n(u[n] - u[n-3])$ and $h_2[n] = -3n(u[n] - u[n-4])$. Use z transforms to find the impulse response of the cascaded system.

6.12 Convert the following transfer functions to standard form:

a. $H(z) = \dfrac{1}{5 - 2z^{-1} + z^{-2}}$

b. $H(z) = \dfrac{1 - z^{-1}}{(1 - 0.8z^{-1})(1 + 0.65z^{-1})}$

6.13 Find the signal $x[n]$ for the z transform $X(z) = 1 + 0.5z^{-1} - 3z^{-4}$.

6.14 Find the signal $y[n]$ that corresponds to the z transform $Y(z) = \dfrac{z^{-1}}{z - 1}$.

6.15 Using the table of z transforms and/or the cover-up method, find the digital signal $x[n]$ whose z transform is:

a. $X(z) = \dfrac{z}{z + 0.12}$

b. $X(z) = \dfrac{5}{1 - z^{-1}}$

c. $X(z) = \dfrac{4}{z^2(z - 0.5)}$

d. $X(z) = \dfrac{2z - 1}{z - 0.9}$

e. $X(z) = \dfrac{1}{(z - 0.2)(z + 0.4)}$

f. $X(z) = \dfrac{1}{z(z + 0.2)(z - 1)}$

g. $X(z) = \dfrac{0.5z}{z^2 + 0.55z + 0.075}$

h. $X(z) = \dfrac{1 - 0.8z^{-2}}{1 - 0.9z^{-1} + 0.03z^{-2}}$

i. $X(z) = \dfrac{z^{-2}}{1 + 0.35z^{-1}}$

j. $X(z) = \dfrac{0.5878z}{z^2 - 1.618z + 1}$

k. $X(z) = \dfrac{1 - 0.5z^{-1}}{1 - z^{-1} + z^{-2}}$

l. $X(z) = \dfrac{1.6829z}{z^2 - 1.0806z + 1}$

6.16 Use long division to find the first eight terms of the inverse z transforms of the following functions:

a. $X(z) = \dfrac{z^2}{z^2 + 0.5z + 0.8}$

b. $X(z) = \dfrac{5z^{-2}}{1 + 0.1z^{-1} + 0.9z^{-2}}$

c. $X(z) = \dfrac{1}{2z^2 - 1.7z + 0.05}$

6.17 Find the impulse response for the system whose transfer function is

$$H(z) = \frac{z^{-1}}{1 - 0.75z^{-1}}$$

6.18 The input to a digital filter is $x[n] = \delta[n] - \delta[n-1]$. The impulse response for the filter is $h[n] = (0.25)^n u[n]$.
  a. Find z transforms for the input and impulse response.
  b. Use z transform methods to find an expression for the output from the system.

6.19 The impulse response for a system is $h[n] = \delta[n] - \delta[n-1]$, and the input is $x[n] = 5nu[n]$. Use z transforms to find the output from the system. Confirm the results using convolution.

6.20 The impulse response of a system is $e^{-2n}u[n]$ and the input is $u[n]$. Use z transform methods to find an expression for the output from the system.

6.21 Use z transforms to find the impulse response for the digital filter with the difference equation

$$1.2y[n] + 0.18y[n-1] - 0.084y[n-2] = 6x[n] - 0.3x[n-1]$$

6.22 For the transfer function

$$H(z) = \frac{5}{(1 - 0.1z^{-1})(1 - 0.9z^{-1})}$$

find the:

a. Impulse response

b. Step response

6.23 Produce a pole-zero plot for each of the following systems:

a. $H(z) = \dfrac{1}{0.71 - z^{-1}}$

b. $H(z) = \dfrac{z + 1.2}{z^2 + 1.2z + 0.5}$

c. $H(z) = \dfrac{z^{-1}}{2 - 3z^{-1} + 0.8z^{-2}}$

6.24 A digital filter has a zero at $z = 0$ and poles at $z = -0.8$ and $z = -0.5 \pm j0.6$. The filter has unity gain.

a. Draw a pole-zero diagram for the filter.

b. Find the transfer function of the filter.

6.25 A filter has unity gain and a zero at $z = 0.5$. Its poles lie at $z = 0.9$ and $z = -0.1$.

a. Find the transfer function of the filter.

b. Which pole has the greatest impact on the shape of the impulse response?

6.26 The poles of a unity gain filter $H(z)$ are $0.7 \pm j0.4$, $0.4 \pm j0.65$, and -0.5. The zeros of the filter are $0.65 \pm j0.3$, and $0.5 \pm j0.5$.

a. How many poles does the filter have?

b. Produce a pole-zero plot for the filter.

c. Find the transfer functions for three first and second order filters $H_1(z), H_2(z)$, and $H_3(z)$ that can be cascaded together to implement the transfer function $H(z)$. Note that the implementation will have the best numerical properties if the poles and zeros that are close together are grouped into the same transfer function.

6.27 A digital filter has the transfer function

$$H(z) = \frac{z - 0.2}{z^2(z - 0.7)}$$

a. Is the system stable?

b. Find the output $y[n]$ for the filter if the input is $x[n] = (0.9)^n u[n]$.

6.28 For each filter, find the poles and zeros and determine whether or not the filter is stable:

a. $H(z) = \dfrac{0.4}{0.2 - 0.1z^{-1}}$

b. $H(z) = \dfrac{1 - 1.44z^{-2}}{3 + 0.2z^{-1} + 1.5z^{-2}}$

c. $H(z) = \dfrac{1}{z^2 - 0.2z - 0.99}$

d. $H(z) = \dfrac{z^{-1} - 0.8z^{-2}}{1 - 0.5z^{-1} - 0.5z^{-2}}$

6.29 The pole-zero diagrams for several filters are given in Figures 6.28 through 6.33. In each case, provide a rough sketch of what the impulse response for each filter looks like.

FIGURE 6.28
Pole-zero plot for Question 6.29.

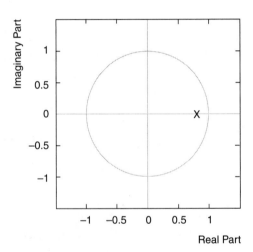

FIGURE 6.29
Pole-zero plot for Question 6.29.

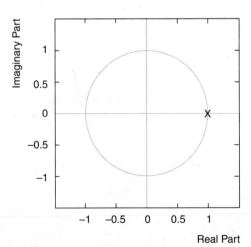

FIGURE 6.30
Pole-zero plot for Question 6.29.

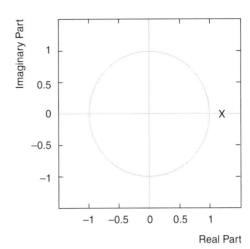

FIGURE 6.31
Pole-zero plot for Question 6.29.

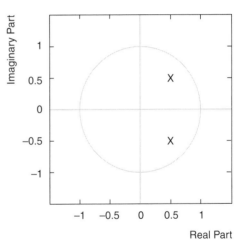

FIGURE 6.32
Pole-Zero plot for Question 6.29.

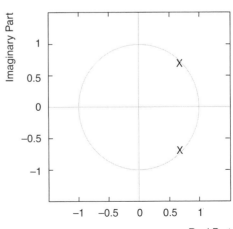

FIGURE 6.33
Pole-zero plot for Question 6.29.

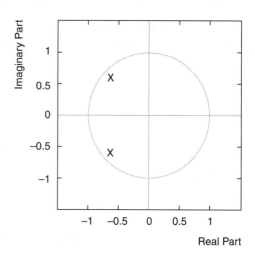

6.30 For the following first order transfer functions:
- **(i)** Find an expression for the impulse response.
- **(ii)** Sketch the first 10 samples of the impulse response.
- **(iii)** Sketch the pole-zero plot.

a. $H(z) = \dfrac{3}{1 - 0.5z^{-1}}$

b. $H(z) = \dfrac{5z}{4z + 1}$

c. $H(z) = \dfrac{1}{1.1 + z^{-1}}$

6.31 For the following second order transfer functions:
- **(i)** Find the poles and zeros and create a pole-zero plot.
- **(ii)** Determine the difference equation and use it to produce 10 samples of the step response.

a. $H(z) = \dfrac{1}{1 - 0.15z^{-1} - 0.4z^{-2}}$

b. $H(z) = \dfrac{1}{1 + 0.15z^{-1} + 0.4z^{-2}}$

c. $H(z) = \dfrac{z^{-1}}{1 - 0.85z^{-1} - 0.6z^{-2}}$

6.32 Find the steady state output when a unit step input is applied to each of the following second order systems:

a. $H(z) = \dfrac{z^{-2}}{1 - 0.5z^{-1} + 0.3z^{-2}}$

b. $H(z) = \dfrac{z^{-2}}{1 - 0.7z^{-1} + 0.7z^{-2}}$

c. $H(z) = \dfrac{1 - 0.5z^{-1}}{1 - 0.4z^{-1} + 0.5z^{-2}}$

d. $H(z) = \dfrac{1 - 0.25z^{-1} + z^{-2}}{1 - 1.4z^{-1} + 0.5z^{-2}}$

6.33 Two transfer functions are described by $H_1(z) = \dfrac{1 - 0.9z^{-1}}{1 - 0.5z^{-1} + 0.3z^{-2}}$ and

$H_2(z) = \dfrac{0.8}{1 + 0.7z^{-1} - 0.2z^{-2}}$.

a. Find the steady state output for each filter's step response.

b. Which filter's step response settles to its steady state value more quickly?

6.34 Place the following stable transfer functions in order according to the number of samples required for the impulse response to settle close to zero:

a. $H_1(z) = \dfrac{z^{-1} - 2z^{-2}}{1 - 0.4z^{-1} + 0.05z^{-2}}$

b. $H_2(z) = \dfrac{1 + z^{-2}}{1 - 1.05z^{-1} + 0.17z^{-2}}$

c. $H_3(z) = \dfrac{z^{-1} + 0.98z^{-2}}{1 + 0.28z^{-1} - 0.686z^{-2}}$

d. $H_4(z) = \dfrac{z^{-1}}{1 + 1.786z^{-1} + 0.903z^{-2}}$

e. $H_5(z) = \dfrac{z^{-1} - 0.8z^{-2}}{1 - 1.31z^{-1} + 0.405z^{-2}}$

6.35 For each of the following first order systems,

    (i) Plot a pole-zero diagram.

    (ii) Compute the impulse response $h[n]$ using inverse z transforms.

    (iii) Plot 15 samples of the impulse response.

a. $H(z) = \dfrac{1 + 0.1z^{-1}}{1 + 0.8z^{-1}}$

b. $H(z) = \dfrac{1 + 0.5z^{-1}}{1 + 0.8z^{-1}}$

c. $H(z) = \dfrac{1 + 0.7z^{-1}}{1 + 0.8z^{-1}}$

7

FOURIER TRANSFORMS AND FILTER SHAPE

The discrete time Fourier transform identifies the shape of a filter, and hence its behavior. This chapter:

- ➤ defines the discrete time Fourier transform (DTFT)
- ➤ defines the frequency response of a digital filter
- ➤ makes connections between frequency responses, transfer functions, difference equations, and impulse responses
- ➤ uses the discrete time Fourier transform to calculate filter outputs for input sinusoids
- ➤ identifies the magnitude response and the phase response of a digital filter
- ➤ relates analog frequency and digital frequency
- ➤ shows how pole and zero locations determine filter shape
- ➤ investigates first and second order system examples

7.1 FOURIER TRANSFORM BASICS

The discrete time Fourier transform (DTFT) is a tool for digital signals that corresponds to the Fourier transform for analog signals, not studied in this text. The DTFT takes the description of a signal or filter from the time domain to the frequency domain. The usual reason for doing this is to study the frequency characteristics of the signal or the filter. Like the z transform introduced in the preceding chapter, the discrete time Fourier transform can be used to make calculations easier. Finding filter shapes and signal spectra is this transform's main contribution. In the case of a signal, the information provided by the DTFT is referred to as the signal's spectrum. Signal spectra will be studied in the next chapter. In the case of a filter, the information provided by the DTFT is referred to as the filter's **frequency response**. A frequency response consists of two parts, the **magnitude response** and the **phase response**, studied further in Section 7.3.2. It is the magnitude response that gives the shape of a filter and the greatest insight into the way the filter behaves.

The definition for the discrete time Fourier transform of a signal $x[n]$ is

$$X(\Omega) = \sum_{n=-\infty}^{\infty} x[n]e^{-jn\Omega} \tag{7.1}$$

where Ω is the digital frequency defined by Equation (3.6). For brevity, the DTFT of a signal $x[n]$ is denoted as:

$$X(\Omega) = \mathbf{F}\{x[n]\}$$

To get some feel for the transform, Euler's identity $e^{j\theta} = \cos\theta + j\sin\theta$, and its alternative form $e^{-j\theta} = \cos(-\theta) + j\sin(-\theta) = \cos\theta - j\sin\theta$, both defined in Appendix A.13.2.5, are helpful. With the second form, the DTFT may be written as

$$X(\Omega) = \sum_{n=-\infty}^{\infty} x[n](\cos(n\Omega) - j\sin(n\Omega))$$

The transform $X(\Omega)$ is different at each different digital frequency Ω and is largest when $x[n]$ "resonates" with the sinusoids $\cos(n\Omega)$ or $\sin(n\Omega)$. That is, $X(\Omega)$ is large when $x[n]$ varies at nearly the frequency Ω. Figure 7.1 illustrates how this occurs. The product of the cosine signal $x[n]$ with other cosines is largest when the signals are most similar. In this way, the DTFT $X(\Omega)$ reports the frequencies present in a signal. In particular, the large product obtained in Figure 7.1(c) indicates that the signal being studied has a digital frequency close to $\Omega = 0.1\pi$ radians.

EXAMPLE 7.1
Find the discrete time Fourier transform of the signal shown in Figure 7.2.

Only the four nonzero samples at $n = 0, 1, 2$, and 4 contribute to the transform.

$$X(\Omega) = \sum_{n=-\infty}^{\infty} x[n]e^{-jn\Omega} = 2 - e^{-j\Omega} + 3e^{-j2\Omega} + e^{-j4\Omega}$$

As is generally the case, this DTFT is complex-valued.

EXAMPLE 7.2
Find the DTFT of the signal $x[n] = 4(u[n] - u[n-3])$.

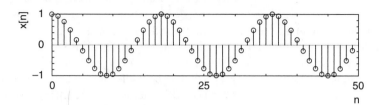

(a) Signal $x[n] = \cos\left(\dfrac{\pi n}{9}\right) = \cos(0.111\pi n)$

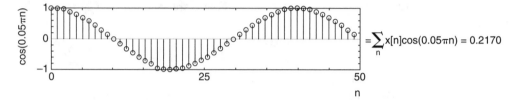

$= \displaystyle\sum_n x[n]\cos(0.05\pi n) = 0.2170$

(b) Comparing against $\cos(0.05\pi n)$

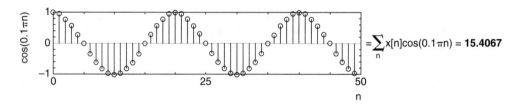

$= \displaystyle\sum_n x[n]\cos(0.1\pi n) = \mathbf{15.4067}$

(c) Comparing against $\cos(0.1\pi n)$

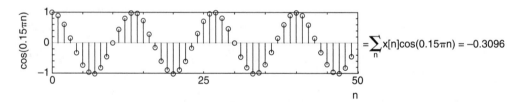

$= \displaystyle\sum_n x[n]\cos(0.15\pi n) = -0.3096$

(d) Comparing against $\cos(0.15\pi n)$

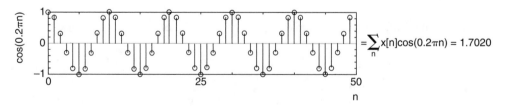

$= \displaystyle\sum_n x[n]\cos(0.2\pi n) = 1.7020$

(e) Comparing against $\cos(0.2\pi n)$

FIGURE 7.1

Signal resonance for the discrete time Fourier transform.

FIGURE 7.2

Signal for Example 7.1.

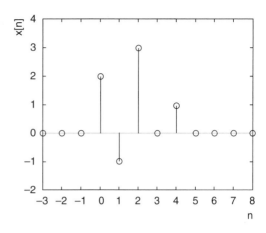

All signal values $n < 0$ and $n \geq 3$ are zero. Therefore,

$$X(\Omega) = \sum_{n=-\infty}^{\infty} x[n]e^{-jn\Omega} = 4 + 4e^{-j\Omega} + 4e^{-j2\Omega}$$

Two important properties of discrete time Fourier transforms are needed. The first is the effect of time delay and the second is periodicity. To find the effect of a time delay on a DTFT, suppose that the DTFT of a signal $x[n]$ exists and is called $X(\Omega)$. The discrete time Fourier transform of a signal $x[n - n_0]$ will then be

$$\sum_{n=-\infty}^{\infty} x[n - n_0]e^{-jn\Omega}$$

Letting $m = n - n_0$, the transform becomes

$$\sum_{m=-\infty}^{\infty} x[m]e^{-j(m + n_0)\Omega} = e^{-jn_0\Omega} \sum_{m=-\infty}^{\infty} x[m]e^{-jm\Omega} = e^{-jn_0\Omega}X(\Omega)$$

Thus, a delay of n_0 in the time domain introduces a complex exponential $e^{-jn_0\Omega}$ in the frequency domain.

The second property is periodicity. Consider $X(\Omega + 2\pi)$:

$$X(\Omega + 2\pi) = \sum_{n=-\infty}^{\infty} x[n]e^{-jn(\Omega + 2\pi)} = \sum_{n=-\infty}^{\infty} x[n]e^{-jn\Omega}e^{-j2\pi n}$$

Using Euler's identity, $e^{-j2\pi n} = \cos(2\pi n) - j\sin(2\pi n) = 1$, for all values of n. Therefore,

$$X(\Omega + 2\pi) = \sum_{n=-\infty}^{\infty} x[n]e^{-jn\Omega} = X(\Omega)$$

Since $X(\Omega + 2\pi) = X(\Omega)$, the DTFT is periodic with period 2π. In other words, all discrete time Fourier transforms repeat every 2π, endlessly, for all Ω.

7.2 FREQUENCY RESPONSES AND OTHER FORMS

7.2.1 Frequency Responses and Difference Equations

The general difference equation

$$\sum_{k=0}^{N} a_k y[n-k] = \sum_{k=0}^{M} b_k x[n-k]$$

or

$$a_0 y[n] + a_1 y[n-1] + a_2 y[n-2] + \ldots + a_N y[n-N]$$
$$= b_0 x[n] + b_1 x[n-1] + b_2 x[n-2] + \ldots + b_M x[n-M]$$

may be transformed into the frequency domain by taking the DTFT of each term. This process requires the time delay property presented in the previous section and gives:

$$a_0 Y(\Omega) + a_1 e^{-j\Omega} Y(\Omega) + a_2 e^{-j2\Omega} Y(\Omega) + \ldots + a_N e^{-jN\Omega} Y(\Omega)$$
$$= b_0 X(\Omega) + b_1 e^{-j\Omega} X(\Omega) + b_2 e^{-j2\Omega} X(\Omega) + \ldots + b_M e^{-jM\Omega} X(\Omega)$$

Factoring out the common terms gives

$$Y(\Omega)(a_0 + a_1 e^{-j\Omega} + a_2 e^{-j2\Omega} + \ldots + a_N e^{-jN\Omega})$$
$$= X(\Omega)(b_0 + b_1 e^{-j\Omega} + b_2 e^{-j2\Omega} + \ldots + b_M e^{-jM\Omega})$$

Thus, the ratio between output and input in the frequency domain is

$$H(\Omega) = \frac{Y(\Omega)}{X(\Omega)} = \frac{b_0 + b_1 e^{-j\Omega} + b_2 e^{-j2\Omega} + \ldots + b_M e^{-jM\Omega}}{a_0 + a_1 e^{-j\Omega} + a_2 e^{-j2\Omega} + \ldots + a_N e^{-jN\Omega}} \qquad (7.2)$$

$H(\Omega)$ is referred to as the frequency response of the filter.

EXAMPLE 7.3

Find an expression for the frequency response of the filter described by the difference equation

$$y[n] = -0.85y[n-1] + 0.5x[n]$$

Taking the DTFT of each term gives

$$Y(\Omega) + 0.85e^{-j\Omega} Y(\Omega) = 0.5X(\Omega)$$

Factoring out common terms,

$$Y(\Omega)(1 + 0.85e^{-j\Omega}) = X(\Omega)(0.5)$$

The frequency response is

$$H(\Omega) = \frac{Y(\Omega)}{X(\Omega)} = \frac{0.5}{1 + 0.85e^{-j\Omega}}$$

The expression is complex-valued.

EXAMPLE 7.4

Find the frequency response that corresponds to the difference equation

$$y[n] + 0.1y[n-1] + 0.85y[n-2] = x[n] - 0.3\,x[n-1]$$

The coefficients are easy to identify as $a_0 = 1$, $a_1 = 0.1$, $a_2 = 0.85$, $b_0 = 1$, and $b_1 = -0.3$. Therefore, from Equation (7.2), the frequency response for this filter is:

$$H(\Omega) = \frac{Y(\Omega)}{X(\Omega)} = \frac{b_0 + b_1 e^{-j\Omega} + b_2 e^{-j2\Omega} + \ldots + b_M e^{-jM\Omega}}{a_0 + a_1 e^{-j\Omega} + a_2 e^{-j2\Omega} + \ldots + a_N e^{-jN\Omega}}$$

$$= \frac{1 - 0.3e^{-j\Omega}}{1 + 0.1e^{-j\Omega} + 0.85e^{-j2\Omega}}$$

7.2.2 Frequency Responses and Transfer Functions

A very close relationship exists between the frequency response $H(\Omega)$ and the transfer function $H(z)$ of a filter. From Equation (6.2), the general transfer function has the form

$$H(z) = \frac{Y(z)}{X(z)} = \frac{b_0 + b_1 z^{-1} + \ldots + b_M z^{-M}}{a_0 + a_1 z^{-1} + \ldots + a_N z^{-N}}$$

The general frequency response is, from Equation (7.2),

$$H(\Omega) = \frac{Y(\Omega)}{X(\Omega)} = \frac{b_0 + b_1 e^{-j\Omega} + b_2 e^{-j2\Omega} + \ldots + b_M e^{-jM\Omega}}{a_0 + a_1 e^{-j\Omega} + a_2 e^{-j2\Omega} + \ldots + a_N e^{-jN\Omega}}$$

Thus, the frequency response may be obtained from the transfer function by replacing every occurrence of z^{-1} by $e^{-j\Omega}$ (or every occurrence of z by $e^{j\Omega}$).

EXAMPLE 7.5

Find an expression for the frequency response of the filter with transfer function

$$H(z) = \frac{1 - 0.2z^{-2}}{1 + 0.5z^{-1} + 0.9z^{-2}}$$

The frequency response is

$$H(\Omega) = \frac{1 - 0.2e^{-j2\Omega}}{1 + 0.5e^{-j\Omega} + 0.9e^{-j2\Omega}}$$

7.2.3 Frequency Responses and Impulse Responses

The DTFT provides yet another way to describe a filter. Figure 7.3 shows all possibilities: difference equation, impulse response, transfer function, and frequency response. Like the other descriptions, the filter's frequency response contains all the information needed to predict how the filter will behave. As indicated in Section 6.2.2, the transfer function $H(z)$ of a filter is the z transform of the impulse response $h[n]$. It is not difficult to see how the frequency response can be obtained from the impulse response. When the input $x[n]$ to a filter is an impulse function $\delta[n]$, its DTFT is

$$X(\Omega) = \sum_{n=-\infty}^{\infty} \delta[n]e^{-jn\Omega} = 1$$

Since the output $y[n]$ for an impulse input is known to be the impulse response $h[n]$, its DTFT must be

$$Y(\Omega) = \sum_{n=-\infty}^{\infty} h[n]e^{-jn\Omega}$$

According to Section 7.2.2, the frequency response $H(\Omega)$ of a filter is given by the ratio of output $Y(\Omega)$ to input $X(\Omega)$. Thus, the frequency response $H(\Omega)$ is identical with the DTFT of the impulse response $h[n]$. That is,

$$H(\Omega) = \mathbf{F}\{h[n]\} = \sum_{n=-\infty}^{\infty} h[n]e^{-jn\Omega} \tag{7.3}$$

EXAMPLE 7.6

The impulse response for a digital filter is

$$h[n] = 5\delta[n] - \delta[n-1] + 0.2\delta[n-2] - 0.04\delta[n-3]$$

Find an expression for the frequency response of the filter.

The frequency response is the DTFT of the impulse response, as shown in Equation (7.3). Thus,

$$H(\Omega) = \sum_{n=-\infty}^{\infty} h[n]e^{-jn\Omega} = 5 - e^{-j\Omega} + 0.2e^{-j2\Omega} - 0.04e^{-j3\Omega}$$

7.3 FREQUENCY RESPONSE AND FILTER SHAPE

7.3.1 Filter Effects on Sine Wave Inputs

As Figure 7.3 suggests, the frequency response of a filter can be used to calculate filter outputs by a method that parallels the one presented for z transforms in Chapter 6. Since

FIGURE 7.3
Finding outputs in time and frequency domains.

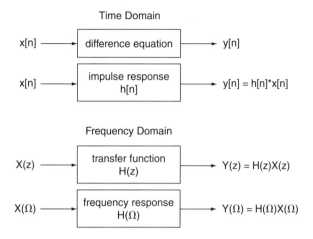

$H(\Omega) = Y(\Omega)/X(\Omega)$, the discrete time Fourier transform (DTFT) of the output from a filter can be found from the product of the frequency response and the DTFT of the input:

$$Y(\Omega) = H(\Omega)X(\Omega) \qquad (7.4)$$

The output signal can then be found by taking an inverse transform: $y[n] = \mathbf{F}^{-1}\{Y(\Omega)\}$. Though this method is suitable for use with any input at all, the z transform method presented in Section 6.2.3 is more convenient to use for the calculation of general outputs. The DTFT approach is normally reserved for finding outputs for sinusoidal inputs only. To understand why this is reasonable, it is important to remember that the frequency response of a filter reports its behavior at every frequency. Since the focus is on individual frequencies, sine waves (or cosine waves), the only real signals that consist of a single frequency, are naturally the inputs of interest. The scope is broadened by the fact that very complex signals can be constructed by adding sine waves of different amplitudes and phases together, and the output of a filter can still be predicted using the DTFT method.

The frequency response $H(\Omega)$ for a filter is a complex number and, as such, may be expressed in polar form:

$$H(\Omega) = \left|H(\Omega)\right|e^{j\theta(\Omega)}$$

Here, $|H(\Omega)|$ is the gain of the filter at the digital frequency Ω, and $\theta(\Omega)$ is the **phase difference** of the filter at the same frequency. The gain $|H(\Omega)|$ is unitless, but is also frequently presented in decibels, or dB, described in Appendix A.10. In this case, the gain is $20\log|H(\Omega)|$. Phase differences $\theta(\Omega)$ may be plotted in either degrees or radians. Digital frequency Ω is measured in radians, as usual. As their notations suggest, the gain $|H(\Omega)|$ and phase difference $\theta(\Omega)$ vary with frequency. At any given frequency, the gain and phase difference may be used to predict the filter's response. The gain provides the amplification factor for the input, while the phase difference determines the phase shift to be applied to the input. Since $H(\Omega)$, $X(\Omega)$, and $Y(\Omega)$ are all complex numbers, Equation (7.4) can be rewritten using polar form as

$$Y(\Omega) = |Y(\Omega)|e^{j\theta_Y(\Omega)}$$

$$= H(\Omega)X(\Omega) = |H(\Omega)|e^{j\theta(\Omega)}|X(\Omega)|e^{j\theta_X(\Omega)} = |H(\Omega)||X(\Omega)|e^{j(\theta(\Omega)+\theta_X(\Omega))}$$

The conclusions are

$$|Y(\Omega)| = |H(\Omega)||X(\Omega)| \tag{7.5a}$$

$$\theta_Y(\Omega) = \theta(\Omega) + \theta_X(\Omega) \tag{7.5b}$$

In other words, at a given frequency Ω, the magnitude of the output is the product of the filter gain and the magnitude of the input. The phase of the output is the sum of the phase difference of the filter and the phase of the input. Notice that the magnitudes themselves must be used for the calculation in Equation (7.5a). Decibel values cannot be used.

The magnitude spectrum $|X(\Omega)|$ of a digital cosine signal, $x[n] = A\cos(n\Omega_0 + \theta_X)$, is established in Appendix I and shown in Figure 7.4(a) for the range $0 \le \Omega \le \pi$. As the figure shows, a cosine signal has a nonzero contribution at only one positive frequency, Ω_0. The magnitudes are zero at all other positive frequencies. When a filter gain $|H(\Omega)|$ and a signal magnitude $|X(\Omega)|$ are multiplied together according to Equation (7.5a) to get the output magnitude $|Y(\Omega)|$, the result is another magnitude spectrum that is zero everywhere except at $\Omega = \Omega_0$. This characteristic means that a sinusoid is obtained at the output when a sinusoid is applied at the input. Suppose that at $\Omega = \Omega_0$ the gain of the filter is H and the phase difference is θ. Then, for the cosine input with peak magnitude $A\pi$ in Figure 7.4(a), the magnitude of the output is

$$|Y(\Omega_0)| = |H(\Omega_0)||X(\Omega_0)| = HA\pi$$

for $\Omega = \Omega_0$ and zero for all other values of Ω. Because the magnitude is nearly always zero, the only phase value that counts is the one at Ω_0. At $\Omega = \Omega_0$, the phase of the input, from Figure 7.4(b), is $\theta_X(\Omega) = \theta_X$. At the same frequency, the phase of the output is, according to Equation (7.5b),

$$\theta_Y(\Omega_0) = \theta(\Omega_0) + \theta_X(\Omega_0) = \theta + \theta_X$$

This information yields the spectrum for the output cosine shown in Figure 7.5.

If the spectrum for the input signal $x[n] = A\cos(n\Omega_0 + \theta_X)$ is as shown in Figure 7.4, with magnitude $A\pi$ and phase θ_X, then, by analogy, the output signal represented by the spectrum in Figure 7.5, with magnitude $HA\pi$ and phase $\theta + \theta_X$, must be $y[n] = HA\cos(n\Omega_0 + \theta + \theta_X)$. Notice that the magnitude $A\pi$ of the DTFT of $x[n]$ at Ω_0 includes a factor π that appears in $|Y(\Omega)|$, but disappears again when the inverse transform of $Y(\Omega)$ is taken to get $y[n]$. For simplicity, this π factor will be omitted from the calculations from now on. Sometimes a short form is used for cosine (or sine) signals. The short form for $x[n] = A\cos(n\Omega_0 + \theta_X)$ is $A|\theta_X$. This short form records the amplitude and phase of the cosine, but not the frequency Ω_0, which is assumed to remain unchanged through the calculation. The short form for the frequency response, including magnitude and phase information, is $H|\theta$. To multiply short forms, magnitudes (or amplitudes) are multiplied and phases are added. Thus,

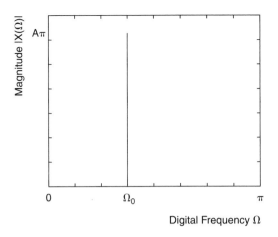 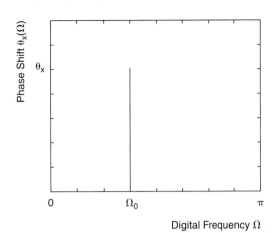

(a) Magnitude Spectrum (b) Phase Spectrum

FIGURE 7.4
Spectrum of input cosine signal.

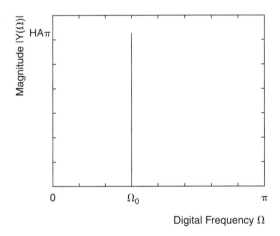 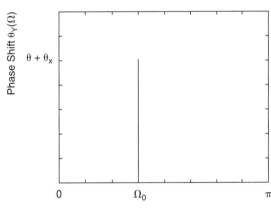

(a) Magnitude Spectrum (b) Phase Spectrum

FIGURE 7.5
Spectrum of output cosine signal.

$$Y(\Omega) = H(\Omega)X(\Omega) = (H\underline{\theta})(A\underline{\theta_X}) = HA\underline{\theta + \theta_X}$$

Expanding the short form into a cosine signal again gives

$$y[n] = HA\cos(n\Omega_0 + \theta + \theta_X)$$

Thus, when a sinusoid passes through a linear filter, its amplitude and phase change, but its frequency remains unchanged.

EXAMPLE 7.7

A cosine input signal with a digital frequency of 1.5 radians is applied to a filter. At this frequency, the filter has a gain of -21 dB and a phase difference of $86°$. If the amplitude of the input is 20 and the phase of the input is $12°$, what are the magnitude and phase of the output?

The short form for the input is $20\underline{|12°}$. This is an abbreviation for the full cosine signal, which would be $20\cos(1.5n + 12°)$. At 1.5 radians, the gain of the filter is -21 dB, but this quantity cannot be used for calculations. It must be converted to a linear quantity, using $10^{-21/20} = 0.0891$. Since the phase difference is $86°$, the short form for the frequency response is $= 0.0891\underline{|86°}$. The output is the product of the frequency response and the input in the Fourier domain, that is,

$$Y(\Omega) = H(\Omega)X(\Omega) = (20\underline{|12°})(0.0891\underline{|86°}) = 1.782\underline{|98°}$$

Expanding this short form gives the output signal

$$y[n] = 1.782\cos(1.5n + 98°)$$

As mentioned in Section 3.3.4, care must be taken when evaluating phase-shifted digital sinusoids of the form $A\sin(\Omega n + \theta)$ or $A\cos(\Omega n + \theta)$. Since Ω is measured in radians, θ must also be expressed in radians in order to evaluate the sinusoid correctly. Example 7.8 addresses this issue.

EXAMPLE 7.8

Find the values of the output signal from the last example,

$$y[n] = 1.782\cos(1.5n + 98°)$$

for $n = 10$ and $n = 20$.

Once θ has been converted to radians, the equation reads

$$y[n] = 1.782\cos(1.5n + 0.5444\pi)$$

Using this equation, $y[10] = -0.9593$ and $y[20] = 1.7054$.

7.3.2 Magnitude Response and Phase Response

The frequency response for a filter furnishes vital information about the filter. As mentioned in the last section, the frequency response $H(\Omega)$ at the digital frequency Ω can be partitioned into a gain $|H(\Omega)|$ and a phase difference $\theta(\Omega)$ using polar form as $H(\Omega) = |H(\Omega)|e^{j\theta(\Omega)}$. The collection of gains across all digital frequencies forms what is called the filter's magnitude response. The collection of phase differences forms the filter's phase response. The frequency response can be plotted conveniently in terms of these two parts. The first part, the magnitude response, is a graph of gain (or magnitude) $|H(\Omega)|$ versus frequency Ω. The second part, the phase response, is a graph of phase difference $\theta(\Omega)$ versus frequency Ω. The magnitude response and the phase response together make up the frequency response for a digital filter.

Magnitude and phase are evaluated most easily with a calculator (or computer) that has the ability to do polar calculations. When this is possible, each form $Ae^{j\theta}$ can be written as $A\underline{|\theta}$ for short. When polar calculations are not available, Euler's identity is necessary; computations are performed using complex numbers in rectangular form, and a conversion to polar form is made at the end. Both types of calculations are illustrated in the next example.

EXAMPLE 7.9

The frequency response of a system is

$$H(\Omega) = \frac{1}{1 - 0.4e^{-j\Omega}} \tag{7.6}$$

Find and sketch the magnitude response and the phase response for the system. For the magnitude response, plot gain versus digital frequency in radians and for the phase response, plot phase in radians versus digital frequency in radians. Cover a range of digital frequencies from $-\pi$ to 3π radians.

Table 7.1 is used to record the results for each digital frequency. Steps of $\pi/4$ radians are used arbitrarily. Greater accuracy can be obtained by using smaller steps. Analysis with and without polar calculations is illustrated for $\Omega = 3\pi/4$ radians (highlighted). Phases are calculated in radians.

With polar calculations:

$$H\left(\frac{3\pi}{4}\right) = \frac{1}{1 - 0.4e^{-j3\pi/4}} = \frac{1}{1 - 0.4\underline{\left|-\frac{3\pi}{4}\right.}} = 0.7612\underline{|-0.2170}$$

Without polar calculations:

$$H\left(\frac{3\pi}{4}\right) = \frac{1}{1 - 0.4e^{-j3\pi/4}} = \frac{1}{1 - 0.4\left(\cos\left(\frac{3\pi}{4}\right) - j\sin\left(\frac{3\pi}{4}\right)\right)}$$

$$= \frac{1}{1 - 0.4(-0.7071 - j0.7071)}$$

$$= \frac{1}{1.2828 + j0.2828} = \frac{1}{1.2828 + j0.2828}\frac{1.2828 - j0.2828}{1.2828 - j0.2828}$$

$$= 0.7434 - j0.1639 = 0.7612\underline{|-0.2170}$$

TABLE 7.1
Calculations for Example 7.9

Ω	$H(\Omega)$	$|H(\Omega)|$	$\theta(\Omega)$ (rads)
$-\pi$	$\dfrac{1}{1 - 0.4e^{j\pi}}$	0.7143	0.0000
$\dfrac{-3\pi}{4}$	$\dfrac{1}{1 - 0.4e^{j3\pi/4}}$	0.7612	0.2170
$\dfrac{-\pi}{2}$	$\dfrac{1}{1 - 0.4e^{j\pi/2}}$	0.9285	0.3805
$\dfrac{-\pi}{4}$	$\dfrac{1}{1 - 0.4e^{j\pi/4}}$	1.2972	0.3757
0	$\dfrac{1}{1 - 0.4e^{j0}}$	1.6667	0.0000
$\dfrac{\pi}{4}$	$\dfrac{1}{1 - 0.4e^{-j\pi/4}}$	1.2972	-0.3757
$\dfrac{\pi}{2}$	$\dfrac{1}{1 - 0.4e^{-j\pi/2}}$	0.9285	-0.3805
$\dfrac{3\pi}{4}$	$\dfrac{1}{1 - 0.4e^{-j3\pi/4}}$	**0.7612**	**-0.2170**
π	$\dfrac{1}{1 - 0.4e^{-j\pi}}$	0.7143	0.0000
$\dfrac{5\pi}{4}$	$\dfrac{1}{1 - 0.4e^{-j5\pi/4}}$	0.7612	0.2170
$\dfrac{3\pi}{2}$	$\dfrac{1}{1 - 0.4e^{-j3\pi/2}}$	0.9285	0.3805
$\dfrac{7\pi}{4}$	$\dfrac{1}{1 - 0.4e^{-j7\pi/4}}$	1.2972	0.3757
2π	$\dfrac{1}{1 - 0.4e^{-j2\pi}}$	1.6667	0.0000
$\dfrac{9\pi}{4}$	$\dfrac{1}{1 - 0.4e^{-j9\pi/4}}$	1.2972	-0.3757
$\dfrac{5\pi}{2}$	$\dfrac{1}{1 - 0.4e^{-j5\pi/2}}$	0.9285	-0.3805
$\dfrac{11\pi}{4}$	$\dfrac{1}{1 - 0.4e^{-j11\pi/4}}$	0.7612	-0.2170
3π	$\dfrac{1}{1 - 0.4e^{-j3\pi}}$	0.7143	0.0000

The magnitude and phase responses in the table are plotted in Figure 7.6. Triangles mark the calculated points. The figure also shows the exact magnitude and phase responses as dotted lines. These responses are obtained by calculating additional intermediate points beyond those listed in the table.

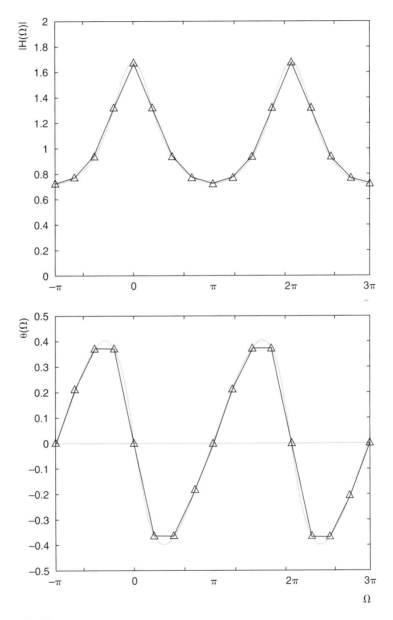

FIGURE 7.6
Frequency response for Example 7.9.

As Example 7.9 showed, the magnitude and phase responses are periodic, and they repeat every 2π radians. As shown in Section 7.1, this periodicity holds for all discrete time Fourier transforms. Both the magnitude and phase responses are continuous functions, meaning they have values for every value of frequency. Example 7.9 also shows that the magnitude response is an even function, since the portion to the left of zero is a perfect mirror image of what lies to the right of zero, that is,

$$|H(-\Omega)| = |H(\Omega)|$$

This is true for all magnitude responses. The phase response, on the other hand, is an odd function, since the portion to the left of zero is an upside-down and mirrored copy of what lies to the right of zero. That is,

$$\theta(-\Omega) = -\theta(\Omega)$$

This is true for all phase responses. Thus, there is normally no need to record the part of the magnitude and phase responses to the left of $\Omega = 0$: This portion can be deduced from the plot for $\Omega > 0$. Furthermore, there is no real need to consider values of Ω above π. Recalling, from Section 3.3.4, that

$$\Omega = 2\pi\frac{f}{f_S}$$

it is easy to see that the case where $\Omega = \pi$ corresponds to the case where the input frequency f is one-half the sampling frequency f_S. This is the Nyquist frequency, the highest filter input frequency that can be permitted without introducing aliasing. Therefore, frequency responses for digital filters are usually plotted for $0 \leq \Omega \leq \pi$ only. A rough sense of the shape of the filter can be obtained when frequency response values are calculated in quarter-π steps, and smaller steps can be used to get finer detail.

In Example 7.9, the magnitude response was plotted as linear gain $|H(\Omega)|$ versus digital frequency Ω. When decibels (dB) are used, as discussed in the last section, the magnitude response is plotted in logarithmic form, as $20\log|H(\Omega)|$ versus Ω. The advantage of using decibels is that an extremely large range of gains, from very large to very small, can be conveniently plotted on a single graph. The use of decibels changes the shape of the plot. For the phase response, $\theta(\Omega)$ versus Ω, phase differences are presented in either radians or degrees. Finally, it is not uncommon to see the digital frequencies Ω along the horizontal axis plotted on a logarithmic scale. Using a log scale permits a wide range of frequencies to be presented on the same graph. The following examples illustrate the use of decibels in frequency response analysis.

EXAMPLE 7.10

The magnitude response and the phase response for a filter appear in Figure 7.7. Identify the gain and phase difference where the digital frequency is:

    a. 2 radians
    b. 3 radians

FIGURE 7.7

Frequency response for Example 7.10.

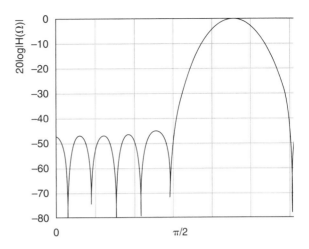

(a) Magnitude Response

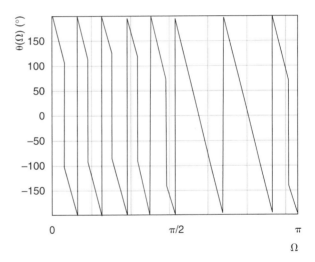

(b) Phase Response

a. The gain at 2 radians is about −4 dB and the phase is about −70°. The exact values are −4.4 dB and −66°.

b. The gain at 3 radians is about −43 dB. The phase is about 80°. The exact values are −42.9 dB and 81°.

EXAMPLE 7.11

The digital signal

$$x[n] = 0.5\cos\left(\frac{\pi}{4}n\right)$$

provides the input to a filter whose frequency response is shown in Figure 7.8. Find the output signal.

FIGURE 7.8
Frequency response for Example 7.11.

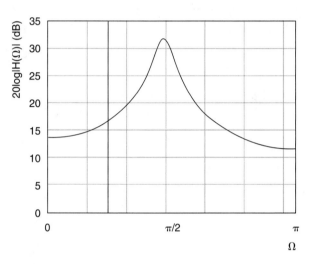

(a) Magnitude Response

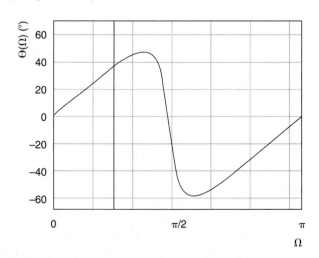

(b) Phase Response

The frequency of the input signal is $\pi/4 = 0.7854$ radians, marked on Figure 7.8 by a solid line. At this frequency, the filter has a gain of 17.4 dB, or 7.41. At the same frequency, the filter has a phase difference of $37.5°$. The short form for the input $x[n]$ is $0.5\underline{|0°}$. Therefore,

$$Y(\Omega) = H(\Omega)X(\Omega) = (7.41\underline{|37.5°})(0.5\underline{|0°}) = 3.71\underline{|37.5°}$$

The output signal is

$$y[n] = 3.71\cos\left(\frac{\pi}{4}n + 37.5°\right)$$

EXAMPLE 7.12

For the system in Example 7.9, Figure 7.9 plots gain and gain in dB against frequency, and also phases, in both radians and degrees versus frequency. Following the analysis presented

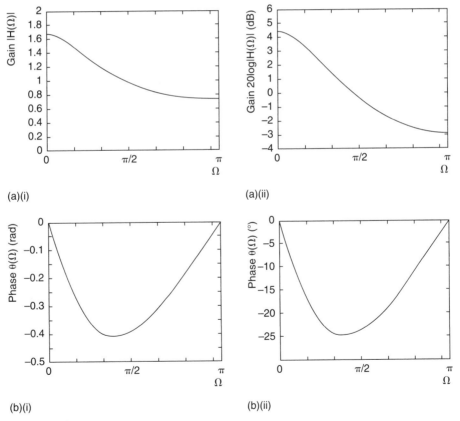

(a)(i)   (a)(ii)

(b)(i)   (b)(ii)

FIGURE 7.9

Frequency response for Example 7.12.

above, only the range of digital frequencies between 0 and π radians need be considered. Figure 7.9(a)(i) displays linear gain versus frequency. It is simply a repeat of Figure 7.6(a), covering only the essential range. Figure 7.9(a)(ii) shows gain in dB versus frequency. In Figure 7.9(b)(i), phase in radians is plotted against frequency, and Figure 7.9(b)(ii) shows phase in degrees plotted against frequency. The gain plots show more clearly than Figure 7.6(a) that the filter with the frequency response described in Equation (7.6) tends to pass low frequency signals better than high frequency ones.

A comparison of Figures 7.9(a)(i) and (ii) shows that conversion to dB has a strong effect on the appearance of the magnitude response. Conversion from radians to degrees, however, does not involve a change of shape, so in Figures 7.9(b)(i) and (ii) only the units along the vertical axis change.

The **filter shape** for a digital filter is furnished by its magnitude response. The magnitude response indicates the general filter type, namely, low pass, high pass, band pass, or band stop. Figure 7.10 summarizes the basic filter shapes that can be expected. Each tends to pass a different range of frequencies. As mentioned in Section 4.1, a filter is considered to pass signals where the gain is at least 0.707 of its maximum value. This is equivalent to a gain of 3 dB below maximum. In other words, when the maximum gain of a filter is 1, or 0 dB, the edge of the pass band occurs where the gain is 0.707, or -3 dB. Figure 7.10 contains, for each filter type, the linear magnitude response, also shown in Figure 4.2, and the logarithmic magnitude response, for which gains are recorded in dB.

EXAMPLE 7.13

Sketch the frequency response

$$H(\Omega) = \frac{1 - 0.3e^{-j\Omega}}{1 + 0.5e^{-j\Omega}}$$

The polar calculations used to compute the frequency response are shown in Table 7.2. The frequency response is graphed in Figure 7.11. The points from the table are marked by triangles and are joined by solid lines. The exact responses are added as dashed lines. The magnitude response reveals a high pass filter.

EXAMPLE 7.14

The impulse response for a filter is

$$h[n] = \delta[n] - 1.25\delta[n-2] \tag{7.7}$$

Find and sketch the frequency response for this filter. Plot phases in degrees and linear gains.

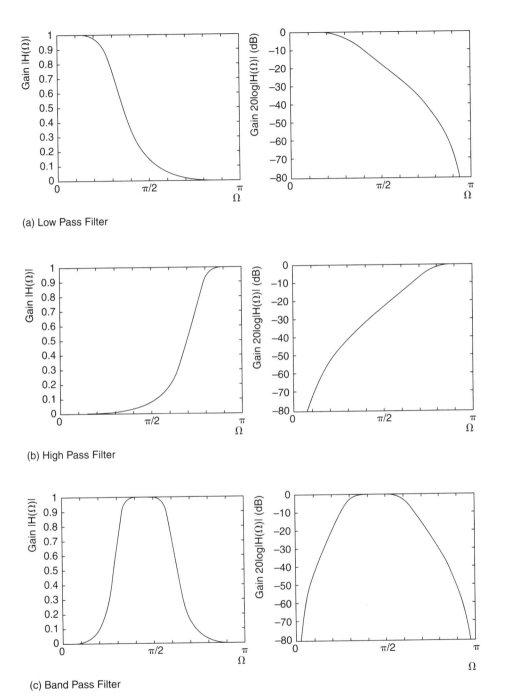

(a) Low Pass Filter

(b) High Pass Filter

(c) Band Pass Filter

FIGURE 7.10

Frequency responses for common filters.

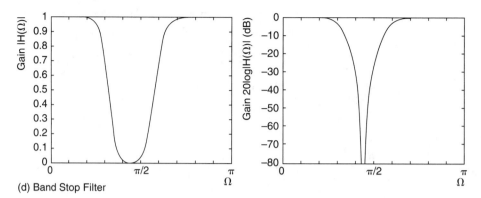

(d) Band Stop Filter

FIGURE 7.10
Continued

TABLE 7.2
Calculations for Example 7.13

Ω	$H(\Omega)$	$|H(\Omega)|$	$\theta(\Omega)$(rads)	$20\log|H(\Omega)|$ (dB)	$\theta(\Omega)$ (°)
0	$\dfrac{1 - 0.3\lfloor 0}{1 + 0.5\lfloor 0}$	0.4667	0.0000	−6.6199	0.0000
$\dfrac{\pi}{4}$	$\dfrac{1 - 0.3\lfloor -\dfrac{\pi}{4}}{1 + 0.5\lfloor -\dfrac{\pi}{4}}$	0.5832	0.5185	−4.6831	29.7082
$\dfrac{\pi}{2}$	$\dfrac{1 - 0.3\lfloor -\dfrac{\pi}{2}}{1 + 0.5\lfloor -\dfrac{\pi}{2}}$	0.9338	0.7551	−0.5948	43.2643
$\dfrac{3\pi}{4}$	$\dfrac{1 - 0.3\lfloor -\dfrac{3\pi}{4}}{1 + 0.5\lfloor -\dfrac{3\pi}{4}}$	1.6701	0.6737	4.4549	38.6017
π	$\dfrac{1 - 0.3\lfloor -\pi}{1 + 0.5\lfloor -\pi}$	2.6000	0.0000	8.2995	0.0000

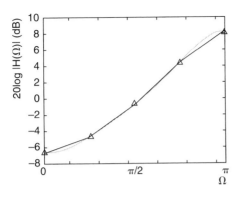

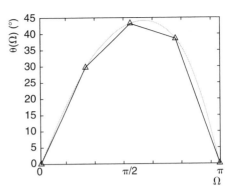

(a) Magnitude Response   (b) Phase Response

FIGURE 7.11
Frequency response for Example 7.13.

TABLE 7.3
Calculations for Example 7.14

Ω	$H(\Omega)$	$|H(\Omega)|$	$\theta(\Omega)$ (°)
0	$1 - 1.25\lfloor 0$	0.2500	180.0000
$\dfrac{\pi}{4}$	$1 - 1.25\left\lfloor -\dfrac{\pi}{2}\right.$	1.6008	51.3402
$\dfrac{\pi}{2}$	$1 - 1.25\lfloor -\pi$	2.2500	0.0000
$\dfrac{3\pi}{4}$	$1 - 1.25\left\lfloor -\dfrac{3\pi}{2}\right.$	1.6008	-51.3402
π	$1 - 1.25\lfloor -2\pi$	0.2500	-180.0000

Using Equation (7.3), the frequency response is given by

$$H(\Omega) = 1 - 1.25e^{-j2\Omega}$$

Magnitude and phase responses are calculated in Table 7.3 and are plotted in Figure 7.12. Clearly the impulse response of Equation (7.7) belongs to a band pass filter.

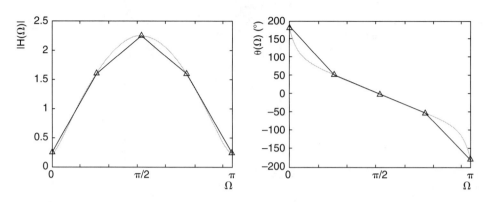

FIGURE 7.12
Frequency response for Example 7.14.

EXAMPLE 7.15

The difference equation describing a digital filter is

$$y[n] = 1.5y[n-1] - 0.85y[n-2] + x[n]$$

Find and sketch the frequency response. Plot gains in decibels and phases in degrees.

Using Equation (7.2), the expression for frequency response is

TABLE 7.4
Calculations for Example 7.15

Ω	$H(\Omega)$	$20\log|H(\Omega)|$	$\theta(\Omega)$ (°)
0	$\dfrac{1}{1-1.5\lfloor 0 + 0.85\lfloor 0}$	9.1186	0.0000
$\dfrac{\pi}{4}$	$\dfrac{1}{1 - 1.5\left\lfloor -\dfrac{\pi}{4} + 0.85\left\lfloor -\dfrac{\pi}{2}\right.\right.}$	13.1824	−106.0639
$\dfrac{\pi}{2}$	$\dfrac{1}{1 - 1.5\left\lfloor -\dfrac{\pi}{2} + 0.85\lfloor -\pi\right.}$	−3.5650	−84.2894
$\dfrac{3\pi}{4}$	$\dfrac{1}{1 - 1.5\left\lfloor -\dfrac{3\pi}{4} + 0.85\left\lfloor -\dfrac{3\pi}{2}\right.\right.}$	−8.9746	−42.8369
π	$\dfrac{1}{1 - 1.5\lfloor -\pi + 0.85\lfloor -2\pi}$	−10.5009	0.0000

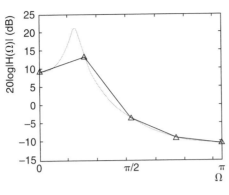

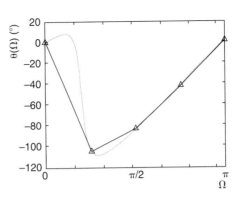

(a) Magnitude Response (b) Phase Response

FIGURE 7.13
Frequency response for Example 7.15.

$$H(\Omega) = \frac{1}{1 - 1.5e^{-j\Omega} + 0.85e^{-j2\Omega}}$$

Table 7.4 reports the results for magnitude and phase responses. The frequency response is plotted in Figure 7.13. For this particular filter, five points are clearly not adequate to get a good picture of the filter's shape. The exact curves show that the filter passes only a narrow band of frequencies.

For the remainder of this chapter, exact magnitude and phase response graphs will be presented without tables of calculations. When a software tool is not available to make the work easy, calculating the responses at quarter-π intervals—as was done in Examples 7.13, 7.14, and 7.15—is a good way to get a rough picture of filter behavior.

EXAMPLE 7.16
Find the filter shape of the recursive filter whose transfer function is

$$H(z) = \frac{1}{1 - 0.5z^{-8}}$$

According to Section 7.2.2, this transfer function becomes the frequency response

$$H(\Omega) = \frac{1}{1 - 0.5e^{-j8\Omega}}$$

The filter shape, or magnitude response, is shown in Figure 7.14. The filter is, for obvious reasons, called a **comb filter**. It has the effect of modeling reverberations, as Figure 7.15 shows. The pulse in Figure 7.15(a) when passed through the comb filter produces the series of pulses in Figure 7.15(b).

FIGURE 7.14
Magnitude response of
comb filter for Example
7.16.

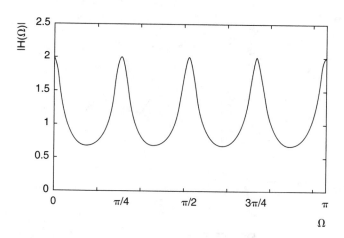

FIGURE 7.15
Pulse passed through
comb filter.

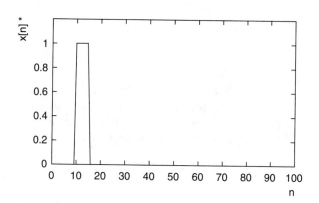

(a) Input

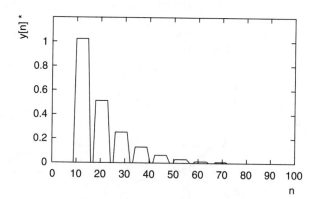

* For clarity, the envelope of the signals rather than their
individual samples is plotted.

(b) Output

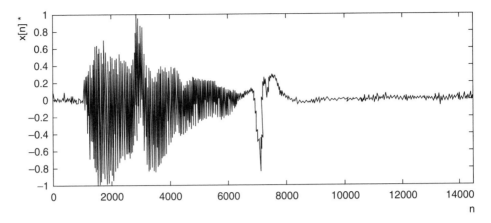

(a) Filter Input ("Hello")

hello.wav

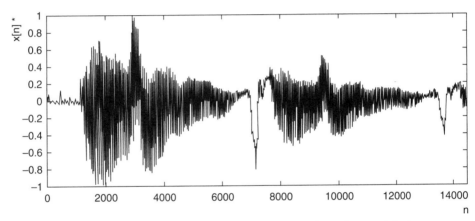

helloo.wav

* For clarity, the envelopes of the signals rather than their individual samples are plotted.

(b) Filter Input ("Hello Hello")

FIGURE 7.16
"Hello" passed through comb filter.

The word "hello" can be passed through a comb filter similar to the one used in Example 7.16. The result is a series of progressively quieter copies of the word. The input and the first two iterations of the output are shown in Figure 7.16. For this illustration, the transfer function

$$H(z) = \frac{1}{1 - 0.5z^{-6499}}$$

is used.

EXAMPLE 7.17

Find the filter shape for the recursive digital filter described by the transfer function

$$H(z) = \frac{1.1368 - 0.7127z^{-10}}{1 - 0.8495z^{-10}}$$

The filter shape is computed from the frequency response

$$H(\Omega) = \frac{1.1368 - 0.7127e^{-j10\Omega}}{1 - 0.8495e^{-j10\Omega}}$$

and is plotted in Figure 7.17. This filter is another example of a comb filter. Section 8.3 will introduce the idea that the spectrum of a periodic digital signal contains spikes at regular intervals. The frequencies that mark the spikes are called **harmonics**. The comb filter provides a means to remove noise from a periodic signal, since it can attenuate all frequencies between the harmonic spikes.

FIGURE 7.17

Magnitude response of comb filter for Example 7.17.

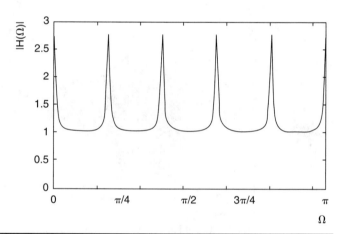

7.3.3 Analog Frequency f and Digital Frequency Ω

The shape of a digital filter, $|H(\Omega)|$, can be designed independent of sampling frequency. The sampling frequency chosen will, however, determine the range of input frequencies over which the filter will act. When the sampling rate is known, the frequency axis can be plotted in terms of analog frequency f in Hertz instead of digital frequency Ω in radians. It is often easier to understand filter behavior in this way. To perform the conversion requires Equation (3.6). Solving for f gives

$$f = \Omega \frac{f_S}{2\pi} \tag{7.8}$$

Nothing about the frequency response diagram changes except for the labeling of the horizontal axis, but the choice of sampling frequency has an enormous impact on the effect of the filter. In fact, the characteristics of a digital filter specified in terms of digital frequency Ω cannot be known exactly until the sampling frequency is selected. For example, according to Equation (7.8), a narrow band pass filter centered at the digital frequency 0.6π rads passes 300 Hz signals when the sampling frequency is 1 kHz and 1.2 kHz when the sampling frequency is 4 kHz. Equation (7.8) permits the range of digital frequencies between 0 and π rads to be replaced by analog frequencies between 0 and $f_s/2$ Hz. Following this change, the magnitude response can be described as $|H(f)|$ versus f (or $20\log|H(f)|$ versus f when plotting gain in dB), and the phase response can be described as $\theta(f)$ versus f. The relationship between f and Ω is clarified in the following example.

EXAMPLE 7.18

A filter with the frequency response plotted in Figure 7.18 is used in a digital filter with a 12 kHz sampling frequency.

FIGURE 7.18

Frequency response for Example 7.18.

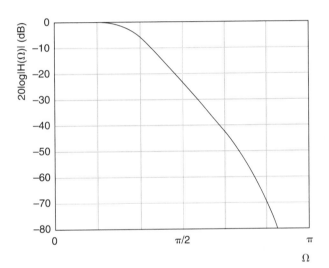

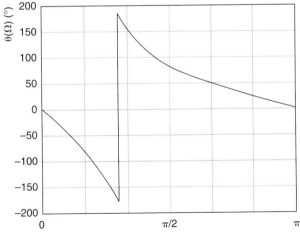

a. Plot the frequency response against analog frequencies in Hz instead of digital frequencies in radians.
b. What is the bandwidth of the low pass filter in Hz?
c. What is the bandwidth if the sampling frequency is changed to 30 kHz?

a. The digital frequencies from 0 to π radians convert to analog frequencies between 0 and 6 kHz, half the sampling rate. The shapes of the magnitude and phase responses do not change. Only the labeling of the horizontal axis changes, to reflect the change from digital to analog frequencies. This is shown in Figure 7.19.

FIGURE 7.19

Frequency response plotted against frequency in Hz.

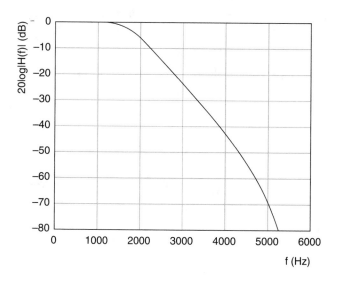

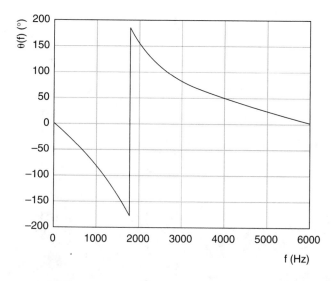

b. The bandwidth is the range of frequencies passed by the filter. For a low pass filter, this corresponds to the frequency where the gain drops by 3 dB from its low frequency value. For the magnitude response in Figure 7.19, this occurs at a frequency of 1800 Hz.

c. If the sampling frequency were changed to 30 kHz, frequencies of interest would range from 0 to 15 kHz, and the bandwidth would become 4500 Hz.

EXAMPLE 7.19

A magnitude response is given in Figure 7.20. Determine the gain that will act on an input signal with analog frequency 1 kHz if the sampling frequency is

    a. $f_S = 4$ kHz
    b. $f_S = 10$ kHz

In Figure 7.20, the magnitude response is plotted from $\Omega = 0$ to $\Omega = \pi$ radians. Once a sampling frequency has been selected, the response may be replotted from $f = 0$ to $f = f_s/2$ Hz.

    a. Figure 7.21 replots the magnitude response for $f_S = 4$ kHz. The gain applied to an input with frequency 1 kHz would be 0 dB, or 1, for this filter.

    b. Figure 7.22 replots the magnitude response for $f_S = 10$ kHz. The gain applied to an input with frequency 1 kHz would be about -63 dB, or 7.0795×10^{-4}, for this filter.

The vastly different results for parts (a) and (b) of this example underline the impact sampling frequency has on digital filter behavior.

FIGURE 7.20
Magnitude response for
Example 7.19.

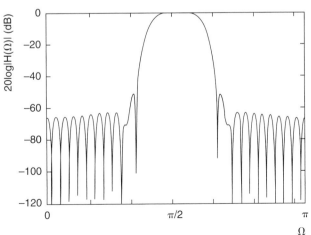

FIGURE 7.21
Magnitude response
for $f_S = 4$ kHz for
Example 7.19.

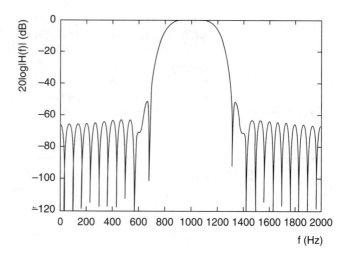

FIGURE 7.22
Magnitude response
for $f_S = 10$ kHz for
Example 7.19.

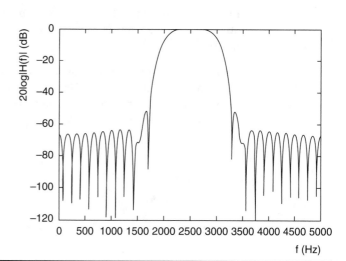

EXAMPLE 7.20

An analog signal $x(t) = 2\cos(2 \times 10^4 t + 15°)$ is sampled at 16 kHz and digitized. Estimate the gain that will be applied to the resulting signal by the filter with the magnitude response presented in Figure 7.23.

There are two ways to solve this problem. Both methods require the frequency of the input signal in Hz, which is

$$f = \frac{\omega}{2\pi} = \frac{20000}{2\pi} = 3183.1 \text{ Hz}$$

FIGURE 7.23

Magnitude response versus Ω for Example 7.20.

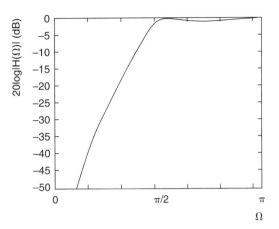

FIGURE 7.24

Magnitude response versus f for Example 7.20.

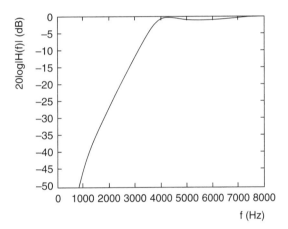

In the first method, the digital frequency for the input signal is calculated as

$$\Omega = 2\pi \frac{f}{f_S} = 2\pi \frac{3183.1}{16000} = \frac{20000}{16000} = 1.25 \text{ rads}$$

Figure 7.23 shows that the gain is about -10 dB at this digital frequency. The exact value is -10.3 dB, or 0.3055.

The second method requires replotting the magnitude response for $f_S = 16$ kHz, as shown in Figure 7.24. On this figure, the gain for a 3183.1 Hz signal is -10.3 dB, or 0.3055, as before.

7.3.4   Filter Shape from Poles and Zeros

Section 7.3.2 described how to obtain plots of filter shape from a filter's frequency response. Another method, based on the locations of the filter's poles and zeros, is more intuitive, especially for low order filters. To show how this method works, consider a transfer function of the form

$$H(z) = \frac{1}{z - p}$$

The frequency response for this filter is

$$H(\Omega) = \frac{1}{e^{j\Omega} - p}$$

and the magnitude response, or filter shape, is given by

$$|H(\Omega)| = \left|\frac{1}{e^{j\Omega} - p}\right| = \frac{1}{|e^{j\Omega} - p|} \tag{7.9}$$

By Euler's identity, the complex number $e^{j\Omega}$ equals $\cos\Omega + j\sin\Omega$. As indicated in Appendix A.13.2.5, this complex number lies on the unit circle in the complex plane for all values of Ω. As shown in Figure 7.25, it may be written as the ordered pair $(\cos\Omega, \sin\Omega)$, with a location that changes with Ω. As Ω increases over its usual range from 0 to π, the ordered pair $(\cos\Omega, \sin\Omega)$ travels along the unit circle from $(1, 0)$ to $(-1, 0)$. For the sake of generality, the pole p is defined as a complex number $\alpha + j\beta$, marked on the figure as the ordered pair (α, β). Using these definitions, the quantity $|e^{j\Omega} - p|$ from Equation (7.9) can be written as

$$|e^{j\Omega} - p| = |(\cos\Omega + j\sin\Omega) - (\alpha + j\beta)| = |(\cos\Omega - \alpha) + j(\sin\Omega - \beta)|$$

According to Equation (A.2), this magnitude is $\sqrt{(\cos\Omega - \alpha)^2 + (\sin\Omega - \beta)^2}$, which is nothing more than the distance between the ordered pairs $(\cos\Omega, \sin\Omega)$ and (α, β), from Appendix A.13.2.1. Therefore, the filter shape in Equation (7.9) is given by

FIGURE 7.25

Graphical view of filter shape.

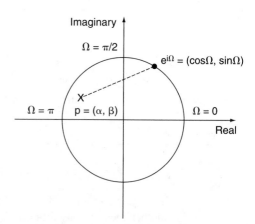

$$|H(\Omega)| = \frac{1}{\text{distance from } e^{j\Omega} \text{ to } p}$$

According to this expression, the smaller the distance between $e^{j\Omega}$ and p for a particular value of Ω, the larger the magnitude response. The magnitude response has its maximum value when $e^{j\Omega}$, along the unit circle, is as close to the pole p as possible. In other words, the maximum magnitude is obtained when Ω matches the phase of the pole p. Furthermore, the closer the location of the pole to the unit circle, the larger this maximum value will be.

The idea developed in the preceding paragraphs may be extended to filters with multiple poles and zeros. For example, for the filter with the transfer function

$$H(z) = \frac{K(z - z_0)}{(z - p_0)(z - p_1)}$$

where K is a gain, the z_j are zeros, and the p_j are poles, the filter shape is

$$|H(\Omega)| = \frac{K|e^{j\Omega} - z_0|}{|e^{j\Omega} - p_0||e^{j\Omega} - p_1|} = \frac{K(\text{distance to } z_0)}{(\text{distance to } p_0)(\text{distance to } p_1)}$$

In general, the filter shape is given by

$$|H(\Omega)| = \frac{K \,(\text{Product of distances to zeros})}{(\text{Product of distances to poles})} \tag{7.10}$$

For a digital frequency Ω between 0 and π radians, the closer $e^{j\Omega}$ is to a pole of a filter, and the farther $e^{j\Omega}$ is from a zero, the larger the magnitude tends to be. Also, a pole close to the unit circle leads to a filter shape with an extremely large magnitude at some frequency, while a zero close to the unit circle leads to a filter shape with a very small magnitude at some frequency. Extremely large and small magnitudes tend to signal filters that are very selective. Thus, thinking about Equation (7.10) can give a good sense of the filter's shape, as the following examples will illustrate.

EXAMPLE 7.21

Deduce the shape of the filter described by the transfer function

$$H(z) = \frac{1}{z - 0.45}$$

This filter has one pole, at $z = 0.45$, no zeros, and unity gain, so Equation (7.10) gives

$$|H(\Omega)| = \frac{1}{\text{distance to } (0.45, 0)}$$

The pole-zero plot for the filter is shown in Figure 7.26, with a sample position for $e^{j\Omega}$ marked. Comparing with Figure 7.25, the distance to the pole should be smallest near $\Omega = 0$,

FIGURE 7.26

Pole-zero plot for Example 7.21.

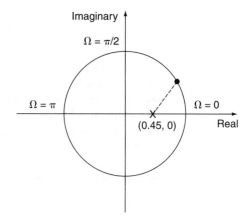

TABLE 7.5

Calculations for Example 7.21

Ω	$e^{j\Omega} = (\cos\Omega, \sin\Omega)$	Distance to Pole at (0.45, 0)	$\lvert H(\Omega)\rvert$
0	$(1, 0)$	0.5500	1.8182
$\dfrac{\pi}{4}$	$\left(\dfrac{1}{\sqrt{2}}, \dfrac{1}{\sqrt{2}}\right)$	0.7524	1.3291
$\dfrac{\pi}{2}$	$(0, 1)$	1.0966	0.9119
$\dfrac{3\pi}{4}$	$\left(-\dfrac{1}{\sqrt{2}}, \dfrac{1}{\sqrt{2}}\right)$	1.3561	0.7374
π	$(-1, 0)$	1.4500	0.6897

and largest Ω near π radians. Since the magnitude for this filter is simply the reciprocal of this distance, the magnitude will be largest near $\Omega = 0$ and smallest Ω near π radians. In other words, this filter shows low pass behavior. Since the pole is not very close to the unit circle, the filter is not particularly selective. This rough analysis is confirmed in Table 7.5, which computes the exact distances and magnitudes for several values of Ω. The magnitudes obtained in this manner match those that would be obtained by frequency response calculations like those found in Example 7.15. The filter shape is plotted in Figure 7.27.

FIGURE 7.27
Filter shape for Example 7.21.

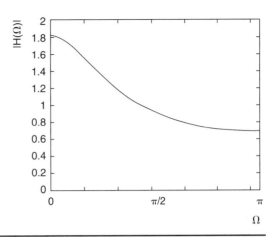

EXAMPLE 7.22

Deduce the filter shape for the filter described by the difference equation

$$y[n] = x[n-1] + x[n-3]$$

The transfer function for the filter is

$$H(z) = z^{-1} + z^{-3} = \frac{z^2 + 1}{z^3}$$

The filter has three poles at $z = 0$, and zeros at $z = j$ and $z = -j$. The pole-zero plot is shown in Figure 7.28, with a sample position for $e^{j\Omega}$ marked. Since the poles are all at the origin, the distances from the poles to $e^{j\Omega}$ are the same for all values of Ω. Thus, only the zero positions affect the overall filter shape as given by Equation (7.10). A magnitude of zero will be obtained for $\Omega = \pi/2$, since the distance to one of the zeros is zero for this frequency. The filter will therefore have a band stop effect, as confirmed by the exact filter shape shown in Figure 7.29.

FIGURE 7.28
Pole-zero plot for Example 7.22.

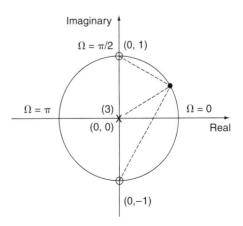

FIGURE 7.29

Filter shape for Example 7.22.

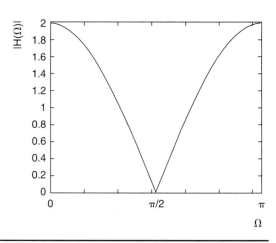

EXAMPLE 7.23

Compare the filter shapes for the filters:

$$\text{a.} \quad H(z) = \frac{z^{-2}}{1 + 1.537z^{-1} + 0.9025z^{-2}}$$

$$\text{b.} \quad H(z) = \frac{z^{-2}}{1 + 1.456z^{-1} + 0.81z^{-2}}$$

$$\text{c.} \quad H(z) = \frac{z^{-2}}{1 + 1.294z^{-1} + 0.64z^{-2}}$$

The pole-zero plots for all three filters are shown in Figure 7.30. None of the filters have zeros. The first filter has the poles $-0.7686 \pm j0.5584$, with magnitude 0.95. The second filter has the poles $-0.7281 \pm j0.5290$, with magnitude 0.9. The third filter has the poles

FIGURE 7.30

Pole-zero plot for Example 7.23.

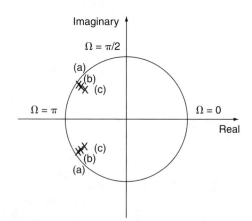

FIGURE 7.31

Filter shapes for
Example 7.23.

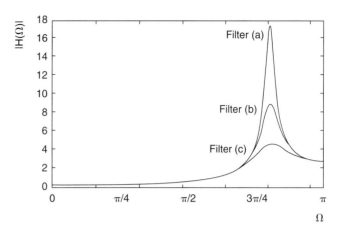

$-0.6472 \pm j0.4702$, with magnitude 0.8. As the figure shows, the value of digital frequency
for which the poles are nearest the unit circle is the same for all three filters. Therefore, the
magnitude responses should all give a spike at this value of Ω, which happens to be $\Omega =
0.8\pi$, the phase from the polar form of each number. Since, of the three filters, the poles for
filter (a) are closest to the unit circle, the spike for this filter should be the largest. The filter
shapes, plotted in Figure 7.31, confirm these hypotheses.

7.3.5   First Order Filters

The discrete time Fourier transform (DTFT) provides the tools to investigate the frequency
response of first order filters. Section 6.4.3 presented the impulse response and the pole-
zero diagram for a first order difference equation of the form

$$y[n] + \alpha y[n-1] = x[n]$$

where α can be positive or negative. Because α is the only variable in this equation, this
structure can support a limited number of filter shapes. Changing the sign of α can, however, have a dramatic effect on the shape, as the following examples show.

EXAMPLE 7.24

Find the frequency response for the filter $y[n] + 0.5y[n-1] = x[n]$.

The frequency response is

$$H(\Omega) = \frac{1}{1 + 0.5e^{-j\Omega}}$$

and is shown in Figure 7.32. The filter has a high pass nature, but is not an effective high
pass filter because the gain difference between the pass band and the stop band is less than

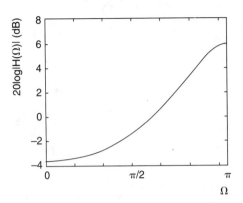 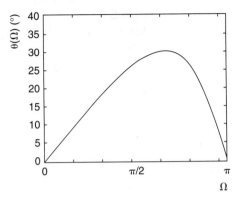

FIGURE 7.32
Frequency response for Example 7.24.

10 dB. The pole-zero plot, impulse response, and step response for the filter appear in the left column of Figure 6.17. Following the discussion in the last section, the single pole of this filter, at $z = -0.5$, explains why the filter is high pass but not very selective.

EXAMPLE 7.25

Find the frequency response for the filter $y[n] - 0.5y[n-1] = x[n]$.

The frequency response is, by Equation (7.2),

$$H(\Omega) = \frac{1}{1 - 0.5e^{-j\Omega}}$$

The frequency response is shown in Figure 7.33. The filter has a shallow low pass characteristic. The pole-zero plot, impulse response, and step response for the filter appear in the right column of Figure 6.17. Changing the sign of the coefficient 0.5 clearly changes the filter's behavior dramatically.

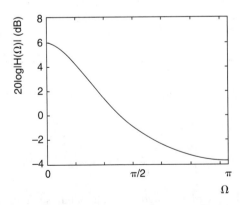 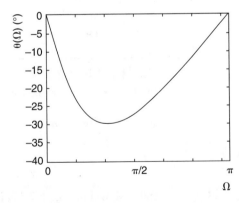

FIGURE 7.33
Frequency response for Example 7.25.

7.3.6   Second Order Filters

Second order filters of the form

$$y[n] + \alpha y[n-1] + \beta y[n-2] = x[n]$$

were studied in Section 6.4.4 of the last chapter. The frequency responses for second order filters are now developed using the discrete time Fourier transform.

EXAMPLE 7.26

Find the frequency response for the second order filter described in the diagram of Figure 7.34.

The difference equation for this filter is

$$y[n] = -0.8y[n-1] - 0.9y[n-2] + x[n]$$

The frequency response is

$$H(\Omega) = \frac{1}{1 + 0.8e^{-j\Omega} + 0.9e^{-j2\Omega}}$$

The magnitude and phase responses are plotted in Figure 7.35.

FIGURE 7.34
Difference equation diagram for
Example 7.26.

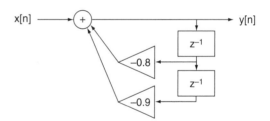

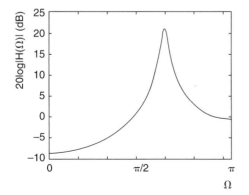
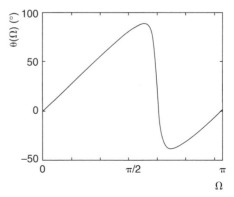

FIGURE 7.35
Frequency response for Example 7.26.

The following second order transfer functions were presented in Section 6.4.4:

a. $H(z) = \dfrac{z^{-2}}{1 + 0.2z^{-1} + 0.01z^{-2}}$

b. $H(z) = \dfrac{z^{-2}}{1 + 0.6z^{-1} + 0.13z^{-2}}$

c. $H(z) = \dfrac{z^{-2}}{1 - 1.2z^{-1} + 0.36z^{-2}}$

d. $H(z) = \dfrac{z^{-2}}{1 - z^{-1} + 0.5z^{-2}}$

e. $H(z) = \dfrac{z^{-2}}{1 - 1.15z^{-1} + 0.28z^{-2}}$

f. $H(z) = \dfrac{z^{-2}}{1 + 1.7z^{-1} + 0.7625z^{-2}}$

g. $H(z) = \dfrac{z^{-2}}{1 + 1.8z^{-1} + 0.81z^{-2}}$

h. $H(z) = \dfrac{z^{-2}}{1 - 1.6z^{-1} + 0.9425z^{-2}}$

The frequency responses for each filter are presented in Figure 7.36. The magnitude responses, or filter shapes, are shown in the left column, and the phase responses in the right column. Varying coefficient values produce a wide variety of filter shapes, including low pass, high pass, and band pass. Figure 6.25 in the last chapter presented the pole-zero plots and impulse responses for the same group of filters. As expected, the filters whose poles are closest to the unit circle are the most selective.

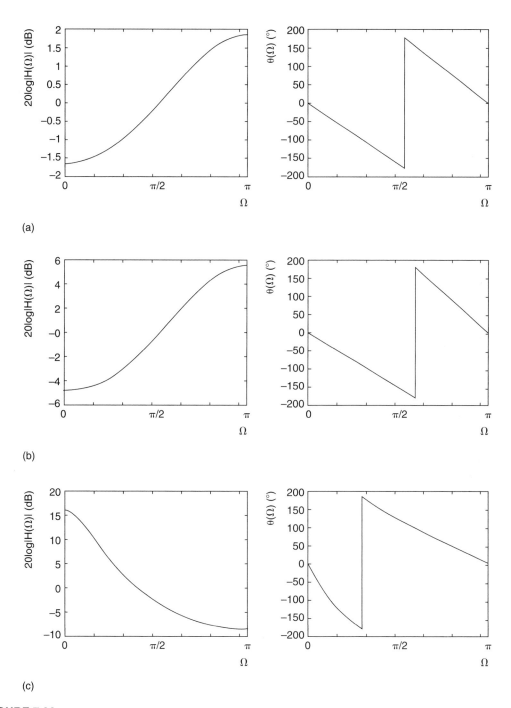

FIGURE 7.36
Second order frequency responses.

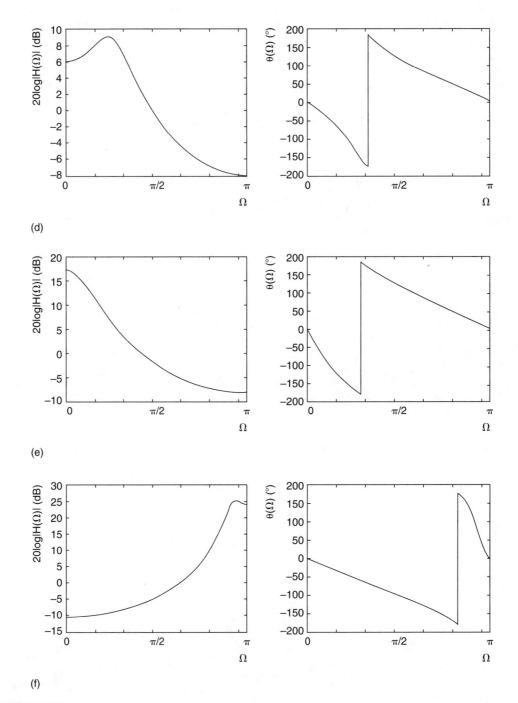

(d)

(e)

(f)

FIGURE 7.36
Continued

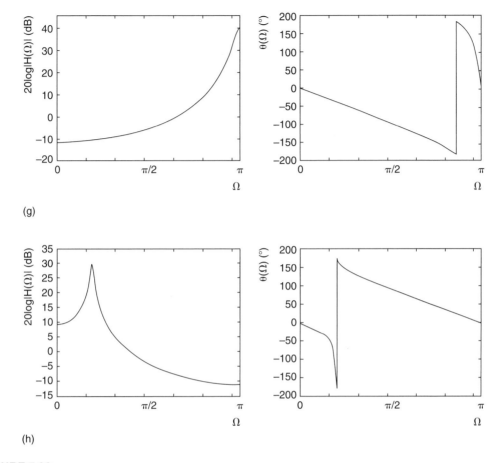

(g)

(h)

FIGURE 7.36
Continued

CHAPTER SUMMARY

Matlab
Support

1. The discrete time Fourier transform (DTFT) of a signal $x[n]$ is given by

$$X(\Omega) = \sum_{n=-\infty}^{\infty} x[n]e^{-jn\Omega}$$

It reports the frequencies present in a signal.

2. The DTFT of a signal $x[n]$ gives the signal's spectrum $X(\Omega)$.

3. The DTFT of a system $h[n]$ gives the system's frequency response $H(\Omega)$.

4. The DTFT is periodic with period 2π.

5. A difference equation can be expressed as a frequency response.

6. A transfer function can be expressed as a frequency response.

7. The frequency response $H(\Omega)$ is the DTFT of the impulse response $h[n]$.

8. A frequency response $H(\Omega)$ is a complex number and may be expressed in polar form in terms of a gain $|H(\Omega)|$ and a phase difference $\theta(\Omega)$ as

$$H(\Omega) = |H(\Omega)|e^{j\theta(\Omega)}$$

9. The frequency response can be used to find a filter's output for a sinusoidal input. The output is a sinusoid with the same frequency as the input, but with an amplitude multiplied by the gain of the filter and a phase shifted by the phase difference of the filter. For the input $x[n] = A\cos(n\Omega_0 + \theta_x)$, with a digital frequency Ω_0, the output is $y[n] = HA\cos(n\Omega_0 + \theta + \theta_x)$, where the gain of the filter is $H = |H(\Omega_0)|$ and the phase difference is $\theta = \theta(\Omega_0)$.

10. The gains applied by the filter at each frequency form the magnitude response, plotted as $|H(\Omega)|$ versus Ω, or $20\log|H(\Omega)|$ versus Ω. The phase differences applied by the filter at each frequency form the phase response, plotted as $\theta(\Omega)$, in degrees or radians, versus Ω.

11. It is sufficient to plot a frequency response for a system for the digital frequencies between 0 and π radians, as $\Omega = \pi$ radians corresponds to the Nyquist frequency or $f_S/2$.

12. The magnitude response shows the shape of a filter: low pass, high pass, band pass, or band stop. Some filters, such as comb filters, have more complicated shapes.

13. The magnitude and phase responses, $|H(\Omega)|$ and $\theta(\Omega)$, are normally plotted against digital frequency Ω in radians. They can be plotted instead against analog frequency f in Hz using the equation

$$f = \Omega\frac{f_S}{2\pi}$$

With this substitution, the magnitude response may be plotted as $|H(f)|$ versus f, and the phase response as $\theta(f)$ versus f.

14. The shape of a filter can be deduced from its pole-zero plot. As Ω increases, the complex number $e^{j\Omega}$ moves around the unit circle. Proximity to a zero tends to reduce the magnitude of the filter, while proximity to a pole tends to increase it. The closer the poles and zeros are to the unit circle, the more selective the filter is.

REVIEW QUESTIONS

7.1 Find an expression for the DTFT for the signal $x[n] = \delta[n] - \delta[n-1] + \delta[n-2]$.

7.2 What is the expression for the DTFT of the signal $x[n] = e^{-0.1n}(u[n] - u[n-4])$?

7.3 Find an expression for the frequency response of the digital filter whose difference equation is

$$2y[n] - y[n-1] + 3y[n-2] = x[n] - 4x[n-1]$$

7.4 A filter's difference equation is

$$y[n] = \frac{1}{4}(x[n] + x[n-1] + x[n-2] + x[n-3])$$

Find an expression for the frequency response.

7.5 Find expressions for the frequency responses of the filters whose transfer functions are given by:

a. $H(z) = \dfrac{1}{1 - 1.1z^{-1} + 0.4z^{-2}}$

b. $H(z) = \dfrac{z - 0.7}{z^2 - 0.5z + 0.3}$

7.6 The impulse response of a filter is

$$h[n] = \cos\left(n\frac{\pi}{2}\right)(u[n] - u[n-5])$$

Find an expression for the filter's frequency response.

7.7 The frequency response of a filter is given by:

$$H(\Omega) = \frac{1}{1 + 0.8e^{-j\Omega}}$$

a. Find and sketch the frequency response of the filter between $\Omega = 0$ and $\Omega = 4\pi$ radians. Use linear gains and phases in radians.

b. With what period does the response repeat?

7.8 Sketch the frequency response

$$H(\Omega) = \frac{3}{2 - e^{-j\Omega}}$$

for $\leq \theta \leq \pi$ rads. Plot the gains in dB and the phases in degrees.

7.9 The frequency response for a high pass filter is given in Figure 7.37. Determine the gain (dB) and the phase (°) of the filter at each of the following digital frequencies:

a. 1 radian

b. 2 radians

c. 3 radians

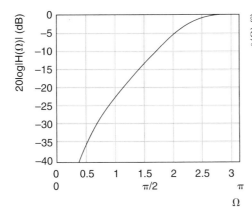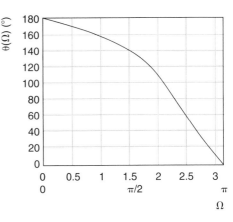

FIGURE 7.37

Frequency response for Question 7.9.

7.10 Sketch the magnitude responses (as linear gains versus digital frequencies) for each filter. Indicate the filter type (low pass, high pass, band pass, band stop) in each case.

a. $H(\Omega) = \dfrac{1.3}{1.3 - e^{-j\Omega}}$

b. $H(\Omega) = \dfrac{0.2 - 0.4e^{-j\Omega} + 0.2e^{-j2\Omega}}{1 + 0.2e^{-j2\Omega}}$

c. $H(\Omega) = \dfrac{0.75(1 + e^{-j2\Omega})}{1 + 0.5e^{-j2\Omega}}$

d. $H(\Omega) = \dfrac{1}{1 - 0.3e^{-j\Omega} + 0.9e^{-j2\Omega}}$

7.11 The impulse response of a filter is $h[n] = (u[n] - u[n-4])/4$.
a. Sketch the shape of this filter in dB versus digital frequency in radians.
b. What kind of filter is this?

7.12 Sketch the frequency response (using decibels and degrees) for the filter

$$y[n] = 1.3y[n-1] - 0.7y[n-2] + x[n] - 0.5x[n-1]$$

7.13 Draw the filter shape as dB versus radians for the filters:

a. $H(z) = \dfrac{1}{1 - 0.6z^{-6}}$

b. $H(z) = \dfrac{1.1 - 0.7z^{-9}}{1 - 0.85z^{-9}}$

7.14 The digital signal $x[n] = 2\cos(n\,4\pi/9 + 20°)$ is applied to the input of a digital filter whose frequency response is plotted in Figure 7.38. Determine the output signal.

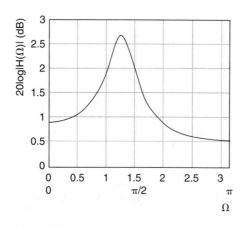
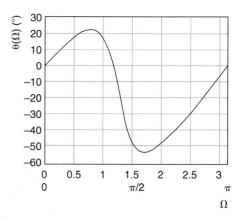

FIGURE 7.38
Frequency response for Question 7.14.

7.15 A filter has the frequency response shown in Figure 7.39.
   a. If the input to the filter is $x[n] = 25\cos(1.5n)$, what is the output?
   b. Calculate and plot the first 10 samples of the input and output.

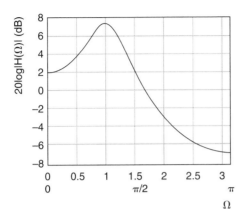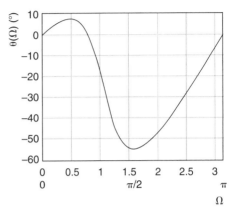

FIGURE 7.39
Frequency response for Question 7.15.

7.16 A filter with the magnitude response shown in Figure 7.40 is used in a 16 kHz sampled system.
   a. What frequency in Hz corresponds to the digital frequency π radians?
   b. Find the bandwidth of the filter in Hz.
   c. Find the gain that will be applied to a 5 kHz input.
   d. What input frequency in Hz will experience a -2 dB gain?

FIGURE 7.40
Magnitude response
for Question 7.16.

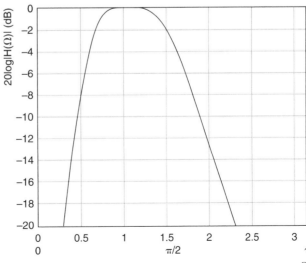

7.17 Assuming a 10 kHz sampling rate, plot the frequency response of the filter

$$H(z) = \frac{1 - 0.95z^{-1}}{1 - 0.9z^{-1}}$$

with magnitudes in dB, phases in degrees, and frequencies in Hz.

7.18 Plot the magnitude and phase responses for the filter with the impulse response $h[n] = \delta[n-8]$. Use steps of digital frequency of $\pi/8$ radians or smaller.

7.19 Draw the magnitude and phase responses against frequency in Hz for the filter defined by the difference equation

$$y[n] = -0.8y[n-1] - 0.9y[n-2] + x[n] - 0.5x[n-1]$$

if the sampling frequency is 8 kHz. Use linear gains and phases in degrees.

7.20 Plot the magnitude response $|H(\Omega)|$ for the filter with the transfer function

$$H(z) = \frac{1 + z^{-4}}{1 - 0.82z^{-4}}$$

Use steps of digital frequency Ω of $\pi/8$ radians or smaller. What type of filter is this?

7.21 Find the filter shape of the filter with the impulse response

$$h[n] = \delta[n] - 0.98\delta[n-6]$$

a. as $|H(\Omega)|$ versus Ω

b. as $|H(f)|$ versus f, if the sampling frequency is 16 kHz

7.22 An analog signal $x(t) = \cos(100\pi t)$ is sampled at 200 Hz. The digital signal produced from this sampled signal acts as an input to a filter whose magnitude response is shown in Figure 7.41.

a. Write an expression for the digital filter input.

b. What gain will be applied to the input by the filter?

c. What are the amplitude and digital frequency of the output signal?

FIGURE 7.41

Magnitude response for Question 7.22.

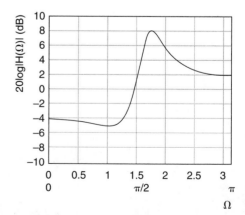

7.23 The frequency response of a filter is given by

$$H(\Omega) = \frac{0.5 - 0.2e^{-j2\Omega}}{1 + 0.7e^{-j\Omega} + 0.1e^{-j2\Omega}}$$

Find the output $y[n]$ from the filter if the input is:
a. $x[n] = \cos(n3\pi/7)$
b. $x[n] = 2\sin(n\pi/6 - \pi/2)$

7.24 A filter has the transfer function

$$H(z) = \frac{z - 0.85}{z + 0.95}$$

Deduce the filter shape without computing the frequency response.

7.25 The pole-zero plot for a filter is shown in Figure 7.42. Deduce the filter shape without computing the frequency response.

FIGURE 7.42
Pole-zero diagram for Question 7.25.

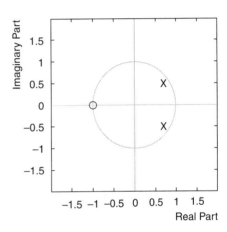

7.26 The transfer function for a filter is

$$H(z) = \frac{(z + 1)}{z^2(z - 1)}$$

Sketch the magnitude response without computing $H(\Omega)$.

7.27 The impulse response $h[n] = \delta[n] - \delta[n-1]$ describes a differencing filter. Without computing $H(\Omega)$, describe the nature of the filter.

7.28 The pole-zero plot for a seven-term moving average filter is shown in Figure 7.43. Use it to predict the filter's shape.

FIGURE 7.43

Pole-zero diagram for Question 7.28.

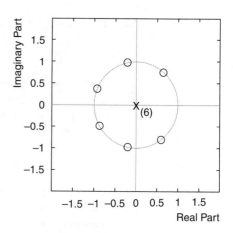

7.29 The pole-zero plot for a filter is shown in Figure 7.44. Use it to predict the filter's shape.

FIGURE 7.44

Pole-zero plot for Question 7.29.

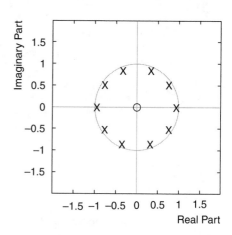

7.30 By choosing locations for a suitable pair of poles, find the transfer function of a narrowband filter centered at $f = 400$ Hz, for a sampling frequency of $f_S = 2400$ Hz.

8

DIGITAL SIGNAL SPECTRA

This chapter investigates digital signal spectra, which provide the frequency content of both nonperiodic and periodic signals. The chapter:

> ➤ provides definitions for the spectrum, the magnitude spectrum, and the phase spectrum of a digital signal
> ➤ demonstrates how to calculate spectra for nonperiodic signals using the discrete time Fourier transform (DTFT)
> ➤ introduces the discrete Fourier series (DFS)
> ➤ shows how to calculate spectra for periodic signals using the discrete Fourier series
> ➤ defines fundamental frequency and harmonic frequency

8.1 THE MEANING OF THE SPECTRUM

This chapter, like the preceding one, is concerned with the frequency domain. While the last chapter examined the frequency characteristics of filters, this one investigates the frequency characteristics of signals. In the frequency domain, every digital signal has a characteristic signature. A sine wave, for example, contains only a single frequency, while **white noise** contains elements at all frequencies. The smooth transitions in a signal come from its low frequency components, while sharp edges and rapid changes come from its high frequency elements. A square wave, for instance, contains low frequency elements that produce the smooth up and down changes, as well as high frequency elements that form the sharp transitions at the edges. Figure 8.1 illustrates why this might be true. The figure begins with a single, low frequency sine wave. In each part of the figure, a higher frequency sine wave is added in. Well-chosen sine waves of suitable frequencies and amplitudes add together to produce a square wave, as part (e) of the figure suggests, according to the equation

$$y(t) = \frac{4}{\pi}\left(\sin(\omega t) + \frac{1}{3}\sin(3\omega t) + \frac{1}{5}\sin(5\omega t) + \frac{1}{7}\sin(7\omega t) + \frac{1}{9}\sin(9\omega t) + \ldots\right)$$

The **spectrum** of a signal is a detailed description of the frequency components the signal contains. For the square wave, for example, the spectrum would show spikes for all

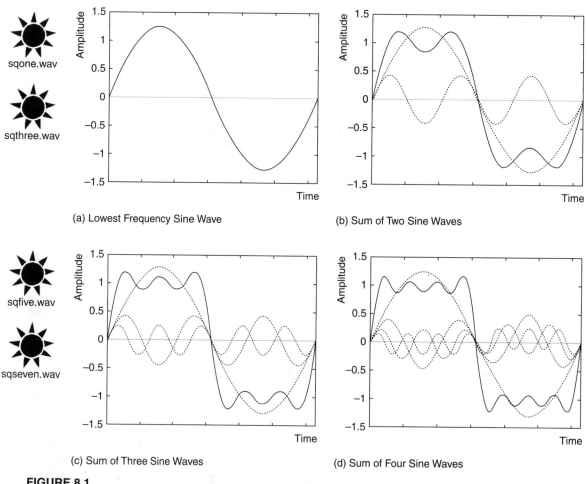

(a) Lowest Frequency Sine Wave

(b) Sum of Two Sine Waves

(c) Sum of Three Sine Waves

(d) Sum of Four Sine Waves

FIGURE 8.1

Building a square wave from sine waves.

FIGURE 8.1
Continued

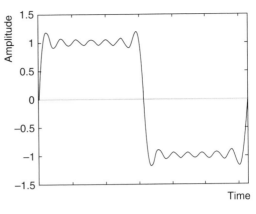

(e) Sum of Seven Sine Waves

the individual sine waves that combine to produce it. This information can be important for many reasons. For example, the frequency content of a piece of music dictates the characteristics of the speakers that must be used to reproduce it faithfully. As another example, a microphone in a speech recognition system must have a wide enough frequency response to be able to pick up all of the important frequencies in the incoming speech. To anticipate the effect a filter will have on a signal, it is necessary to know not only the nature of the filter but also the spectrum of the signal.

The spectrum of a signal has two parts, the magnitude spectrum and the phase spectrum, which parallel perfectly the magnitude and phase responses that make up the frequency responses of the last chapter. The **magnitude spectrum** relates to the size or amplitude of the components at each frequency. The **phase spectrum** gives the phase relationships between the components at different frequencies. All digital signals have spectra, but different tools must be used to compute them, depending on whether the signal is nonperiodic (nonrepeating) or periodic (repeating). These tools are investigated in Sections 8.2 and 8.3.

8.2   NONPERIODIC DIGITAL SIGNALS

Nonperiodic signals are those that do not repeat at regular intervals. Some examples are shown in Figure 8.2. The tool required to compute the spectra for nonperiodic signals is the discrete time Fourier transform (DTFT). In Section 7.1, the DTFTs for several nonperiodic digital signals were calculated. This calculation is the first step in obtaining the spectrum of a nonperiodic signal. The next steps follow the method for finding magnitude and phase responses for digital filters in Section 7.3.2. In that section, the magnitude and phase of the DTFT were used to produce the magnitude response and phase response of the filter. Exactly the same process is used to find magnitude and phase spectra for nonperiodic signals.

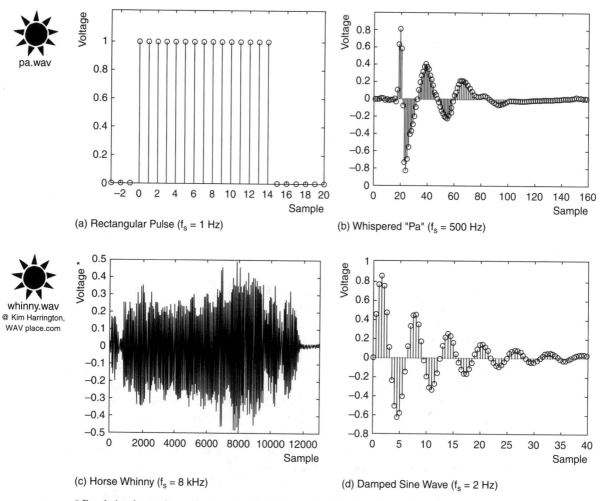

(a) Rectangular Pulse (f_s = 1 Hz)

(b) Whispered "Pa" (f_s = 500 Hz)

(c) Horse Whinny (f_s = 8 kHz)

(d) Damped Sine Wave (f_s = 2 Hz)

pa.wav

whinny.wav
@ Kim Harrington,
WAV place.com

* For clarity, the envelope rather than the individual samples of the signal is plotted.

FIGURE 8.2
Nonperiodic digital signals.

The DTFT for a nonperiodic signal gives the signal's spectrum as

$$X(\Omega) = \sum_{n=-\infty}^{\infty} x[n]e^{-jn\Omega}$$

As the equation indicates, calculation of the DTFT requires all samples of the nonperiodic signal $x[n]$. When the signal has an infinite number of nonzero samples that decrease in size,

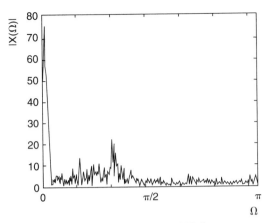

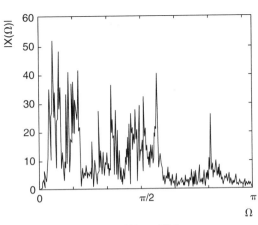

(a) Magnitude Spectrum of Whispered "Pa"

(b) Magnitude Spectrum of Horse Whinny

FIGURE 8.3

Magnitude spectra of nonperiodic signals.

the DTFT may be approximated by truncating the signal where its amplitude drops below some suitably low threshold. The DTFT spectrum $X(\Omega)$ is a complex number and may be expressed in polar form

$$X(\Omega) = |X(\Omega)|e^{j\theta(\Omega)} \tag{8.1}$$

The magnitude spectrum is given by $|X(\Omega)|$ versus Ω, and the phase spectrum by $\theta(\Omega)$ versus Ω. Figure 8.3 shows the magnitude spectra for two of the signals in Figure 8.2. Because they are obtained by using the DTFT, the spectra of nonperiodic signals are continuous and periodic with period 2π, as shown in Section 7.3.2 for frequency responses. The same section reviews the units used for the magnitude and phase plots. Magnitude spectra are plotted either as linear gains, $|X(\Omega)|$, or gains in dB, $20\log|X(\Omega)|$, against digital frequency Ω in radians. Phase spectra are plotted as $\theta(\Omega)$, in degrees or radians, versus Ω.

As was done for frequency responses in Section 7.3.3 to make spectra easier to interpret, the digital frequencies Ω from 0 to π radians can be converted to analog frequencies f between 0 and $f_s/2$ if the sampling frequency f_s that was used to collect the signal samples is known. Equation (7.8),

$$f = \Omega\frac{f_S}{2\pi}$$

is used to make the conversion. The following examples investigate the spectra of three nonperiodic signals.

EXAMPLE 8.1

Find the magnitude and phase spectra for the rectangular pulse $x[n] = u[n] - u[n-4]$ as functions of Ω. Plot linear gains, and phases in radians.

The rectangular pulse is plotted in Figure 8.4. The spectrum is found in a manner analogous to that used in Example 7.2. It is

$$X(\Omega) = \sum_{n=-\infty}^{\infty} x[n]e^{-jn\Omega} = 1 + e^{-j\Omega} + e^{-j2\Omega} + e^{-j3\Omega}$$

As in Section 7.3.2, the easiest way to evaluate this expression is to construct a table. Quarter-π intervals for frequency Ω are convenient. For this example only, a range of digital frequencies from $-\pi$ to 3π radians is examined. Table 8.1 shows four places in every 2π interval where the magnitude is zero. At these positions, phase is irrelevant, since a zero magnitude makes the spectrum $X(\Omega)$ identically zero, according to Equation (8.1).

The magnitude and phase spectra are plotted in Figure 8.5 for the entire range of frequencies shown in the table. Triangles mark the points calculated at quarter-π intervals, while the dotted lines show the exact spectral shapes. The magnitude spectrum in (a) has a shape that is characteristic for all rectangular pulses, called a **sinc function**, defined in Section 9.4. The phase spectrum in (b) is particularly poorly predicted by the points from the table. As expected, the spectra are continuous and periodic with period 2π. The same characteristic was evident for the filter shapes in Section 7.3.2. It comes as no surprise that the magnitude spectrum is even and the phase spectrum is odd. These properties mean that all spectral information can be deduced from the information between 0 and π radians.

The spectrum is replotted in Figure 8.6 for the range $0 \leq \Omega \leq \pi$. The two zeros uncovered in the table, at $\Omega = \pi/2$ and $\Omega = \pi$, are the only two zeros in the magnitude spectrum. The magnitude spectrum in Figure 8.6(a) shows that the signal contains

FIGURE 8.4

Rectangular pulse for Example 8.1.

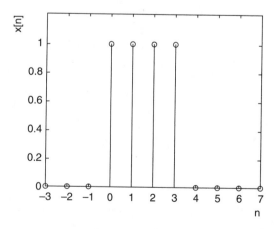

TABLE 8.1
Calculations for Example 8.1

Ω	$X(\Omega)$	$\lvert X(\Omega)\rvert$	$\theta(\Omega)$ (rads)
$-\pi$	$1 + 1\angle\pi + 1\angle 2\pi + 1\angle 3\pi$	0.0000	0.0000
$-\dfrac{3\pi}{4}$	$1 + 1\angle\dfrac{3\pi}{4} + 1\angle\dfrac{3\pi}{2} + 1\angle\dfrac{9\pi}{4}$	1.0824	0.3927
$-\dfrac{\pi}{2}$	$1 + 1\angle\dfrac{\pi}{2} + 1\angle\pi + 1\angle\dfrac{3\pi}{2}$	0.0000	0.0000
$-\dfrac{\pi}{4}$	$1 + 1\angle\dfrac{\pi}{4} + 1\angle\dfrac{\pi}{2} + 1\angle\dfrac{3\pi}{4}$	2.6131	1.1781
0	$1 + 1\angle 0 + 1\angle 0 + 1\angle 0$	4.0000	0.0000
$\dfrac{\pi}{4}$	$1 + 1\angle -\dfrac{\pi}{4} + 1\angle -\dfrac{\pi}{2} + 1\angle -\dfrac{3\pi}{4}$	2.6131	-1.1781
$\dfrac{\pi}{2}$	$1 + 1\angle -\dfrac{\pi}{2} + 1\angle -\pi + 1\angle -\dfrac{3\pi}{2}$	0.0000	0.0000
$\dfrac{3\pi}{4}$	$1 + 1\angle -\dfrac{3\pi}{4} + 1\angle -\dfrac{3\pi}{2} + 1\angle -\dfrac{9\pi}{4}$	1.0824	-0.3927
π	$1 + 1\angle -\pi + 1\angle -2\pi + 1\angle -3\pi$	0.0000	0.0000
$\dfrac{5\pi}{4}$	$1 + 1\angle -\dfrac{5\pi}{4} + 1\angle -\dfrac{5\pi}{2} + 1\angle -\dfrac{15\pi}{4}$	1.0824	0.3927
$\dfrac{3\pi}{2}$	$1 + 1\angle -\dfrac{3\pi}{2} + 1\angle -3\pi + 1\angle -\dfrac{9\pi}{2}$	0.0000	0.0000
$\dfrac{7\pi}{4}$	$1 + 1\angle -\dfrac{7\pi}{4} + 1\angle -\dfrac{7\pi}{2} + 1\angle -\dfrac{21\pi}{4}$	2.6131	1.1781
2π	$1 + 1\angle -2\pi + 1\angle -4\pi + 1\angle -6\pi$	4.0000	0.0000
$\dfrac{9\pi}{4}$	$1 + 1\angle -\dfrac{9\pi}{4} + 1\angle -\dfrac{9\pi}{2} + 1\angle -\dfrac{27\pi}{4}$	2.6131	-1.1781
$\dfrac{5\pi}{2}$	$1 + 1\angle -\dfrac{5\pi}{2} + 1\angle -5\pi + 1\angle -\dfrac{15\pi}{2}$	0.0000	0.0000
$\dfrac{11\pi}{4}$	$1 + 1\angle -\dfrac{11\pi}{4} + 1\angle -\dfrac{11\pi}{2} + 1\angle -\dfrac{33\pi}{4}$	1.0824	-0.3927
3π	$1 + 1\angle -3\pi + 1\angle -6\pi + 1\angle -9\pi$	0.0000	0.0000

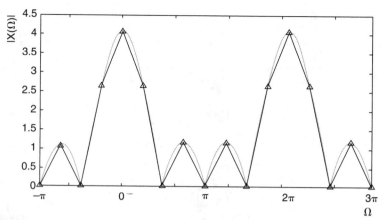

(a) Magnitude Spectrum

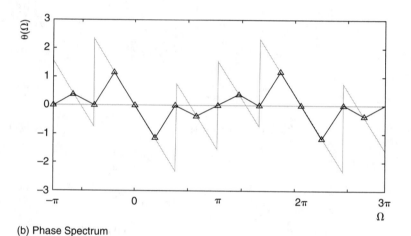

(b) Phase Spectrum

FIGURE 8.5

Extended spectra of signal for Example 8.1.

predominantly low frequencies, that is, below $\Omega = \pi/2$ rads. Without knowing the sampling frequency f_S used to collect the signal samples in Figure 8.4, though, it is impossible to say what the signal frequencies f are, since these frequencies are related to the digital frequencies Ω through $f = \Omega f_S/2\pi$. For example, a sampling frequency of 10 kHz would put the first zero at $f = (\pi/2)(10000/2\pi) = 2.5$ kHz. A sampling frequency of 400 Hz would, however, put the first 0 at 100 Hz. The sampling frequency clearly has a huge impact on how the spectrum must be interpreted.

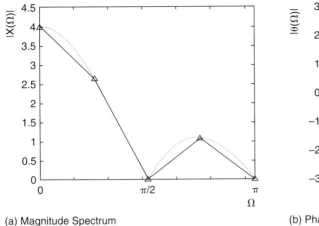

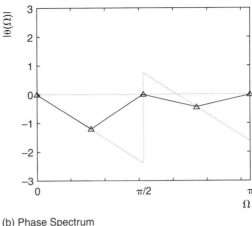

(a) Magnitude Spectrum   (b) Phase Spectrum

FIGURE 8.6
Spectrum of signal for Example 8.1.

EXAMPLE 8.2

Find the magnitude and phase spectra for the signal $x[n] = (0.1)^n u[n]$, sampled at 15 kHz. Plot the magnitude spectrum in dB and the phase spectrum in degrees, both against frequency in Hz.

The first five samples of the signal are listed in Table 8.2. Since the sample amplitudes drop off quickly, the first three samples are sufficient to obtain a good approximation to the DTFT spectrum.

The spectrum for the signal is given by

$$X(\Omega) = \sum_{n=-\infty}^{\infty} x[n]e^{-jn\Omega} \approx 1 + 0.1e^{-j\Omega} + 0.01e^{-j2\Omega}$$

Magnitude and phase spectrum values are calculated at quarter-π intervals and are shown in Table 8.3. In the last column, digital frequency values are converted to analog frequencies using $f = \Omega f_S/2\pi$. The spectral values are not affected by this conversion, as the labels for the fifth and sixth columns attest. The spectrum is plotted in Figure 8.7. The magnitude spectrum shows less than a 2 dB difference in magnitude between the very low and very

TABLE 8.2
Signal Samples for Example 8.2

n	0	1	2	3	4
$x[n]$	1.0000	0.1000	0.0100	0.0010	0.0001

TABLE 8.3
Calculations for Example 8.2

Ω (rads)	$X(\Omega)$	$\lvert X(\Omega)\rvert$	$\theta(\Omega)$ (rads)	$20\log\lvert X(\Omega)\rvert$ (dB) $20\log\lvert X(f)\rvert$ (dB)	$\theta(\Omega)$ (°) $\theta(f)$ (°)	f (Hz)
0	$1 + 0.1\angle 0 + 0.01\angle 0$	1.1100	0.0000	0.9065	0.0000	0
$\dfrac{\pi}{4}$	$1 + 0.1\angle-\dfrac{\pi}{4} + 0.01\angle-\dfrac{\pi}{2}$	1.0737	−0.0752	0.6181	−4.3108	1875
$\dfrac{\pi}{2}$	$1 + 0.1\angle-\dfrac{\pi}{2} + 0.01\angle-\pi$	0.9950	−0.1007	−0.0432	−5.7679	3750
$\dfrac{3\pi}{4}$	$1 + 0.1\angle-\dfrac{3\pi}{4} + 0.01\angle-\dfrac{3\pi}{2}$	0.9313	−0.0652	−0.6185	−3.7378	5625
π	$1 + 0.1\angle-\pi + 0.01\angle-2\pi$	0.9100	0.0000	−0.8192	0.0000	7500

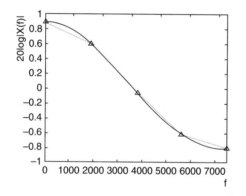
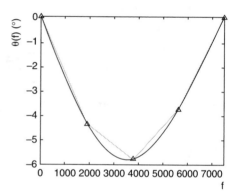

FIGURE 8.7
Magnitude and phase spectra for Example 8.2.

high frequencies, so the signal contains all frequencies in nearly equal measure. Because the sampling frequency is known, it is possible to see that spectral magnitudes are largest in the 0 to 2 kHz range.

EXAMPLE 8.3

A piece of the spoken vowel "eee," sampled at 8 kHz, is shown in Figure 8.8. The magnitude spectrum for this signal is shown in Figure 8.9. What are the main frequency components of $x[n]$?

It is obvious from Figure 8.8 that this vowel sound is quite regular, but not perfectly periodic. As the magnitude spectrum shows, the "eee" consists almost exclusively of 200 Hz and 400 Hz components.

eee.wav

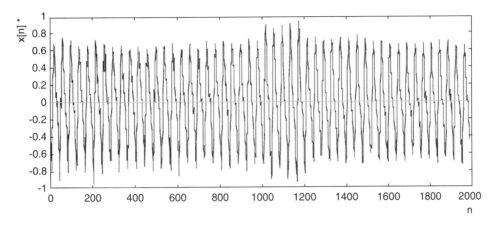

* For clarity, the envelope rather than the individual samples of the signal is plotted.

FIGURE 8.8
The vowel "eee" (2000 samples, $f_S = 8$ kHz) for Example 8.3.

FIGURE 8.9
Magnitude spectrum of "eee"
sound for Example 8.3.

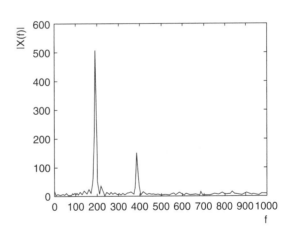

8.3   PERIODIC DIGITAL SIGNALS

Periodic signals are those that repeat at regular intervals for all time. Two examples are provided in Figure 8.10. The number of samples that occur in each interval is called the digital period of the signal. Because the same sequence repeats over all time, the discrete time Fourier transform is not an appropriate tool for calculating the spectrum: The infinite sum that is part of the DTFT would give an infinite result. The tool needed to find the spectrum for a periodic signal is the Fourier series for digital signals, otherwise known as the **discrete Fourier series (DFS)**. This series is particularly worth studying because of its close relationship with the discrete Fourier transform, to be studied in Chapter 11.

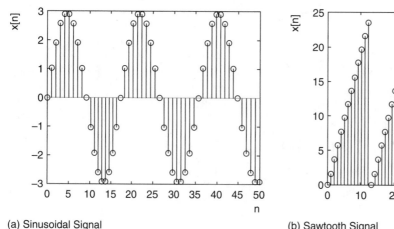

(a) Sinusoidal Signal

(b) Sawtooth Signal

FIGURE 8.10
Periodic digital signals.

According to Fourier theory, every periodic signal can be expressed as the sum of sines and cosines, or, more compactly, as the sum of complex exponentials. The Fourier series representation for a periodic digital signal $x[n]$ with period N is

$$x[n] = \frac{1}{N}\sum_{k=0}^{N-1} c_k e^{j2\pi \frac{k}{N}n} \tag{8.2}$$

where n is the sample number. The index k runs from 0 to $N-1$, but, in fact, any N consecutive values can be used. The **Fourier coefficients** c_k are calculated from the signal samples using

$$c_k = \sum_{n=0}^{N-1} x[n] e^{-j2\pi \frac{k}{N}n} \tag{8.3}$$

where k is the coefficient number. The multiplier $1/N$ in Equation (8.2) is a scaling factor that ensures the signal $x[n]$ can be reconstructed from its Fourier expansion, as shown in Appendix E. Since $x[n]$ has a period of N samples, only N samples of the signal need to be used to find the coefficients c_k for all k. The index n normally runs from 0 to $N-1$, but, since $x[n]$ is periodic, any N consecutive samples can be used. The coefficients c_k also have only N unique values, repeating thereafter. To see this, k can be replaced with $k + N$ in Equation (8.2), as follows:

$$c_{k+N} = \sum_{n=0}^{N-1} x[n] e^{-j2\pi \frac{k+N}{N}n} = \sum_{n=0}^{N-1} x[n] e^{-j2\pi \frac{k}{N}n} e^{-j2\pi n} = \sum_{n=0}^{N-1} x[n] e^{-j2\pi \frac{k}{N}n} = c_k$$

since $e^{-j2\pi n} = \cos(2\pi n) - j\sin(2\pi n) = 1$. This proves that, after N values have been calculated, the coefficients begin to repeat. Because there are only N unique coefficients, no more than N coefficient values are needed in Equation (8.2) to reconstruct $x[n]$ for all n.

In general, the c_k coefficients are complex numbers. They can be expressed in polar form as

$$c_k = |c_k| e^{j\phi_k} \tag{8.4}$$

Equation (8.2) then becomes

$$x[n] = \frac{1}{N} \sum_{k=0}^{N-1} |c_k| e^{j\phi_k} e^{j2\pi \frac{k}{N} n} = \sum_{k=0}^{N-1} \frac{|c_k|}{N} e^{j(2\pi \frac{k}{N} n + \phi_k)} \tag{8.5}$$

which is an alternative expression of the Fourier series expansion. The general digital sinusoid of the form $x[n] = A\cos(n\Omega + \phi)$ can be written, using Euler's identity $e^{j\theta} = \cos\theta + j\sin\theta$, as

$$x[n] = \text{Re}\{Ae^{j(\Omega n + \phi)}\}$$

This model and the terms of Equation (8.5) have several points of similarity. A comparison of the position of the amplitude A and the phase ϕ suggests that magnitude information is carried by $|c_k|/N$ and phase information is carried by ϕ_k. Comparing frequencies shows that the index k is proportional to frequency f. Specifically,

$$2\pi \frac{k}{N} n = \Omega n$$

$$2\pi \frac{k}{N} n = 2\pi \frac{f}{f_S} n$$

or,

$$f = \frac{k}{N} f_S \tag{8.6}$$

where the index k takes the values 0 to $N-1$. This equation provides the frequencies for all the components of the Fourier series. The **DC component** of the signal $|c_0|/N$ is given by $k = 0$; it is the average value of the signal. The other frequencies, given by $k > 0$, are called harmonics of the periodic signal, where $k = 1$ gives the signal's first harmonic, or fundamental frequency, f_S/N. This frequency is important because its reciprocal is the time in seconds to complete one full cycle of the signal. This time equals NT_S, where T_s is the sampling interval. The harmonics fall at frequencies that are integer multiples of the fundamental, according to Equation (8.6). As the equation suggests, the discrete Fourier series covers the frequencies from 0 to $(N-1)f_S/N$, so N discrete Fourier series coefficients address frequencies from zero up to nearly the sampling frequency.

Both the magnitude and phase of $x[n]$ are provided by the Fourier coefficients c_k. The magnitude spectrum for a periodic digital signal is obtained by plotting $|c_k|/N$ versus k, while the phase spectrum is obtained by plotting ϕ_k versus k. Both spectra are periodic with period N. As with the frequency responses studied in Chapter 7, the magnitude spectrum is always even and the phase spectrum is always odd. That is,

$$|c_{-k}| = |c_k|$$
$$\phi_{-k} = -\phi_k$$

There is an important difference between the spectra produced by the discrete Fourier series for periodic digital signals and those produced by the discrete time Fourier transform for nonperiodic digital signals. As seen in Section 8.2, the DTFT produces continuous spectra, which means that the spectra have values for every value of frequency. As a result, the magnitude and phase spectra for nonperiodic signals are smooth, unbroken lines. By contrast, the DFS produces only N points of a spectrum, covering the finite list of frequencies described by Equation (8.6). Thus, the magnitude and phase spectra for periodic signals are line spectra, vertical lines spaced at equal intervals. Equation (8.6) indicates that the spectral lines are f_S/N Hz apart when the spectra are plotted against frequency f instead of index k. Figure 8.11 compares and contrasts the two Fourier tools.

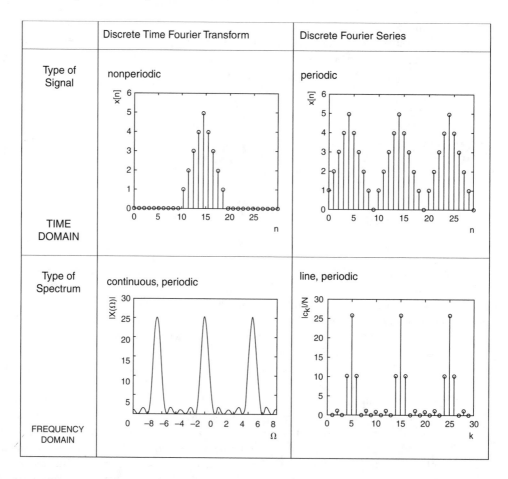

FIGURE 8.11

Comparing DTFTs and DFSs for digital signals.

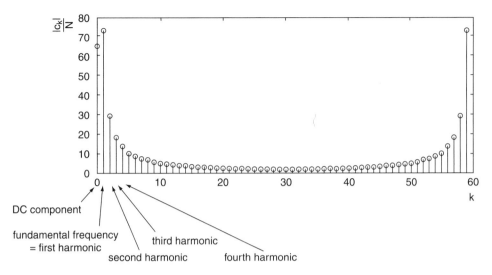

FIGURE 8.12
Magnitude spectrum for a periodic signal.

A sample magnitude spectrum for a periodic signal of period $N = 60$ is shown in Figure 8.12. The horizontal axis is marked with k, the index for the DFS. When a digital signal repeats every N samples, its fundamental frequency is f_S/N, or $f_S/60$ Hz. This frequency, also considered the first harmonic frequency, corresponds to the first non-DC line in the spectrum, marked in the diagram. All of the other lines in the spectrum mark the other harmonics, at integer multiples of the fundamental frequency. In the spectrum of Figure 8.12, all harmonics are present. For example, the frequency of the second harmonic is $2f_S/N$, or $f_S/30$ Hz, and the frequency of the $(N-1)$st harmonic is $(N-1)f_S/N$, or $59f_S/60$ Hz. Not all periodic signals show all harmonics. For example, some spectra contain odd harmonics only, with even components set to zero, while others have zero contributions at just a few of the harmonic frequencies.

EXAMPLE 8.4

Find the magnitude and phase spectra for the periodic square wave signal shown in Figure 8.13. Since this digital signal is periodic, the discrete Fourier series is the tool of choice for finding the magnitude and phase spectra. The period $N = 8$. The Fourier coefficients are calculated from Equation (8.3):

$$c_k = \sum_{n=0}^{N-1} x[n]e^{-j2\pi\frac{k}{N}n} = \sum_{n=0}^{7} x[n]e^{-j2\pi\frac{k}{8}n}$$

$$= x[0] + x[1]e^{-j2\pi\frac{k}{8}} + x[2]e^{-j2\pi\frac{k}{8}2} + x[3]e^{-j2\pi\frac{k}{8}3} + x[4]e^{-j2\pi\frac{k}{8}4}$$

$$+ x[5]e^{-j2\pi\frac{k}{8}5} + x[6]e^{-j2\pi\frac{k}{8}6} + x[7]e^{-j2\pi\frac{k}{8}7}$$

$$= 1 + e^{-j\frac{\pi k}{4}} + e^{-j\frac{\pi k}{2}} + e^{-j\frac{3\pi k}{4}}$$

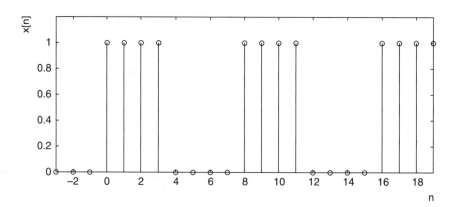

FIGURE 8.13
Digital signal for Example 8.4.

Table 8.4 shows c_k for k from 0 to 7, and the corresponding values of the magnitude spectrum $|c_k|/N$ and the phase spectrum ϕ_k. Other values of k can be used as well, but the same magnitude and phase values will repeat, with period 8. The spectra are plotted in Figure 8.14 for $k = -4 \ldots 12$. The signal contains a DC component ($k = 0$) and odd harmonics ($k = 1, 3, 5, 7$), but no even harmonics. The signal in this example has a

TABLE 8.4
Calculations for Example 8.4

k	c_k	$|c_k|/8$	ϕ_k
0	$1 + 1\lfloor 0 + 1\lfloor 0 + 1\lfloor 0$	0.5000	0.0
1	$1 + \left\lfloor -\dfrac{\pi}{4} + 1 \right\rfloor -\dfrac{\pi}{2} + 1 \left\lfloor -\dfrac{3\pi}{4} \right.$	0.3266	−1.1781
2	$1 + 1\left\lfloor -\dfrac{\pi}{2} + 1\lfloor -\pi + 1 \right\rfloor -\dfrac{3\pi}{2}$	0.0	0.0
3	$1 + 1\left\lfloor -\dfrac{3\pi}{4} + 1 \right\rfloor -\dfrac{3\pi}{2} + 1\left\lfloor -\dfrac{3\pi}{2} \right.$	0.1353	−0.3927
4	$1 + 1\lfloor -\pi + 1\lfloor -2\pi + 1\lfloor -3\pi$	0.0	0.0
5	$1 + 1\left\lfloor -\dfrac{5\pi}{4} + 1 \right\rfloor -\dfrac{5\pi}{2} + 1\left\lfloor -\dfrac{15\pi}{4} \right.$	0.1353	0.3927
6	$1 + 1\left\lfloor -\dfrac{3\pi}{2} + 1\lfloor -3\pi + 1 \right\rfloor -\dfrac{9\pi}{2}$	0.0	0.0
7	$1 + 1\left\lfloor -\dfrac{7\pi}{4} + 1 \right\rfloor -\dfrac{7\pi}{2} + 1\left\lfloor -\dfrac{21\pi}{4} \right.$	0.3266	1.1781

square wave nature like the one in Example 8.1. Therefore, its magnitude spectrum shows the same sinc function shape noted there. Notice that the signal in Figure 8.13 and the spectra in Figure 8.14 repeat with the same period.

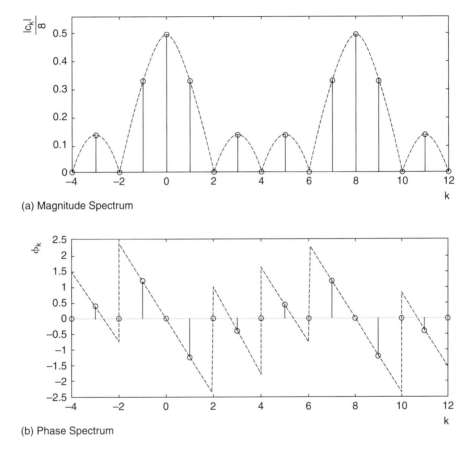

(a) Magnitude Spectrum

(b) Phase Spectrum

FIGURE 8.14
Spectrum of signal for Example 8.4.

The signal in Example 8.4 is a periodic version of the nonperiodic signal studied in Example 8.1. A comparison of the spectra for the two signals shows that the line spectra of the periodic signal in Figure 8.14 are sampled versions of the continuous spectra of the nonperiodic signal in Figure 8.5. This is a general relationship, as is suggested by Figure 8.11.

EXAMPLE 8.5

Find magnitude and phase spectra for $x[n] = \sin(n\pi/5)$. The sampling rate is 1 kHz.

The sample values are listed in Table 8.5, and the signal is plotted in Figure 8.15. Recall from Section 3.3.4 that the digital sinusoid will repeat if $2\pi/\Omega$ can be written as a ratio of integers N/M. In this case, $\Omega = \pi/5$, so $2\pi/\Omega = 10$, or 10/1, so the digital signal is periodic. The integer in the numerator gives the digital period N, so $N = 10$ for the discrete Fourier series calculations.

When the sample values are substituted into Equation (8.3), values of c_k may be found for any value of k, and, from c_k, the magnitude spectrum values $|c_k|/N$ and phase spectrum

TABLE 8.5
Signal Sample Values for Example 8.5

n	$x[n]$
0	0.0000
1	0.5878
2	0.9511
3	0.9511
4	0.5878
5	0.0000
6	−0.5878
7	−0.9511
8	−0.9511
9	−0.5878

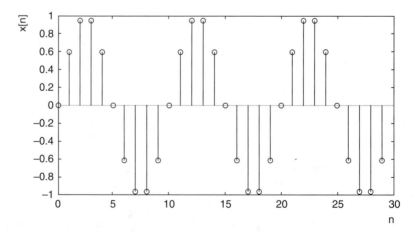

FIGURE 8.15
Digital sine wave for Example 8.5.

values ϕ_k may be determined from Equation (8.4) using polar calculations. Table 8.6 shows c_k, as well as the corresponding spectrum values, computed from

$$c_k = \sum_{n=0}^{N-1} x[n]e^{-j2\pi\frac{k}{N}n} = \sum_{n=0}^{9} x[n]e^{-j2\pi\frac{k}{10}n}$$

Plotting more than N values demonstrates the periodicity of the discrete Fourier series. In Figure 8.16, values for k from -4 to 25 are shown. This figure confirms that both the magnitude and phase spectra are periodic with period N, where $N = 10$ in this case. Since the sampling rate f_S is known to be 1 kHz, the 10 unique discrete Fourier series coefficients correspond according to Equation (8.6) to the stepped frequencies kf_S/N, or $100k$ Hz, where k

TABLE 8.6
Fourier Coefficients for Example 8.5

k	c_k	$|c_k|/N$	ϕ_k
0	0.0000	0.0000	0.0000
1	$-j5.0000$	0.50000	-1.5708
2	0.0000	0.0000	0.0000
3	0.0000	0.0000	0.0000
4	0.0000	0.0000	0.0000
5	0.0000	0.0000	0.0000
6	0.0000	0.0000	0.0000
7	0.0000	0.0000	0.0000
8	0.0000	0.0000	0.0000
9	$j5.0000$	0.5000	1.5708

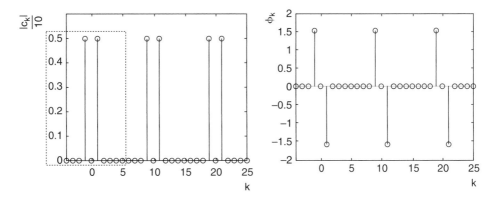

FIGURE 8.16
Spectrum of sine wave for Example 8.5.

ranges from 0 to $N-1$, that is, from 0 to 9. As the magnitude spectrum shows, the only nonzero components of this signal occur for $k = 1$ and $k = 9$. These k values correspond to frequencies 100 and 900 Hz. In a system employing a 1 kHz sampling rate, signal frequencies should be limited to 500 Hz. Thus, the 900 Hz spike is a spectral image of the 100 Hz signal spike. Only one true frequency remains in the spectrum, a basic feature of all sinusoids. In fact, the 100 Hz finding is not unexpected. The signal $x[n] = \sin(n\pi/5)$ has a digital frequency $\Omega = \pi/5$ radians. With a sampling frequency f_S of 1 kHz, Equation (7.8) identifies the analog frequency that corresponds to this digital frequency as $f = \Omega f_S/(2\pi) = 100$ Hz.

To verify that the Fourier coefficients correctly specify the signal, the spectrum values from Table 8.6 can be substituted into the Fourier expansion in Equation (8.5). Most of the spectrum values are zero in this example, which makes the calculations straightforward. The expansion is

$$x[n] = \sum_{k=0}^{N-1} \frac{|c_k|}{N} e^{j(2\pi \frac{k}{N} n + \phi_k)} = \frac{|c_1|}{10} e^{j(\frac{2\pi}{10} n + \phi_1)} + \frac{|c_9|}{10} e^{j(\frac{2\pi 9}{10} n + \phi_9)}$$

$$= 0.5 e^{j(\frac{\pi n}{5} - \frac{\pi}{2})} + 0.5 e^{j(\frac{9\pi n}{5} + \frac{\pi}{2})} = 0.5 e^{j(\frac{\pi n}{5} - \frac{\pi}{2})} + 0.5 e^{j\frac{10\pi n}{5}} e^{j(-\frac{\pi n}{5} + \frac{\pi}{2})}$$

$$= 0.5(e^{j(\frac{\pi n}{5} - \frac{\pi}{2})} + e^{j2\pi n} e^{-j(\frac{\pi n}{5} - \frac{\pi}{2})})$$

$$= 0.5(e^{j(\frac{\pi n}{5} - \frac{\pi}{2})} + e^{-j(\frac{\pi n}{5} - \frac{\pi}{2})})$$

since, by Euler's identity, $e^{j2\pi n} = \cos(2\pi n) + j\sin(2\pi n) = 1$ for any value of n. Using Euler's form of the cosine, $\cos\theta = (e^{j\theta} + e^{-j\theta})/2$ and the trigonometric identity $\cos(x-\pi/2) = \sin(x)$ gives

$$x[n] = 0.5(e^{j(\frac{n\pi}{5} - \frac{\pi}{2})} + e^{-j(\frac{n\pi}{5} - \frac{\pi}{2})}) = \cos\left(\frac{n\pi}{5} - \frac{\pi}{2}\right) = \sin\left(\frac{n\pi}{5}\right)$$

which matches the original function.

EXAMPLE 8.6

Find magnitude and phase spectrum for $x[n] = 1 + \sin(n\pi/2) + \cos(n\pi/4)$.

This problem may be handled in two ways. The most straightforward, but most tedious, way is to follow the method of Example 8.5. In this method, the period N of $x[n]$ is found, N sample values are computed for $x[n]$, and N coefficients c_k are calculated. An alternative method is available for signals that contain only sinusoids and constants. It involves rewriting the signal in terms of complex exponentials and finding coefficients by inspection. Both methods give the same results.

The digital period of $x[n]$ can be deduced by checking each term. The first term, 1, can be thought of as having a digital period of 1, since it is a constant and thus repeats every sample. The period for each sinusoidal term can be found by computing the ratio $2\pi/\Omega$. If the ratio can be expressed in the form N/M, then N is the digital period. For the second term, the ratio $2\pi/\Omega = 2\pi/(\pi/2) = 4$, which marks a periodic signal with period 4. For the third term, the ratio $2\pi/\Omega = 2\pi/(\pi/4) = 8$, indicating a digital period of 8. Thus, the second sequence repeats twice as often as the third. The lowest common multiple of the digital periods for each term gives the digital period for the signal $x[n]$. In this case, the signal repeats every 8 samples.

Using the Euler forms of the sine and cosine,

$$x[n] = 1 + \frac{e^{j\frac{n\pi}{2}} - e^{-j\frac{n\pi}{2}}}{2j} + \frac{e^{j\frac{n\pi}{4}} + e^{-j\frac{n\pi}{4}}}{2}$$

To make the signal as similar as possible to the Fourier representation of Equation (8.2), the exponents of all exponential terms must have denominator N, that is, 8. With the exponents written in this form, and using the fact that $1/j = -j$, $x[n]$ becomes

$$x[n] = \frac{j}{2}e^{-j\frac{n4\pi}{8}} + \frac{1}{2}e^{-j\frac{n2\pi}{8}} + 1e^{j0} + \frac{1}{2}e^{j\frac{n2\pi}{8}} - \frac{j}{2}e^{j\frac{n4\pi}{8}}$$

$$= \frac{1}{8}\left(4je^{-j2\pi\frac{2}{8}n} + 4e^{-j2\pi\frac{1}{8}n} + 8e^{j2\pi\frac{0}{8}n} + 4e^{j2\pi\frac{1}{8}n} - 4je^{j2\pi\frac{2}{8}n}\right) \qquad (8.7)$$

In this form, $x[n]$ has a lot in common with the Fourier series expansion

$$x[n] = \frac{1}{N}\sum_{k=0}^{N-1}c_k e^{j2\pi\frac{k}{N}n} = \frac{1}{8}\sum_{k=0}^{7}c_k e^{j2\pi\frac{k}{8}n}$$

Though N consecutive coefficients c_k must be used, the range need not be 0 through 7. Since the values of k in the exponents of Equation (8.7) range from -2 to 2, a more convenient choice in this case is

$$x[n] = \frac{1}{8}\sum_{k=-3}^{4}c_k e^{j2\pi\frac{k}{8}n} \qquad (8.8)$$

Equations (8.7) and (8.8) have exactly parallel structures. By inspection, the coefficients can be identified:

$$c_{-3} = 0 \quad c_{-2} = 4j \quad c_{-1} = 4 \quad c_0 = 8 \quad c_1 = 4 \quad c_2 = -4j \quad c_3 = 0 \quad c_4 = 0$$

Magnitude and phase values can now be computed according to Equation (8.4). They are listed in Table 8.7. Plots are presented in Figure 8.17.

TABLE 8.7

Table of Spectral Values for Example 8.6

k	c_k	$|c_k|/8$	ϕ_k
-3	0	0.0	0
-2	$4j$	0.5	$\pi/2$
-1	4	0.5	0
0	8	1.0	0
1	4	0.5	0
2	$-4j$	0.5	$-\pi/2$
3	0	0.0	0
4	0	0.0	0

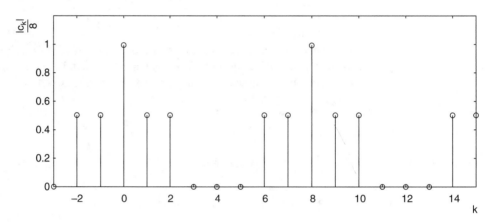

(a) Magnitude Spectrum

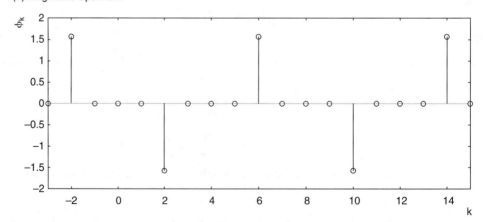

(b) Phase Spectrum

FIGURE 8.17

Spectrum of signal for Example 8.6.

EXAMPLE 8.7

The vowel sound "eee," of Example 8.3, has mostly 200 Hz and 400 Hz content for the female speaker who recorded it. The 200 Hz component is about four times as strong as the 400 Hz component. Speech synthesis, discussed in more detail in Section 14.3, means artificially producing speech sounds that sound like natural speech. An artificial "eee" sound can be approximated with a signal that contains a strong 200 Hz component and a weaker 400 Hz component. With an $f_S = 8$ kHz sampling rate, the expression $\Omega = 2\pi f/f_S$ gives the digital frequencies for each sine wave term, $2\pi(200/8000) = \pi/20$ and $2\pi(400/8000) = \pi/10$. With amplitudes that make the 200 Hz component strongest, the digital signal becomes

$$x[n] = \frac{2}{3}\sin\left(n\frac{\pi}{20}\right) + \frac{1}{6}\sin\left(n\frac{\pi}{10}\right)$$

For the first term, $2\pi/\Omega = 2\pi/(\pi/20) = 40$, indicating a digital period of 40 samples. For the second term, the digital period is 20 samples, exactly half that of the first. Thus, the signal $x[n]$ that combines the two terms is purely periodic and repeats every $N = 40$ samples. It is plotted in Figure 8.18 as a smooth curve rather than a digital signal for ease of comparison with Figure 8.8. The spectrum can be found with the discrete Fourier series, following the same method as used in Example 8.6. The magnitude spectrum is plotted in Figure 8.19 against frequencies in Hz calculated from $f = kf_S/N = k(8000/40) = 200k$ for $k = 0$ to 5. It is clearly very similar to the spectrum of the original sound in Figure 8.9. Important sound qualities are missing, however, as is evident upon listening to the reconstructed signal. This simple exercise gives a hint of the difficulties attending the synthesis of natural-sounding speech.

fakeee.wav

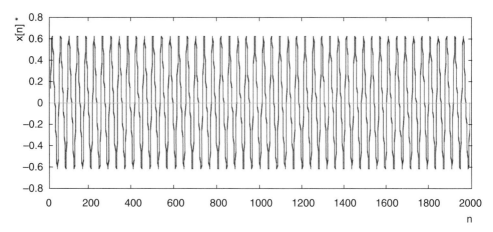

FIGURE 8.18
Artificial "eee" signal for Example 8.7.

FIGURE 8.19
Magnitude spectrum of artificial
"eee" sound for Example 8.7.

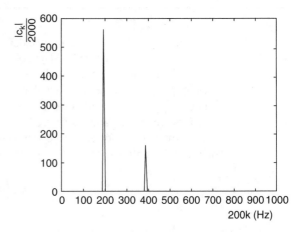

EXAMPLE 8.8

Figure 8.21 displays the magnitude spectrum for the repetitive square wave of Figure 8.20, periodic with a period of 256 samples. The spectrum is obtained by finding the magnitudes for the c_k using Equation (8.3). Remember that the section of the spectrum appearing in Figure 8.21 repeats endlessly in both directions, since the DFS is periodic with period $N = 256$, producing the expected even magnitude spectrum.

Figure 8.22 plots a narrower range of indices to allow a closer view of the frequency content of the square wave. The lines indicate that specific, individual frequencies make up the square wave. Furthermore, only odd harmonics are present, and these are present at

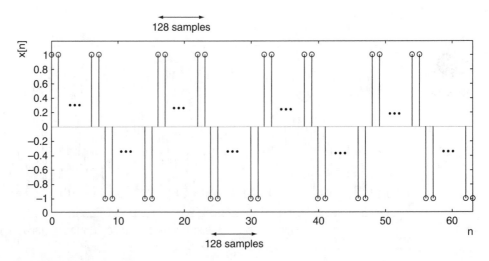

FIGURE 8.20
Periodic square wave for Example 8.8.

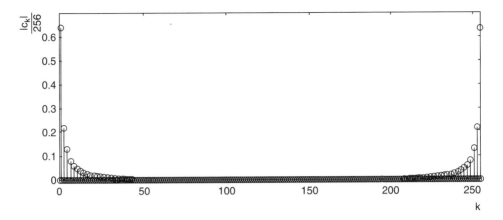

FIGURE 8.21
Magnitude spectrum of square wave for Example 8.8.

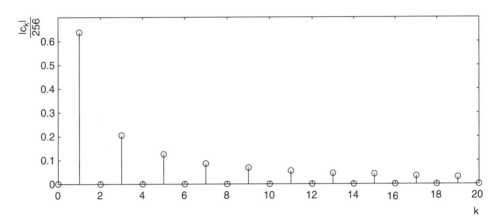

FIGURE 8.22
First few harmonic components of square wave for Example 8.8.

ever-decreasing strengths. Specifically, the third, fifth, and seventh harmonics are 1/3, 1/5, and 1/7 of the amplitude of the first. This confirms the analysis of Section 8.1 of this chapter, which showed that a square wave can be constructed from odd harmonic sine waves.

EXAMPLE 8.9

A digital system collects samples at the rate of 20 ksamples/sec. The digital signal obtained is described by the function

$$x[n] = \cos\left(n\frac{\pi}{6}\right) + \cos\left(n\frac{3\pi}{7}\right) + \cos\left(n\frac{2\pi}{3}\right)$$

Plot the magnitude spectrum for this digital signal.

The digital frequencies of the terms are $\Omega_1 = \pi/6$, $\Omega_2 = 3\pi/7$, and $\Omega_3 = 2\pi/3$ rads. With $2\pi/\Omega_1 = 12$, $2\pi/\Omega_2 = 14/3$, and $2\pi/\Omega_3 = 3$, all ratios of integers N/M, the periods of the three terms may be identified as 12, 14, and 3 samples. Because each component is periodic, the signal is periodic as well. The period may be found by finding the lowest common multiple of the three periods. For the digital periods 12, 14, and 3, the lowest common multiple is 84, which is the digital period N of $x[n]$. Equation (7.8) provides the analog frequencies for each term in $x[n]$. For the first term,

$$f = \Omega \frac{f_S}{2\pi} = \frac{\pi}{6} \frac{20000}{2\pi} = 1667 \text{ Hz}$$

For the second and third terms, the analog frequencies are 4286 and 6667 Hz.

Whether Equation (8.3) is used to find the spectrum, as in Example 8.5, or a Euler expansion is used, as in Example 8.6, the result is the same. The 84 points of the spectrum are shown in Figure 8.23. Because the signal $x[n]$ comprises three sinusoids, the three peaks in the spectrum are no surprise. Recall that the frequency of each Fourier series component is given by Equation (8.6) as $f = kf_S/N$, where $k = 0, \ldots, N-1$, so the discrete Fourier series covers the frequencies from 0 to $(N-1)f_S/N$ Hz. In this example, $N = 84$ and $f_S = 20$ kHz, so the $k = 0$ component corresponds to 0 Hz and the $k = 84$ component corresponds to the frequency $(N-1)f_S/N = 83(20)/84 = 19.762$ kHz. Frequencies within the Nyquist band, in the range 0 to $f_S/2 = 10$ kHz, are represented by c_0 to c_{42}. The peaks in the magnitude spectrum occur at $k = 7$, 18, and 28. These spectral peaks correspond to frequencies of $f = kf_S/N = k(20/84)$, or 1.667, 4.286, and 6.667 kHz, identical with the true frequencies, calculated above. The remaining three peaks, at $k = 84 - 7$, $84 - 18$, and $84 - 28$, or $k = 77$, 66, and 56, are spectral images of the three frequencies already identified.

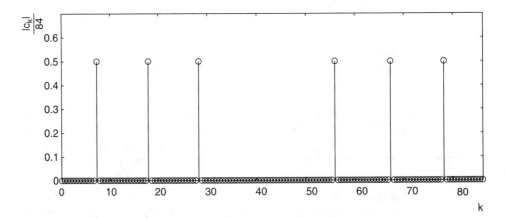

FIGURE 8.23
Magnitude spectrum for Example 8.9.

CHAPTER SUMMARY

Matlab
Support

1. The spectrum of a signal presents the signal's frequency content.
2. A spectrum comprises a magnitude spectrum and a phase spectrum.
3. The spectrum for a nonperiodic signal $x[n]$ is calculated using the discrete time Fourier transform (DTFT). The magnitude spectrum may be plotted as $|X(\Omega)|$ versus digital frequency Ω, or as $|X(f)|$ versus analog frequency f. The phase spectrum may be plotted as $\theta(\Omega)$ versus Ω, or as $\theta(f)$ versus f.
4. The magnitude and phase spectra for nonperiodic signals are smooth, continuous functions of frequency.
5. It is sufficient to plot a nonperiodic signal's spectrum for the digital frequencies between 0 and π radians, as $\Omega = \pi$ radians corresponds to the Nyquist frequency.
6. The magnitude spectrum of a rectangular pulse has the shape of a sinc function.
7. The spectrum of a periodic signal $x[n]$ with period N is calculated using the discrete Fourier series (DFS). The DFS coefficients are calculated as

$$c_k = \sum_{n=0}^{N-1} x[n] e^{-j2\pi \frac{k}{N} n}$$

The coefficients c_k are complex numbers and may be written in the form

$$c_k = |c_k| e^{j\phi_k}$$

The magnitude spectrum is plotted as $|c_k|$ versus k, and the phase spectrum is plotted as ϕ_k versus k.

8. The DFS is periodic with period N.
9. The DFS index k corresponds to the frequency

$$f = k \frac{f_S}{N}$$

The $k = 0$ coefficient gives the DC component of the signal. The $k > 0$ coefficients give the harmonic elements of the signal. The fundamental frequency is given by $k = 1$.

10. The magnitude and phase spectra for periodic signals are line functions, with contributions only at DC and harmonic frequencies.
11. The envelope of the magnitude spectrum for a square wave has the shape of a sinc function.

REVIEW QUESTIONS

8.1 Find the magnitude and phase spectra for the signal $x[n] = \delta[n] + 0.5\delta[n-1] + 0.25\delta[n-2]$. Plot linear magnitudes, and phases in radians, against digital frequencies in radians.

8.2 Plot the magnitude spectrum in dB versus Hz for the signal

$$x[n] = -0.5\delta[n-1] + u[n-3] - u[n-5]$$

sampled at 8 kHz.

8.3 Find the magnitude spectrum of the nonperiodic signal shown in Figure 8.24.

FIGURE 8.24

Signal for Question 8.3.

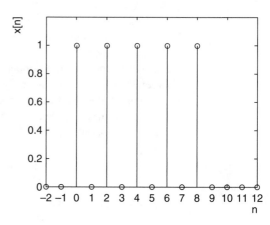

8.4 Find approximate magnitude and phase spectra for the signal $x[n] = (-4)^{-n}u[n]$.

8.5 Compare the magnitude and phase spectra for the signals $x_1[n]$, $x_2[n]$, and $x_3[n]$ shown in Figure 8.25.

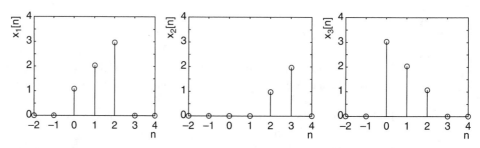

FIGURE 8.25

Signals for Question 8.5.

8.6 Plot the magnitude spectrum of the signal $x[n] = \sin(n2\pi/3)(u[n] - u[n-5])$.

8.7 Find the spectrum of the signal whose repeating pattern is shown in Figure 8.26.

8.8 **a.** Plot 12 points of the magnitude and phase spectra for the signal $x[n]$ shown in Figure 8.27. Assume the pattern evident in the figure continues. Plot gains in dB and phases in degrees.

b. With what digital period do the spectra in (a) repeat?

FIGURE 8.26

Signal for Question 8.7.

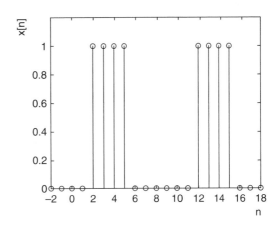

FIGURE 8.27

Signal for Question 8.8.

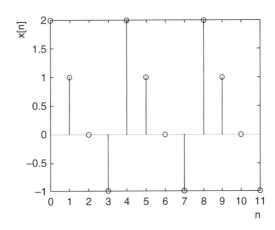

8.9 A group of samples $[1, 0, -1, 0]$ repeats with a period of 4 to produce a signal $x[n]$. Plot magnitude and phase spectra for the signal.

8.10 What are the major differences between the spectra of nonperiodic and periodic signals?

8.11 a. Find the magnitude and phase spectra for the signal in Figure 8.28(a).
   b. How do the magnitude and phase spectra for the signal in Figure 8.28(b) compare to the result in (a)?

8.12 Identify the first five harmonic frequencies in Hz for a signal sampled at 12 kHz that repeats every 72 samples.

8.13 Plot and interpret the magnitude spectrum of the signal

$$x[n] = 2\cos\left(n\frac{2\pi}{3}\right) + \cos\left(n\frac{\pi}{3}\right)$$

for a sampling frequency of 4 kHz. Use linear gains.

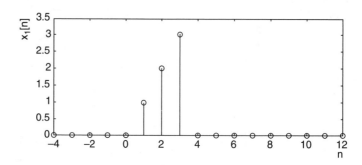

(a) Signal $x_1[n]$

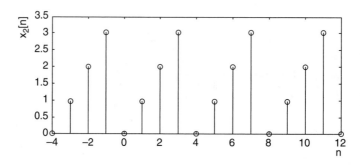

(b) Signal $x_2[n]$

FIGURE 8.28
Signals for Question 8.11.

8.14 A digital signal is created by sampling a sum of three sine waves. Its magnitude spectrum is given in Figure 8.29.

    a. What is the period of the digital signal (measured in number of samples)?

FIGURE 8.29
Magnitude spectrum for Question 8.14.

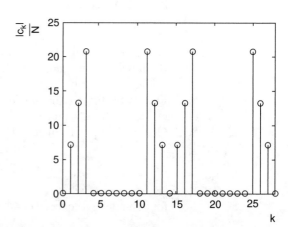

    b. If the sampling rate is 16 kHz, what are the analog frequencies of the three sine waves?

8.15 A periodic digital signal is obtained by sampling an analog signal at 12 kHz. One period of its magnitude spectrum is shown in Figure 8.30.

    a. What is the period of the digital signal, measured in number of samples?

    b. What frequencies in Hz were present in the analog signal?

    c. What are the digital frequencies of the components of the digital signal?

FIGURE 8.30

Magnitude spectrum for Question 8.15.

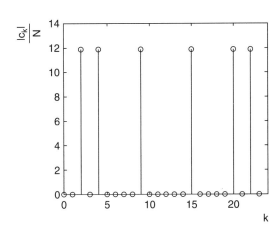

8.16 The complex Fourier series expansion for a periodic signal with period $N = 16$ is

$$x[n] = e^{-j\frac{\pi n}{4}} - \frac{j}{2}e^{-j\frac{\pi n}{8}} + 1 + \frac{j}{2}e^{j\frac{\pi n}{8}} + e^{j\frac{\pi n}{4}}$$

Draw magnitude and phase spectra for $x[n]$. Use linear gains, and phases in radians.

8.17 For each of the following signals, sampled at 10 kHz,

    (i) Find the digital period.

    (ii) Without doing Fourier series calculations, sketch an approximate magnitude spectrum. Use linear gains.

    a. $x[n] = \sin\left(n\frac{6\pi}{7}\right)$

    b. $x[n] = \cos\left(n\frac{3\pi}{5}\right)$

    c. $x[n] = \sin\left(\frac{n\pi}{3}\right) + \cos\left(n\frac{\pi}{8}\right)$

    d. $x[n] = 2\sin\left(n\frac{\pi}{3}\right) + \cos\left(n\frac{\pi}{8}\right)$

8.18 The DTFT magnitude spectrum for a signal $x[n]$ sampled at 20 kHz is shown in Figure 8.31. Describe the signal.

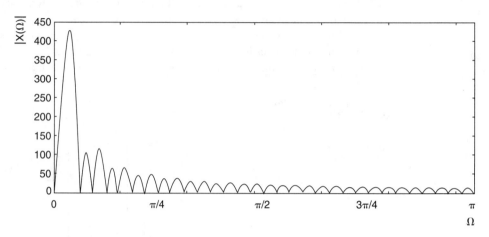

FIGURE 8.31
Magnitude spectrum for Question 8.18.

8.19 One period of the DFS magnitude spectrum for a signal $x[n]$ sampled at 20 kHz is shown in Figure 8.32. Describe the signal. Compute the period of the signal in both samples and seconds, and list the harmonic frequencies.

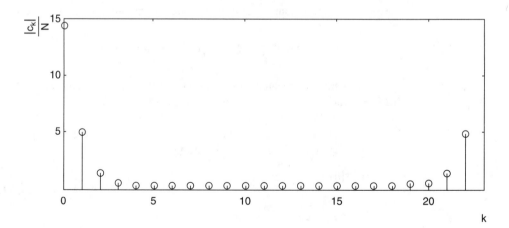

FIGURE 8.32
Magnitude spectrum for Question 8.19.

8.20 A digital signal sampled at 16 kHz is described as $x[n] = -0.5\delta[n] + 0.5\delta[n-1]$.

    a. Draw the magnitude spectrum for the signal.

    b. An ideal low pass filter with a cut-off frequency of 2 kHz is applied to the signal. Draw the spectrum of the filter output.

    c. An ideal band pass filter with cut-off frequencies of 3 and 6 kHz is applied to the signal. Draw the spectrum of the filter output.

    d. An ideal high pass filter with a cut-off frequency of 7 kHz is applied to the signal. Draw the spectrum of the filter output.

8.21 A digital signal is described as $x[n] = \sin(n\pi/7) + \sin(n2\pi/3) + \sin(n\pi/8)$, for a sampling frequency of 16 kHz.

    a. Draw the magnitude spectrum for the signal without performing Fourier series calculations.

    b. An ideal low pass filter with a cut-off frequency of 2 kHz is applied to the signal. Draw the spectrum of the filter output.

    c. An ideal band pass filter with cut-off frequencies of 3 and 6 kHz is applied to the signal. Draw the spectrum of the filter output.

    d. An ideal high pass filter with a cut-off frequency of 7 kHz is applied to the signal. Draw the spectrum of the filter output.

8.22 A periodic square wave is sampled at 4 kHz. One period of its magnitude spectrum is shown in Figure 8.33.

    a. What is the fundamental frequency of the square wave?

    b. What is the period of the square wave in seconds?

    c. What is the average value of the square wave?

    d. If the square wave is filtered by a high order low pass filter with a cut-off frequency of 500 Hz, what frequencies will be present in the filtered output?

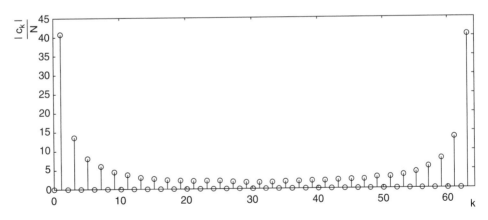

FIGURE 8.33

Magnitude spectrum for Question 8.22.

9

FINITE IMPULSE RESPONSE FILTERS

This chapter introduces a major class of digital filters: finite impulse response, or FIR, filters. The chapter:

- ➤ reviews the difference equation, impulse response, transfer function, and frequency response for a nonrecursive filter
- ➤ shows how to design moving average filters
- ➤ discusses phase delay and phase distortion
- ➤ presents the impulse response for an ideal low pass filter
- ➤ defines important filter features, including pass band ripple, stop band ripple, transition width, and bandwidth
- ➤ defines FIR windows, including rectangular, Hanning, Hamming, Blackman, and Kaiser windows
- ➤ provides guidelines for choosing window type and filter order
- ➤ lists low pass FIR filter design steps
- ➤ describes how to design band pass and high pass FIR filters
- ➤ provides a band stop FIR filter design method
- ➤ completes low pass, band pass, high pass, and band stop FIR filter designs
- ➤ discusses equiripple FIR filter design
- ➤ highlights practical considerations, including quantization and overflow

9.1 FINITE IMPULSE RESPONSE FILTER BASICS

The majority of the important underlying theory for DSP has now been covered, and in this chapter, attention is turned to the design of digital filters. **Finite impulse response filters**, also known as nonrecursive filters, rely only on past input information, never on past output information. They represent a special case of the general difference equation introduced in Section 4.4. For a nonrecursive filter, the difference equation takes the form

$$y[n] = b_0 x[n] + b_1 x[n-1] + b_2 x[n-2] + \ldots + b_M x[n-M]$$

Written more compactly,

$$y[n] = \sum_{k=0}^{M} b_k x[n-k] \tag{9.1}$$

Implementations for this type of filter were discussed in Section 4.6.1. Because each new filter output $y[n]$ does not rely on previous outputs, the impulse response for a nonrecursive filter is guaranteed to have a finite number of terms. The impulse response is

$$h[n] = \sum_{k=0}^{M} b_k \delta[n-k]$$

It consists of $M+1$ impulse functions, weighted by the coefficients b_k. Because nonrecursive filters have finite impulse responses, they are frequently known as finite impulse response, or FIR, filters. Example 9.1 reviews this point.

EXAMPLE 9.1

Find the impulse response for the nonrecursive filter

$$y[n] = x[n] - 0.5\, x[n-1] + 0.3\, x[n-2]$$

The impulse response is found using

$$h[n] = \delta[n] - 0.5\, \delta[n-1] + 0.3\, \delta[n-2]$$

Table 9.1 shows its first five samples. The difference equation uses past inputs up to two steps in the past to compute a new output. At the time of the third step, the impulse input,

TABLE 9.1
Impulse Response for Example 9.1

n	$h[n]$
0	1
1	−0.5
2	0.3
3	0
4	0

which occurred at $n = 0$, is more than two steps behind and is no longer referenced by the difference equation. Thus, the impulse response persists only while the impulse input is within range of the difference equation.

The design of FIR filters requires choosing the b_k coefficients in Equation (9.1) to give the desired filter behavior with as few coefficients as possible. The FIR structure provides a great deal of flexibility in the choice of filter shape. Generally, the steeper the desired filter roll-off, the larger the number of coefficients required. FIR filters quite often require 100 to 200 coefficients for acceptable performance. Recursive filters, which will be studied in the next chapter, generally require fewer coefficients, but nonrecursive filters do offer certain advantages that cannot be provided by recursive filters. Specifically, nonrecursive filters are guaranteed to be stable, and they eliminate pass band phase distortion, making them ideal for audio applications.

A z transform expression for an FIR filter can be found by using the time shift property of the z transform established in Section 6.1, that is, $\mathbf{Z}\{x[n - k]\} = z^{-k}X(z)$. The z transform of the difference equation in Equation (9.1) is

$$Y(z) = \sum_{k=0}^{M} b_k z^{-k} X(z)$$

Rearranging this equation gives the transfer function for the filter in the z domain:

$$H(z) = \frac{Y(z)}{X(z)} = \sum_{k=0}^{M} b_k z^{-k} = b_0 + b_1 z^{-1} + b_2 z^{-2} + \ldots + b_{M-1} z^{-(M-1)} + b_M z^{-M}$$

Multiplying top and bottom by z^M:

$$H(z) = \frac{b_0 z^M + b_1 z^{M-1} + b_2 z^{M-2} + \ldots + b_{M-1} z + b_M}{z^M}$$

Thus, an FIR filter with $M+1$ coefficients b_k has M zeros and M poles, since both the numerator and denominator polynomials have degree M. The expression makes it easy to see that nonrecursive filters can have poles only at $z = 0$. In fact, there are exactly M poles at $z = 0$. Since all of these poles are inside the unit circle, all filters of this form are stable.

Filter shape may be determined, as usual, by finding the magnitude of the frequency response for the filter, using the discrete time Fourier transform (DTFT):

$$H(\Omega) = \sum_{k=0}^{M} b_k e^{-jk\Omega}$$

In Section 9.3, the DTFT is used to demonstrate the phase properties of nonrecursive filters.

9.2 MOVING AVERAGE FILTERS REVISITED

Moving average filters, discussed in Section 5.3, are nonrecursive filters that compute a running average of a sequence of digital samples. This section will examine their transfer

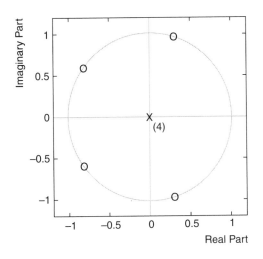

(a) Pole-Zero Diagram

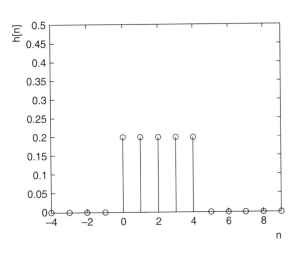

(b) Impulse Response

FIGURE 9.1

Five-term moving average filter.

functions and frequency responses, as well as suggest a design method. As an example, the pole-zero diagram and the impulse response for a moving average filter with five terms are shown in Figure 9.1. The difference equation for the filter is

$$y[n] = 0.2(x[n] + x[n-1] + x[n-2] + x[n-3] + x[n-4])$$

Therefore, replacing $y[n]$ by $h[n]$ and $x[n]$ by $\delta[n]$, the equation for the impulse response is

$$h[n] = 0.2(\delta[n] + \delta[n-1] + \delta[n-2] + \delta[n-3] + \delta[n-4])$$

The general form of the difference equation for a moving average filter with M terms is

$$y[n] = \frac{1}{M}(x[n] + x[n-1] + \ldots + x[n-(M-1)]) = \frac{1}{M}\sum_{k=0}^{M-1} x[n-k]$$

with the impulse response

$$h[n] = \frac{1}{M}(\delta[n] + \delta[n-1] + \ldots + \delta[n-(M-1)]) = \frac{1}{M}\sum_{k=0}^{M-1} \delta[n-k]$$

In Example 5.9, a five-term moving average filter smoothed the sharp changes from a signal. In other words, the high frequency variations in the input signal were removed by the filter. Examination of the filter shape will confirm that the moving average filter is in fact a low pass filter. The transfer function of the five-term moving average filter is

$$H(z) = 0.2(1 + z^{-1} + z^{-2} + z^{-3} + z^{-4})$$

FIGURE 9.2
Filter shape for five-term moving average filter.

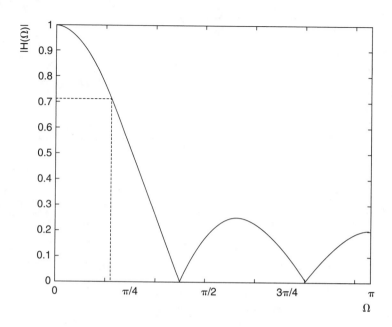

and its frequency response, replacing z by $e^{j\Omega}$, is

$$H(\Omega) = 0.2(1 + e^{-j\Omega} + e^{-j2\Omega} + e^{-j3\Omega} + e^{-j4\Omega})$$

A plot of the magnitude response for the five-term moving average filter is shown in Figure 9.2, for Ω values between 0 and π. The figure confirms the low pass characteristic of the moving average filter. It shows that the lowest digital frequency where the gain is zero occurs at $2\pi/M = 2\pi/5 = 1.257$ radians. This is one instance of an observation that applies to all moving average filters: The first zero occurs at $2\pi/M$ radians, where M is the number of terms in the impulse response. Furthermore, Figure 9.2 indicates that the cut-off frequency, where the gain drops to 0.707 of its DC value, lies about halfway to the first zero gain point, at π/M radians. Again, this observation is true for all moving average filters. It means that the higher the number of terms in the filter, the stronger the low pass filtering effect of the filter, and the more high frequency information that is removed. The general expressions for the transfer function and the frequency response of a moving average filter are

$$H(z) = \frac{1}{M}(1 + z^{-1} + z^{-2} + \dots + z^{-(M-1)})$$

$$H(\Omega) = \frac{1}{M}(1 + e^{-j\Omega} + \dots + e^{-j(M-1)\Omega})$$

EXAMPLE 9.2

Design a moving average filter with a -3 dB frequency of 480 Hz if the sampling frequency is 10 kHz.

The -3 dB point for the digital filter is required to be $f = 480$ Hz. With a sampling frequency of $f_S = 10$ kHz, this means a digital frequency of

$$\Omega = 2\pi \frac{f}{f_S} = 2\pi \frac{480}{10000} = 0.096\pi = 0.302 \text{ radians}$$

For a moving average filter with M terms, the magnitude first becomes zero at a digital frequency of $2\pi/M$ radians and the -3 dB point occurs at a frequency of approximately half this, or π/M rads. Since the -3 dB point is required to be 0.302 radians, M can be found from

$$\frac{\pi}{M} = 0.302$$

This gives $M = 10.4$. Thus, a moving average filter with 11 terms will give a -3 dB frequency slightly below that required. The impulse response for this filter is

$$h[n] = \frac{1}{11} \sum_{k=0}^{10} \delta[n - k]$$

and the difference equation is

$$y[n] = \frac{1}{11} \sum_{k=0}^{10} x[n - k]$$

The pole-zero plot for the filter is shown in Figure 9.3(a) and the filter shape, given by the magnitude response, in Figure 9.3(b). As required, the -3 dB point occurs close to $\pi/M = \pi/11 = 0.286$.

FIGURE 9.3
Eleven-term moving average filter for Example 9.2.

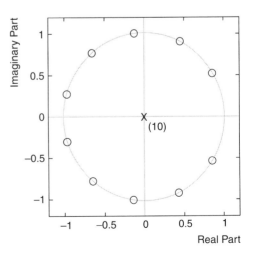

(a) Pole-Zero Diagram

FIGURE 9.3
Continued

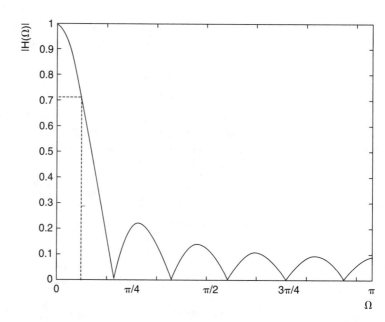

(b) Filter Shape

9.3 PHASE DISTORTION

When a sinusoidal signal passes through a linear digital filter, both its amplitude and phase are modified, as discussed in Section 7.3.1. The gain of the filter, $|H(\Omega)|$, and the phase difference of the filter, $\theta(\Omega)$, both vary with frequency. The gain is dimensionless; the phase is measured in radians or degrees. For an input sinusoid $A\cos(n\Omega_0)$, the output will be $HA\cos(n\Omega_0 + \theta)$, where $H = |H(\Omega_0)|$ is the gain of the filter at $\Omega = \Omega_0$ and $\theta = \theta(\Omega_0)$ is the phase difference of the filter at the same frequency. The digital frequency of the output is the same as that of the input, but both the amplitude and phase are different. Figure 9.4 plots an input and an output signal, with the output lagging the input by three samples. In general this lag can be calculated from the output function by setting

$$n\Omega_0 + \theta = 0 \tag{9.2}$$

as described at the end of Section 3.3.4. This equation determines the value of n for which a zero angle is obtained in $\cos(n\Omega_0 + \theta)$; when $\theta = 0$, a zero angle is obtained for $n = 0$. Solving Equation (9.2) for n gives

$$n = -\frac{\theta}{\Omega_0} = -\frac{\theta(\Omega_0)}{\Omega_0}$$

which is the **phase delay** of the filter, measured in samples, at the digital frequency Ω_0 radians. For the example in Figure 9.4, the digital frequency is $\Omega_0 = \pi/9$. At this frequency,

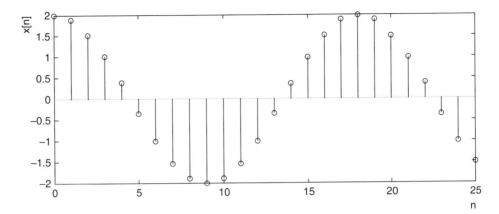

(a) $x[n] = 2\cos\left(n\dfrac{\pi}{9}\right)$

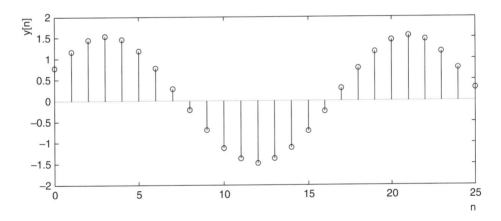

(b) $y[n] = 1.5\cos\left(n\dfrac{\pi}{9} - \dfrac{\pi}{3}\right)$

FIGURE 9.4
Phase delay.

$n\Omega_0 + \theta = n\pi/9 - \pi/3 = 0$ for $n = 3$, which gives a phase delay of 3 samples. This is confirmed by the plots: The signal $y[n]$ is delayed from the signal $x[n]$ by 3 samples.

Phase distortion occurs when different frequency components experience different phase delays through a filter. The only way to ensure that phase distortion will not occur is to make sure that signals of different frequencies have the same delay as they pass through the filter. Requiring a constant phase shift for all frequencies, however, will not have the effect intended. In Section 8.1, analog sine waves were added to produce a square wave using the equation

FIGURE 9.5
Effect of phase
distortion.

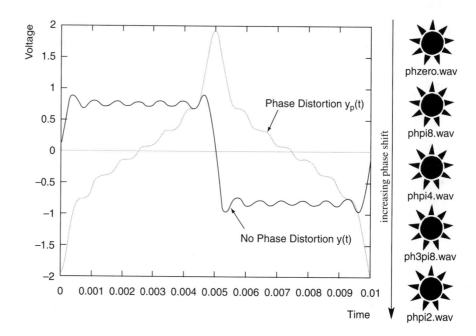

$$y(t) = \frac{4}{\pi}\left(\sin(\omega t) + \frac{1}{3}\sin(3\omega t) + \frac{1}{5}\sin(5\omega t) + \frac{1}{7}\sin(7\omega t) + \frac{1}{9}\sin(9\omega t) + \cdots\right)$$

If a $\pi/2$ radian phase shift were applied to each sine wave, the result would be

$$y_P(t) = \frac{4}{\pi}\left(\sin\left(\omega t - \frac{\pi}{2}\right) + \frac{1}{3}\sin\left(3\omega t - \frac{\pi}{2}\right) + \frac{1}{5}\sin\left(5\omega t - \frac{\pi}{2}\right)\right.$$

$$\left. + \frac{1}{7}\sin\left(7\omega t - \frac{\pi}{2}\right) + \frac{1}{9}\sin\left(9\omega t - \frac{\pi}{2}\right) + \cdots\right)$$

$$= -\frac{4}{\pi}\left(\cos(\omega t) + \frac{1}{3}\cos(3\omega t) + \frac{1}{5}\cos(5\omega t) + \frac{1}{7}\cos(7\omega t) + \frac{1}{9}\cos(9\omega t) + \cdots\right)$$

since $\sin(x - \pi/2) = -\cos x$. As Figure 9.5 shows, the sum of the shifted sine waves no longer resembles a square wave at all. The visible effect is also audible.

One way to ensure that all phase delays are equal is to require a zero phase difference for all frequencies, but this is not a very practical condition. Another way is to change the phase with frequency so that the phase delay remains constant for all frequencies. This strategy can be implemented by making phase difference a linear function of frequency. Specifically, when $\theta(\Omega) = -k\Omega$, the phase delay $-\theta(\Omega)/\Omega$ becomes equal to k samples. Because this result is independent of frequency, input components of all frequencies appear at the output of the filter at the same time. Phase changes that are approximately linear in frequency also work fairly well to reduce phase distortion. Small differences in phase delay across frequency cannot be detected by most listeners, but dramatic differences are definitely noticeable.

To see how a filter may be constructed to have a linear phase characteristic, and therefore no phase distortion, consider an impulse response that is symmetrical about zero. The equation for any impulse response with an equal number of terms on either side of zero is

$$h_1[n] = h_1[-M]\delta[n+M] + \ldots + h_1[-1]\delta[n+1] + h_1[0]\delta[n]$$

$$+ h_1[1]\delta[n-1] + \ldots + h_1[M]\delta[n-M]$$

Note that this equation requires that the impulse response have an odd number of terms. When the impulse response is symmetrical, as shown in Figure 9.6 for $M = 6$, $h_1[-M] = h_1[M]$ and so on, so that

$$h_1[n] = h_1[M]\delta[n+M] + \ldots + h_1[1]\delta[n+1] + h_1[0]\delta[n]$$

$$+ h_1[1]\delta[n-1] + \ldots + h_1[M]\delta[n-M]$$

Taking the z transform of this expression gives the transfer function

$$H_1(z) = h_1[M]z^M + \ldots + h_1[1]z + h_1[0] + h_1[1]z^{-1} + \ldots + h_1[M]z^{-M}$$

Finally, the frequency response is given by the DTFT:

$$H_1(\Omega) = h_1[M]e^{jM\Omega} + \ldots + h_1[1]e^{j\Omega} + h_1[0] + h_1[1]e^{-j\Omega} + \ldots + h_1[M]e^{-jM\Omega}$$

$$= h_1[0] + h_1[1](e^{j\Omega} + e^{-j\Omega}) + \ldots + h_1[M](e^{jM\Omega} + e^{-jM\Omega})$$

Using Euler's form for the cosine function from Equation (A.6), this becomes

$$H_1(\Omega) = h_1[0] + 2h_1[1]\cos\Omega + \ldots + 2h_1[M]\cos M\Omega$$

It is clear from this expression that, when the impulse response is symmetrical around zero, the Fourier transform is purely real. Purely real numbers can only have two possible phases: zero (for positive real numbers) and $180°$ (for negative real numbers). Thus, as long as $H(\Omega)$ is positive in the filter pass band, the phase $\theta_1(\Omega)$ is guaranteed to be zero in this range, and no phase distortion will occur. Unfortunately, the filter whose impulse response is pictured in Figure 9.6 has a practical problem: Because there are samples that occur prior to $n = 0$, the filter is noncausal. This means that the filter is capable of producing outputs before inputs are applied. All practical filters are causal, so the impulse response must be shifted in time, as shown in Figure 9.7.

Notice that the new causal impulse response is symmetrical about M, which has the value 6 in Figure 9.7. This impulse response, $h[n]$, is a time-shifted version of the original:

$$h[n] = h_1[n - M]$$

Time-shifting has no effect on the filter coefficients, but does affect how the filter is implemented. Prior to time-shifting, the noncausal difference equation for the impulse response $h_1[n]$ in Figure 9.6 is

FIGURE 9.6
Noncausal impulse response.

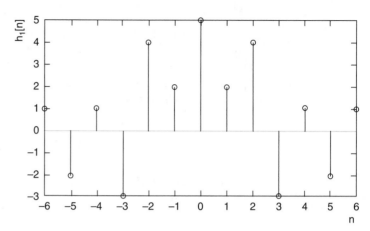

FIGURE 9.7
Causal impulse response.

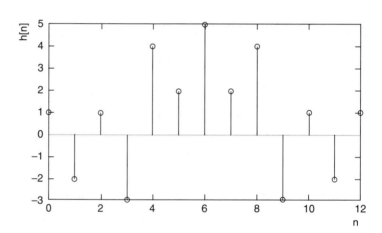

$$y[n] = h[0]x[n+6] + h[1]x[n+5] + \ldots + h[11]x[n-5] + h[12]x[n-6]$$

After time-shifting, the causal difference equation corresponding to the impulse response $h[n]$ in Figure 9.7 is

$$y[n] = h[0]x[n] + h[1]x[n-1] + \ldots + h[11]x[n-11] + h[12]x[n-12]$$

Time-shifting has an impact on the frequency response of the filter as well. Using the time-shift property of the Fourier transform,

$$H(\Omega) = \mathbf{F}\{h[n]\} = \mathbf{F}\{h_1[n - M]\} = e^{-jM\Omega}H_1(\Omega)$$

Both $H(\Omega)$ and $H_1(\Omega)$ can be written in magnitude-phase form, so

$$|H(\Omega)|e^{j\theta(\Omega)} = e^{-jM\Omega}|H_1(\Omega)|e^{j\theta_1(\Omega)}$$

$$= |H_1(\Omega)|e^{j(\theta_1(\Omega) - M\Omega)}$$

Time-shifting, therefore, does not have any effect on the magnitude response of the filter, since $|H(\Omega)| = |H_1(\Omega)|$, but does add $-M\Omega$ to the phase response, since $\theta(\Omega) = \theta_1(\Omega) - M\Omega$. If the phase of $H_1(\Omega)$ is zero in the pass band, then the phase of $H(\Omega)$ is $-M\Omega$ in the pass band. In other words, the phase in the pass band is linear in frequency Ω. As discussed above, this condition ensures that no phase distortion will occur during filtering. To summarize, a filter whose impulse response is symmetrical around its midpoint is guaranteed to be free from phase distortion. In addition, the symmetry of such filters allows very efficient implemenations.

EXAMPLE 9.3

Consider a nine-term moving average filter with impulse response

$$h[n] = \frac{1}{9}\sum_{k=0}^{8}\delta[n - k]$$

Because this impulse response is symmetrical around a midpoint, the phase response should be linear in frequency within the pass band. The magnitude response for this filter is plotted in Figure 9.8, and the phase response in Figure 9.9. The pass band, where the magnitude response is at least 0.707 of the value at DC, is indicated on the figure. Note the straight line nature of the phase within the pass band. In fact, the phase is linear in many other regions also, with shifts of π whenever $H(\Omega)$ changes sign. This causal impulse response can easily be implemented by using the difference equation

$$y[n] = \frac{1}{9}\sum_{k=0}^{8}x[n - k]$$

FIGURE 9.8
Magnitude
response for
Example 9.3.

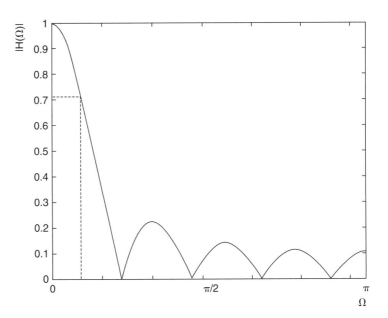

FIGURE 9.9
Phase response
for Example 9.3.

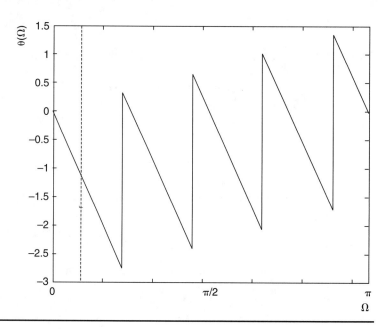

9.4 APPROXIMATING AN IDEAL LOW PASS FILTER

Moving average filters provide only crude low pass filtering. Higher performance FIR filters can be obtained by approximating ideal filter characteristics. Figure 9.10 shows the one-sided magnitude response for an ideal low pass filter with a cut-off frequency of Ω_1 radians. This ideal filter provides the starting point for one common FIR filter design technique, which is explained step by step in Section 9.6. The impulse response that will produce the ideal low pass filter shape can be found using the inverse discrete time Fourier transform, defined in Appendix F. Appendix G presents the mathematical development of the impulse response, with the final result

$$h_1[n] = \frac{1}{n\pi} \sin(n\Omega_1)$$

defined for all positive and negative integers n. This expression can be evaluated easily for all values except $n = 0$. For $n = 0$, it is worth reexpressing the impulse response in terms of the sinc function, defined by

$$\text{sinc } x = \frac{\sin x}{x}$$

The sinc function has the shape shown in Figure 9.11. Note that $\text{sinc}(0) = 1$ and that zeros of the function occur when $x = \pm n\pi$ ($n \neq 0$). Using the sinc function in the expression for the impulse response of the ideal low pass filter,

FIGURE 9.10

Ideal low pass filter magnitude response.

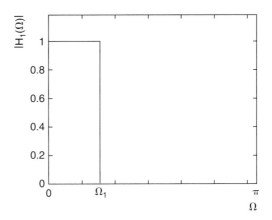

FIGURE 9.11

Sinc function.

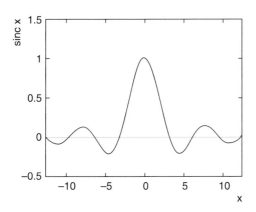

(a) Sinc Function

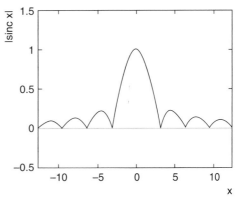

(b) Absolute Value of Sinc Function

$$h_1[n] = \frac{1}{n\pi} \sin(n\Omega_1) = \frac{1}{n\pi} (n\Omega_1) \frac{\sin(n\Omega_1)}{n\Omega_1} = \frac{\Omega_1}{\pi} \operatorname{sinc}(n\Omega_1)$$

Since $\operatorname{sinc}(0) = 1$, $h_1[0] = \Omega_1/\pi$. A quick alternative for evaluating $h_1[0]$ without using the sinc function is to use a very small value for n (such as 0.0001) instead of 0: The result will be a close approximation to the correct answer.

Thus, the impulse response that produces an ideal low pass filter is computed from the equation

$$h_1[n] = \frac{\sin(n\Omega_1)}{n\pi}$$

where Ω_1 is the cut-off for the filter. The impulse response for a filter with a cut-off of $\Omega_1 = 0.25\pi$ radians, for example, is depicted in Figure 9.12 for values of n from -15 to 15.

FIGURE 9.12

Impulse response of ideal low pass filter.

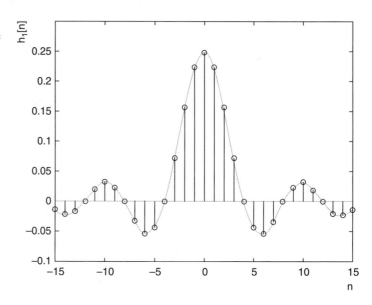

This impulse response has an infinite number of samples in each direction, and its envelope has the expected sinc shape.

Unfortunately, creating an ideal low pass filter is not quite this easy. The first problem is that the impulse response is noncausal, since it has terms prior to $n = 0$. It cannot be shifted in time, as was suggested for moving average filters, because the number of nonzero terms for negative values of n is infinite. Second, the infinite number of terms means that the impulse response cannot be converted directly into a nonrecursive difference equation. A simple solution is to truncate from both ends of the impulse response in Figure 9.12 those samples for which the response is small. The impulse response becomes finite in length and can be shifted for causality, which makes the filter characterized by the impulse response usable for the first time. In Figure 9.13, for example, all but the middle 21 terms have been eliminated. Figure 9.14 shows the impulse response that results after shifting by 10 samples for causality.

Naturally, truncating terms from an impulse response will affect its frequency response. After truncation, the filter shape loses its ideal rectangular shape. For example, Figure 9.15 shows the magnitude response of the 21-term causal impulse response shown in Figure 9.14. The ideal low pass filter shape has been superimposed to show how the filter shape is smeared by truncation in the time domain. Of course, the more terms that are retained, the more closely the filter shape approximates the ideal. It is important to notice that the Ω_1 cut-off no longer marks the point where the gain drops below unity. It now marks the point where the gain is 0.5. For this reason, the design choice for Ω_1 must be larger than the desired pass band edge frequency. This correction is described in Section 9.6.1.

When filter shapes become nonideal, some means of specifying key features must be identified. Figure 9.16 shows the most common features used. The filter pass band determines the range of frequencies passed by the filter. In the filter stop band, signals are

FIGURE 9.13
Truncated impulse
response.

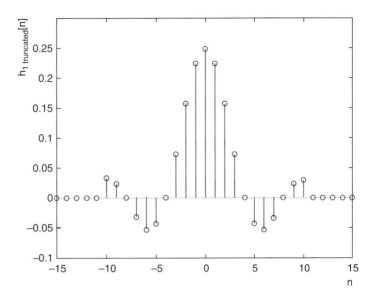

FIGURE 9.14
Truncated and
shifted impulse
response.

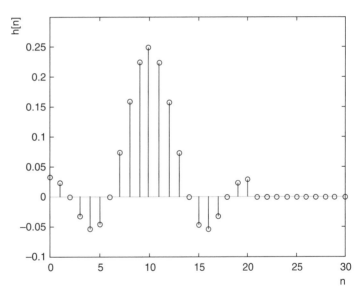

strongly attenuated by the filter. The parameter δ_p defines the **pass band ripple**, which is the maximum deviation from unity gain within the pass band of the filter. The gain at the edge of the pass band is defined as $1 - \delta_p$. The parameter δ_s defines the **stop band ripple**, the maximum deviation from zero gain experienced in the stop band of the filter. At the stop band edge, the gain of the filter is δ_s. The **transition width** is the distance in Hz between the stop band and pass band edges, or

$$\text{Transition width} = \text{Stop band edge frequency} - \text{Pass band edge frequency}$$

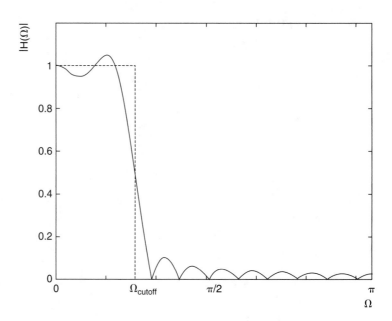

FIGURE 9.15
Nonideal low pass filter magnitude response.

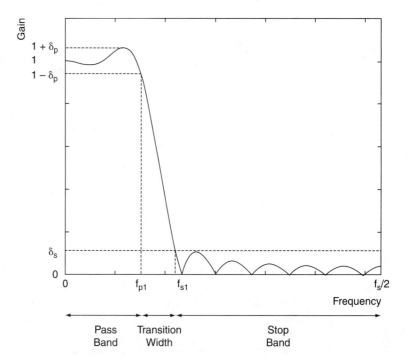

FIGURE 9.16
Filter features.

Note that filter bandwidth normally refers to the range of frequencies defined by a gain of 0.707 (or −3 dB), which may not correspond to the FIR pass band edge frequency, defined as the frequency where the gain drops to $1 - \delta_p$, or $20\log(1 - \delta_p)$ dB.

The terms gain and attenuation, when they are used to refer to filter performance at a certain frequency, must be differentiated. Suppose the pass band ripple $\delta_p = 0.1$ and the stop band ripple $\delta_s = 0.05$. If the DC gain is 1, or 0 dB, then the gain at the edge of the pass band is $1 - 0.1 = 0.9$, or $20\log(0.9) = -0.915$ dB. The gain at the edge of the stop band is 0.05, or $20\log(0.05) = -26.02$ dB. Measured in dB, attenuation is simply the negative of gain. Thus, the attenuation at the edge of the pass band is 0.915 dB, and the stop band attenuation is 26.02 dB. Stop band attenuation values are often quoted with reference to pass band gain. For example, a filter with a quoted stop band attenuation of 40 dB has a stop band gain of −40 dB when the pass band gain is about 0 dB, but a stop band gain of −20 dB when the pass band gain is about 20 dB.

EXAMPLE 9.4

For the low pass filter shown in Figure 9.17, identify the pass band ripple, the frequency at the edge of the pass band, the stop band ripple, the frequency at the edge of the stop band, the transition width, the bandwidth, and the −3 dB, or cut-off, frequency.

The largest excursion from unity in the pass band is determined by the positive-going ripple in this case. The maximum gain in the pass band is 1.0905, so the pass band ripple δ_p is 0.0905. Therefore, the edge of the pass band is located where the gain drops to $1 - \delta_p = 1 - 0.0905 = 0.9095$. The frequency at this point is 1060 Hz, the edge of the pass band. The

FIGURE 9.17
Magnitude
response for
Example 9.4.

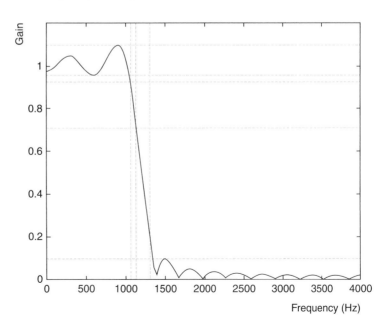

maximum gain in the stop band, at the first side lobe, is 0.0911, which is the stop band ripple δ_s. The gain of the filter first drops to this value at a frequency of 1337 Hz, the edge of the stop band. The transition width is the difference between the stop band and pass band edges, or $1337 - 1060 = 277$ Hz. For this low pass filter, the -3 dB or cut-off frequency, where the gain drops to 0.707, is 1136 Hz. This frequency also determines the bandwidth of the filter, which is 1136 Hz as well.

9.5   WINDOWS

9.5.1   Rectangular Window

The purpose of a window function is to select a finite number of samples from the infinite number of samples in the ideal low pass impulse response $h_1[n] = \sin(n\Omega_1)/n\pi$. This important step allows a real filter to be implemented from the impulse response samples. Truncation, the solution suggested in the previous section, can in fact be viewed as multiplication by a rectangular window. On the left side of Figure 9.18, the finite impulse response $h[n]$ in (c)(i) is obtained by multiplying the ideal response $h_1[n] = \sin(n\Omega_1)/n\pi$ in (a)(i) by the finite rectangular window $w[n]$ shown in (b)(i), that is, $h[n] = h_1[n]w[n]$. The N-term rectangular window is equal to

$$w[n] = 1$$

for $|n| \le (N-1)/2$, and is zero elsewhere.

Windowing creates a usable filter impulse response from an unusable one, but has attendant side effects in the frequency domain. The right-hand side of Figure 9.18 shows the DTFT magnitude responses for each of the time domain signals at the left. As expected, the impulse response $h_1[n]$ gives an ideal low pass filter shape, as shown in (a)(ii). The magnitude response for the window function produces the sinc shape that is characteristic of spectra for rectangular time functions, as seen in Figure 9.18(b)(ii). Figure 9.18(c)(ii) shows the nonideal filter shape produced by the finite impulse response $h[n]$ in (c)(i).

As indicated in Section 9.4, eliminating terms from the ideal low pass filter impulse response degrades the shape of the filter, but is necessary to limiting the number of terms. One other problem must be corrected before the filter may be implemented: The impulse response shown in Figure 9.18(c)(i) may be finite, but it is also noncausal. As suggested in Section 9.4, and as illustrated in Figures 9.13 and 9.14, the simple solution to this problem is to shift the impulse response to the right so that its first nonzero term aligns with $n = 0$. This action creates a causal FIR filter with the N-term impulse response

$$h[n] = h[0] + h[1]z^{-1} + h[2]z^{-2} + \ldots + h[N-1]z^{-(N-1)}$$

The impulse response samples can then be used to find the filter shape. Using the DTFT:

$$H(\Omega) = h[0] + h[1]e^{-j\Omega} + h[2]e^{-j2\Omega} + \ldots + h[N-1]e^{-j(N-1)\Omega}$$

The filter shape, or magnitude response, is shown in Figure 9.18(c)(ii), scaled to give a DC gain of unity. Recalling from Section 4.7 that the impulse response samples also give

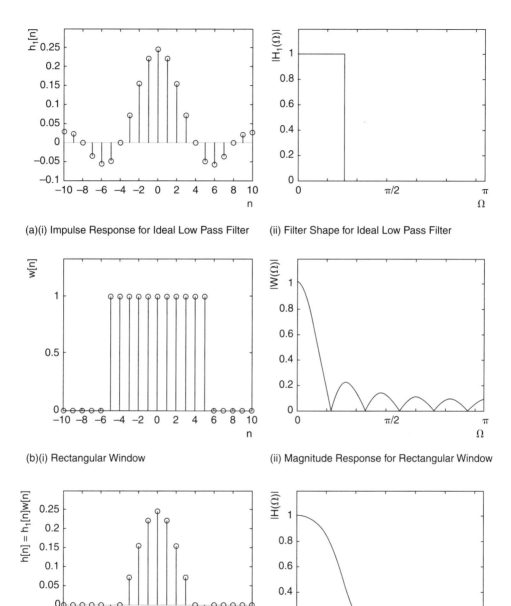

(a)(i) Impulse Response for Ideal Low Pass Filter

(ii) Filter Shape for Ideal Low Pass Filter

(b)(i) Rectangular Window

(ii) Magnitude Response for Rectangular Window

(c)(i) Impulse Response for Non-Ideal Low
       Pass Filter

(ii) Filter Shape for Non-Ideal Low Pass Filter

FIGURE 9.18

Creating a Non-Ideal low pass filter.

FIGURE 9.19

Impulse response for rectangular window.

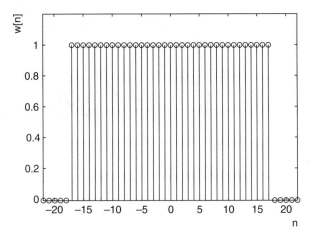

FIGURE 9.20

Magnitude response of rectangular window.

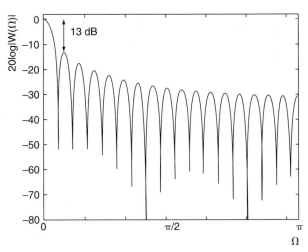

the difference equation coefficients for an FIR filter, the difference equation becomes

$$y[n] = h[0]x[n] + h[1]x[n-1] + h[2]x[n-2] + \ldots + h[N-1]x[n-(N-1)]$$

This equation permits the filter to be implemented.

While the low pass filter produced using the rectangular window may appear to behave reasonably well, it has problems that must be addressed. As Figure 9.18(c)(ii) shows, the gain for the biggest side lobe is about 10% of the pass band gain. For many applications, this is unacceptably large. The side lobes are evidence of the **ringing** that occurs because of the sharp vertical edges of the rectangular window. When the gains are plotted in dB, the side lobes become even more obvious. A 35-term rectangular window is shown in Figure 9.19. The biggest side lobe is 13 dB below the DC magnitude, as shown by Figure 9.20. A low pass filter produced using the rectangular window shows an approximate difference of 21 dB between its pass band and stop band gains, as seen in Figure 9.21.

FIGURE 9.21

Filter shape for filter made with rectangular window.

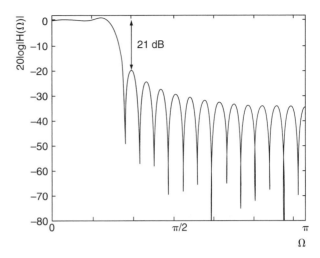

---

sig.wav

sig-21.wav

EXAMPLE 9.5

If the gain in the pass band is unity, or 0 dB, then an attenuation of 21 dB corresponds to a gain of $10^{-21/20} = 0.089125$. Though this may seem a small number, a signal multiplied by this gain is still quite audible. Therefore, a filter that attenuates signals in its stop band by only 21 dB is a poor one. A better filter would apply greater attenuation.

Ringing problems can be reduced by choosing a window with smoother edges. Common choices for windows $w[n]$ include Hanning, Hamming, Blackman, and Kaiser windows, investigated in the following sections. All of these window functions play the same role: to make the number of terms in the impulse response finite. To achieve this, the impulse response for a low pass filter is calculated as $h[n] = h_1[n]w[n]$, where $h_1[n]$ is the impulse response of an ideal low pass filter and $w[n]$ is a window function. After windowing, the impulse response must be shifted to the right before it can be implemented as a causal difference equation. The details of these FIR filter design steps are covered in Section 9.6.

9.5.2   Hanning Window

The N-term Hanning window is defined by the equation

$$w[n] = 0.5 + 0.5\cos\frac{2\pi n}{N-1}$$

for $|n| \leq (N-1)/2$, and is zero elsewhere. Figure 9.22 shows the window function for $N = 35$, and Figure 9.23 shows its magnitude response. Notice how the smoother contours in the window function lead to smaller side lobes than for the rectangular window. The side lobes for the Hanning window are 31 dB below its DC magnitude. This translates into filter designs with better stop band attenuation. Figure 9.24 shows the magnitude response for a low

FIGURE 9.22
Impulse response for
Hanning window.

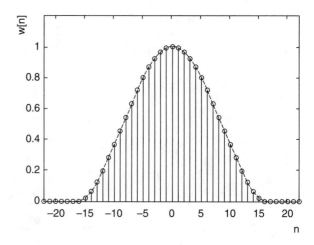

FIGURE 9.23
Magnitude response for
Hanning window.

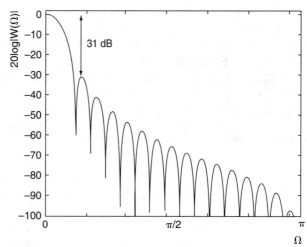

FIGURE 9.24
Filter shape for filter made
with Hanning window.

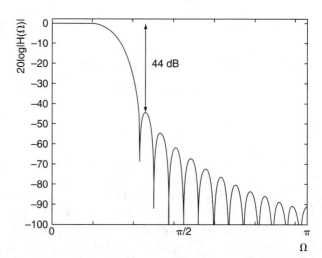

pass FIR filter designed with a Hanning rather than a rectangular window. Stop band gains for the biggest side lobe are 44 dB below the pass band gain, compared to 21 dB for the rectangular window design.

9.5.3   Hamming Window

The N-term Hamming window is defined by the equation

$$w[n] = 0.54 + 0.46\cos\frac{2\pi n}{N-1}$$

for $|n| \leq (N-1)/2$, and is zero elsewhere. Figure 9.25 shows the window function for $N = 35$, and Figure 9.26 shows its magnitude response. As with the Hanning window, the smooth contours in the window function lead to small side lobes. The side lobes for the Hamming window are 41 dB below its DC magnitude. For a low pass filter designed with a Hamming window, the largest side lobe in the stop band is 55 dB below the pass band gain, as shown in Figure 9.27.

FIGURE 9.25

Impulse response for Hamming window.

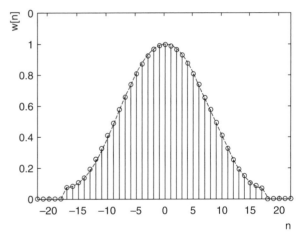

FIGURE 9.26

Magnitude response for Hamming window.

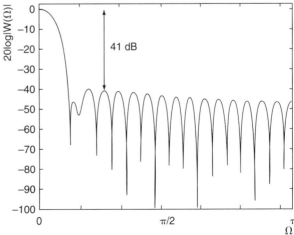

FIGURE 9.27
Filter shape for filter made with Hamming window.

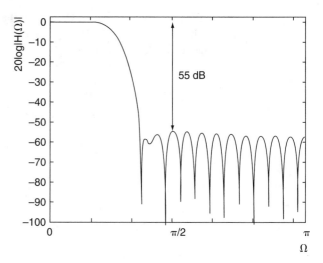

9.5.4 Blackman Window

The N-term Blackman window is defined by the equation

$$w[n] = 0.42 + 0.5\cos\frac{2\pi n}{N-1} + 0.08\cos\frac{4\pi n}{N-1}$$

for $|n| \le (N-1)/2$, and is zero elsewhere. The 35-term Blackman window is shown in Figure 9.28. The magnitude response for this window, shown in Figure 9.29, indicates a 57 dB difference between the DC magnitude and the largest side lobe. A low pass filter designed with a Blackman window gives side lobes that are 75 dB below the pass band gain, as seen in Figure 9.30.

FIGURE 9.28
Impulse response for Blackman window.

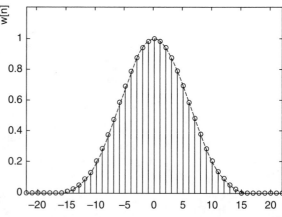

FIGURE 9.29
Magnitude response for Blackman window.

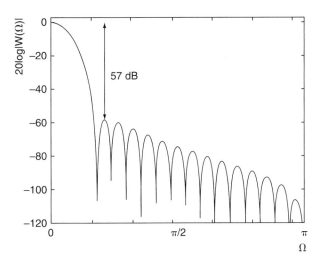

FIGURE 9.30
Filter shape for filter made with Blackman window.

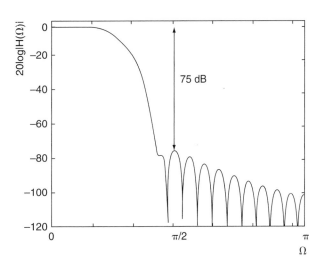

9.5.5 Kaiser Window

The expression for the N-term Kaiser window is

$$w[n] = \frac{I_0\left[\beta\sqrt{1 - \left(\frac{2n}{N-1} - 1\right)^2}\right]}{I_0[\beta]}$$

defined for $|n| \leq (N-1)/2$. As for the other windows, the Kaiser window is zero outside of this range. $I_0(x)$ is the so-called zero order modified Bessel function of the first kind, defined by

$$I_0(x) = 1 + \sum_{j=1}^{\infty} \left[\frac{(x/2)^j}{j!} \right]^2$$

and normally approximated by summing the first 20 or 25 terms. Most software packages with signal processing capability provide n^{th} order Bessel functions of the first kind for any value of x.

Kaiser windows may be designed to match a desired stop band attenuation. The shape of a Kaiser window is determined by the parameter β in the equation for $w[n]$, which is estimated according to the stop band requirements. Kaiser showed that, for a desired stop band attenuation A in dB, an approximate value for β can be chosen using the empirical relation

$$\beta = 0.1102A - 0.9587 \tag{9.3}$$

as long as A is greater than 50 dB. Figure 9.31 shows an example of 35-term Kaiser window for $\beta = 8$, which, according to Equation (9.3), corresponds to a stop band attenuation of 81 dB for a filter designed with this window. The magnitude response for the window is shown in Figure 9.32. For the value of β chosen, the biggest side lobe is 58 dB below the DC gain. When this window is used to design a low pass filter, the resulting filter has side lobe gains approximately 81 dB below the pass band gain, as illustrated in Figure 9.33. Table 9.2 shows the filter stop band attenuations that result for various values of β, along with the approximations that Equation (9.3) provides. Using either the table or Equation (9.3), a value for β may be chosen to give the desired filter response.

FIGURE 9.31

Impulse response for Kaiser window, $\beta = 8$.

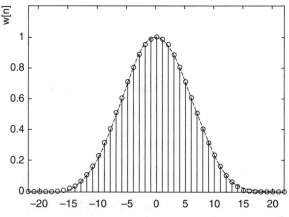

FIGURE 9.32
Magnitude response for Kaiser window, β = 8.

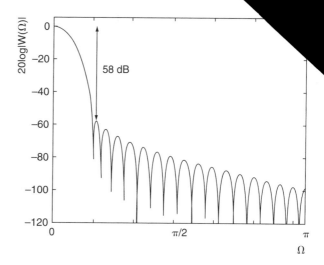

FIGURE 9.33
Filter shape for filter made with Kaiser window, β = 8.

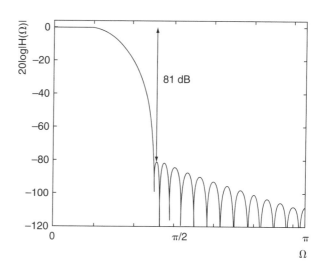

TABLE 9.2
Choosing the Kaiser Filter Parameter β

β	Predicted Stop Band Attenuation (dB)	Actual Filter Stop Band Attenuation (dB)
5.0	54	56
6.0	63	64
7.0	72	72
8.0	81	81
9.0	90	90
10.0	99	100

R FILTER DESIGN

Design Guidelines

FIR filter can be designed from filter specifications by choosing an appropriate
unction with sufficient terms. Table 9.3 contains guidelines, obtained empirically,
g these choices. Normally a window type is chosen from the shapes shown in Fig-
by first selecting the required stop band attenuation in Table 9.3. The best choice
adow that meets the stop band requirements as closely as possible. Windows that
exceed the requirements will also demand a larger number of terms to accomplish the task.
Once the window choice has been made, the number of terms needed may be calculated using the equations provided in the third column of the table. The steeper the required roll-off
of the filter, that is, the smaller the required transition width, the larger the number of terms
required. Choosing an odd number of terms will guarantee a symmetrical impulse response
can be constructed, which will eliminate phase distortion.

TABLE 9.3
FIR Filter Guidelines

Window Type	Window Function $\|n\| \leq \dfrac{N-1}{2}$	Number of Terms, N^*	Filter Stop Band Attenuation (dB)	Gain at Edge of Pass Band $20\log(1 - \delta_p)$ (dB)
Rectangular	1	$0.91\dfrac{f_S}{\text{T.W.}}$	21	-0.9
Hanning	$0.5 + 0.5\cos\left(\dfrac{2\pi n}{N-1}\right)$	$3.32\dfrac{f_S}{\text{T.W.}}$	44	-0.06
Hamming	$0.54 + 0.46\cos\left(\dfrac{2\pi n}{N-1}\right)$	$3.44\dfrac{f_S}{\text{T.W.}}$	55	-0.02
Blackman	$0.42 + 0.5\cos\left(\dfrac{2\pi n}{N-1}\right)$ $+ 0.08\cos\left(\dfrac{4\pi n}{N-1}\right)$	$5.98\dfrac{f_S}{\text{T.W.}}$	75	-0.0014
Kaiser	$\dfrac{I_0\left(\beta\sqrt{1 - \left(\dfrac{2n}{N-1}\right)^2}\right)}{I_0(\beta)}$	$4.33\dfrac{f_S}{\text{T.W.}}$ ($\beta = 6$)	64	-0.0057
		$5.25\dfrac{f_S}{\text{T.W.}}$ ($\beta = 8$)	81	-0.00087
		$6.36\dfrac{f_S}{\text{T.W.}}$ ($\beta = 10$)	100	-0.000013

$^*N$ = number of terms in window, f_S = sampling frequency, T.W. = transition width.

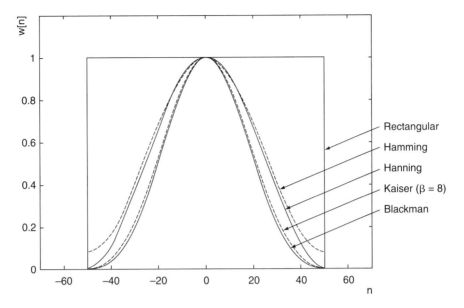

FIGURE 9.34
Window shape envelopes ($N = 101$).

The window $w[n]$ is applied to the ideal low pass filter impulse response $h_1[n]$ given in Section 9.4 to obtain a nonideal low pass filter, according to $h[n] = h_1[n]w[n] = (\sin(n\Omega_1)/n\pi)w[n]$. The pass band edge frequency Ω_1 chosen for the low pass filter must take into account the smearing that is a result of the finite number of terms used, as discussed in Section 9.4 and shown in Figure 9.15. As a rule of thumb, instead of using the desired pass band edge, the middle of the transition width, halfway between the pass band and stop band edges, should be used, as illustrated in Figure 9.35. This will put the actual pass band edge at the desired location. In equation form,

$$\begin{matrix} \text{Pass band edge frequency} \\ \text{for design} \end{matrix} = \begin{matrix} \text{Desired pass band} \\ \text{frequency} \end{matrix} + \frac{\text{Transition width}}{2}$$

When the pass band edge frequency and the number of window terms are determined from Table 9.3, the gain $20\log(1-\delta_p)$ at the pass band edge will be close to the value listed in the final column of the table.

Finally, when the impulse response of the ideal low pass filter is windowed, a noncausal low pass filter is obtained because the impulse response for the filter is symmetrical around zero. As discussed in Section 9.4, for the filter to be practical, it must be causal. To make the filter causal, its impulse response must be shifted to the right to begin at zero. This

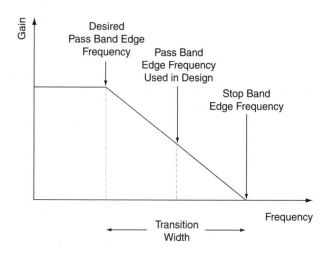

FIGURE 9.35
Choosing a pass band edge for design.

occurs in the last step of the design. The following section lists all of the design steps and provides examples of complete designs.

EXAMPLE 9.6

A filter must have a stop band attenuation of 75 dB and a transition width of 1 kHz for 16 kHz sampling. Which window should be selected, and how many terms should be used?

The two windows in Table 9.3 that best satisfy the stop band requirement are the Blackman window and the $\beta = 8$ Kaiser window. The Blackman window requires $5.98(16000/1000) = 95.68 \cong 95$ terms. The $\beta = 8$ Kaiser window requires $5.25(16000/1000) = 84 \cong 85$ terms. In general, the number of terms required is rounded to the nearest odd integer. Rounding up produces a filter that exceeds specifications slightly, while rounding down gives a filter that falls a little short of the requirements.

9.6.2 Steps for Low Pass FIR Filter Design

The steps needed for low pass FIR filter design are listed in Table 9.4. These steps provide the impulse response $h[n]$ for the desired low pass filter. The transfer function $H(z)$ for the filter, if needed, can be found, using Equation (6.3), as

$$H(z) = h[0] + h[1]z^{-1} + h[2]z^{-2} + h[3]z^{-3} + \ldots + h[N - 1]z^{-(N-1)}$$

This transfer function has $(N-1)$ zeros and $(N-1)$ poles, and is specified by the N coefficients $h[0], h[1], \ldots,$ and $h[N-1]$. From the transfer function, the difference equation

$$y[n] = h[0] + h[1]x[n-1] + h[2]x[n-2] + h[3]x[n-3] + \ldots$$
$$+ h[N-1]x[n - (N-1)]$$

TABLE 9.4
Design Steps for Windowed Low Pass FIR Filter Design

1. Choose a pass band edge frequency in Hz for the filter in the middle of the transition width:

$$f_1 = \text{Desired pass band edge frequency} + \frac{\text{Transition width}}{2}$$

2. Calculate $\Omega_1 = 2\pi f_1/f_S$ and substitute this value into $h_1[n]$, the infinite impulse response for an ideal low pass filter:

$$h_1[n] = \frac{\sin(n\Omega_1)}{n\pi}$$

3. Choose a window from Table 9.3 that will satisfy the stop band attenuation and other filter requirements. Calculate the number of nonzero window terms required using the formula for N in the table. Choose an odd number so that the impulse response can be perfectly symmetrical, thereby avoiding phase distortion in the final filter. Evaluate the window function $w[n]$ for $|n| \le (N-1)/2$.
4. Calculate the (finite) impulse response $h[n]$ for the filter from $h[n] = h_1[n]w[n]$ for $|n| \le (N-1)/2$, with $h[n] = 0$ for other values of n. This impulse response is noncausal.
5. Shift the impulse response values to the right by $(N-1)/2$ steps to ensure that the first nonzero value occurs at $n = 0$, thereby making the low pass filter causal.

and the frequency response

$$H(\Omega) = h[0] + h[1]e^{-j\Omega} + h[2]e^{-j2\Omega} + h[3]e^{-j3\Omega} + \ldots + h[N-1]e^{-j(N-1)\Omega}$$

may be obtained. The frequency response provides the magnitude response $|H(\Omega)|$ versus Ω, which gives the filter shape.

EXAMPLE 9.7

A low pass filter must be designed according to the following specifications:

Pass band edge	2 kHz
Stop band edge	3 kHz
Stop band attenuation	40 dB
Sampling frequency	10 kHz

For this filter, from Figure 9.16,

$$\text{Transition width} = \text{Stop band edge frequency} - \text{Pass band edge frequency}$$
$$= 3 - 2 = 1 \text{ kHz}$$

Step 1 of the design method then yields

$$f_1 = 2000 + \frac{1000}{2} = 2500 \text{ Hz}$$

For Step 2,

$$\Omega_1 = 2\pi \frac{f_1}{f_S} = 2\pi \frac{2500}{10000} = 0.5\pi$$

$$h_1[n] = \frac{\sin(n\Omega_1)}{n\pi} = \frac{\sin(0.5\pi n)}{n\pi}$$

Step 3 demands a choice of window. For a stop band attenuation of 40 dB, Table 9.3 suggests the Hanning window, with

$$N = 3.32\frac{f_S}{\text{T.W.}} = 3.32\frac{10}{1} = 33.2$$

Choosing $N = 33$, the window function becomes

$$w[n] = 0.5 + 0.5 \cos\left(\frac{2\pi n}{N-1}\right) = 0.5 + 0.5 \cos\left(\frac{2\pi n}{32}\right)$$

for $-16 \leq n \leq 16$. The first four columns in Table 9.5 show the calculations that give the impulse response, as required by Step 4. The impulse response is given by $h[n] = h_1[n]w[n]$ for $|n| \leq 16$ and $h[n] = 0$ for all other values of n. The renumbering in the last column of Table 9.5 records the shifting of the impulse response values for causality, as indicated in Step 5. To obtain a causal realization of the filter, the $h[n]$ values must be associated with these new values of n.

The impulse response for the filter is shown in Figure 9.36. Its values can be used to find the difference equation for the filter:

$$\begin{aligned}
y[n] = &- 0.0002x[n-1] + 0.0021x[n-3] - 0.0064x[n-5] + 0.0142x[n-7] \\
&- 0.0272x[n-9] + 0.0495x[n-11] - 0.0972x[n-13] \\
&+ 0.3153x[n-15] + 0.5x[n-16] + 0.3153x[n-17] - 0.0972x[n-19] \\
&+ 0.0495x[n-21] - 0.0272x[n-23] + 0.0142x[n-25] \\
&- 0.0064x[n-27] + 0.0021x[n-29] - 0.0002x[n-31]
\end{aligned}$$

The impulse response values can also be used to find the transfer function $H(z)$, from which the poles and zeros may be computed. The low pass filter's pole-zero diagram is shown in Figure 9.37(a). Since the filter has $N = 33$ coefficients, it has 32 zeros and 32 poles. Only 30 of the 32 zeros are shown, because the impulse response is negligible for the first and

TABLE 9.5
Calculations for Example 9.7

n	$h_1[n]$	$w[n]$	$h[n]$	new n
−16	0.0000	0.0000	0.0000	0
−15	−0.0212	0.0096	−0.0002	1
−14	0.0000	0.0381	0.0000	2
−13	0.0245	0.0843	0.0021	3
−12	0.0000	0.1464	0.0000	4
−11	−0.0289	0.2222	−0.0064	5
−10	0.0000	0.3087	0.0000	6
−9	0.0354	0.4025	0.0142	7
−8	0.0000	0.5000	0.0000	8
−7	−0.0455	0.5975	−0.0272	9
−6	0.0000	0.6913	0.0000	10
−5	0.0637	0.7778	0.0495	11
−4	0.0000	0.8536	0.0000	12
−3	−0.1061	0.9157	−0.0972	13
−2	0.0000	0.9619	0.0000	14
−1	0.3183	0.9904	0.3153	15
0	0.5000	1.0000	0.5000	16
1	0.3183	0.9904	0.3153	17
2	0.0000	0.9619	0.0000	18
3	−0.1061	0.9157	−0.0972	19
4	0.0000	0.8536	0.0000	20
5	0.0637	0.7778	0.0495	21
6	0.0000	0.6913	0.0000	22
7	−0.0455	0.5975	−0.0272	23
8	0.0000	0.5000	0.0000	24
9	0.0354	0.4025	0.0142	25
10	0.0000	0.3087	0.0000	26
11	−0.0289	0.2222	−0.0064	27
12	0.0000	0.1464	0.0000	28
13	0.0245	0.0843	0.0021	29
14	0.0000	0.0381	0.0000	30
15	−0.0212	0.0096	−0.0002	31
16	0.0000	0.0000	0.0000	32

last samples in the table. All 32 poles lie at the origin. Therefore, the filter is stable, as for all FIR filters. The magnitude response $|H(f)|$ is shown in Figure 9.37(b). The pass band edge, defined by Table 9.3 to be at a gain of −0.06 dB, occurs at a frequency 2013 Hz. A stop band attenuation of 40 dB occurs at a frequency of 2976 Hz. These measures closely match the specifications.

FIGURE 9.36
Causal impulse
response for
Example 9.7.

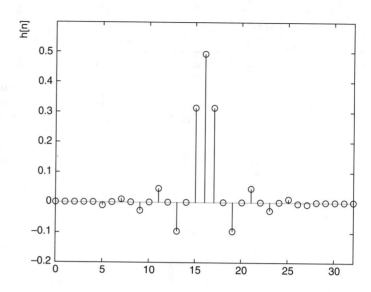

FIGURE 9.37
Pole-zero diagram and filter
shape for Example 9.7.

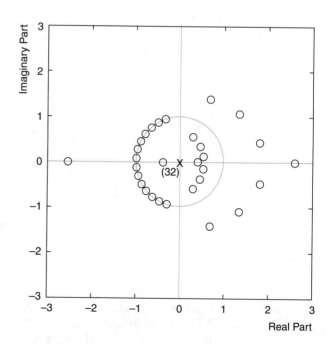

(a) Pole-Zero Diagram

FIGURE 9.37
Continued

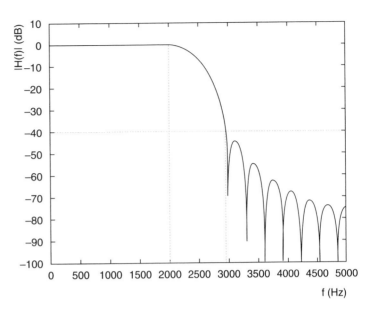

(b) Filter Shape

EXAMPLE 9.8

A low pass filter must be designed according to the following specifications:

Pass band edge	10 kHz
Stop band edge	22 kHz
Stop band attenuation	75 dB
Sampling frequency	50 kHz

Following the approach of the last example:

$$\text{Transition width} = 12 \text{ kHz}$$

$$f_1 = 10000 + \frac{12000}{2} = 16 \text{ kHz}$$

$$\Omega_1 = 2\pi\frac{f_1}{f_S} = 0.64\pi$$

$$h_1[n] = \frac{\sin(n\Omega_1)}{n\pi} = \frac{\sin(0.64\pi n)}{n\pi}$$

According to Table 9.3, both the Blackman and the Kaiser windows meet the requirement for a stop band attenuation of 75 dB. Though the Kaiser window would meet the objectives using the fewest terms, the Blackman window is selected to avoid the complexities of the Bessel function. The number of window terms required is

$$N = 5.98 \frac{f_S}{\text{T.W.}} = 5.98 \frac{50}{12} = 24.9$$

Choosing $N = 25$, the window function is

$$w[n] = 0.42 + 0.5 \cos\left(\frac{2\pi n}{24}\right) + 0.08 \cos\left(\frac{4\pi n}{24}\right)$$

The impulse response for the filter is given by $h[n] = h_1[n]w[n]$ for $|n| \le 12$ and $h[n] = 0$ for other values of n. Table 9.6 shows the calculations that generate the impulse response, which is depicted in Figure 9.38. The resulting low pass filter shape is shown in Figure 9.39. The pass band edge, defined in Table 9.3 by a gain of -0.0014 dB, occurs at 10.11 kHz, and a gain of -75 dB occurs at 21.685 kHz. Thus, the filter closely matches the specifications.

TABLE 9.6

Calculations for Example 9.8

n	$h_1[n]$	$w[n]$	$h[n]$	new n
-12	-0.022	0.000	0.000	0
-11	-0.004	0.006	0.000	1
-10	0.030	0.027	0.001	2
-9	-0.024	0.066	-0.002	3
-8	-0.015	0.130	-0.002	4
-7	0.045	0.221	0.010	5
-6	-0.026	0.340	-0.009	6
-5	-0.037	0.480	-0.018	7
-4	0.078	0.630	0.049	8
-3	-0.026	0.774	-0.020	9
-2	-0.123	0.893	-0.110	10
-1	0.288	0.972	0.280	11
0	0.640	1.000	0.640	12
1	0.288	0.972	0.280	13
2	-0.123	0.893	-0.110	14
3	-0.026	0.774	-0.020	15
4	0.078	0.630	0.049	16
5	-0.037	0.480	-0.018	17
6	-0.026	0.340	-0.009	18
7	0.045	0.221	0.010	19
8	-0.015	0.130	-0.002	20
9	-0.024	0.066	-0.002	21
10	0.030	0.027	0.001	22
11	-0.004	0.006	0.000	23
12	-0.022	0.000	0.000	24

FIGURE 9.38

Causal impulse response for Example 9.8.

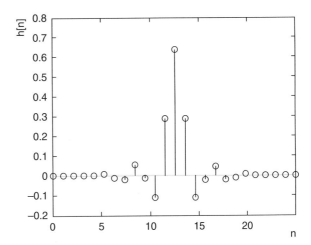

FIGURE 9.39

Filter shape for Example 9.8.

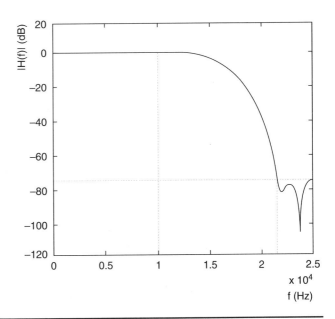

EXAMPLE 9.9

A recording of a song is severely degraded by annoying high frequency noise. The spectrum of the signal plus noise is shown in Figure 9.40. The sampling rate for the system is 16 kHz. Design a filter that will improve the quality of the recording.

Removing the noise is not an easy task. From the spectrum in Figure 9.40, the noise appears to act starting at about 2 kHz. Since this frequency is quite low, removing the noise will cause portions of the music to be lost as well. The best trade-off retains as much of the

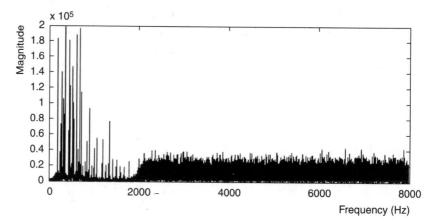

noisy.wav

FIGURE 9.40
Spectrum of signal plus noise for Example 9.9.

FIGURE 9.41
Impulse response for Example 9.9.

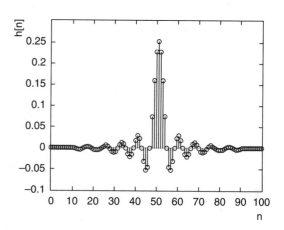

music content as possible, while removing the most distracting noise. A low pass filter with its pass band edge at 2 kHz should serve. Since no particular stop band attenuation is specified, any window that gives reasonable performance will do. In this case, a Hamming window is chosen. No transition width is specified for this problem either. To obtain a reasonably steep roll-off, $N = 101$ terms are selected. From Table 9.3, this number of terms will give a transition width of 3.44(16000/101), or about 545 Hz.

The impulse response, and then the difference equation, for the filter may be calculated by following the method in Example 9.7. The impulse response is plotted in Figure 9.41. When the difference equation is used to filter the noisy signal, the result is a filtered signal whose spectrum is shown in Figure 9.42. The filtered signal is almost free of noise, though slightly muffled, as the high frequency elements in the song are eliminated along with the noise during filtering. It should be noted that the noise would have been much more

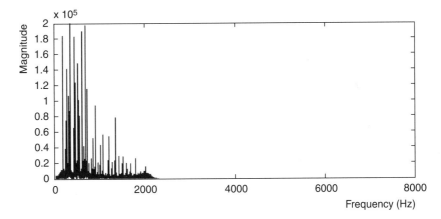

filtered.wav

FIGURE 9.42
Spectrum of signal plus noise after low pass filtering for Example 9.9.

difficult to remove from the song had it not been conveniently restricted to frequencies above 2 kHz. When noise, as is more typical, is spread across all frequencies, the portions that lie in the same ranges as the signal cannot be eliminated without drastically reducing signal quality.

9.7   BAND PASS AND HIGH PASS FIR FILTERS

Band pass and high pass FIR filters can be designed by starting with a low pass filter and shifting in frequency to produce the desired filter. This section describes the process of conversion. The one-sided magnitude response for a low pass FIR filter is shown in Figure 9.43, while Figure 9.44 shows the two-sided filter response. As discussed in Section 2.2.2, sampling causes spectral images of the two-sided response to appear at every multiple of the sampling frequency. Appendix K shows that this is a result of convolving the filter shape with a series of impulse functions in the frequency domain. Such convolution places one copy of the filter shape at every impulse function location. The same trick can be used to produce the frequency shift needed to convert a low pass prototype for a filter into a band pass or high pass filter. In this case, a single impulse function in the frequency domain must be located at the desired filter center frequency. Convolving a low pass filter shape with this impulse function has the effect of shifting a copy of the two-sided low pass filter shape to a new location, as explained in Appendix K. The result is a band pass or high pass filter with the same shape as the low pass prototype. The right column of Figure 9.45 illustrates these relationships.

The necessary impulse in Figure 9.45(b)(ii) can be conveniently provided by a cosine function since its spectrum has a single spike, as noted in Section 7.3.1 and derived in Appendix I. This cosine is shown in Figure 9.45(b)(i). Appendix D explains that, when convolution occurs in the frequency domain, multiplication must occur in the time domain. To convert a noncausal low pass filter $h_1[n]w[n]$ as designed in the last section into a band pass

FIGURE 9.43

One-sided magnitude response of low pass filter.

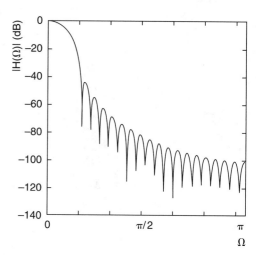

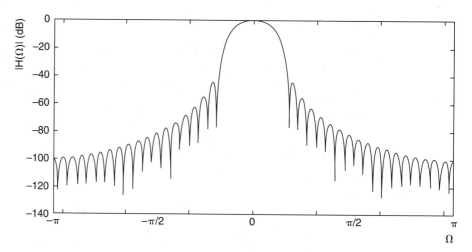

FIGURE 9.44

Two-sided magnitude response of low pass filter.

or high pass filter, then, the impulse response of the low pass filter must be multiplied by the cosine function

$$q[n] = \cos(n\Omega_0)$$

to produce

$$h[n] = h_1[n]w[n]\cos(n\Omega_0) \qquad \textbf{(9.4)}$$

where Ω_0 is chosen to equal the desired center frequency of the two-sided filter shape. This multiplication is illustrated in the left column of Figure 9.45. As is the case for low pass

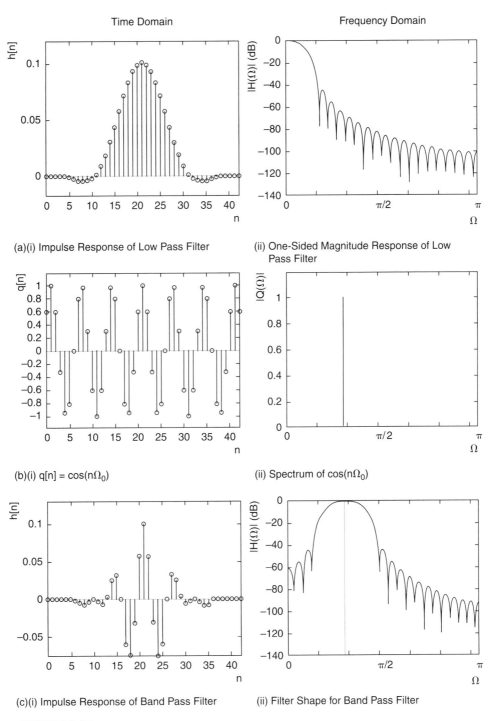

Time Domain

Frequency Domain

(a)(i) Impulse Response of Low Pass Filter

(ii) One-Sided Magnitude Response of Low Pass Filter

(b)(i) $q[n] = \cos(n\Omega_0)$

(ii) Spectrum of $\cos(n\Omega_0)$

(c)(i) Impulse Response of Band Pass Filter

(ii) Filter Shape for Band Pass Filter

FIGURE 9.45

Construction of a band pass filter.

designs, the last step in the design process is to shift the impulse response $h[n]$ for causality. The final, causal impulse response can be used to compute a frequency response, which in turn gives the new filter shape shown in Figure 9.45(c)(ii).

Thus, two small modifications must be made to the steps for low pass filter design given in Table 9.4: First, between Steps 3 and 4, the digital frequency Ω_0 must be calculated as

$$\Omega_0 = 2\pi \frac{f_0}{f_S}$$

where f_0 is the desired center frequency of the filter in Hz. For a band pass filter, this center frequency will lie between 0 and $f_S/2$ Hz. For a high pass filter, the center frequency f_0 is chosen to equal the Nyquist limit $f_S/2$ Hz, so that $\Omega_0 = \pi$ radians. Second, Step 4 must include the factor $\cos(n\Omega_0)$, so that the impulse response is calculated according to Equation (9.4). Note that these facts imply that the impulse response for a high pass filter is given by

$$h_{\text{high}}[n] = \cos(n\pi)h_{\text{low}}[n] = (-1)^n h_{\text{low}}[n]$$

which means sign changes on alternating samples of the low pass impulse response. As Figure 9.45 suggests, and as Examples 9.10 and 9.11 will show, specifications for a band pass or high pass filter must first be transformed into specifications for a low pass filter. The low pass design is then shifted in the frequency domain to obtain the filter required.

EXAMPLE 9.10

An FIR band pass filter must be designed for a 22 kHz system. The center frequency must be 4 kHz, with pass band edges at 3.5 and 4.5 kHz. Transition widths should be 500 Hz, and the stop band attenuation should be 50 dB. Design the filter.

The specifications must first be translated to describe a low pass filter. The band pass filter is shown in Figure 9.46. Its low pass equivalent is shown as a dashed line in the same figure. Only the frequencies between 0 and 11 kHz are shown because the sampling rate is 22 kHz.

The pass band edge frequencies of the band pass filter are 3.5 and 4.5 kHz, and the center frequency is 4 kHz, so the low pass filter must have its pass band edge at 500 Hz. Since the band pass filter must have a transition width of 500 Hz, the low pass filter must as well. Therefore, the pass band edge frequency f_1 and the equivalent digital frequency Ω_1 become, according to Table 9.4:

$$f_1 = 500 + \frac{500}{2} = 750 \text{ Hz}$$

$$\Omega_1 = 2\pi \frac{f_1}{f_S} = 0.06818\pi$$

The impulse response for the ideal low pass filter with this pass band edge is

$$h_1[n] = \frac{\sin(n\Omega_1)}{n\pi} = \frac{\sin(0.06818\pi n)}{n\pi}$$

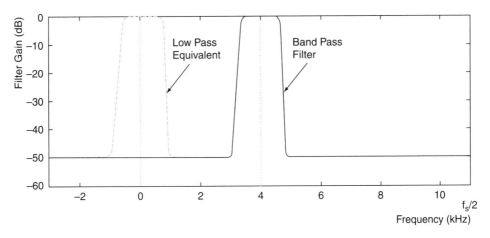

FIGURE 9.46
Band pass filter and low pass filter equivalent for Example 9.10.

From Table 9.3, the required stop band attenuation dictates that a Hamming window be used in the design. According to the table,

$$N = 3.44 \frac{f_S}{\text{T.W.}} = 3.44 \frac{22000}{500} = 151.4$$

The nearest odd integer is $N = 151$, so the window function becomes

$$w[n] = 0.54 + 0.46 \cos\left(\frac{2\pi n}{150}\right)$$

Since the center frequency f_0 for the band pass filter must be 4 kHz, the center digital frequency Ω_0 for the cosine function $\cos(n\Omega_0)$ must be located at

$$\Omega_0 = 2\pi \frac{f_0}{f_S} = 2\pi \frac{4000}{22000} = 0.3636\pi$$

The impulse response samples are calculated from $h[n] = h_1[n]w[n]\cos(n\Omega_0)$. The table of calculations is reproduced in part in Table 9.7, and the impulse response is illustrated in Figure 9.47. As expected, the impulse response is symmetrical around its central axis, $n = 75$ in this case.

As seen in Section 4.7, the difference equation may be constructed from the impulse response for this nonrecursive FIR filter. The impulse response is

$$h[n] = h[0]\delta[n] + h[1]\delta[n-1] + h[2]\delta[n-2] + \ldots + h[150]\delta[n-150]$$

so the difference equation is

$$y[n] = h[0]x[n] + h[1]x[n-1] + h[2]x[n-2] + \ldots + h[150]x[n-150]$$

TABLE 9.7
Calculations of Band Pass Impulse Response for Example 9.10

n	$h_1[n]$	$w[n]$	$\cos(n\Omega_0)$	$h[n]$	new n
-75	-0.001	0.0800	-0.661	0.00008	0
-74	-0.001	0.0804	-0.957	0.00005	1
-73	0.000	0.0816	-0.134	0.00000	2
. . . . .	. . .	. . .	. . .	. . .	
-5	0.056	0.9899	0.841	0.04651	70
-4	0.060	0.9936	-0.143	-0.00853	71
-3	0.064	0.9964	-0.960	-0.06079	72
-2	0.066	0.9984	-0.655	-0.04321	73
-1	0.068	0.9996	0.416	0.02810	74
0	0.068	1.0000	1.000	0.06818	75
1	0.068	0.9996	0.416	0.02810	76
2	0.066	0.9984	-0.655	-0.04321	77
3	0.064	0.9964	-0.960	-0.06079	78
4	0.060	0.9936	-0.143	-0.00853	79
5	0.056	0.9899	0.841	0.04651	80
. . .	. . .	. . .	. . .	. . .	. . .
73	0.000	0.0816	-0.134	0.00000	148
74	-0.001	0.0804	-0.957	0.00005	149
75	-0.001	0.0800	-0.661	0.00008	150

FIGURE 9.47
Band pass filter
impulse response
for Example 9.10.

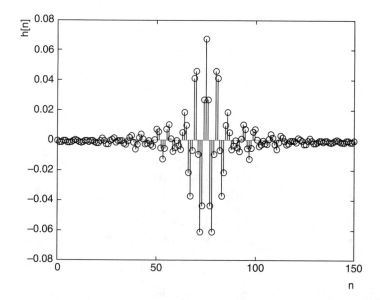

FIGURE 9.48
Filter shape of band pass filter response for Example 9.10.

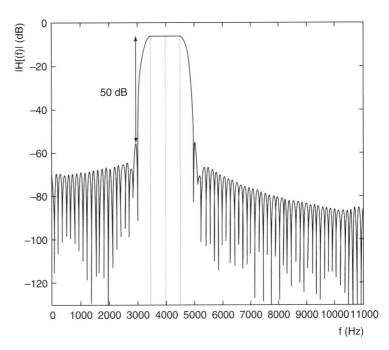

The impulse response $h[n]$ may also be used to compute the frequency response $H(\Omega)$. The magnitude response $|H(f)|$, computed from the frequency response, is plotted in Figure 9.48. The gain in the pass band is -6 dB, or 0.5. If desired, this gain may be changed to 0 dB, or 1, by multiplying all $h[n]$ values by $1/0.5 = 2$. According to Table 9.3, the pass band edges will lie 0.02 dB below the maximum gain. In this case, then, the gain at the pass band edges is $-6 - 0.02 = -6.02$ dB. These edges occur at 3501 and 4499 Hz, giving the required bandwidth of 1 kHz. An attenuation of 50 dB (or a gain of -56 dB, since the gain in the pass band is -6 dB) is obtained at 3.005 and 4.994 kHz, producing the specified 500 Hz transition widths.

EXAMPLE 9.11

A high pass filter is required with a pass band edge at 8 kHz and a stop band edge at 6 kHz. The stop band gain should be at least 40 dB below the pass band gain. The sampling rate is 22 kHz. Design the filter and provide its impulse response.

The first step in the design is to find the low pass equivalent of the desired high pass filter. Recall that, for the high pass filter, the center frequency f_0 is chosen as half the sample rate, 11 kHz. As Figure 9.49 suggests, the high pass filter is designed to be symmetrical about this frequency. For a sampling frequency of 22 kHz, the pass band of the high pass filter lies between 8 kHz and 11 kHz. The transition width must be $8 - 6 = 2$ kHz. For the

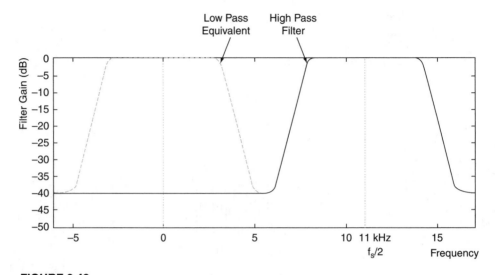

FIGURE 9.49
High pass filter and equivalent low pass filter for Example 9.11.

prototype low pass filter, then, the pass band edge must occur at 3 kHz and the stop band edge at 5 kHz, giving a required transition width.

As outlined in Section 9.6.2, the pass band edge frequency f_1 and the corresponding digital frequency Ω_1 for the low pass filter become

$$f_1 = 3000 + \frac{2000}{2} = 4000 \text{ Hz}$$

$$\Omega_1 = 2\pi \frac{f_1}{f_S} = 0.3636\pi$$

The impulse response for the ideal low pass filter is

$$h_1[n] = \frac{\sin(n\Omega_1)}{n\pi} = \frac{\sin(0.3636\pi n)}{n\pi}$$

From Table 9.3, a Hanning window can provide the required stop band attenuation, with

$$N = 3.32\frac{f_S}{\text{T.W.}} = 3.32\frac{22000}{2000} = 36.5$$

Rounding up to the next odd integer, $N = 37$ is the number of terms selected. The window function is

$$w[n] = 0.5 + 0.5 \cos\left(\frac{2\pi n}{36}\right).$$

The center frequency of the high pass filter is $f_0 = 11$ kHz, so digital center frequency Ω_0 for $\cos(n\Omega_0)$ must be

$$\Omega_0 = 2\pi \frac{f_0}{f_S} = 2\pi \frac{11000}{22000} = \pi$$

This is true for all high pass filters, as noted earlier. Table 9.8 shows the calculations of the impulse response as $h[n] = h_1[n]w[n]\cos(n\Omega_0)$, and Figure 9.50 displays its symmetrical nature. A pole-zero diagram for the filter, computed from the transfer function $H(z)$, is shown in Figure 9.51(a). Since the filter has $N = 37$ coefficients, it has 36 zeros and 36 poles. The first and last samples of the impulse response are zero, so the system has $36 - 2 = 34$ zeros. One of these zeros lies at $z = -6.93$ and therefore does not appear in Figure 9.51(a). All poles lie at the origin. The shape of the filter produced from the impulse

TABLE 9.8
Calculation of High Pass Impulse Response for Example 9.11

n	$h_1[n]$	$w[n]$	$\cos(n\Omega_0)$	$h[n]$	new n
−18	0.0175	0.0000	1.000	0.0000	0
−17	0.0101	0.0076	−1.000	−0.0001	1
...	...	...	...	...	...
−5	−0.0344	0.8214	−1.000	0.0283	13
−4	−0.0788	0.8830	1.000	−0.0695	14
−3	−0.0299	0.9330	−1.000	0.0279	15
−2	0.1203	0.9698	1.000	0.1167	16
−1	0.2895	0.9924	−1.000	−0.2873	17
0	0.3636	1.0000	1.000	0.3636	18
1	0.2895	0.9924	−1.000	−0.2873	19
2	0.1203	0.9698	1.000	0.1167	20
3	−0.0299	0.9330	−1.000	0.0279	21
4	−0.0788	0.8830	1.000	−0.0695	22
5	−0.0344	0.8214	−1.000	0.0283	23
...	...	...	...	...	...
17	0.0101	0.0076	−1.000	−0.0001	35
18	0.0175	0.0000	1.000	0.0000	36

FIGURE 9.50
Impulse response of high pass filter for Example 9.11.

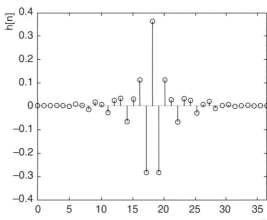

FIGURE 9.51
Pole-zero plot and
filter shape of high
pass filter for
Example 9.11.

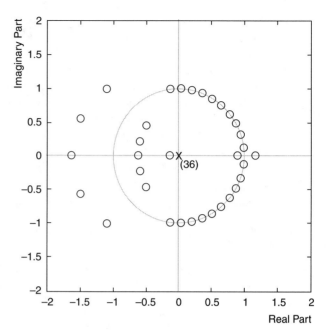

(a) Pole-Zero Diagram

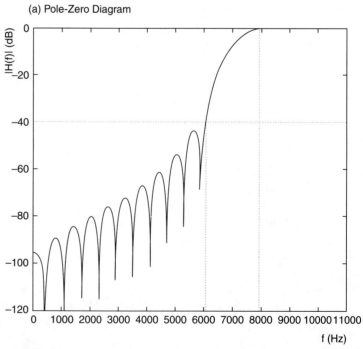

(b) Magnitude Response

response in Table 9.8 is presented in Figure 9.51(b). The edge of the pass band is found at 7.948 kHz, marked according to Table 9.3 by a gain of -0.06 dB. An attenuation of 40 dB occurs at a frequency of 6.075 kHz. Thus, all filter specifications are met.

EXAMPLE 9.12

there.wav
lowthere.wav
highthere.wav

A low pass filter and a high pass filter are applied to a speech sample. Both filters have a pass band edge at 1.5 kHz. Each uses a Hanning window with $N = 125$ terms. The low pass filter produces a muffled version of the original speech that consists of low frequency elements only. The high pass filter passes only the tinny, high frequency portions of the signal. The outputs of the two filters summed together give a close approximation to the original signal, differing only because the filters are not ideal.

EXAMPLE 9.13

A cellular phone channel with a bandwidth of 30 kHz must be extracted from a band of channels between 900 and 900.1 MHz. The center of the channel lies at 900.03 MHz. Design a filter for the task if a sampling rate of 200 kHz is used.

The cell phone channel lies between $900.03 - 0.015 = 900.015$ and $900.03 + 0.015 = 900.045$ MHz. The sampling rate is much lower than these signal frequencies, so aliasing definitely occurs. Since 900 MHz is an integer multiple of the sampling frequency, important spectral information that lies between 900.015 MHz and 900.045 MHz is mirrored across 900 MHz to the range 899.955 MHz to 899.985 MHz, and similar spectral copies appear around all other multiples of the sampling frequency, as shown in Figure 9.52. As a result, the cell phone spectrum aliases into the baseband between 15 and 45 kHz, but with the same shape as the high frequency original.

From Figure 9.52, the filtering task reduces to extracting the signal between 15 and 45 kHz for a 200 kHz sampling rate. This task may be addressed with a band pass filter with a center frequency of 30 kHz, whose low pass equivalent has a pass band edge at 15 kHz. For reasonable isolation of the desired channel from others, a transition width of 1 kHz is selected, and the Hamming window is chosen for good stop band attenuation. A design pass band edge frequency of $f_1 = 15 + 1/2 = 15.5$ kHz will place the actual pass band edge at the desired frequency of 15 kHz, so $\Omega_1 = 2\pi f_1/f_S = 2\pi(15500/200000) = 0.155\pi$ radians. The impulse response for the ideal low pass filter with this pass band edge frequency is

$$h_1[n] = \frac{\sin(n\Omega_1)}{n\pi} = \frac{\sin(0.155\pi n)}{n\pi}$$

From Table 9.3,

$$N = 3.44\frac{f_S}{\text{T.W.}} = 3.44\frac{200000}{1000} = 688$$

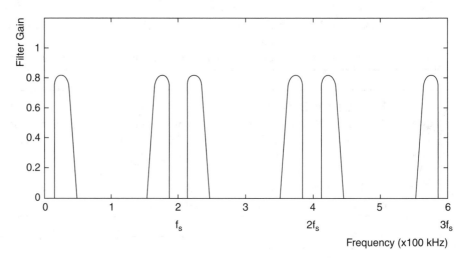

FIGURE 9.52
Aliasing of signal due to undersampling in Example 9.13.

Selecting $N = 689$ as the next higher odd number, the window function becomes

$$w[n] = 0.54 + 0.46 \cos\left(\frac{2\pi n}{688}\right)$$

The final element of the design is the cosine function $\cos(n\Omega_0)$ that shifts the low pass prototype to the desired band pass center frequency, 30 kHz. The digital center frequency is

$$\Omega_0 = 2\pi\frac{30000}{200000} = 0.3\pi$$

The impulse response for the band pass filter is given by $h[n] = h_1[n]w[n]\cos(0.3\pi n)$, where $h_1[n]$ and $w[n]$ are defined above. The magnitude response for the filter is plotted against frequency in Hz in Figure 9.53. The filter has extremely steep sides as a result of the large number of terms used. A possible obstacle to implementing a high quality filter like this in practice is that a DSP processor that is fast enough to multiply and add 688 terms in a single clock cycle must be available.

9.8 BAND STOP FIR FILTERS

Band stop filters suppress one range of frequencies, the stop band, while passing all others. They cannot be designed according to the method presented in the previous section, but a simple method is available. Provided the pass band edge frequencies are chosen correctly, a band stop filter can be constructed by combining a low pass filter with a high pass filter. The low pass filter sets the pass band edge frequency for the lower end of the stop band, and the high pass filter sets the pass band edge frequency for the higher end, as depicted in

FIGURE 9.53

Magnitude response of band pass filter for Example 9.13.

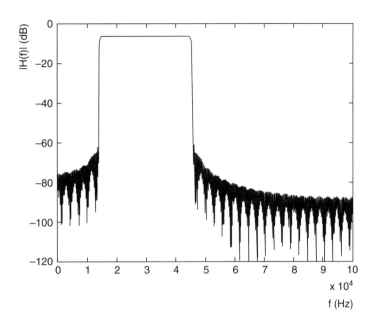

FIGURE 9.54

Constructing a band stop filter.

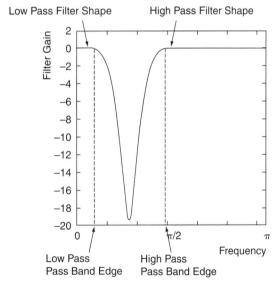

Figure 9.54. Thus, to create a band stop characteristic, the pass band edge frequency for the low pass filter must lie below that of the high pass filter.

The impulse responses for the two filters, $h_{low}[n]$ and $h_{high}[n]$ can be found using the methods of the previous section. To understand how they must be combined to produce the impulse response for the band stop filter, it is useful to consider the sum in the z domain

FIGURE 9.55

Summing low and high pass filters
to create a band stop filter.

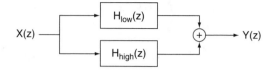

shown in Figure 9.55. Recall from Figure 6.4 that an input $X(z)$ applied to a low pass filter produces the output $H_{low}(z)X(z)$, and that the same input applied to a high pass filter gives the output $H_{high}(z)X(z)$. The desired output is the combination of these two outputs:

$$Y(z) = H_{low}(z)X(z) + H_{high}(z)X(z)$$

and the overall transfer function is

$$H_{band\ stop}(z) = H_{low}(z) + H_{high}(z)$$

Because the sum of the high and low pass transfer functions produces the transfer function for a band stop filter, the impulse responses of the three filters are related in the same way:

$$h_{band\ stop}[n] = h_{low}[n] + h_{high}[n] \tag{9.5}$$

Thus, the impulse response for a band stop filter is simply the sum of the impulse responses for the high pass and low pass filters.

Band pass filters may be designed in a similar fashion. Both a low pass and a high pass filter are required, but, in this case, the output of one filter is passed to the next as input, as shown in Figure 9.56. Figure 9.57 illustrates the relationship between the band pass filter and the low pass and high pass filters used to construct it. To obtain the required band pass behavior, the low pass filter's pass band edge frequency must be greater than that of the high pass filter, and the frequency difference between the two pass band edges defines the width of the pass band. The transfer function of the band pass filter is described by the product

$$H_{band\ pass}(z) = H_{low}(z)H_{high}(z)$$

As discussed in Appendix D, multiplication in the frequency domain is equivalent to convolution in the time domain. Therefore, the impulse responses of the three filters are related by the equation

$$h_{band\ pass}[n] = h_{low}[n]*h_{high}[n] \tag{9.6}$$

FIGURE 9.56

Cascading low and high pass
filters to create a band pass
filter.

FIGURE 9.57

Constructing a band pass filter.

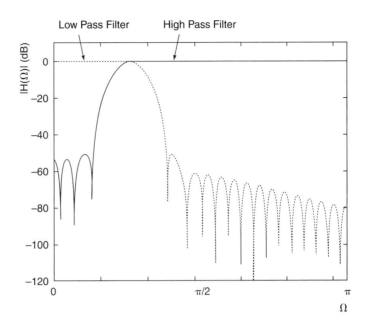

EXAMPLE 9.14

A selection of samples from a digital voltage signal is shown in Figure 9.58, collected using a 1 kHz sampling rate. A 60 Hz disturbance is suspected of influencing the measurements. A band stop filter must be designed to remove the disturbance.

To be sure the 60 Hz disturbance is removed, a band stop filter with a stop band between 40 and 80 Hz will be designed. To achieve this, the low pass filter used in the construction of the band stop filter must have a pass band edge at 40 Hz, and the high pass filter at 80 Hz. Hamming windows will be used for both designs, with $N = 151$ to ensure a reasonably steep roll-off. The window function is

$$w[n] = 0.54 + 0.46 \cos\left(\frac{2\pi n}{150}\right)$$

Since $N = 3.44 \, f_s/\text{T.W.}$, 151 terms give a transition width of 23 Hz. For the low pass filter, the pass band edge frequency is

$$\Omega_{1L} = 2\pi \frac{f_{1L}}{f_S} = 2\pi \frac{\left(40 + \dfrac{23}{2}\right)}{1000} = 0.103\pi$$

This gives the ideal impulse response

$$h_{1L}[n] = \frac{\sin(n\Omega_{1L})}{n\pi} = \frac{\sin(0.103\pi n)}{n\pi}$$

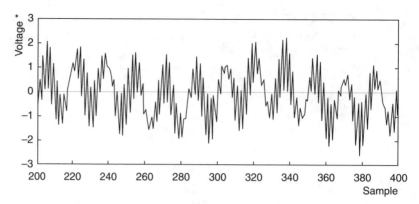

* For clarity, the envelope, rather than the individual samples of the digital signal is plotted.

FIGURE 9.58

Voltage signal for Example 9.14.

The impulse response for the low pass filter is $h_L[n] = h_{1L}[n]w[n]$. The high pass filter has its pass band edge at 80 Hz. The low pass prototype for this filter has its pass band edge at $f_S/2 - 80 = 500 - 80 = 420$ Hz. Thus,

$$\Omega_{1H} = 2\pi \frac{f_{1H}}{f_S} = 2\pi \frac{\left(420 + \dfrac{23}{2}\right)}{1000} = 0.863\pi$$

and

$$h_{1H}[n] = \frac{\sin(n\Omega_{1H})}{n\pi} = \frac{\sin(0.863\pi n)}{n\pi}$$

To convert the low pass filter described by $h_{1H}[n]w[n]$ to a high pass filter requires multiplying by $\cos(n\Omega_0)$, as directed in Section 9.7. For all high pass filters, $\Omega_0 = \pi$. The impulse response for the high pass filter, then, is $h_H[n] = h_{1H}[n]w[n]\cos(n\pi)$. According to Equation (9.5), the impulse response for the band stop filter is the sum of the low pass and high pass filter responses, or $h_{BP}[n] = h_L[n] + h_H[n]$. The filter shape obtained from this impulse response is shown in Figure 9.59.

It is instructive to see the effects of the filters, alone and together. In each case, filtering may be implemented through either convolution or difference equations. Figure 9.60 shows what happens when the low pass filter alone is applied to the voltage signal. A slow variation can be observed. Figure 9.61 shows the output from the high pass filter alone. Finally, Figure 9.62 shows the output when the band stop filter is applied to the signal. This signal is the sum of the signals in Figure 9.60 and Figure 9.61. It is clearly very different from the original in Figure 9.58, so some element has been removed by band stop filtering.

To verify that it was indeed a 60 Hz disturbance that was removed, the voltage signal in Figure 9.58 may be applied to a band pass filter with a pass band between 40 and 80 Hz.

FIGURE 9.59

Band stop filter shape for Example 9.14.

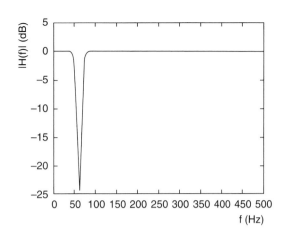

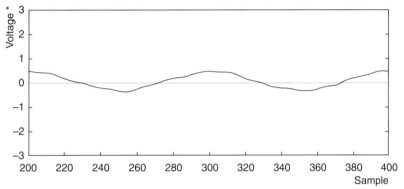

* For clarity, the envelope of the signal, rather than its individual samples, is plotted.

FIGURE 9.60

Low pass filter output for Example 9.14.

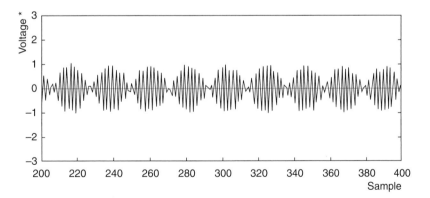

* For clarity, the envelope of the signal, rather than its individual samples, is plotted.

FIGURE 9.61

High pass filter output for Example 9.14.

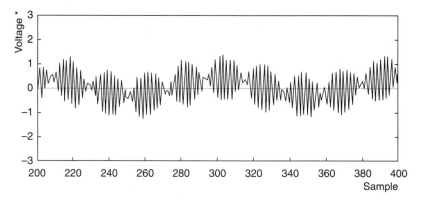

* For clarity, the envelope of the signal, rather than its individual samples, is plotted.

FIGURE 9.62

Band stop filter output for Example 9.14.

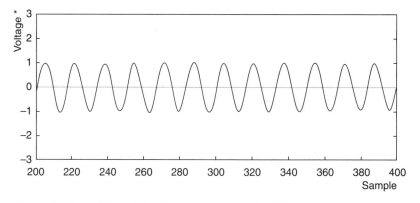

* For clarity, the envelope of the signal, rather than its individual samples, is plotted.

FIGURE 9.63

Band pass filter output for Example 9.14.

This time, the pass band edge frequency for the low pass filter is 80 Hz and that for the high pass filter is 40 Hz, so $f_{1L} = 80 + 23/2 = 91.5$ Hz and $f_{1H} = (500-40) + 23/2 = 471.5$ Hz. It follows that $\Omega_{1L} = 0.183\pi$ and $\Omega_{1H} = 0.943\pi$, and also that $h_{1L}[n] = \sin(0.183\pi n)/n\pi$ and $h_{1H}[n] = \sin(0.943\pi n)/n\pi$. The impulse response for the low pass filter is $h_L[n] = h_{1L}[n]w[n]$, and that for the high pass filter is $h_H[n] = h_{1H}[n]w[n]\cos(n\pi)$. According to Equation (9.6), the impulse response for the band pass filter is the convolution of $h_L[n]$ and $h_H[n]$. Figure 9.63 shows the output of the band pass filter when the voltage signal is applied. The hypothesis of a 60 Hz disturbance is confirmed: 12 cycles in 200 samples at a 1 kHz sampling rate correspond to (1000 samples/second)(12 cycles)/(200 samples) = 60 cycles/second.

9.9   EQUIRIPPLE FIR FILTER DESIGN

The windowed FIR filter designs examined in the preceding sections have one thing in common: Their pass bands and stop bands show rippling that varies in amplitude. As Figure 9.64(a) illustrates, the pass ripple and the stop band ripple are largest near the pass band and stop band edges. When the filter shape is plotted in decibels, as in Figure 9.64(b), changing ripple amplitude is especially apparent in the stop band. The stop band attenuation specification is met at the first side lobe, but far exceeded by the side lobes situated at higher frequencies. The idea behind **equiripple filter** design is to even out the ripple amplitudes, as shown in Figure 9.65, so that a better approximation to an ideal filter response may be achieved. As it happens, equiripple FIR filters also generally require fewer terms than the windowed FIR filters studied in the previous sections to achieve the same objectives.

Some aspects of equiripple design are familiar: To obtain phase distortion-free operation, for example, the requirement for a symmetrical impulse response is maintained. Section 9.3 showed that the frequency response of a symmetrical, noncausal impulse response with $2M + 1$ terms is given by

$$H_1(\Omega) = h_1[0] + 2h_1[1]\cos\Omega + 2h_1[2]\cos2\Omega + \ldots + 2h_1[M]\cos M\Omega$$

The frequency response $H_1(\Omega)$ is a purely real function of frequency Ω. For a given value of M, the shape of the function changes with the coefficients $h_1[n]$. Mathematically, the equiripple design relies on minimizing the maximum (weighted) error that occurs between the proposed filter shape $H_1(\Omega)$ and its ideal counterpart, which are compared in Figure 9.66. The weighted error is calculated from

$$E(\Omega) = W(\Omega)(H_{\text{ideal}}(\Omega) - H_1(\Omega))$$

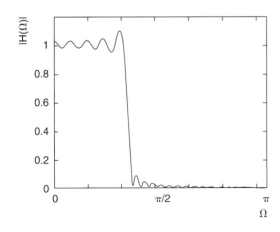

(a) Linear Filter Shape

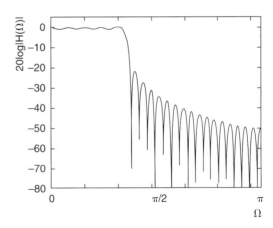

(b) Logarithmic Filter Shape

FIGURE 9.64
Varying ripple windowed FIR filter shape ($N = 51$).

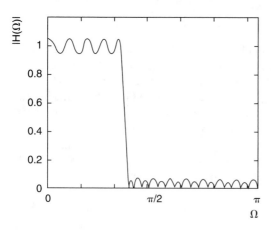

(a) Linear Filter Shape

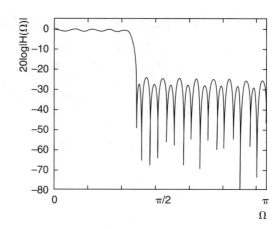

(b) Logarithmic Filter Shape

FIGURE 9.65

Equiripple FIR filter shape ($N = 51$).

FIGURE 9.66

Features of
equiripple filters.

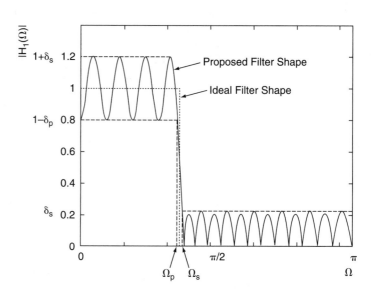

where $W(\Omega)$ is a weighting function that permits the ripples in the pass band to differ in size from the ripples in the stop band. $H_{\text{ideal}}(\Omega)$ is the ideal filter response, with unity gain in the pass band and zero gain in the stop band, and $H_1(\Omega)$ is the frequency response of the proposed filter. In practice, the optimal solution is found by modeling $H_1(\Omega)$ as an Mth order polynomial in powers of $\cos\Omega$.

In finding the frequency response $H_1(\Omega)$ that minimizes the maximum weighted error, $\max_{\Omega}|E(\Omega)|$, two requirements must be met: The gain at the edge of the pass band must equal the minimum gain in the pass band, and the gain at the edge of the stop band must equal the maximum gain in the stop band. These requirements are embodied by the equations:

$$H_1(\Omega_p) = 1 - \delta_p \tag{9.7a}$$

$$H_1(\Omega_s) = \delta_s \tag{9.7b}$$

It turns out that, for a given choice of M, the function $H_1(\Omega)$ that minimizes the maximum weighted error subject to the constraints in Equation (9.7) is a function that has an **equiripple** characteristic. That is, the size of the pass band ripple is consistent throughout the pass band, and the size of the stop band ripple is consistent throughout the stop band. The method that identifies the optimal choice for $H_1(\Omega)$ is called the Parks-McClellan algorithm (Oppenheim & Schafer, 1999) or the Remez exchange algorithm (Ifeachor & Jervis, 1993). Once the best filter shape $H_1(\Omega)$ has been identified, it must be converted to an impulse response $h_1[n]$ to allow implementation of the filter. This is achieved in two steps: by first sampling the filter shape, and then taking an inverse discrete Fourier transform of the samples to obtain a set of impulse response values. This process will be illustrated in Example 11.3.

9.10   HAZARDS OF PRACTICAL FIR FILTERS

Difference equation diagrams for implementing FIR filters were introduced in Section 4.6.1. Practical implementations experience difficulties that theoretical designs do not. First, the inputs to the filters are obtained from analog sensors whose measurements have been converted to digital, with the attendant quantization noise and aliasing errors. Second is the quantization of the filter coefficients themselves. The coefficients must be represented using the number of bits the processor makes available, and large errors can occur when only a small number of bits is used. Errors in the coefficient values change the final shape of the filter. The specific side effects include lower attenuation (that is, higher gain) in the stop band of the filter, as well as undesirable ripple in the pass band.

EXAMPLE 9.15

A 41-term equiripple filter is designed, with coefficients listed in the first column of Table 9.9. The filter shape is shown as a solid line in Figure 9.67. When the filter coefficients are quantized, for implementation in DSP software, significant changes occur. Each original coefficient C is quantized according to the formula

$$C_{\text{quant}} = \text{sign}(C)\frac{\text{truncate}(2^Q|C| + 0.5)}{2^Q} \tag{9.8}$$

where Q is the number of bits used to quantize. This formula ensures that the closest Q-bit representation possible is used for each coefficient. For this example, $Q = 8$. Coefficients quantized to $Q = 8$ bits are listed in the second column of Table 9.9. The dramatic effect on the filter shape is evident in Figure 9.67, where the quantized filter has been added with a dashed line.

TABLE 9.9
Unquantized and Quantized Filter Coefficients for Example 9.15

Coefficient	Quantized Coefficient
−0.00037217	0.00000000
−0.00068571	0.00000000
0.00008766	0.00000000
0.00183039	0.00000000
0.00187290	0.00000000
−0.00177549	0.00000000
−0.00530586	−0.00390625
−0.00167450	0.00000000
0.00793053	0.00781250
0.01003589	0.01171875
−0.00438293	−0.00390625
−0.02021976	−0.01953125
−0.01025327	−0.01171875
0.02383469	0.02343750
0.03710594	0.03515625
−0.00707476	−0.00781250
−0.06998092	−0.07031250
−0.05223657	−0.05078125
0.09752810	0.09765625
0.29812590	0.29687500
0.39176594	0.39062500
0.29812590	0.29687500
0.09752810	0.09765625
−0.05223657	−0.05078125
−0.06998092	−0.07031250
−0.00707476	−0.00781250
0.03710594	0.03515625
0.02383469	0.02343750
−0.01025327	−0.01171875
−0.02021976	−0.01953125
−0.00438293	−0.00390625
0.01003589	0.01171875
0.00793053	0.00781250
−0.00167450	0.00000000
−0.00530586	−0.00390625
−0.00177549	0.00000000
0.00187290	0.00000000
0.00183039	0.00000000
0.00008766	0.00000000
−0.00068571	0.00000000
−0.00037217	0.00000000

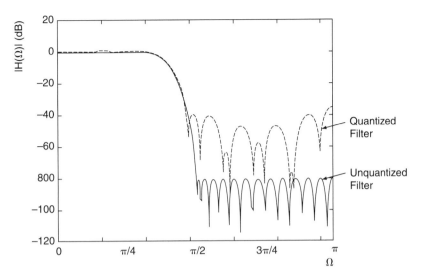

FIGURE 9.67
Unquantized and quantized FIR filter shapes for Example 9.15.

Other sources of error are arithmetic in nature. Repetitive multiplication and addition makes arithmetic overflow a serious possibility. When it occurs, a number that should be extremely large may appear as a very small number, causing the output at that step to be distorted. Overflow can be avoided by scaling down the coefficients and the inputs before calculations occur and scaling them back up at the output, but these operations can result in loss of precision as well. Many processors multiply using extra-wide hardware registers. The results of multiplication, though, must eventually be transmitted on normal-width buses, which means the least significant bits have to be eliminated through truncation or rounding. This adds errors to intermediate results, which accumulate in the final answer. Quantization and arithmetic errors that are a result of the limited number of bits used to represent values before, during, and after computations are classed as finite word length effects. Lessening the impact of these effects is a major consideration in the design of DSP algorithms.

CHAPTER SUMMARY

Matlab
Support

1. The difference equation for an FIR filter is nonrecursive.
2. All FIR filters are stable.
3. A moving average filter is one example of a low pass FIR filter.
4. Phase distortion causes some frequency elements of a signal to arrive at the output of a filter ahead of others. When extreme, the effect is audible.
5. An FIR filter whose impulse response is symmetrical about its midpoint is free of phase distortion. A symmetrical impulse response produces a linear phase relationship in the

pass band of the filter. Thus, a linear phase relationship also guarantees no phase distortion will occur.

6. Low pass filter specifications include pass band edge frequency, stop band edge frequency, transition width, stop band attenuation, pass band ripple, stop band ripple, and sampling frequency.

7. One FIR filter design method windows the impulse response for an ideal low pass filter. After windowing, the filter shape is no longer ideal, but has a finite transition width and ripples in its pass and stop bands. The larger the number of terms in the window, the closer to ideal the filter shape will be. Choices of window functions include rectangular, Hanning, Hamming, Blackman, and Kaiser. Each offers a characteristic stop band attenuation. Band pass and high pass filters are produced by shifting a low pass filter prototype to the desired location in the frequency domain.

8. Band stop filters are created by summing a low pass filter and a high pass filter: The pass band edge frequency for the high pass filter exceeds that of the low pass filter. Band pass filters may be created by cascading a low pass and a high pass filter. The pass band edge frequency for the low pass filter exceeds that of the high pass filter.

9. Equiripple FIR filter design determines a filter shape that minimizes the maximum difference from the ideal. The ripples are equal through each of the pass and stop bands.

10. When FIR filter coefficients are quantized for use with a DSP processor, quantization errors can cause dramatic changes to the designed shape of the filter.

REVIEW QUESTIONS

9.1 Show that the filter described by the difference equation

$$y[n] = 0.1x[n] + 0.5x[n-1] + 0.9x[n-2] + 0.5x[n-3] + 0.1x[n-4]$$

has a finite impulse response. What is the length of the response, and how does it relate to the maximum delay in the difference equation?

9.2 For a seven-term moving average filter, write an expression for the
   a. Difference equation
   b. Impulse response
   c. Transfer function
   d. Frequency response

9.3 a. Sketch the frequency response for a seven-term moving average filter using dB for magnitudes and degrees for phases. Use digital frequency steps of $\pi/8$ radians or smaller.
   b. With reference to the frequency response sketch, explain why this filter can guarantee that no pass band phase distortion will occur.

9.4 Filter the signal shown in Figure 9.68 using a three-term moving average filter.
   a. Plot the first 15 samples of the output.
   b. Plot the filter shape as $|H(\Omega)|$ versus Ω.
   c. Demonstrate that the phase response, $\theta(\Omega)$ versus Ω, is linear in the pass band.

FIGURE 9.68

Signal for Question 9.4.

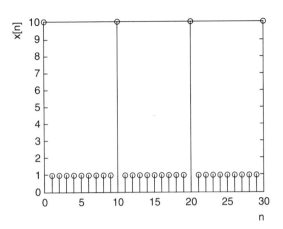

9.5 Draw the impulse response between $n = -8$ and $n = 8$ for an ideal low pass filter with a pass band edge at $\Omega_1 = 0.5$ rads.

9.6 The impulse response for an ideal low pass filter with a pass band edge at $\Omega_1 = \pi/4$ radians is truncated outside $-3 \leq n \leq 3$.

  a. Plot the truncated impulse response.

  b. Shift the truncated impulse response to make it causal. Write the equation for this new impulse response and plot it.

  c. Draw the magnitude response $|H(\Omega)|$ for the causal impulse response and show the magnitude response for the ideal low pass filter on the same graph.

9.7 For the magnitude response in Figure 9.69, determine the:

  a. Pass band ripple

  b. Stop band ripple

FIGURE 9.69

Filter shape for
Question 9.7.

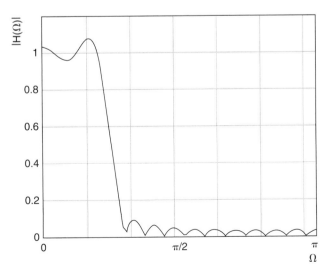

      c. Gain at edge of pass band (dB)

      d. Gain at edge of stop band (dB)

      e. Stop band attenuation (dB)

      f. Transition width (radians)

9.8 The magnitude response for a filter in a 22 kHz sampled system appears in Figure 9.70. Determine the:

      a. Pass band ripple

      b. Stop band ripple

      c. Bandwidth (Hz)

      d. Stop band attenuation

      e. Center frequency (Hz)

      f. Transition width (Hz)

FIGURE 9.70

Filter shape for Question 9.8.

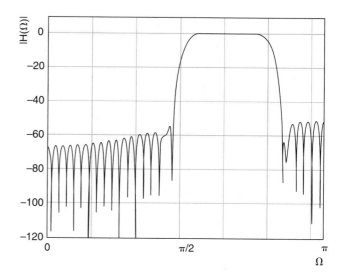

9.9 Draw a linear magnitude response for a filter with the characteristics:

      a. Maximum pass band gain = 1.03

          Pass band ripple = 0.05

          Transition width = 1500 Hz

          Stop band edge frequency = 3 kHz

          Stop band ripple = 0.075

          Sampling frequency = 8 kHz

      b. Pass band ripple = 0.08

          Stop band attenuation = 25 dB

          Gain at pass band edge = −0.72 dB

          Pass band edge at 2 kHz

          Stop band edge at 4 kHz

          Sampling frequency 20 kHz

9.10 Draw nine-term windows, $w[n]$ versus n, for each of the following window types:
  a. Rectangular
  b. Hanning
  c. Hamming
  d. Blackman

9.11 Draw magnitude responses, $|W(\Omega)|$ versus Ω, for nine-term windows of each of the following window types:
  a. Rectangular
  b. Hanning
  c. Hamming
  d. Blackman

9.12 For each of the following low pass filter specifications, select an FIR window type and identify the number of terms needed to meet the requirements:
  a. Stop band attenuation 20 dB; transition width 1 kHz; sampling frequency 12 kHz
  b. Stop band attenuation 50 dB; transition width 2 kHz; sampling frequency 5 kHz
  c. Stop band attenuation 50 dB; transition width 500 Hz; sampling frequency 5 kHz
  d. Pass band gain 10 dB; stop band gain -30 dB; pass band edge frequency 5 kHz; stop band edge frequency 6.5 kHz; sampling frequency 22 kHz

9.13 For each set of the FIR specifications in (a) and (b):
  (i) Draw the filter shape.
  (ii) Choose a window and calculate the number of terms needed.
  (iii) Choose the pass band edge frequency that must be used for the design.
  a. Pass band gain 0 dB; stop band gain -40 dB; pass band edge frequency 1 kHz; stop band edge frequency 2.5 kHz; sampling frequency 12 kHz
  b. Pass band gain 10 dB; stop band gain -10 dB; pass band edge frequency 500 Hz; transition width 330 Hz; sampling frequency 2 kHz

9.14 Find the number of zeros, poles, and coefficients that characterize an FIR filter with
  a. $N = 51$ terms
  b. $N = 101$ terms

9.15 Design a low pass FIR filter to meet the following specifications. Provide all equations needed to produce the filter's impulse response. Specifications: pass band gain 0 dB, stop band attenuation 40 dB, pass band ripple < 0.008, stop band ripple < 0.01, pass band edge 3 kHz, stop band edge 3.5 kHz, sampling frequency 12 kHz.

9.16 A rectangular window with 25 terms is used to design a low pass FIR filter. The pass band edge is located at 2 kHz and the pass band gain is about 0 dB. For a sampling frequency of 20 kHz:
  a. Determine the transition width in Hz.
  b. Draw the filter shape as $|H(f)|$ versus f in Hz.

9.17 Find and plot the impulse response for a filter with the following specifications:
  Low pass
  Linear phase in pass band
  Sampling frequency 16 kHz
  Pass band edge frequency 4.5 kHz
  Stop band edge frequency 6 kHz
  Pass band gain 0 dB

 Stop band attenuation 75 dB

(Use software to assist with repetitive calculations.)

9.18 A low pass filter with 79 terms uses the Kaiser window with $\beta = 6$. At the pass band edge, the gain is 0 dB and the frequency is 6 kHz. If the sampling frequency is 11.025 kHz:

a. What is the transition width?

b. What is the stop band edge frequency?

c. What is the gain at the stop band edge?

9.19 Design a low pass FIR filter for 10 kHz sampling, with a pass band edge at 2 kHz, a stop band edge at 3 kHz, and 20 dB stop band attenuation. Find the impulse response and the difference equation for the filter.

9.20 For the low pass filter with the specifications:

 Finite impulse response

 Stop band attenuation 50 dB

 Pass band edge 1.75 kHz

 Transition width 1.5 kHz

 Sampling frequency 8 kHz

a. Write the difference equation for the filter.

b. Plot the magnitude response for the filter in dB versus Hz, and verify that the specifications are met.

9.21 To smooth the dynamic mass measurements produced by a strain gauge in a 10 Hz sampled system, all frequencies above 1 Hz must be removed. Design a suitable filter.

9.22 The impulse response of a high pass filter is $h[n] = 0.0101\delta[n] - 0.2203\delta[n-1] + 0.5391\delta[n-2] - 0.2203\delta[n-3] + 0.0101\delta[n-4]$. Find the impulse response of the matching low pass filter.

9.23 An FIR filter must be designed to the following specifications:

 Band pass

 Sampling frequency 16 kHz

 Center frequency 4 kHz

 Pass band edges at 3 kHz and 5 kHz

 Transition widths 900 Hz

 Stop band attenuation 40 dB

Find and plot the impulse response for the filter. (Use software to assist with repetitive calculations.)

9.24 Signals between 2 kHz and 8 kHz must be extracted from a sensor signal in a 24 kHz sampled system. The required filter must have a stop band attenuation of at least 50 dB, and the transition widths must be no greater than 500 Hz. Determine all the equations that are necessary to specify the impulse response for the filter.

9.25 A voice signal sampled at 8 kHz must be filtered to eliminate elements outside the range from 300 to 3400 Hz before it is coded for transmission. Design the filter.

9.26 A high pass filter with a pass band edge frequency of 5.5 kHz must be designed for a 16 kHz sampled system. The stop band attenuation must be at least 40 dB, and the transition width must be no greater than 3.5 kHz. Write the difference equation for the filter.

9.27 The signal transitions below 200 Hz must be eliminated from a temperature sensor signal in a 1 kHz sampled system. Design the filter.

9.28 Design a band stop filter according to the following specifications:

Pass band edges at 2 kHz and 5 kHz

Transition widths 1 kHz

Stop band attenuation $\geq$ 40 dB

Sampling rate 12 kHz

9.29 Use a low pass filter and a high pass filter to design a band pass filter passing frequencies between 7 and 8 kHz, if the sampling frequency is 24 kHz. Stop band attenuation must be at least 70 dB, and transition widths must not exceed 500 Hz.

9.30 Design a filter to remove a 1200 Hz tone from a 4 kHz sampled signal.

9.31 Compare the filter shape for the filter described by the transfer function

$$H(z) = 0.0152 + 0.2263z^{-1} + 0.5171z^{-2} + 0.2263z^{-3} + 0.0152z^{-4}$$

to the shape obtained after the coefficients are quantized to:

a. 5 bits

b. 4 bits

10

INFINITE IMPULSE RESPONSE FILTERS

This chapter introduces a second major class of digital filters: infinite impulse response, or IIR, filters. The chapter:

> reviews the difference equation and transfer function for a recursive filter
> defines simple low pass analog filter types
> introduces the bilinear transformation as a means of converting from an analog to a digital filter
> describes how to choose IIR filter order given filter specifications
> presents low pass Butterworth filter designs
> presents low pass Chebyshev Type I filter designs
> describes the impulse invariance design method
> discusses methods for creating band pass, high pass, and band stop filters from low pass filters
> highlights practical considerations

10.1 INFINITE IMPULSE RESPONSE FILTER BASICS

The last chapter was concerned with the design of finite impulse response, or nonrecursive, filters. In this chapter, the design of **infinite impulse response filters**, otherwise known as recursive filters, is considered. These require past and present inputs as well as past outputs to generate each new output. If the general difference equation

$$\sum_{k=0}^{N} a_k y[n-k] = \sum_{k=0}^{M} b_k x[n-k]$$

is rearranged assuming $a_0 = 1$, a convenient expression for the filter is obtained:

$$y[n] = -\sum_{k=1}^{N} a_k y[n-k] + \sum_{k=0}^{M} b_k x[n-k]$$

$$= -a_1 y[n-1] - a_2 y[n-2] - \cdots - a_N y[n-N]$$

$$+ b_0 x[n] + b_1 x[n-1] + b_2 x[n-2] + \cdots + b_M x[n-M]$$

The filter equation is recursive because new outputs depend on older outputs. There is no loss of generality in assuming $a_0 = 1$, as established in Section 4.4. Implementations for this type of filter were examined in Section 4.6.2.

Nonrecursive filters use only past inputs, and not past outputs, to compute new outputs, and therefore rely only on the coefficients b_k. Recursive filters, on the other hand, make use of both the a_k and b_k coefficients, thereby permitting a great deal of flexibility for filter shape design. The presence of a_k coefficients means that recursive filters have impulse responses with an infinite number of terms. For this reason, recursive filters are also referred to as infinite impulse response or IIR filters.

EXAMPLE 10.1
Find the impulse response for the simple recursive filter

$$y[n] = 0.8y[n-1] + x[n]$$

Replacing $y[n]$ by $h[n]$ and $x[n]$ by $\delta[n]$, the impulse response may be found from the equation

$$h[n] = 0.8h[n-1] + \delta[n]$$

TABLE 10.1
Impulse Response Calculations for Example 10.1

n	0	1	2	3	4	5
$\delta[n]$	1	0	0	0	0	0
$h[n]$	1.0	0.8	0.64	0.512	0.4096	0.32768

Table 10.1 and Figure 10.1 show the first six impulse response samples. In the difference equation, each new output depends on the preceding output. Because the output $h[0]$ has

FIGURE 10.1

Impulse response for Example 10.1.

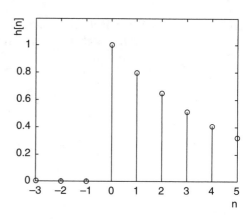

the value 1, the output will be nonzero forever. Because the system is stable, the output $h[n]$ diminishes as n increases.

The transfer function for a general recursive filter is:

$$H(z) = \frac{b_0 + b_1 z^{-1} + b_2 z^{-2} + \cdots + b_M z^{-M}}{1 + a_1 z^{-1} + a_2 z^{-2} + \cdots + a_N z^{-N}}$$

Assuming $N > M$ and multiplying every term by z^N gives:

$$H(z) = \frac{b_0 z^N + b_1 z^{N-1} + b_2 z^{N-2} + \cdots + b_M z^{N-M}}{z^N + a_1 z^{N-1} + a_2 z^{N-2} + \cdots + a_N}$$

Where the nonrecursive filters of the last chapter had poles only at zero, recursive filters have poles determined by the denominator polynomial. This means that recursive filters cannot be guaranteed to be stable. In fact, checking stability is an important part of many software programs for recursive filter design.

There are several other important differences between recursive and nonrecursive filters. With recursive filters, it is not easy to achieve the linear phase that was guaranteed for nonrecursive filters. However, recursive filters do have the advantage that they require far fewer coefficients than nonrecursive filters to achieve similar performance specifications. The design approach that will be used for recursive filters is to choose a prototype analog filter that has the desired characteristics and then convert it to a digital filter.

10.2 LOW PASS ANALOG FILTERS

In the continuous domain, filters are described in terms of s rather than z. The transfer function for a simple low pass analog filter is

$$H(s) = \frac{1}{s + 1}$$

FIGURE 10.2

Shape of simple low pass analog filter.

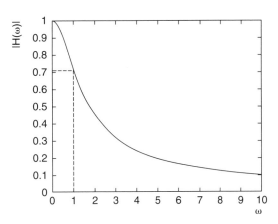

In Section 7.2.2, the frequency response for digital filters was found by replacing every z in the transfer function $H(z)$ with $e^{j\Omega}$, where Ω is the digital frequency in radians. For analog filters, the frequency response may generally be found by replacing every s in the transfer function $H(s)$ with $j\omega$, where ω is the analog frequency in radians per second, related to the analog frequency f in Hz through the relationship $\omega = 2\pi f$. Using this rule,

$$H(\omega) = \frac{1}{j\omega + 1}$$

$H(\omega)$ is called the frequency response of the analog filter.

The frequency response $H(\omega)$ is a complex number with magnitude

$$\left|H(\omega)\right| = \frac{1}{\sqrt{\omega^2 + 1}}$$

as shown in Appendix A.13.2.4. For large values of ω, $|H(\omega)|$ approaches zero. For small values of ω, $|H(\omega)|$ approaches one. This means that the filter has low pass behavior. When $\omega = 1$, $|H(\omega)|$ equals $1/\sqrt{2}$. Since $20 \log(1/\sqrt{2}) = -3$, $\omega = 1$ rad/sec is the -3 dB frequency for this low pass filter, which also marks its bandwidth. In Hz, the -3 dB frequency is $f = \omega/2\pi = 1/2\pi$ Hz. The shape of the filter is shown in Figure 10.2.

To make the filter more useful, the -3 dB frequency can be set to a frequency other than $\omega = 1$ rad/sec by altering the transfer function slightly:

$$H(s) = \frac{\omega_{p1}}{s + \omega_{p1}} \tag{10.1}$$

For this analog transfer function, the Fourier transform is

$$H(\omega) = \frac{\omega_{p1}}{j\omega + \omega_{p1}} = \frac{1}{j\dfrac{\omega}{\omega_{p1}} + 1} \tag{10.2}$$

FIGURE 10.3
Low pass filter with cut-off
frequency ω_{p1}.

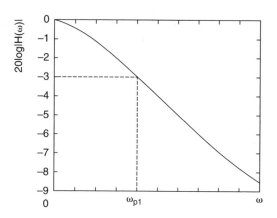

and the magnitude is

$$|H(\omega)| = \frac{1}{\sqrt{\left(\dfrac{\omega}{\omega_{p1}}\right)^2 + 1}} \tag{10.3}$$

This is still a low pass filter, but the -3 dB frequency now occurs when $\omega = \omega_{p1}$ rad/sec or $f = \omega_{p1}/2\pi$ Hz. This filter is shown in Figure 10.3.

Common analog filter types include Butterworth, Chebyshev Type I, Chebyshev Type II, and elliptic filters. They differ mainly in their general shape, as Figure 10.4 shows. Butterworth filters are monotonic through the pass band and stop band, meaning they change smoothly in a single direction. Chebyshev Type I filters are monotonic through the stop band, but ripple in the pass band. Chebyshev Type II filters are monotonic in the pass band, but ripple in the stop band. Finally, elliptic filters ripple in both the pass and stop bands. Butterworth and Chebyshev Type I filters will be examined in detail in Sections 10.4 and 10.5.

10.3 BILINEAR TRANSFORMATION

The bilinear transformation provides a means of converting between an analog filter and a digital filter. The bilinear transformation is defined by:

$$s \Leftrightarrow 2f_S\frac{z - 1}{z + 1} \tag{10.4}$$

where f_S is the sampling frequency in Hz. Replacing z by $e^{j\Omega}$ gives the discrete time Fourier transform of the right-hand side. Manipulating the expression gives

$$2f_S\frac{e^{j\Omega} - 1}{e^{j\Omega} + 1} = 2f_S\frac{e^{j\frac{\Omega}{2}}(e^{j\frac{\Omega}{2}} - e^{-j\frac{\Omega}{2}})}{e^{j\frac{\Omega}{2}}(e^{j\frac{\Omega}{2}} + e^{-j\frac{\Omega}{2}})} = 2f_S\frac{e^{j\frac{\Omega}{2}} - e^{-j\frac{\Omega}{2}}}{e^{j\frac{\Omega}{2}} + e^{-j\frac{\Omega}{2}}}$$

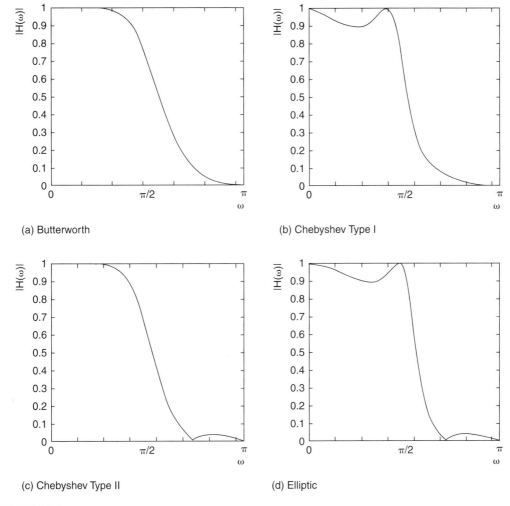

(a) Butterworth

(b) Chebyshev Type I

(c) Chebyshev Type II

(d) Elliptic

FIGURE 10.4
Common analog filter types.

Using Euler's identities

$$\sin\theta = \frac{e^{j\theta} - e^{-j\theta}}{2j} \quad \text{and} \quad \cos\theta = \frac{e^{j\theta} + e^{-j\theta}}{2}$$

the right-hand side can be further simplified as

$$2f_S \frac{2j\sin\left(\dfrac{\Omega}{2}\right)}{2\cos\left(\dfrac{\Omega}{2}\right)} = j2f_S \tan\left(\frac{\Omega}{2}\right)$$

In the Fourier domain, the s on the left-hand side of Equation (10.4) becomes $j\omega$. Then, comparing the left and right sides after canceling a j factor from each shows

$$\omega \Leftrightarrow 2f_S \tan\left(\frac{\Omega}{2}\right) \tag{10.5}$$

Equation (10.5) is called the **prewarping equation**. The range of interest for digital frequency Ω is normally from 0 to π radians, as discussed in Section 7.3.2. According to Equation (10.5), this range for Ω maps to the range 0 to ∞ rad/sec for the analog frequency ω, that is, to all possible frequency values in the analog domain. In fact, this is the reason the bilinear transformation is used. It expands the digital frequency scale from 0 to π rads to the analog frequency scale from 0 to ∞ rad/sec. Conversely, the inverse bilinear transformation compresses analog frequencies from 0 to ∞ to digital frequencies from 0 to π according to

$$\Omega \Leftrightarrow 2 \tan^{-1}\left(\frac{\omega}{2f_S}\right)$$

Figure 10.5 shows the relationship between digital frequency Ω and analog frequency ω given by Equation (10.5). Since $\omega = 2\pi f$ and $\Omega = 2\pi f/f_S$, ω/f_S should equal Ω, as indicated by the line $\omega/f_S = \Omega$ marked on the diagram. But, as the figure shows, the bilinear transformation warps this relationship between digital frequencies and analog frequencies. This warping must be taken into account during the design process. The $2f_S$ factor in the bilinear transformation in Equation (10.5) ensures that the transformation distorts the frequency axis as little as possible. For small Ω values, $\tan \Omega/2 \approx \Omega/2$, so, in Equation (10.5), $\omega \approx 2f_S (\Omega/2) = f_S\Omega = f_S(2\pi f/f_S) = 2\pi f$, as required.

The bilinear transformation makes one other useful connection between the digital and analog domains. As discussed in Section 6.4.2, digital filters are stable if their poles lie

FIGURE 10.5

Warping between digital and analog frequencies.

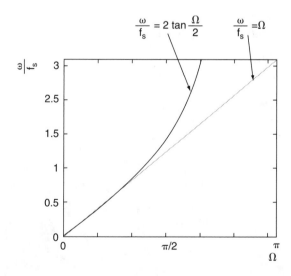

FIGURE 10.6
Stability
regions.

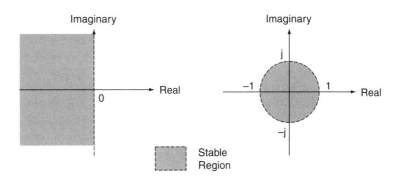

Stability Region for Analog Filters
on Complex s Plane

Stability Region for Digital Filters
on Complex z Plane

inside the unit circle on the complex z plane. The area inside the unit circle is sometimes called the region of stability for digital filters. The corresponding region of stability for analog filters can be found by sending points on the unit circle through the bilinear transformation. As illustrated in Example 10.2 below, any point lying on the unit circle will, after transformation, lie on the imaginary axis of the complex s plane. Any point lying inside the unit circle maps to a point to the left of the imaginary axis. Any point lying outside the unit circle maps to a point in the right half plane. The bilinear transformation in Equation (10.4) causes these mappings to occur. Thus, the region of stability for analog filters is the left half of the complex plane, while the region of stability for digital filters is the interior of the unit circle. Figure 10.6 shows these stability regions for analog and digital filters.

EXAMPLE 10.2

This example provides some sample calculations to show how points are mapped from the z domain complex plane to the s domain complex plane by the bilinear transformation.

a. $z = \dfrac{1}{\sqrt{2}} + j\,\dfrac{1}{\sqrt{2}}$ (on unit circle)

Performing the bilinear transformation:

$$s = 2f_s\frac{z-1}{z+1} = 2f_s\frac{\dfrac{1+j}{\sqrt{2}}-1}{\dfrac{1+j}{\sqrt{2}}+1} = 2f_s\frac{1+j-\sqrt{2}}{1+j+\sqrt{2}}$$

$$= 2f_s\left(\frac{1-\sqrt{2}+j}{1+\sqrt{2}+j}\right)\left(\frac{1+\sqrt{2}-j}{1+\sqrt{2}-j}\right) = 2f_s\frac{j2\sqrt{2}}{4+2\sqrt{2}} = jf_s\frac{2\sqrt{2}}{2+\sqrt{2}}$$

This complex number is purely imaginary, so it lies on the imaginary axis: A point on the unit circle in the z domain maps to a point on the imaginary axis in the s domain.

b. $z = 0$ (inside the unit circle)

Using the bilinear transformation:

$$s = 2f_S \frac{z - 1}{z + 1} = -2f_S$$

This complex number lies in the left half plane: A point inside the unit circle in the z domain maps to a point in the left half plane in the s domain.

c. $z = 2 + j$ (outside the unit circle)

Transforming:

$$s = 2f_S \frac{z - 1}{z + 1} = 2f_S \frac{2 + j - 1}{2 + j + 1} = 2f_S \frac{1 + j}{3 + j}$$

$$= 2f_S \frac{1 + j3 - j}{3 + j3 - j} = 2f_S \frac{4 + j2}{10} = \frac{4f_S}{5} + j\frac{2f_S}{5}$$

This complex number lies in the right half plane: A point outside the unit circle in the z domain maps to a point in the right half plane in the s domain.

To illustrate the bilinear transformation, the next example transforms a simple low pass analog filter introduced in Section 10.2 into a digital filter.

EXAMPLE 10.3

The transfer function for a first order analog low pass filter is described by Equation (10.1). The filter has a -3 dB frequency of 2000 rad/sec, or $2000/2\pi = 318.31$ Hz. Find the transfer function $H(z)$ for a digital filter that matches this analog filter, for a 1500 Hz sampling rate.

The original transfer function $H(s)$ in the analog domain is

$$H(s) = \frac{2000}{s + 2000} \tag{10.6}$$

Following Equation (10.3), the shape of the filter is given by

$$|H(\omega)| = \frac{1}{\sqrt{\left(\dfrac{\omega}{2000}\right)^2 + 1}}$$

FIGURE 10.7
Magnitude response
of analog filter.

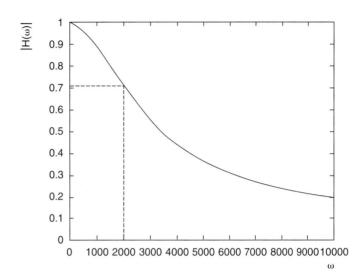

The magnitude response for the analog filter is shown in Figure 10.7.

A digital filter must be designed to match this analog filter. The sampling rate is 1500 Hz, and the bilinear transformation requires that each s in the analog transfer function of Equation (10.6) be replaced by the ratio $2f_S (z - 1)/(z + 1)$. The transfer function in the digital domain becomes:

$$H(z) = \frac{2000}{3000\dfrac{z - 1}{z + 1} + 2000}$$

$$= \frac{2000(z + 1)}{3000(z - 1) + 2000(z + 1)} = \frac{2000(z + 1)}{5000z - 1000}$$

$$= \frac{0.4(z + 1)}{z - 0.2} = \frac{0.4(1 + z^{-1})}{1 - 0.2z^{-1}}$$

The shape of this filter may be verified using the methods of Section 7.3.2 for the frequency response. The frequency response is

$$H(\Omega) = \frac{0.4(1 + e^{-j\Omega})}{1 - 0.2e^{-j\Omega}}$$

Figure 10.8 shows a plot of $|H(\Omega)|$ versus Ω. The -3 dB point coincides with the point where the gain drops to 0.707 of its maximum value. For the digital filter, this occurs where $\Omega = 1.17$ radians, which corresponds according to Equation (7.8) to a frequency of $f = \Omega f_S/2\pi = 279.3$ Hz, nearly 40 Hz below the cut-off of 318.3 Hz for the analog filter.

FIGURE 10.8

Magnitude response of digital filter.

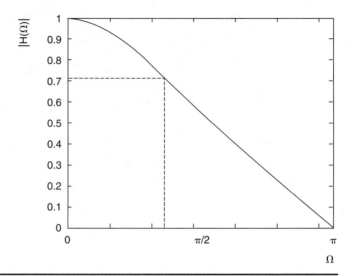

Figure 10.9 compares the analog and digital filter shapes for Example 10.3. The gains for both are plotted against frequency in Hz. The plot is produced from those of $|H(\Omega)|$ versus Ω and $|H(\omega)|$ versus ω in Figures 10.7 and 10.8, using the relationships $\Omega = 2\pi f / f_S$ and $\omega = 2\pi f$ to convert Ω and ω to frequencies in Hz. Notice that the roll-off for the digital filter is steeper than that for the analog filter. Also, the bandwidth of the digital filter does not agree with that of the analog filter. This error occurs because of the warping effect of the bilinear transformation. Fortunately, warping errors can be overcome by prewarping the -3 dB frequency using Equation (10.5). When this step is added to the design process, the correct bandwidth is obtained, as will be shown in Example 10.4.

FIGURE 10.9

Effect of frequency warping.

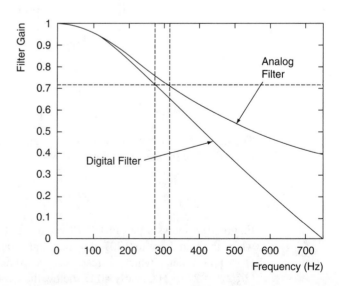

EXAMPLE 10.4

A simple analog low pass filter described by the analog transfer function in Equation (10.1) must be converted into a digital filter with a -3 dB frequency of 318.3 Hz. The sampling rate is 1500 Hz. Find expressions for the transfer function and the frequency response of the digital filter.

The same goal was stated, but not achieved, in Example 10.3. In this example, steps are taken to ensure the goal is reached. With a -3 dB frequency of $f_{p1} = 318.3$ Hz and a sampling frequency of 1500 Hz, the digital cut-off frequency is

$$\Omega_{p1} = 2\pi \frac{f_{p1}}{f_S} = 2\pi \frac{318.3}{1500} = 0.4244\pi \text{ radians}$$

To allow for warping by the bilinear transformation, prewarp the analog frequency using Equation (10.5):

$$\omega_{p1} = 2f_S \tan\left(\frac{\Omega_{p1}}{2}\right) = 2f_S \tan\left(\frac{0.4244\pi}{2}\right) = 2360.4 \text{rad/sec}$$

This frequency will be used in place of the unwarped $\omega_{p1} = 2\pi(318.31) = 2000$ rad/sec. The transfer function for the prototype analog filter is then

$$H(s) = \frac{\omega_{p1}}{s + \omega_{p1}} = \frac{2360.4}{s + 2360.4} \tag{10.7}$$

Converting to a digital filter using the bilinear transformation $s = 3000\,(z-1)/(z+1)$ from Equation (10.4) gives the transfer function

$$H(z) = \frac{0.4403(1 + z^{-1})}{1 - 0.1193z^{-1}}$$

The frequency response for this prewarped digital filter is

$$H(\Omega) = \frac{0.4403(1 + e^{-j\Omega})}{1 - 0.1193e^{-j\Omega}}$$

The magnitude response for the prewarped digital filter is shown in Figure 10.10. The analog filter and the digital filter produced in Example 10.3 without warping are also presented, for comparison purposes. The cut-off frequency for the prewarped filter is much closer to the required value of 318.3 Hz.

There is a quick way to find the shape of a digital filter when it results from a bilinear transformation of an analog filter. If the filter shape $|H(\omega)|$ for the analog filter is known, then the filter shape $|H(\Omega)|$ for the digital filter can be found by substituting $\omega = 2f_S \tan(\Omega/2)$, from Equation (10.5). The following example illustrates how this idea works. Unfortunately, this short cut gives filter shape only; it cannot provide a transfer function or a difference equation, which are needed to implement the filter.

FIGURE 10.10

Effect of prewarping.

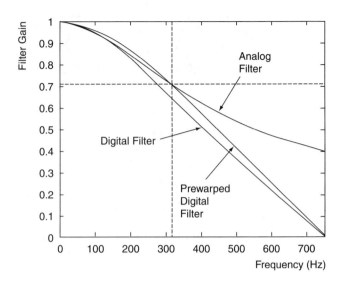

EXAMPLE 10.5

In Example 10.4, a prewarped analog filter was converted into a digital filter for a 1500 Hz sampled system. Find an expression for the shape of the digital filter using an expression for the prewarped analog filter shape.

From Equation (10.3), the filter shape for the analog filter whose transfer function appears in Equation (10.7) is given by

$$|H(\omega)| = \frac{1}{\sqrt{\left(\dfrac{\omega}{\omega_{p1}}\right)^2 + 1}} = \frac{1}{\sqrt{\left(\dfrac{\omega}{2360.4}\right)^2 + 1}}$$

Substituting $\omega = 2f_S \tan(\Omega/2)$, with a sampling frequency of 1500 Hz, into the analog filter shape expression gives the digital filter shape

$$|H(\Omega)| = \frac{1}{\sqrt{\left(\dfrac{3000\tan\left(\dfrac{\Omega}{2}\right)}{2360.4}\right)^2 + 1}}$$

This function is plotted in Figure 10.11. The prewarped digital filter shape that was computed from the frequency response

$$H(\Omega) = \frac{0.4403(1 + e^{-j\Omega})}{1 - 0.1193e^{-j\Omega}}$$

FIGURE 10.11
Filter shape for
Example 10.5.

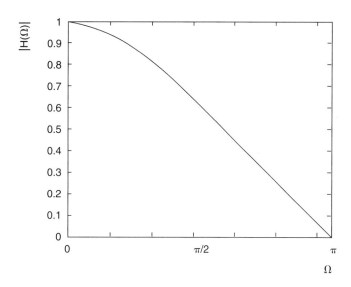

in Example 10.4, and which was plotted in Figure 10.10, is repeated here for comparison. The two curves superimpose exactly.

10.4 BUTTERWORTH FILTER DESIGN

Butterworth filters are the simplest filters in the IIR family. The analog filter that was introduced in Section 10.2, with transfer function

$$H(s) = \frac{\omega_{p1}}{s + \omega_{p1}} \tag{10.8}$$

and filter shape

$$|H(\omega)| = \frac{1}{\sqrt{\left(\dfrac{\omega}{\omega_{p1}}\right)^2 + 1}} \tag{10.9}$$

is actually a first order analog Butterworth filter. An nth order analog Butterworth filter has a rather complicated transfer function $H(s)$. Because of this, the computational burden of finding the transfer function and its bilinear transformation is normally eased with the help of filter design software.

The filter shape $|H(\omega)|$ for the nth order analog Butterworth filter is similar to that of the first order filter. It is given by

$$|H(\omega)| = \frac{1}{\sqrt{\left(\dfrac{\omega}{\omega_{p1}}\right)^{2n} + 1}} \tag{10.10}$$

FIGURE 10.12
Analog Butterworth
filter shapes, order *n*.

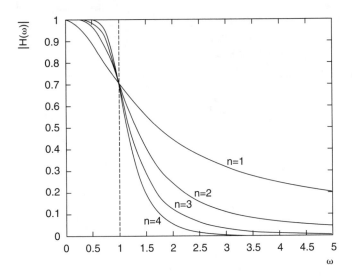

This filter still has a low pass nature, and its roll-off is steeper the higher the order *n*. Figure 10.12 compares a group of analog Butterworth filters of orders 1, 2, 3, and 4 for a -3 dB frequency of 1 rad/sec. All of the filters have a maximum gain of unity, and all show smooth transitions from the pass band through the -3 dB point to the stop band. Two characteristics are important for analog Butterworth filter design: the cut-off (or -3 dB) frequency and the steepness of the roll-off. These two characteristics are chosen independently. The objective of digital Butterworth filter design is to produce a digital filter that has the same characteristics as the analog filter already designed.

The terminology for describing digital IIR filter characteristics is similar to that for FIR filters, introduced in Section 9.4. Important filter characteristics for Butterworth filters are marked in Figure 10.13. Figure 10.13(a) plots gain versus frequency in Hz, and Figure 10.13(b) plots gain in dB versus frequency in Hz. For the low pass Butterworth filter, the -3 dB point occurs, by definition, at f_{p1} Hz, also the bandwidth of the filter. The difference between the maximum and minimum gains in the pass band is called the pass band ripple, or δ_p. Thus, if the maximum gain in the pass band is one, or 0 dB, then the gain at the edge of the pass band is $1-\delta_p$, or $20\log(1-\delta_p)$ dB. In the case of the Butterworth filter, the parameter δ_p must be chosen so that the edge of the pass band lies at the -3 dB boundary. That is, $20\log(1-\delta_p)$ is always equal to -3, so δ_p is always 0.292. The gain at the edge of the stop band, located at the frequency f_{s1} Hz, is called the stop band ripple, or δ_s. In dB, the gain at the stop band edge is $20\log\delta_s$ dB, and the attenuation is equal to $-20\log\delta_s$ dB. For design, the stop band attenuation and the stop band edge frequency must be specified.

Digital Butterworth filters do not offer the phase linearity characteristic of the FIR filters in the last chapter. To illustrate, Figure 10.14 shows the magnitude and phase responses for a fourth order digital Butterworth filter. Not even the phase in the pass band, marked on the diagram, has a linear, or straight line, relationship to frequency. A nonlinear phase response is typical for IIR filters, which means phase distortion is present to some degree in filtered signals.

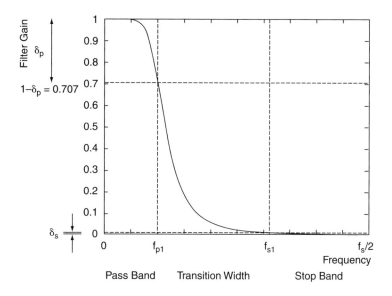

(a) Linear Gain Versus Frequency

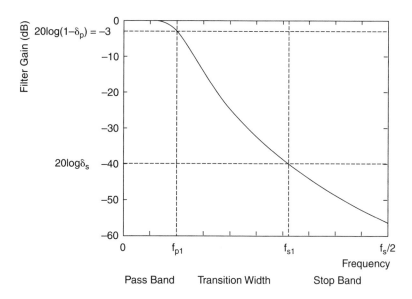

(b) Logarithmic Gain Versus Frequency

FIGURE 10.13
Butterworth filter parameters.

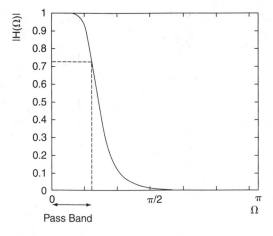

(a) Magnitude Response

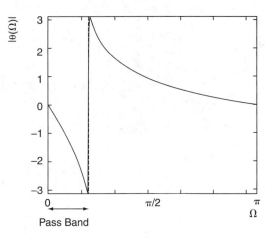

(b) Phase Response

FIGURE 10.14

Frequency response of fourth order Butterworth filter.

Filter design normally begins with the choice of a filter order for the analog filter that will satisfy the specifications. For an nth order analog Butterworth filter, the order required to satisfy the design specifications on the pass band and the stop band is

$$n \geq \frac{\log\left(\dfrac{1}{\delta_s^2} - 1\right)}{2 \log\left(\dfrac{\omega_{s1}}{\omega_{p1}}\right)} \qquad (10.11)$$

where ω_{p1} is the prewarped pass band edge frequency for the analog filter and ω_{s1} is the prewarped stop band edge frequency, both in radians per second. The inequality is derived in Appendix L. Once the order of the analog filter has been selected, and a transfer function $H(s)$ for the analog filter has been specified, a bilinear transformation can be used to find the transfer function of the corresponding digital filter. The transfer function will have the form

$$H(z) = \frac{b_0 + b_1 z^{-1} + b_2 z^{-2} + \cdots + b_n z^{-n}}{1 + a_1 z^{-1} + a_2 z^{-2} + \cdots + a_n z^{-n}}$$

with $2n + 1$ coefficients, n zeros, and n poles. The transfer function can then be converted to a difference equation for filter implementation. The design steps for a Butterworth filter are listed in Table 10.2.

TABLE 10.2
Design Steps for Low Pass Butterworth Filter Design

1. Identify the desired pass band edge frequency f_{p1} Hz, the desired stop band edge frequency f_{s1} Hz, and the desired stop band attenuation $-20\log\delta_s$ dB (or the desired stop band gain $20\log\delta_S$ dB) for the digital filter. The pass band edge frequency must correspond to a gain of -3 dB.
2. Convert the desired edge frequencies in Hz into digital frequencies in radians, using $\Omega = 2\pi f/f_S$, to obtain Ω_{p1} and Ω_{s1}.
3. Calculate prewarped analog frequencies to avoid distortions due to bilinear transformation. Use $\omega = 2f_S \tan(\Omega/2)$ to obtain ω_{p1} and ω_{s1}, in radians per second.
4. Determine the gain δ_s at the edge of the stop band from the specified attenuation $-20\log\delta_s$ (or gain $20\log\delta_s$).
5. Calculate the filter order required using

$$n \geq \frac{\log\left(\frac{1}{\delta_s^2} - 1\right)}{2 \log\left(\frac{\omega_{s1}}{\omega_{p1}}\right)}$$

An integer value for n must be chosen.
6. Substitute ω_{p1} into the nth order analog Butterworth transfer function $H(s)$, and perform a bilinear transformation on $H(s)$ to obtain the nth order digital transfer function $H(z)$.[1] The difference equation needed for filter implementation can be found directly from the transfer function $H(z)$. Filter shape $|H(\Omega)|$ can be obtained by substituting $\omega = 2f_S \tan(\Omega/2)$ into

$$|H(\omega)| = \frac{1}{\sqrt{\left(\frac{\omega}{\omega_{p1}}\right)^{2n} + 1}}$$

EXAMPLE 10.6

A low pass IIR filter with a Butterworth characteristic is to be designed to have a -3 dB frequency of 1200 Hz. The gain must drop to -25 dB by 1500 Hz. The sampling rate is 8000 Hz. Choose a suitable order for the filter and plot the filter shape.

These specifications are fairly demanding since the gain must drop by $25 - 3 = 22$ dB in only 300 Hz, rather a small proportion of the Nyquist frequency of $f_S/2 = 4000$ Hz. The required order will be relatively high. Following the design steps in Table 10.2:

Step 1:   The analog edge frequencies are $f_{p1} = 1200$ Hz and $f_{s1} = 1500$ Hz. The gain at the edge of the stop band is $20\log\delta_s = -25$ dB.

Step 2:   The digital edge frequencies are:

$$\Omega_{p1} = 2\pi \frac{f_{p1}}{f_S} = 2\pi \frac{1200}{8000} = 0.3\pi \text{ radians}$$

[1]   Because the general form of the nth order analog Butterworth transfer function $H(s)$ is rather complicated, this step is usually performed with the assistance of filter design software.

$$\Omega_{s1} = 2\pi\frac{f_{s1}}{f_S} = 2\pi\frac{1500}{8000} = 0.375\pi \text{ radians}$$

Step 3: The prewarped analog edge frequencies are:

$$\omega_{p1} = 2f_S \tan\frac{\Omega_{p1}}{2} = 8152.4 \text{ rad/sec}$$

$$\omega_{s1} = 2f_S \tan\frac{\Omega_{s1}}{2} = 10690.9 \text{ rad/sec}$$

Step 4: Since, from Step 1, $20\log\delta_s = -25$, then $\log\delta_s = -25/20$, and $\delta_s = 10^{-25/20} = 0.0562$.

Step 5: The required order of the filter is

$$n \geq \frac{\log\left(\frac{1}{\delta_s^2} - 1\right)}{2\log\left(\frac{\omega_{s1}}{\omega_{p1}}\right)} = \frac{\log\left(\frac{1}{(0.0562)^2} - 1\right)}{2\log\left(\frac{10690.9}{8152.4}\right)} = 10.6$$

A Butterworth filter of order 11 will slightly exceed the specifications given.

Step 6: Filter design software is best used to produce the digital transfer function $H(z)$ for this 11th order Butterworth filter. The filter shape for the analog prototype is:

$$|H(\Omega)| = \frac{1}{\sqrt{\left(\frac{\omega}{\omega_{p1}}\right)^{2n} + 1}} = \frac{1}{\sqrt{\left(\frac{\omega}{8152.4}\right)^{22} + 1}}$$

Substituting $\omega = 2f_S \tan(\Omega/2)$ into this analog prototype, the filter shape for the digital filter is:

$$|H(\Omega)| = \frac{1}{\sqrt{\left(\frac{2f_S\tan\left(\frac{\Omega}{2}\right)}{\omega_{p1}}\right)^{2n} + 1}} = \frac{1}{\sqrt{\left(\frac{16000\tan\left(\frac{\Omega}{2}\right)}{8152.4}\right)^{22} + 1}}$$

The shape of the filter is presented using linear magnitudes in Figure 10.15(a). Instead of plotting the filter shape against Ω from 0 to π, the shape is plotted against frequencies from 0 to $f_S/2$, as discussed in Section 7.3.3. Figure 10.15(b), which plots the magnitudes in dB, demonstrates that the specifications have been satisfied. The cut-off point is in the expected location, and a gain of -25 dB is reached just below 1500 Hz.

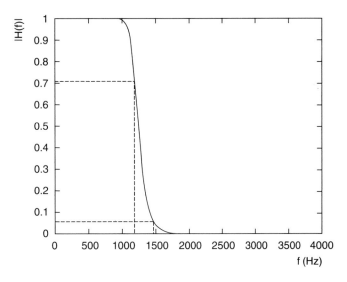

(a) Linear Magnitude Response

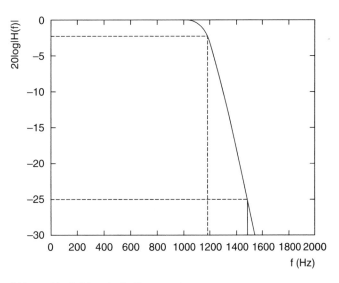

(b) Logarithmic Magnitude Response

FIGURE 10.15
Filter shape for Example 10.6.

EXAMPLE 10.7

A low pass Butterworth filter has a gain of -3 dB at the edge of its pass band, at 1 kHz, and a stop band attenuation at 12 kHz of 30 dB. Find the difference equation and plot the frequency response for the filter, if the sampling rate is 25 kHz.

Following the design steps in Table 10.2:

Step 1: The analog edge frequencies are:

$$f_{p1} = 1000 \text{ Hz}$$

$$f_{s1} = 12000 \text{ Hz}$$

The attenuation at the edge of the stop band is $-20\log\delta_s = 30$ dB.

Step 2: The digital edge frequencies are:

$$\Omega_{p1} = 2\pi\frac{f_{p1}}{f_S} = 2\pi\frac{1000}{25000} = 0.08\pi \text{ radians}$$

$$\Omega_{s1} = 2\pi\frac{f_{s1}}{f_S} = 2\pi\frac{12000}{25000} = 0.96\pi \text{ radians}$$

Step 3: The prewarped analog edge frequencies are:

$$\Omega_{p1} = 2f_S\tan\frac{\Omega_{p1}}{2} = 6316.5 \text{ rad/sec}$$

$$\Omega_{s1} = 2f_S\tan\frac{\Omega_{s1}}{2} = 794727.2 \text{ rad/sec}$$

Step 4: Since, from Step 1, $-20\log\delta_s = 30$, then $\log\delta_s = -30/20$, and $\delta_s = 10^{-30/20} = 0.03162$.

Step 5: The required order of the filter is

$$n \geq \frac{\log\left(\frac{1}{\delta_s^2} - 1\right)}{2\log\left(\frac{\omega_{s1}}{\omega_{p1}}\right)} = \frac{\log\left(\frac{1}{(0.03162)^2} - 1\right)}{2\log\left(\frac{794727.2}{6316.5}\right)} = 0.714$$

Therefore, a first order Butterworth filter is adequate to meet the specifications.

Step 6: The transfer function for the first order analog Butterworth filter, as noted early in this section, is

$$H(s) = \frac{\omega_{p1}}{s + \omega_{p1}} = \frac{6316.5}{s + 6316.5}$$

The transfer function for the digital filter is found using the bilinear transformation defined in Equation (10.4).

$$H(z) = \frac{6316.5}{50000 \dfrac{z-1}{z+1} + 6316.5} = \frac{6316.5(z+1)}{50000(z-1) + 6316.5(z+1)}$$

$$= \frac{0.1122(1 + z^{-1})}{1 - 0.7757z^{-1}}$$

Therefore, the difference equation is

$$y[n] = 0.7757\, y[n-1] + 0.1122\, x[n] + 0.1122\, x[n-1]$$

Replacing z in $H(z)$ by $e^{j\Omega}$, the frequency response

$$H(\Omega) = \frac{0.1122(1 + e^{-j\Omega})}{1 - 0.7757e^{-j\Omega}}$$

is obtained, from which the magnitude and phase can be computed for any value of Ω. The magnitude and phase responses are plotted in Figure 10.16. Note that the phase response in the pass band, where the filter gains are over 0.707, is not linear.

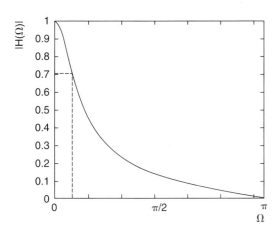

(a) Magnitude Response

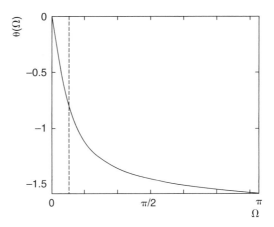

(b) Phase Response

FIGURE 10.16
Magnitude and phase response for Example 10.7.

EXAMPLE 10.8

The transfer function for a second order low pass analog Butterworth filter with -3 dB frequency ω_{p1} rad/sec is

$$H(s) = \frac{\omega_{p1}^2}{s^2 + \sqrt{2}\omega_{p1}s + \omega_{p1}^2}$$

Design a second order low pass digital Butterworth filter with a bandwidth of 500 Hz for a sampling frequency of 4 kHz.

For a low pass filter, the bandwidth is identical with the -3 dB frequency, 500 Hz. The digital frequency that corresponds to this frequency is

$$\Omega_{p1} = 2\pi \frac{f_{p1}}{f_S} = 2\pi \frac{500}{4000} = 0.25\pi \text{ radians}$$

Prewarping to prepare for the bilinear transformation gives:

$$\omega_{p1} = 2f_S\tan\frac{\Omega_{p1}}{2} = 3313.7 \text{ rad/sec}$$

Substituting this value into the analog transfer function gives

$$H(s) = \frac{10980607.7}{s^2 + 4686.3s + 10980607.7}$$

From Equation (10.4), the bilinear transformation is $s = 8000(z - 1)/(z + 1)$ in this case. After transforming, the digital transfer function $H(z)$ becomes

$$H(z) = \frac{0.09763 + 0.19526z^{-1} + 0.09763z^{-2}}{1 - 0.9428z^{-1} + 0.3333z^{-2}}$$

which furnishes the difference equation

$$y[n] = 0.9428y[n-1] - 0.3333y[n-2] + 0.09763x[n]$$
$$+ 0.19526x[n-1] + 0.09763x[n-2]$$

10.5 CHEBYSHEV TYPE I FILTER DESIGN

A Chebyshev filter can often accomplish the same job as a Butterworth filter of higher order, at the cost of slightly greater complexity. The Chebyshev transfer function is rather complicated, but the expression for filter shape is quite straightforward. For an nth order analog Chebyshev Type I filter, the filter shape is defined as:

$$|H(\omega)| = \frac{1}{\sqrt{1 + \varepsilon^2 C_n^2\left(\dfrac{\omega}{\omega_{p1}}\right)}}$$

where ε is a parameter that depends on pass band ripple, and the function $C_n(x)$ is defined as:

$$C_n(x) = \begin{cases} \cos(n\cos^{-1}(x)) & \text{for } |x| \le 1 \\ \cosh(n\cosh^{-1}(x)) & \text{for } |x| > 1 \end{cases}$$

where $\cosh(x)$ refers to the hyperbolic cosine function. The function $C_n(x)$ is known as the Chebyshev polynomial because it can be expanded as a polynomial. Chebyshev filter shapes for several values of n are shown in Figure 10.17. A comparison with Figure 10.12 shows that, as long as the order is greater than one, the Chebyshev filters offer steeper roll-offs than the Butterworth filters.

FIGURE 10.17
Analog Chebyshev
Type I filter shapes,
order n.

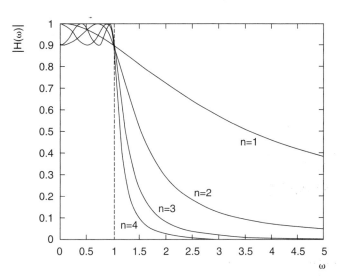

Parameters for Chebyshev filters are displayed in Figure 10.18. The main difference with the Butterworth filter is the fact that the edge of the pass band need not occur at -3 dB. The parameter δ_p, which defines the gain at the edge of the pass band, can be selected by the designer (although bandwidth is still normally measured between -3 dB points). The other important difference is the shape of the filter, which has a ripple in the pass band.

The required order for the analog Chebyshev Type I filter can be found from the expression

$$n \ge \frac{\cosh^{-1}\left(\dfrac{\delta}{\varepsilon}\right)}{\cosh^{-1}\left(\dfrac{\omega_{s1}}{\omega_{p1}}\right)} \tag{10.12}$$

which is derived in Appendix M. The parameters in Equation (10.12) are described in Table 10.3, which contains the steps necessary to design a low pass Chebyshev Type I filter. As in Section 10.4, an nth order filter will have a transfer function of the form

$$H(z) = \frac{b_0 + b_1 z^{-1} + b_2 z^{-2} + \cdots + b_n z^{-n}}{1 + a_1 z^{-1} + a_2 z^{-2} + \cdots + a_n z^{-n}}$$

with $2n + 1$ coefficients, n zeros, and n poles.

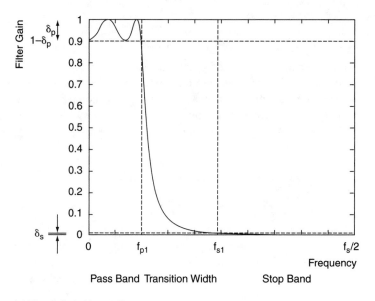

(a) Linear Gain Versus Frequency

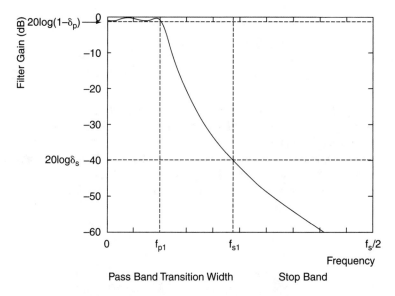

(b) Logarithmic Gain Versus Frequency

FIGURE 10.18
Chebyshev filter parameters.

TABLE 10.3
Design Steps for Low Pass Chebyshev Type I Filter Design

1. Identify the desired pass band edge and stop band edge frequencies, f_{p1} Hz and f_{s1} Hz, as well as the desired gain at the edge of the pass band $20\log(1-\delta_p)$ and the desired stop band attenuation $-20\log\delta_s$ (or the desired stop band gain $20\log\delta_s$).
2. Convert the desired edge frequencies to digital frequencies in radians, using $\Omega = 2\pi f/f_S$, to obtain Ω_{p1} and Ω_{s1}.
3. Prewarp the digital frequencies to avoid errors due to bilinear transformation. Use $\omega = 2f_S \tan(\Omega/2)$ to obtain ω_{p1} and ω_{s1}, in radians per second.
4. Determine the gain at the edge of the pass band, $1-\delta_p$, from the specified pass band edge gain $20\log(1-\delta_p)$. Compute the parameter ε from

$$\varepsilon = \sqrt{\frac{1}{(1-\delta_p)^2} - 1}$$

5. Determine the gain at the edge of the stop band, δ_s, from the specified attenuation $-20\log\delta_s$ or gain $20\log\delta_s$ dB. Compute

$$\delta = \sqrt{\frac{1}{\delta_s^2} - 1}$$

6. Calculate the order required using

$$n \geq \frac{\cosh^{-1}\left(\dfrac{\delta}{\varepsilon}\right)}{\cosh^{-1}\left(\dfrac{\omega_{s1}}{\omega_{p1}}\right)}$$

An integer value for n must be chosen.
7. Substitute ω_{p1} and δ_p into the nth order analog Chebyshev Type I transfer function $H(s)$, and perform a bilinear transformation on $H(s)$ to obtain the nth order digital transfer function $H(z)$.[2] The difference equation needed for filter implementation can be found directly from the transfer function $H(z)$. Filter shape $|H(\Omega)|$ can be obtained by substituting $\omega = 2f_S \tan(\Omega/2)$ into

$$|H(\omega)| = \frac{1}{\sqrt{1 + \varepsilon^2 C_n^2\left(\dfrac{\omega}{\omega_{p1}}\right)}}$$

EXAMPLE 10.9

An IIR filter with a Chebyshev Type 1 characteristic must be designed for a system with a 20 kHz sampling rate. The maximum pass band gain is 0 dB. The pass band edge occurs at 5 kHz, with a gain of -1 dB. The stop band edge begins at 7.5 kHz, with a gain of -32 dB.

[2]   Because the general form of the nth order analog Chebyshev Type I transfer function $H(s)$ is rather complicated, this step is usually performed with the assistance of filter design software.

Following the design steps in Table 10.3:

Step 1: The analog edge frequencies are:

$$f_{p1} = 5000 \text{ Hz}$$

$$f_{s1} = 7500 \text{ Hz}$$

The gain at the edge of the pass band is $20\log(1-\delta_p) = -1$ dB. The gain at the edge of the stop band is $20\log\delta_s = -32$ dB.

Step 2: The digital edge frequencies are:

$$\Omega_{p1} = 2\pi\frac{f_{p1}}{f_S} = 2\pi\frac{5000}{20000} = 0.5\pi \text{ radians}$$

$$\Omega_{s1} = 2\pi\frac{f_{s1}}{f_S} = 2\pi\frac{7500}{20000} = 0.75\pi \text{ radians}$$

Step 3: The prewarped analog edge frequencies are:

$$\omega_{p1} = 2f_S \tan\frac{\Omega_{p1}}{2} = 40000 \text{ rad/sec}$$

$$\omega_{s1} = 2f_S\tan\frac{\Omega_{s1}}{2} = 96568.5 \text{rad/sec}$$

Step 4: Since, from Step 1, $20\log(1-\delta_p) = -1$, then $\log(1 - \delta_p) = -1/20$, and $1-\delta_p = 10^{-1/20} = 0.89125$. Therefore,

$$\varepsilon = \sqrt{\frac{1}{(1 - \delta_p)^2} - 1} = \sqrt{\frac{1}{0.89125^2} - 1} = 0.5088$$

Step 5: Since, from Step 1, $20\log\delta_s = -32$, then $\log\delta_s = -32/20$, and $\delta_s = 10^{-32/20} = 0.0251$. Therefore,

$$\delta = \sqrt{\frac{1}{\delta_s^2} - 1} = \sqrt{\frac{1}{0.0251^2} - 1} = 39.8$$

Step 6: The required order of the filter is:

$$n \geq \frac{\cosh^{-1}\left(\dfrac{\delta}{\varepsilon}\right)}{\cosh^{-1}\left(\dfrac{\omega_{s1}}{\omega_{p1}}\right)} = \frac{\cosh^{-1}\left(\dfrac{39.8}{0.5088}\right)}{\cosh^{-1}\left(\dfrac{96568.5}{40000}\right)} = 3.31$$

To meet the specifications, a filter of order 4 must be chosen.

Step 7: Filter design software can be used to find the transfer function $H(z)$. The filter shape for the analog prototype is

$$\left|H(\omega)\right| = \frac{1}{\sqrt{1 + \varepsilon^2 C_n^2\left(\dfrac{\omega}{\omega_{p1}}\right)}} = \frac{1}{\sqrt{1 + 0.2589 C_4^2\left(\dfrac{\omega}{40000}\right)}}$$

The filter shape for the digital filter is found by substituting $\omega = 2f_S \tan(\Omega/2)$ into the analog prototype:

$$\left|H(\Omega)\right| = \frac{1}{\sqrt{1 + 0.2589 C_4^2\left(\dfrac{40000\tan\left(\dfrac{\Omega}{2}\right)}{40000}\right)}} = \frac{1}{\sqrt{1 + 0.2589 C_4^2\left(\tan\left(\dfrac{\Omega}{2}\right)\right)}}$$

where

$$C_4(x) = \begin{cases} \cos(4\cos^{-1}(x)) & |x| \le 1 \\ \cosh(4\cosh^{-1}(x)) & |x| > 1 \end{cases}$$

The shape of the filter is plotted in Figure 10.19, where frequencies from 0 to $f_S/2$ Hz have been used along the frequency axis in place of 0 to π rads, following the discussion in Section 7.3.3. Notice the characteristic shape of the Chebyshev Type I filter in Figure 10.19(a). The logarithmic plot in Figure 10.19(b) shows that the specifications have been satisfied. A gain of -1 dB is reached at 5 kHz, but, be-

FIGURE 10.19

Filter shape for Example 10.9.

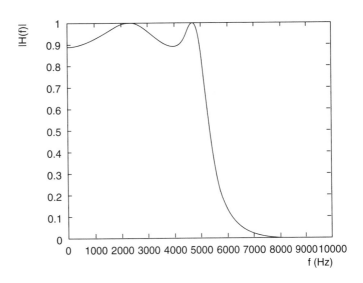

(a) Linear Filter Shape

FIGURE 10.19
Continued

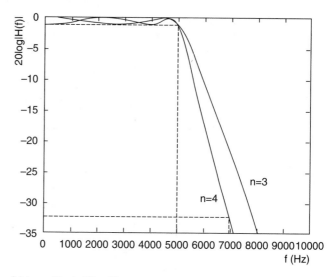

(b) Logarithmic Filter Shape

cause the order was rounded up to 4 from 3.13, the gain drops to -32 dB below 7000 Hz instead of the specified 7500 Hz. Unfortunately, order 3, also presented in Figure 10.19(b), is too low to meet the requirements.

EXAMPLE 10.10

A filter must meet the following specifications:

> Low pass
> Pass band edge 1.8 kHz
> Stop band edge 2.3 kHz
> Pass band ripple 0.292
> Stop band ripple 0.02
> Sampling frequency 8 kHz

Calculate the required order for the filter if it has a
   a. Butterworth characteristic
   b. Chebyshev Type I characteristic

The pass band ripple δ_p is 0.292, and the stop band ripple δ_s is 0.02. Note that the pass band ripple leads to a gain at the edge of the pass band of $1 - 0.292 = 0.708$, or -3 dB, as required for the Butterworth filter. The digital pass band and stop band edge frequencies are

$$\Omega_{p1} = 2\pi \frac{f_{p1}}{f_S} = 2\pi \frac{1800}{8000} = 0.45\pi \text{ radians}$$

$$\Omega_{s1} = 2\pi\frac{f_{s1}}{f_S} = 2\pi\frac{2300}{8000} = 0.575\pi \text{ radians}$$

Prewarping these yields the frequencies to be used for the design of the analog prototype:

$$\omega_{p1} = 2f_S\tan\frac{\Omega_{p1}}{2} = 13665.3 \text{ rad/sec}$$

$$\omega_{s1} = 2f_S\tan\frac{\Omega_{s1}}{2} = 20295.9 \text{ rad/sec}$$

a. For the Butterworth version of the filter,

$$n \geq \frac{\log\left(\frac{1}{\delta^2} - 1\right)}{2\log\left(\frac{\omega_{s1}}{\omega_{p1}}\right)} = \frac{\log\left(\frac{1}{0.02^2} - 1\right)}{2\log\left(\frac{20295.9}{13665.3}\right)} = 9.9$$

A tenth order Butterworth filter is needed to meet the specifications.

b. For the Chebyshev Type I filter,

$$\delta = \sqrt{\frac{1}{\delta_s^2} - 1} = \sqrt{\frac{1}{0.02^2} - 1} = 49.99$$

$$\varepsilon = \sqrt{\frac{1}{(1 - \delta_p)^2} - 1} = \sqrt{\frac{1}{0.708^2} - 1} = 0.9975$$

$$n \geq \frac{\cosh^{-1}\left(\frac{\delta}{\varepsilon}\right)}{\cosh^{-1}\left(\frac{\omega_{s1}}{\omega_{p1}}\right)} = \frac{\cosh^{-1}\left(\frac{49.99}{0.9975}\right)}{\cosh^{-1}\left(\frac{20295.9}{13665.3}\right)} = 4.9$$

A fifth order Chebyshev Type I filter is needed to meet the specifications. Note that the Chebyshev filter has a much lower order than the Butterworth filter, for the same requirements.

EXAMPLE 10.11

A low pass filter must be designed to meet the following specifications:

Pass band edge at 12 kHz
Transition width 4 kHz
Attenuation at pass band edge 0.06 dB
Stop band attenuation 44 dB
Sampling frequency 44 kHz

Determine how many coefficients in the transfer function $H(z)$ are needed to meet the specifications for each of the following filter types:

a. FIR

b. IIR, Chebyshev Type I characteristic

a. According to Table 9.3, to achieve a stop band attenuation of 44 dB, this filter requires a Hanning window. This window also provides the required attenuation at the edge of the pass band. The number of terms required for the specified transition width and sampling frequency is

$$N = 3.32\frac{f_S}{\text{T.W.}} = 3.32\frac{44}{4} = 36.5$$

Thus, $N = 37$ terms are needed. As indicated in Section 9.6.2, a complete design for an FIR filter with N terms has an impulse response with N samples and a difference equation with N coefficients. Therefore, this filter design produces 37 impulse response values, and also a difference equation with 37 coefficients.

b. As shown in Figure 10.18, the gain at the edge of the pass band, -0.06 dB, is given by $20\log(1-\delta_p)$ dB. Thus, $\log(1-\delta_p) = -0.003$, $(1-\delta_p) = 10^{-0.003}$, and $\delta_p = 0.00688$. A stop band attenuation of 44 dB gives a stop band gain of -44 dB, which equals $20\log\delta_s$ dB. Solving $20\log\delta_s = -44$ gives $\delta_s = 0.00631$. Thus, for the Chebyshev filter,

$$\varepsilon = \sqrt{\frac{1}{(1-\delta_p)^2} - 1} = \sqrt{\frac{1}{(1-0.00688)^2} - 1} = 0.1179$$

$$\delta = \sqrt{\frac{1}{\delta_s^2} - 1} = \sqrt{\frac{1}{0.00631^2} - 1} = 158.48$$

The stop band edge is found by adding the transition width to the pass band edge, giving 16 kHz. Thus the digital frequencies at the edges of the pass band and stop band are

$$\Omega_{p1} = 2\pi\frac{f_{p1}}{f_S} = 2\pi\frac{12000}{44000} = 0.5455\pi \text{ radians}$$

$$\Omega_{s1} = 2\pi\frac{f_{s1}}{f_S} = 2\pi\frac{16000}{44000} = 0.7273\pi \text{ radians}$$

Prewarping these yields the frequencies to be used in the design of the analog prototype:

$$\omega_{p1} = 2f_S\tan\frac{\Omega_{p1}}{2} = 101572 \text{ rad/sec}$$

$$\omega_{s1} = 2f_S\tan\frac{\Omega_{s1}}{2} = 192715 \text{ rad/sec}$$

The order of the filter, by Equation (10.12), must be

$$n \geq \frac{\cosh^{-1}\left(\dfrac{\delta}{\varepsilon}\right)}{\cosh^{-1}\left(\dfrac{\omega_{s1}}{\omega_{p1}}\right)} = \frac{\cosh^{-1}\left(\dfrac{158.48}{0.1179}\right)}{\cosh^{-1}\left(\dfrac{192715}{101572}\right)} = 6.3$$

Rounding up, the required order for the Chebyshev filter is $n = 7$. The transfer function for this filter has $2n + 1$ coefficients, so it requires 15 coefficients in all. The Chebyshev Type I filter is able to meet the filter specifications with fewer than half of the coefficients required by the FIR filter.

10.6   IMPULSE INVARIANCE IIR FILTER DESIGN

Every analog filter has an impulse response $h(t)$, just as every digital filter has an impulse response $h[n]$. The impulse invariance method of IIR filter design chooses a digital impulse response $h[n]$ that is a sampled version of the impulse response $h(t)$ of the analog filter that meets the required specifications, as illustrated in Figure 10.20. That is,

$$h[n] = h(nT) \tag{10.13}$$

where T is the sampling interval used. The idea is that a sampled impulse response produced in this manner will give a filter shape that is close to that of the original analog filter.

The impulse response $h(t)$ for an analog filter with the transfer function $H(s)$ can be found using methods outside the scope of this text (Ambardar, 1999). For the simple case of a first order Butterworth filter, for example, the transfer function is, from Equation (10.1),

$$H(s) = \frac{\omega_{p1}}{s + \omega_{p1}}$$

where ω_{p1} is the -3 dB frequency in radians per second. This transfer function gives the impulse response

$$h(t) = \omega_{p1} e^{-\omega_{p1} t} u(t)$$

where $u(t)$ is an analog step function, zero for $t < 0$ and one for $t \geq 0$. When this analog impulse response is sampled every T seconds, the sampled times are given by $t = nT$, and, using Equation (10.13), the digital impulse response becomes

$$h[n] = h(nT) = \omega_{p1} e^{-\omega_{p1} nT} u(nT) = \omega_{p1} e^{-\omega_{p1} Tn} u[n]$$

since a sampled analog step function $u(nT)$ is identical with a digital step function $u[n]$. From Table 6.1, the z transform of $\beta^n u[n]$ is $z/(z - \beta)$. Therefore, the transfer function for

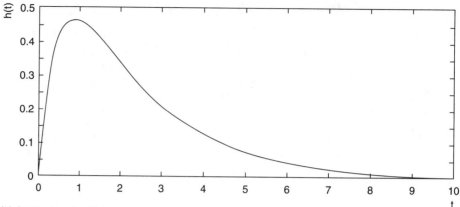

(a) Analog Impulse Response

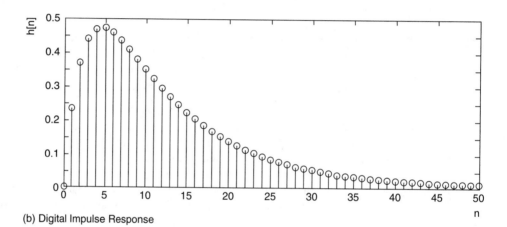

(b) Digital Impulse Response

FIGURE 10.20

Impulse invariance.

the digital filter obtained through impulse invariance from a first order analog Butterworth filter may be found with $\beta = e^{-\omega_{p1}T}$ as

$$H(z) = \frac{\omega_{p1}z}{z - e^{-\omega_{p1}T}} = \frac{\omega_{p1}}{1 - e^{-\omega_{p1}T}z^{-1}} \tag{10.14}$$

This transfer function leads to the difference equation

$$y[n] - e^{-\omega_{p1}T}y[n-1] = \omega_{p1}x[n]$$

The digital filter described by Equation (10.14) has the frequency response

$$H(\Omega) = \frac{\omega_{p1}}{1 - e^{-\omega_{p1}T}e^{-j\Omega}} \tag{10.15}$$

Through the impulse invariance design method, the digital filter should have the same filter shape as the analog filter that provides the source impulse response $h(t)$. According to Equation (10.9), the shape of the first order analog Butterworth filter is given by

$$|H(\omega)| = \frac{1}{\sqrt{\left(\dfrac{\omega}{\omega_{p1}}\right)^2 + 1}}$$

Provided that the sampling interval T used to implement the impulse-invariant digital filter is small, to minimize aliasing, then the shapes of the digital and analog filters are very close, as Example 10.12 shows.

EXAMPLE 10.12

Use the impulse invariance method to design a first order Butterworth filter with a cut-off frequency of 750 Hz.

From Equation (10.8), the first order Butterworth transfer function $H(s)$ for a 750 Hz cut-off frequency is

$$H(s) = \frac{\omega_{p1}}{s + \omega_{p1}} = \frac{2\pi(750)}{s + 2\pi(750)} = \frac{1500\pi}{s + 1500\pi}$$

Note that prewarping is not required because the bilinear transformation is not used. The filter shape for this analog filter, from Equation (10.9), is

$$|H(\omega)| = \frac{1}{\sqrt{\left(\dfrac{\omega}{1500\pi}\right)^2 + 1}}$$

This shape is to be duplicated by a digital filter designed with the impulse invariance method. The transfer function of the digital filter, from Equation (10.14), is

$$H(z) = \frac{\omega_{p1}}{1 - e^{-\omega_{p1}T}z^{-1}} = \frac{1500\pi}{1 - e^{-1500\pi T}z^{-1}}$$

From Equation (10.15), the frequency response of the digital filter is

$$H(\Omega) = \frac{\omega_{p1}}{1 - e^{-\omega_{p1}T}e^{-j\Omega}} = \frac{1500\pi}{1 - e^{-1500\pi T}e^{-j\Omega}}$$

This expression provides the magnitude response, or filter shape, for a sampling interval T.
    The analog and digital filter shapes are most readily compared if both are presented in terms of frequency in Hz. For the analog filter, the expression $\omega = 2\pi f$ may be used to relate analog frequency f in Hz to analog frequency ω in radians per second. For the digital filter, the expression $\Omega = 2\pi f/f_s = 2\pi fT$ relates analog frequency f to digital frequency Ω in radians, for a given sampling interval T. Figure 10.21 compares the analog and digital filter

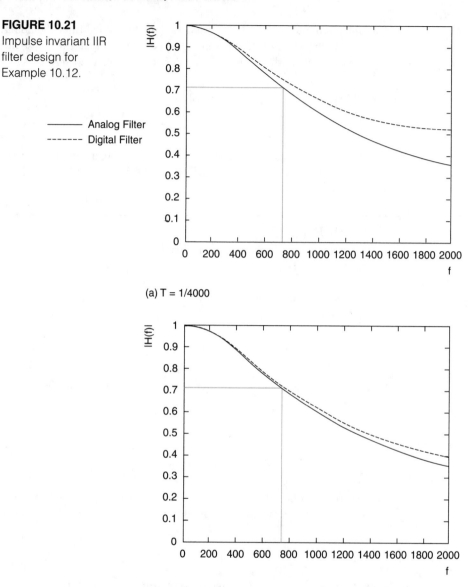

FIGURE 10.21

Impulse invariant IIR filter design for Example 10.12.

——— Analog Filter

------ Digital Filter

(a) T = 1/4000

(b) T = 1/8000

shapes for three different sampling intervals, where all filters have been scaled to give unity DC gain. The desired cut-off frequency of 750 Hz is marked on the diagrams. For the slowest sampling, in (a), the filter shapes diverge at a fairly low frequency. This is evidence of aliasing and indicates the sampling interval is too long. As the sampling interval is shortened, in (b) and (c), the digital filter shape approaches the analog filter shape more and more closely. The closest match is obtained when the sampling rate $1/T$ is many times higher than the desired cut-off frequency.

FIGURE 10.21
Continued

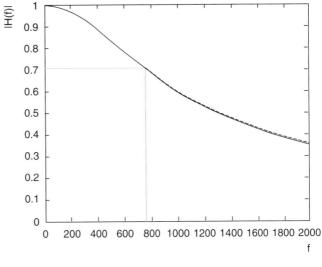

(c) T = 1/16000

10.7 "BEST FIT" FILTER DESIGN

As the examples have shown, design of recursive filters involves manipulating rather complicated equations. For this reason, software support is nearly always needed to complete these designs. One way to automate the design process is to create a program that implements the equations of the previous sections. When the equations become excessively complex, or when no convenient analog prototype exists for a desired filter, approaching the correct filter through iteration can become the best choice for design. Programs that use this method begin by guessing a transfer function, then calculate the filter shape and compare it mathematically to the filter shape. When mismatches occur, the filter coefficients are adjusted and the filter shape is recalculated. The process continues until a transfer function is found that produces the required filter shape.

This "best fit" method is used in many commercial filter design packages, even in cases when other techniques, like those presented in the previous sections, could be used. When filter coefficients do not need to be updated on the fly, as is required by some adaptive systems, speedy coefficient calculation becomes less of an issue, and the extra time needed to go through many iterative cycles is unimportant.

10.8 BAND PASS, HIGH PASS, AND BAND STOP IIR FILTERS

The order for band pass and high pass filters is chosen by working from a low pass prototype. This prototype must have the same shape as the desired filter. Once the transfer

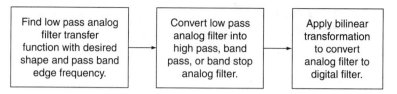

FIGURE 10.22

Steps for design of high pass, band pass, and band stop filters.

function $H_L(s)$ for the low pass analog filter has been obtained, it can be transformed into a high pass, band pass, or band stop filter. Figure 10.22 suggests the steps to be followed. The first step requires choosing a filter type and calculating a minimum filter order, as in Examples 10.6 and 10.9. Naturally this includes prewarping the digital filter requirements. The third step applies the bilinear transformation, as in Example 10.4. The conversions required in the second step can be accomplished by one of a group of simple tricks, described in the following paragraph.

To convert a low pass filter with cut-off frequency ω_p rad/sec, such as

$$H_L(s) = \frac{\omega_p}{s + \omega_p}$$

into a high pass filter with cut-off frequency ω_p' rad/sec, every s must be replaced by $\omega_p \omega_p'/s$, that is,

$$H_H(s) = H_L\left(\frac{\omega_p \omega_p'}{s}\right) \tag{10.16}$$

where $H_L(s)$ is the transfer function of the low pass analog filter and $H_H(s)$ is the transfer function of the high pass analog filter. The transfer function of a low pass filter with cut-off frequency ω_p rad/sec may be converted into the transfer function of a band pass filter with lower cut-off frequency ω_l and upper cut-off frequency ω_u with the equation

$$H_{BP}(s) = H_L\left(\omega_p \frac{s^2 + \omega_l \omega_u}{s(\omega_u - \omega_l)}\right) \tag{10.17}$$

where $H_{BP}(s)$ is the transfer function of the band pass analog filter. A band stop analog filter can be constructed from the low pass analog filter using

$$H_{BS}(s) = H_L\left(\omega_p \frac{s(\omega_u - \omega_l)}{s^2 + \omega_l \omega_u}\right) \tag{10.18}$$

where $H_{BS}(s)$ is the transfer function of the band stop analog filter. Each cut-off frequency ω_p, ω_p', ω_l, and ω_u must be prewarped using Equation (10.5) to avoid distortion of the filter shape. Note that the band pass and band stop filters have twice the order of their low pass prototypes. The bilinear transformation described by Equation (10.4) converts each of these analog filters into its digital counterpart.

EXAMPLE 10.13

In Example 10.4, a first order low pass digital filter was designed for a -3 dB frequency of 318.3 Hz with a sampling rate of 1500 Hz. With the prewarped analog frequency $\omega_p =$ 2360.4 rad/sec, the transfer function for the prototype analog Butterworth filter was

$$H_L(s) = \frac{\omega_p}{s + \omega_p} = \frac{2360.4}{s + 2360.4}$$

The transfer function for the corresponding digital low pass filter was

$$H_L(z) = \frac{0.4403(1 + z^{-1})}{1 - 0.1193z^{-1}}$$

Using the first order Butterworth prototype $H_L(s)$ and a sampling frequency of 1500 Hz, determine a transfer function $H(z)$ for each of the following digital filters:

    a. A high pass filter with a -3 dB frequency of 318.3 Hz.
    b. A band pass filter with -3 dB frequencies of 318.3 and 636.6 Hz.
    c. A band stop filter with -3 dB frequencies of 318.3 and 636.6 Hz.

    a. With a 1500 Hz sampling rate, an analog frequency of 318.3 Hz corresponds to a digital frequency of $\Omega_p' = 2\pi f/f_S = 2\pi(318.3/1500) = 0.4244\pi$ rads. Prewarping gives $\omega_p' = 2f_S \tan(\Omega_p'/2) = 2360.4$ rad/sec, which happens to be the same as the cut-off for the low pass filter. The analog transfer function for the high pass filter is obtained using Equation (10.16):

$$H_H(s) = H_L\left(\frac{\omega_p \omega_p'}{s}\right) = H_L\left(\frac{(2360.4)(2360.4)}{s}\right)$$

$$= \frac{2360.4}{\left(\dfrac{(2360.4)(2360.4)}{s}\right) + 2360.4} = \frac{s}{s + 2360.4}$$

The digital counterpart is given by the bilinear transformation of Equation (10.4):

$$H_H(z) = \frac{3000\dfrac{z - 1}{z + 1}}{3000\dfrac{z - 1}{z + 1} + 2360.4} = \frac{0.5597(1 - z^{-1})}{1 - 0.1193z^{-1}}$$

The magnitude responses for this high pass filter and the original low pass filter $H_L(z)$ are plotted in Figure 10.23. Because the cut-off frequencies are the same for the two filters, their responses are symmetrical across 318.3 Hz.

FIGURE 10.23
High pass filter
from low pass
prototype for
Example 10.13.

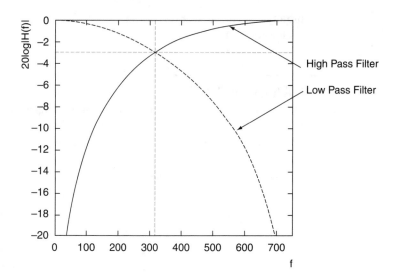

b. As in (a), the prewarped lower cut-off frequency is $\omega_l = 2360.4$ rad/sec. The upper cut-off frequency will be $\Omega_u = 2\pi(636.6/1500) = 0.8488\pi$ rads, or $\omega_u = 2f_S \tan(0.8488\pi/2) = 12392.9$ rad/sec after prewarping. From Equation (10.17),

$$H_{BP}(s) = H_L\left(\omega_P \frac{s^2 + \omega_l\omega_u}{s(\omega_u - \omega_l)}\right)$$

$$= H_L\left(2360.4\frac{s^2 + (2360.4)(12392.9)}{10032.5s}\right)$$

$$= \frac{2360.4}{2360.4\dfrac{s^2 + (2360.4)(12392.9)}{10031.5s} + 2360.4}$$

$$= \frac{10032.5s}{s^2 + 10032.5s + 29252201.2}$$

The bilinear transformation gives the digital transfer function:

$$H_{BP}(z) = \frac{10032.5\left(3000\dfrac{z - 1}{z + 1}\right)}{\left(3000\dfrac{z - 1}{z + 1}\right)^2 + 10032.5\left(3000\dfrac{z - 1}{z + 1}\right) + 29252201.2}$$

$$= \frac{0.4404(1 - z^{-2})}{1 + 0.5926z^{-1} + 0.1193z^{-2}}$$

The magnitude response for this filter is shown in Figure 10.24.

FIGURE 10.24
Band pass and
band stop
filters from low
pass prototype
for Example
10.13.

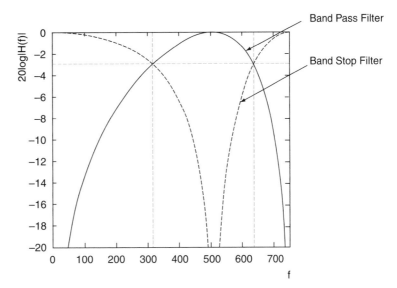

c. As in (b), the lower prewarped cut-off frequency is $\omega_l = 2360.4$ rad/sec and the upper prewarped cut-off frequency $\omega_u = 12392.9$ rad/sec. According to Equation (10.18),

$$H_{BS}(s) = H_L\left(\omega_p \frac{s(\omega_u - \omega_l)}{s^2 + \omega_l\omega_u}\right)$$

$$= H_L\left(2360.4\frac{10032.5s}{s^2 + (2360.4)(12392.9)}\right)$$

$$= \frac{2360.4}{2360.4\dfrac{10032.5s}{s^2 + (2360.4)(12392.9)} + 2360.4}$$

$$= \frac{s^2 + 29252201.2}{s^2 + 10032.5s + 29252201.2}$$

The transfer function for the digital filter is found using the bilinear transformation:

$$H_{BS}(z) = \frac{\left(3000\dfrac{z-1}{z+1}\right)^2 + 29252201.2}{\left(3000\dfrac{z-1}{z+1}\right)^2 + 10032.5\left(3000\dfrac{z-1}{z+1}\right) + 29252201.2}$$

$$= \frac{0.5597 + 0.5926z^{-1} + 0.5597z^{-2}}{1 + 0.5926z^{-1} + 0.1193z^{-2}}$$

The magnitude response for this filter is shown in Figure 10.24. It has the same bandwidth as the band pass filter, also plotted in the figure, but the band stop filter attenuates signals between 318.3 and 636.6 Hz, while the band pass filter passes them.

One good application of IIR filter technology is the generation and recovery of dual tone multifrequency (DTMF) signals used by Touch-Tone telephones (Mock, 1985). This application was alluded to in Section 4.1. The diagram of the Touch-Tone keypad in Figure 4.1 is repeated here in Figure 10.25.

Each time a key is pressed, a pair of tones is generated, one to code the row and the other the column of the key pressed. Each pure tone that forms part of a DTMF tone pair can be generated digitally by applying an impulse function to an IIR filter. As described in Section 6.2.3, the z transform of the output from the filter will be the product of the transfer function $H(z)$ and the z transform of an input $X(z)$:

$$Y(z) = H(z)X(z)$$

The z transform of an impulse function is $X(z) = 1$, so $Y(z) = H(z)$ in this case. This also means that $Y(\Omega) = H(\Omega)$. Thus, the spectrum of the output signal will be identical with the

FIGURE 10.25
Touch-Tone keypad.

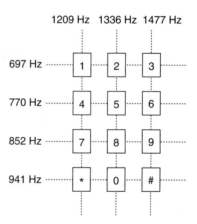

frequency response of the filter. A single pure tone is produced by a sinusoidal signal, so a filter that is capable of creating a pure tone will have a transfer function that is identical with the z transform of a sine signal:

$$H(z) = \frac{z\sin\Omega_0}{z^2 - 2z\cos\Omega_0 + 1} \qquad (10.19)$$

from Table 6.1. Ω_0 is the digital frequency for the desired tone. The difference equation for this tone generator will be

$$y[n] = 2\cos\Omega_0\, y[n-1] - y[n-2] + \sin\Omega_0 x[n-1] \qquad (10.20)$$

FIGURE 10.26
Tone generation
filter.

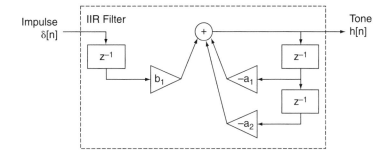

with $a_1 = -2\cos\Omega_0$, $a_2 = 1$, and $b_1 = \sin\Omega_0$. It is presented in diagram form in Figure 10.26. For simplicity, a direct form 1 representation is used. Two tone generation filters of this type are needed to produce a single DTMF signal. Example 10.14 illustrates how to design a tone generator to produce a specified tone.

EXAMPLE 10.14

Design a filter to generate a single 1209 Hz tone with a sampling frequency of 8 kHz.

   a. Sketch the filter shape.

   b. Find the impulse response of the filter.

   c. Sketch the magnitude spectrum of the impulse response.

   a. The required digital frequency $\Omega = 2\pi f/f_S = 2\pi(1209/8000) = 0.30225\pi = 0.9495$ rads. From Equation (10.20), the difference equation for the tone generator should be

$$y[n] = 1.1642y[n-1] - y[n-2] + 0.8131x[n-1]$$

The frequency response for this filter is, by replacing z with $e^{j\Omega}$ in Equation (10.19),

$$H_{\text{filter}}(\Omega) = \frac{0.8131e^{-j\Omega}}{1 - 1.1642e^{-j\Omega} + e^{-j2\Omega}}$$

The filter shape is given by the magnitude response, where the magnitudes $|H(\Omega)|$ can be found for a range of Ω values by using the methods of Section 7.3.2. The results are plotted in Figure 10.27.

   b. By replacing $y[n]$ and $x[n]$ in the difference equation by $h[n]$ and $\delta[n]$, the impulse response is seen to be

$$h[n] = 1.1642h[n-1] - h[n-2] + 0.8131\delta[n-1]$$

The first 256 samples are plotted in Figure 10.28. The impulse response is indeed a sinusoidal tone.

FIGURE 10.27
Magnitude response of tone generator filter for Example 10.14.

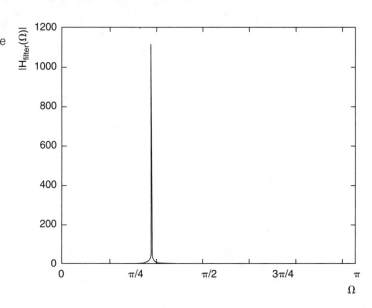

dtmftone.wav

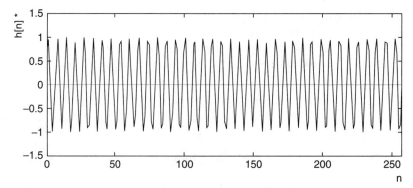

* For clarity, the envelope rather than the individual samples of the signal is plotted.

FIGURE 10.28
Tone produced by tone generator for Example 10.14.

    c. The spectrum of the tone, as distinct from the frequency response of the filter that produced it, may be found using the discrete time Fourier transform as described in Section 7.1. The DTFT for this impulse response is, from Equation (7.3),

$$H_{signal}(\Omega) = \sum_{n=-\infty}^{\infty} h[n]e^{-jn\Omega}$$

The impulse response begins at $n = 0$, so the frequency response can be approximated by summing over some large number of samples, say from $n = 0$ to $n = 2047$.

FIGURE 10.29
Magnitude spectrum
of tone for Example
10.14.

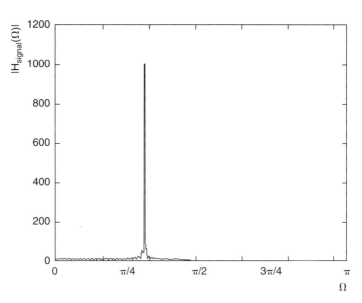

The magnitude spectrum is given by the values of $|H(\Omega)|$ plotted against Ω. Its plot in Figure 10.29 is nearly identical with the plot of the magnitude response in Figure 10.27. In other words, the frequency content of the impulse response matches the frequency response of the filter, as expected.

Dual tone signals for the Touch-Tone telephone can be produced by using two IIR filters, as shown in Figure 10.30. For each tone of a different frequency, the coefficients a_1, a_2, and b_1 in Figure 10.26 change. A low and a high tone are summed to produce each dual frequency tone. For example, a 1336 Hz tone combined with a 770 Hz tone produces the code for the "5" key on the Touch-Tone keypad.

The dual frequency tones produced by the DTMF generator must be decoded by a receiver. A possible digital DTMF receiver is shown in Figure 10.31. Each received tone pair is passed through low pass and high pass filters. As Figure 10.32 illustrates, since the frequencies of the rows in Figure 10.25 lie below the frequencies of the columns, the low and high pass filters separate row and column information. A group of band pass filters operate on the low pass portion to identify the row, and a second group of band pass filters operate

FIGURE 10.30
DTMF generator.

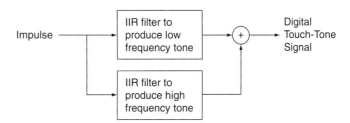

FIGURE 10.31
DTMF receiver.

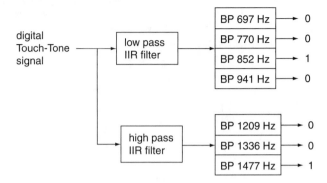

FIGURE 10.32
DTMF row and column frequencies.

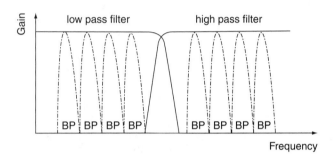

on the high pass portion to identify the column. Each band pass filter reports either a large or a small amplitude output, depending on whether or not a signal is detected in its range. For the set of filter responses shown in Figure 10.31, for example, key "9" on the Touch-Tone keypad would be received.

10.9 HAZARDS OF PRACTICAL IIR FILTERS

Difference equation diagrams for various implementations of IIR filter were presented in Section 4.6.2. All of the implementation issues affecting FIR filters, mentioned in Section 9.10, affect IIR filters as well. That is, coefficient quantization can change the shape of a filter quite dramatically. In addition, though, looms the possibility that quantization can actually cause a filter to become unstable, as Example 10.15 illustrates.

EXAMPLE 10.15

A filter is described by the stable transfer function

$$H(z) = \frac{1}{1 + a_1 z^{-1} + a_2 z^{-2}} = \frac{1}{1 - 1.812 z^{-1} + 0.813 z^{-2}}$$

The poles of this system lie at $z = 0.99452118390532$ and $z = 0.81747881609468$. Both of these poles lie inside the unit circle, so the system is stable. Study the effects of quantizing the coefficients.

Table 10.4 lists the coefficients quantized with between 1 and 14 bits to the right of the binary point. Following Equation (9.7), the quantized values of the coefficients are calculated as

$$q_1 = \frac{-\text{truncate}(2^Q(1.812) + 0.5)}{2^Q}$$

$$q_2 = \frac{\text{truncate}(2^Q(0.813) + 0.5)}{2^Q}$$

where Q is the number of bits used in the quantization. The transfer function defined by these coefficients is

$$H(z) = \frac{1}{1 + q_1 z^{-1} + q_2 z^{-2}}$$

The magnitudes of the poles can now be calculated. Table 10.4 contains the results. At least 10 bits are necessary to implement this filter reliably. Below this number of bits, the poles either fail to lie inside the unit circle or fall too far from the intended locations.

TABLE 10.4
Coefficient Quantization for Example 10.15

Number of Bits Q	Quantized a_1	Quantized a_2	Stable
1	-2.0	1.0	No
2	-1.75	0.75	No
3	-1.75	0.875	Yes
4	-1.8125	0.8125	No
5	-1.8125	0.8125	No
6	-1.8125	0.8125	No
7	-1.8125	0.8125	No
8	-1.8125	0.8125	No
9	-1.8125	0.8125	No
10	-1.8115234375	0.8134765625	Yes
11	-1.81201171875	0.81298828125	Yes
12	-1.81201171875	0.81298828125	Yes
13	-1.81201171875	0.81298828125	Yes
14	-1.81201171875	0.81298828125	Yes

Even when instability does not occur, quantization can dramatically affect the shape of the filter. The stable result for 3 bits in Example 10.15, for instance, is not of much use, since the coefficients quantized at this level produce a very different filter shape from the one intended. Example 10.16 provides a further example of this problem.

EXAMPLE 10.16

The coefficients for the narrow band pass filter

$$H(z) = \frac{1}{1 - 0.17z^{-1} + 0.965z^{-2}}$$

are quantized using 4 bits. Evaluate the effects of quantization.

The new coefficients are

$$q_1 = \frac{-\text{truncate}(2^4(0.17) + 0.5)}{2^4} = -0.1875$$

$$q_2 = \frac{\text{truncate}(2^4(0.965) + 0.5)}{2^4} = 0.9375$$

which gives the transfer function

$$H_q(z) = \frac{1}{1 - 0.1875z^{-1} + 0.9375z^{-2}}$$

The pole-zero diagram in Figure 10.33 shows the poles and zeros for $H(z)$ and $H_q(z)$. The poles for the quantized filter are closer to the origin. The effects of this pole position shift are evident in Figure 10.34, which plots the magnitude responses for both filters. The quantized filter lacks the sharpness and high peak gain of the original because the quantized poles are farther from the unit circle.

FIGURE 10.33

Pole-zero plot for Example 10.16.

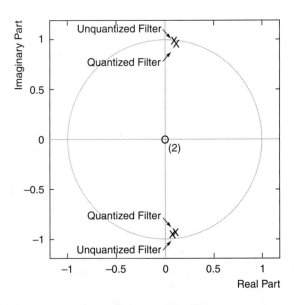

FIGURE 10.34
Magnitude
responses for
Example 10.16.

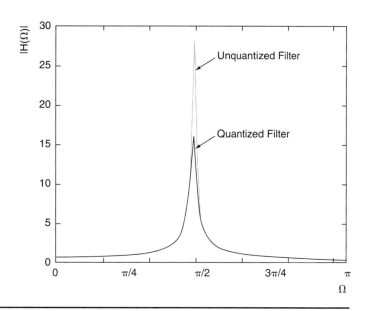

CHAPTER SUMMARY

Matlab
Support

1. The difference equation for an IIR filter is recursive.
2. IIR filters are not guaranteed to be free of phase distortion.
3. For a given filter shape, an IIR filter will require fewer coefficients than an FIR filter.
4. IIR filters are usually designed by converting analog filters into digital filters.
5. The four most common analog prototypes are Butterworth, Chebyshev Type I, Chebyshev Type II, and elliptic.
6. The bilinear transformation is one method of designing IIR filters. It converts an analog transfer function $H(s)$ into a digital transfer function $H(z)$ using the relation $s = 2f_S (z - 1)/(z + 1)$. To avoid the distortion of filter shape that is caused by the bilinear transformation, the frequencies that characterize the analog transfer function must be prewarped.
7. In the impulse invariance IIR design method, the impulse response of an analog filter with the desired behavior is sampled to obtain the impulse response for the equivalent digital filter.
8. Transfer functions for band pass, high pass, and band stop IIR filters can be obtained from the transfer function for a low pass filter.
9. When IIR filter coefficients are quantized for use with a DSP processor, quantization errors can change the behavior of the filter and even cause it to become unstable.

REVIEW QUESTIONS

10.1 A recursive filter has the difference equation

$$y[n] = -0.8y[n-1] + 0.1y[n-2] + x[n]$$

    a. Find the impulse response for the filter.
    b. How many nonzero terms does the impulse response contain?

10.2 Show the effect of prewarping by plotting analog frequency $\omega = 2f_s \tan(\Omega/2)$, in rad/sec, against digital frequency Ω, in rads, for $\Omega = 0$, $\pi/8$, $\pi/4$, $3\pi/8$, $\pi/2$, $5\pi/8$, $3\pi/4$, $7\pi/8$ rads, for a 16 kHz sampling rate.

10.3 A first order analog low pass Butterworth filter has a -3 dB frequency of 2.5 kHz and a DC gain of unity.

    a. Use a bilinear transformation to find a digital filter $H(z)$ to match the analog filter for a sampling rate of 8 kHz:
        (i)   Without prewarping
        (ii) With prewarping
    b. Plot the magnitude responses for both filters obtained in (a).

10.4 The first order analog low pass filter $H(s) = 1500/(s + 1500)$ must be transformed into a digital filter operating with an 8 kHz sampling rate.

    a. Prewarp the cut-off frequency for the filter and modify the analog transfer function accordingly.
    b. Find the transfer function for the digital filter.
    c. Find the difference equation for the filter.
    d. Find the frequency response and draw the filter shape.
    e. Find the digital filter shape directly from the magnitude $|H(\omega)|$ of the analog filter, and show that the shape matches the filter shape obtained in (d).

10.5 Choose an order for the Butterworth IIR filter with the specifications shown in Figure 10.35. The sampling rate is 8 kHz.

FIGURE 10.35
Butterworth filter
specification for
Question 10.5.

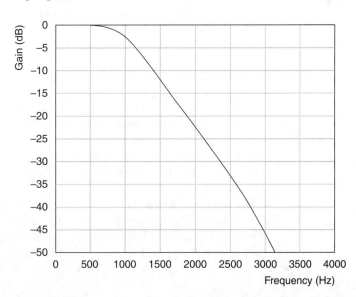

10.6 A low pass Butterworth filter design requires a cut-off frequency of 6.5 kHz and a stop band attenuation of 28 dB at 8 kHz. The sampling rate is 24 kHz.
   a. Choose an order for the filter.
   b. Find an expression for filter shape $|H(\Omega)|$ and then plot filter shape, $|H(f)|$ versus f.

10.7 The -3 dB frequency for a low pass Butterworth filter is 2.5 kHz. The stop band begins at 3.8 kHz with an attenuation of 35 dB. Determine the minimum order of the filter if the sampling rate is 8 kHz.

10.8 A fifth order low pass Butterworth filter in a 10 kHz sampled system has a pass band edge at 1.3 kHz and a transition width of 600 Hz. What is the stop band attenuation for this filter?

10.9 Find the difference equation for a first order high pass digital Butterworth filter with a cut-off frequency of 3.5 kHz, for a 16 kHz sampling frequency.

10.10 A design for a high pass filter with a Butterworth characteristic is needed. Filter specifications require a cut-off frequency of 9 kHz for a sampling rate of 44 kHz, and a stop band attenuation of 40 dB. Find the transition width if the order of the filter is:
   a. $n = 3$
   b. $n = 6$

10.11 Find the transfer function $H(z)$ for a second order low pass digital Butterworth filter with a cut-off frequency of 2.5 kHz and a sampling frequency of 8 kHz.

10.12 A design is required for a low pass filter meeting the following specifications: -3 dB frequency 1 kHz, stop band attenuation 44 dB, stop band edge frequency 1.5 kHz, sampling frequency 8 kHz. How many filter coefficients will be required if the filter is:
   a. Windowed FIR
   b. Butterworth IIR

10.13 Identify the order of the Chebyshev Type I filter shown in Figure 10.36. The sampling rate is 10 kHz.

FIGURE 10.36
Chebyshev Type I filter specification for Question 10.13.

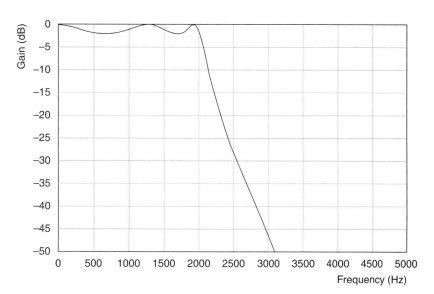

10.14 A low pass Chebyshev Type I filter has a pass band edge gain of -0.5 dB at 10 kHz, and a stop band edge gain of -20 dB at 12 kHz, with a 32 kHz sampling frequency.
a. Choose an order for the filter.
b. Find and plot an expression for filter shape $|H(f)|$ versus f.

10.15 Filter specifications are pass band ripple 0.292, stop band ripple 0.08, pass band edge frequency 4.8 kHz, stop band edge frequency 5.4 kHz, sampling frequency 15 kHz. Design a filter to meet these specifications that has a:
a. Butterworth characteristic
b. Chebyshev Type I characteristic

10.16 A band pass filter has a center frequency of 5 kHz. Its maximum gain is 0 dB, and its pass band edges lie at -0.5 dB, 1.6 kHz apart. The transition widths are 400 Hz, and the stop band attenuation is 35 dB. Choose an order for the Chebyshev Type I filter that satisfies these specifications for a 15 kHz sampling rate.

10.17 Use a first order low pass Butterworth filter prototype to find the transfer function $H(z)$ for a high pass Butterworth filter with a cut-off frequency of 3 kHz, for an 8 kHz sampling rate.

10.18 Use a first order low pass Butterworth filter prototype to find the transfer function $H(z)$ for a band pass Butterworth filter with a lower cut-off frequency of 1 kHz and an upper cut-off frequency of 1.5 kHz, for an 8 kHz sampling frequency.

10.19 Use a first order low pass Butterworth filter prototype to find the transfer function for a band stop Butterworth filter with lower cut-off frequency 55 Hz and upper cut-off frequency 65 Hz, for a 2 kHz sampling rate. Draw the filter shape, $|H(f)|$ versus f.

10.20 Find the transfer function for a second order high pass digital Butterworth filter with a bandwidth of 30 Hz. The sampling rate is 200 Hz.

10.21 **a.** Write the difference equation for a 1 kHz tone generator for a 4 kHz sampled system.
b. Sketch the filter shape for the filter in (a).

10.22 The impulse response for an analog filter is given by $h(t) = e^{-2t}u(t)$. An equivalent digital filter is obtained using the impulse invariance method.
a. Find an expression for the digital filter shape $|H(\Omega)|$ if the sampling interval is:
(i) $T = 1.0$ sec
(ii) $T = 0.5$ sec
b. Plot the digital filter shapes from (a) and compare them to the analog filter shape, given by $|H(\omega)| = 1/\sqrt{4 + \omega^2}$, from 0 to 4 Hz. Scale all filters to give unity DC gain. (Note: Taking small frequency steps will give the best pictures. Use software to ease the computational burden.)
c. What sampling interval will produce a digital filter that closely matches the analog filter shape up to 4 Hz?

10.23 Using the impulse invariance method, find the digital equivalent to a first order analog Butterworth filter with a cut-off frequency of 4 kHz for a 32 kHz sampling rate. Compare the digital filter shape with the analog filter shape up to 20 kHz, scaling both filters to give unity DC gain.

10.24 An analog filter has the impulse response

$$h(t) = \frac{1}{300}e^{-200t}\sin(300t)u(t)$$

Find the impulse-invariant digital filter obtained with a sampling interval of 2 msec. Plot the digital filter shape $|H(f)|$. Compare the digital filter shape with the analog filter shape, given by

$$|H(\omega)| = \frac{1}{\sqrt{(130000 - \omega^2)^2 + (400\omega)^2}}$$

up to 400 Hz. Scale both filters to give unity DC gain.

10.25 The transfer function for a digital IIR filter is

$$H(z) = \frac{0.3525 + 1.0576z^{-1} + 1.0576z^{-2} + 0.3525z^{-3}}{1 + 1.0487z^{-1} + 0.6920z^{-2} + 0.0797z^{-3}}$$

Find the transfer function and plot the shape of the filter after its coefficients are quantized to:

a. 4 bits

b. 5 bits

11

DFT AND FFT PROCESSING

The DFT identifies the frequency content in a signal from a set of the signal's samples. Its most efficient implementation, the FFT, is one of the most well-used of all DSP tools. It provides a practical means of identifying a signal's spectrum. This chapter:

- ➤ defines and interprets the discrete Fourier transform (DFT)
- ➤ provides a graphical interpretation of the DFT
- ➤ links the DFT and the discrete time Fourier transform (DTFT)
- ➤ links the DFT and the discrete Fourier series (DFS)
- ➤ identifies the limited resolution of the DFT and problems such as smearing that arise from it
- ➤ applies the DFT to nonperiodic and periodic signals
- ➤ discusses the effect of using windows in DFT calculations
- ➤ introduces spectrograms as a means of presenting time and frequency information about a signal
- ➤ develops the fast Fourier transform (FFT) as a computationally efficient alternative to the DFT
- ➤ defines the two-dimensional DFT for images and other two-dimensional data

11.1 DFT BASICS

In this chapter, attention turns to the development of efficient computational algorithms for practical analysis. In Chapter 7, the discrete time Fourier transform (DTFT) was introduced. It was used in Chapter 7 to find filter shape and in Chapter 8 to find signal spectra. The definition of the DTFT was given by Equation (7.1) as:

$$X(\Omega) = \sum_{n=-\infty}^{\infty} x[n]e^{-jn\Omega} \tag{7.1}$$

One of the problems with implementing this transform on a computer is that an infinite number of samples may be required. The **discrete Fourier transform (DFT)** solves this problem by making the number of samples required finite. The definition for the DFT is:

$$X[k] = \sum_{n=0}^{N-1} x[n]e^{-j2\pi\frac{k}{N}n} \qquad \text{for } k = 0, 1, \cdots, N-1 \tag{11.1}$$

This equation transforms a time domain signal $x[n]$ into a frequency domain signal $X[k]$. Like the DTFT, the DFT reports the frequency content of a signal. For the DFT, the index k marks the frequency of each element, in a manner to be detailed shortly. The number of time samples $x[n]$ used as input to the DFT is the same as the number of frequency samples $X[k]$ produced as output: Both are equal to N. Thus, the DFT solves another problem connected with the DTFT: $X(\Omega)$ is defined at an infinite number of frequencies Ω, whereas $X[k]$ is defined at a finite number of indices k, making it amenable to processing. The N time samples used in the DFT are said to lie within the **DFT window**. The length of the window is measured in numbers of samples and equals N. Samples in the signal that lie outside the window do not affect the analysis.

The quantity $X[k]$ is in general complex, and so can be written as

$$X[k] = |X[k]|e^{j\theta[k]}$$

Its magnitude $|X[k]|$ is plotted against k to produce the **DFT magnitude spectrum**. The **DFT phase spectrum** plots the phase $\theta[k]$ against k. It is easy to show that $X[k]$ is periodic with a period of N samples:

$$X[k+N] = \sum_{n=0}^{N-1} x[n]e^{-j2\pi\frac{(k+N)}{N}n} = \sum_{n=0}^{N-1} x[n]e^{-j2\pi\frac{k}{N}n}e^{-j2\pi n}$$

$$= \sum_{n=0}^{N-1} x[n]e^{-j2\pi\frac{k}{N}n} = X[k]$$

since, by Euler's identity, $e^{-j2\pi n} = \cos(2\pi n) - j\sin(2\pi n) = 1$. Since the DFT is periodic with period N, the magnitudes and phases of the DFT need to be calculated only for $k = 0$ to $N-1$, and repeat every N points after that. As is the case for the DTFT, the magnitude spectrum for the DFT is always even and the phase spectrum is always odd. When frequency samples $X[k]$ must be converted back into time samples $x[n]$, the **inverse DFT (IDFT)** is required. This inverse transform, also periodic with period N, is defined as:

$$x[n] = \frac{1}{N}\sum_{k=0}^{N-1} X[k]e^{j2\pi\frac{k}{N}n} \qquad \text{for } n = 0, 1, \cdots, N-1 \tag{11.2}$$

The scaling factor $1/N$, explained in Appendix E, ensures that the original signal samples $x[n]$ are correctly reproduced from the DFT $X[k]$.

Equation (11.1) for the DFT has a similar form to that of the DTFT. In fact, when the DFT and the DTFT operate on the same set of time samples, the DFT matches the DTFT exactly, but at a finite number of points only. The DTFT produces magnitude and phase spectra that are smooth functions of frequency, and the DFT produces sampled versions of the same spectra. The continuous variable Ω in the DTFT is replaced in the DFT by the discrete variable k, and the smooth spectra of the DTFT become line spectra for the DFT. Figure 11.1 illustrates the links between the signal, its DTFT, its DFT, and the inverse DFT. Figure 11.1(a) shows a nonperiodic signal. Its exact DTFT magnitude spectrum is provided in Figure 11.1(b). Notice its 2π periodicity, characteristic for all nonperiodic digital signals.

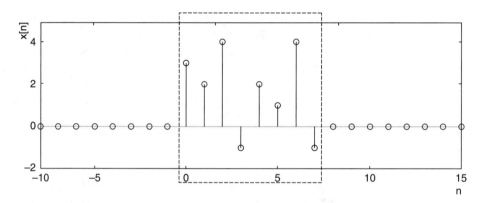

(a) Original Signal

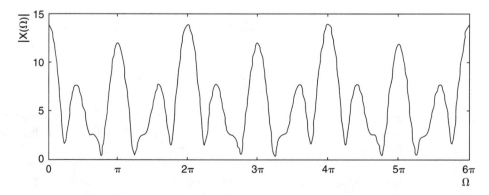

(b) Discrete Time Fourier Transform (DTFT)

FIGURE 11.1
Relationship between DTFT and DFT, signal and IDFT.

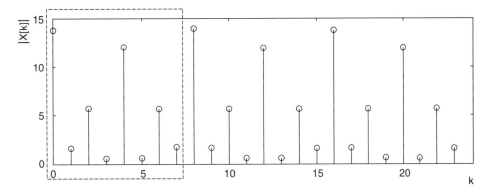

(c) Discrete Fourier Transform (DFT)

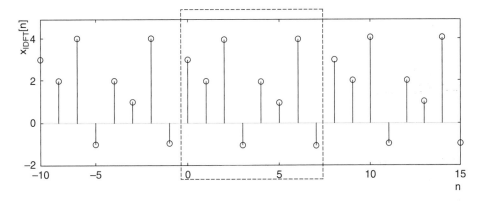

(d) Inverse Discrete Fourier Transform (IDFT)

FIGURE 11.1
Continued

The DFT magnitude spectrum of the windowed portion of the signal is given in Figure 11.1(c). Each DFT point matches the DTFT spectrum at the same location. In other words, the DFT is a sampled version of the DTFT. Figure 11.1(d) shows the inverse DFT. Note that it does not match the original, nonperiodic signal. The inverse DFT produces a signal that is a periodic version of the windowed portion in Figure 11.1(a). The N unique values of the DFT magnitudes and of the inverse DFT are highlighted in boxes.

EXAMPLE 11.1
Find the DFT magnitude and phase spectra for samples of the signal selected in Figure 11.2, and verify that the inverse DFT reproduces these samples.

FIGURE 11.2
Signal for Example 11.1.

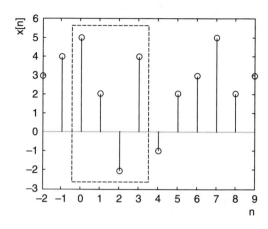

The DFT window selects four samples for analysis, that is, $N = 4$. Using the DFT in Equation (11.1),

$$X[k] = \sum_{n=0}^{N-1} x[n]e^{-j2\pi\frac{k}{N}n} = x[0] + x[1]e^{-j2\pi\frac{k}{4}} + x[2]e^{-j2\pi\frac{k}{4}2} + x[3]e^{-j2\pi\frac{k}{4}3}$$

$$= 5 + 2e^{-j\frac{\pi k}{2}} - 2e^{-j\pi k} + 4e^{-j\frac{\pi k3}{2}}$$

Each complex number in polar form $re^{j\theta}$ can be written for short as $r\underline{|\theta}$, which records magnitude $\underline{|phase}$ for the number. Terms of the form $e^{j\theta}$ may be written, simply, as $1\underline{|\theta}$. Using this notation, $X[k]$ may be written for calculation as

$$X[k] = 5\underline{|0} + 2\left|-\frac{\pi k}{2}\right. - 2\underline{|-\pi k} + 4\left|-\frac{\pi k3}{2}\right.$$

with angles in radians. Table 11.1 evaluates $X[k]$ for $k = 0, 1, 2, 3$. Other values of k can be chosen, but values of $X[k]$ other than those in the table do not arise. The DFT magnitude and phase spectra are shown in Figure 11.3. The DFT is periodic with period $N = 4$.

TABLE 11.1
Calculations of $X[k]$ for Example 11.1

k	$X[k]$	$|X[k]|$	$\theta[k]$ (rads)
0	$5\underline{|0} + 2\underline{|0} - 2\underline{|0} + 4\underline{|0}$	9.0000	0
1	$5\underline{|0} + 2\left|-\frac{\pi}{2}\right. - 2\underline{|-\pi} + 4\left|-\frac{3\pi}{2}\right.$	7.2801	0.2783
2	$5\underline{|0} + 2\underline{|-\pi} - 2\underline{|-2\pi} + 4\underline{|-3\pi}$	3.0000	−3.1416
3	$5\underline{|0} + 2\left|-\frac{3\pi}{2}\right. - 2\underline{|-3\pi} + 4\left|-\frac{9\pi}{2}\right.$	7.2801	−0.2783

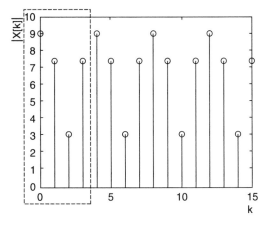

(a) DFT Magnitude Spectrum

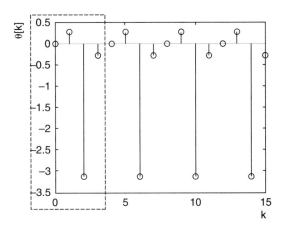

(b) DFT Phase Spectrum

FIGURE 11.3
DFT magnitude and phase spectra for Example 11.1.

According to Equation (11.2), the signal samples can be reproduced using the inverse DFT:

$$x[n] = \frac{1}{N}\sum_{k=0}^{N-1} X[k]e^{j2\pi\frac{k}{N}n} = \frac{1}{4}\left(X[0] + X[1]e^{j2\pi\frac{1}{4}n} + X[2]e^{j2\pi\frac{2}{4}n} + X[3]e^{j2\pi\frac{3}{4}n}\right)$$

$$= \frac{1}{4}\left(X[0] + X[1]e^{j\frac{\pi n}{2}} + X[2]e^{j\pi n} + X[3]e^{j\frac{3\pi n}{2}}\right)$$

Writing the $X[k]$ values from the table in magnitude $\lfloor\underline{phase}$ form, this becomes

$$x[n] = \frac{1}{4}\left(9\lfloor\underline{0} + (7.2801\lfloor\underline{0.2783})(1\lfloor\underline{1.5708n})\right.$$

$$\left. + \left(3\lfloor\underline{-3.1416})(1\lfloor\underline{3.1416n})\right) + \left(7.2801\lfloor\underline{-0.2783})(1\lfloor\underline{4.7124n})\right)$$

The values of $x[n]$ for $n = 0, 1, 2, 3$ are shown in Table 11.2, and the samples for a wider range of n values are plotted in Figure 11.4. When rounding errors are neglected, the inverse DFT produces exactly the original sample values within the DFT window. As Figure 11.4 shows, the complete inverse DFT repeats every $N = 4$ samples. Notice that the inverse DFT samples have no relationship to the original signal samples outside the window.

TABLE 11.2
Calculations of $x[n]$ for Example 11.1

n	0	1	2	3
$x[n]$	5	2	−2	4

FIGURE 11.4
Inverse DFT for Example 11.1.

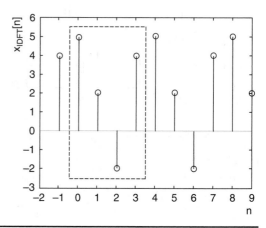

A comparison of the DTFT in Equation (7.1) and the DFT in Equation (11.1) shows that, when they operate on the same finite group of samples, it must be the case that

$$\Omega = 2\pi\frac{k}{N} \tag{11.3}$$

As k ranges from 0 to $N-1$, it maps to digital frequencies Ω between 0 and $2\pi(N-1)/N$ radians. N points of the DFT, then, cover essentially 2π radians. Since $\Omega = 2\pi f/f_S$, it follows that the first N points of the DFT cover the analog frequencies between zero and the sampling frequency f_S. Example 11.2 compares the DTFT and DFT for a signal with four nonzero samples.

While the equation for the DFT may be similar to that of the DTFT in Equation (7.1), it is absolutely identical with Equation (8.3) for the discrete Fourier series (DFS) coefficients. Thus, the calculations are exactly the same for both methods. The relationships between the DFT and the DFS, and the reasons why they have so much in common, are explored further in Section 11.3.

EXAMPLE 11.2

Find the magnitude spectrum using both the DTFT and the DFT for the signal shown in Figure 11.5.

The DTFT is

$$X(\Omega) = \sum_{n=-\infty}^{\infty} x[n]e^{-jn\Omega} = 2 - e^{-j\Omega} + 3e^{-j2\Omega} + 3e^{-j3\Omega}$$

FIGURE 11.5

Signal for Example 11.2.

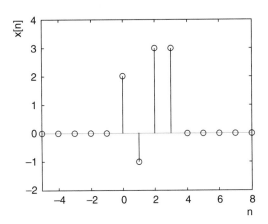

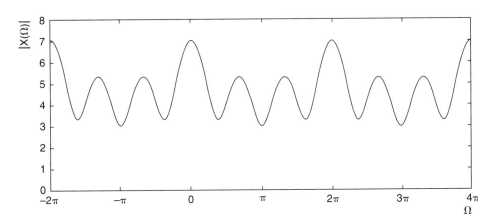

FIGURE 11.6

DTFT magnitude spectrum for Example 11.2.

The magnitude of this expression for the range $\Omega = -2\pi, ..., 4\pi$ is presented in Figure 11.6. Note that the portion between $\Omega = 0$ and $\Omega = \pi$ is enough to deduce the remainder of the data. First, the magnitude spectrum is even, which provides the part between $\Omega = -\pi$ and $\Omega = 0$. Second, the magnitude spectrum repeats every 2π.

For the four nonzero samples in Figure 11.5, a DFT with $N = 4$ is computed from

$$X[k] = \sum_{n=0}^{N=1} x[n] e^{-j2\pi \frac{k}{N}n} = 2 - e^{-j2\pi \frac{k}{4}} + 3e^{-j2\pi \frac{k}{4}2} + 3e^{-j2\pi \frac{k}{4}3}$$

$$= 2 - e^{-j\frac{\pi k}{2}} + 3e^{-j\pi k} + 3e^{-j\frac{\pi 3k}{2}}$$

The magnitude of this expression is shown in Figure 11.7. Since N points of the DFT cover 2π radians for the DTFT, the range $k = -4$ to $k = 8$ is plotted to match the range to that of Figure 11.6. Note that any N values of the sequence, for example, from $k = 0$ to $k = 3$, or

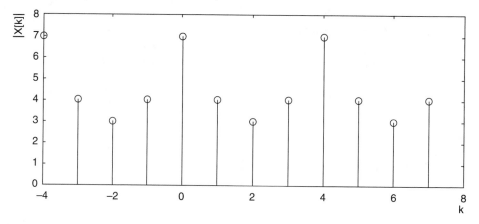

FIGURE 11.7
DFT magnitude spectrum for Example 11.2.

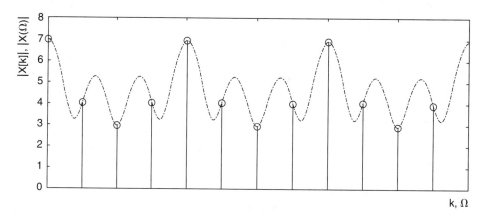

FIGURE 11.8
DTFT and DFT results superimposed for Example 11.2.

from $k = 2$ to $k = 5$, are sufficient to generate the entire sequence, since the sequence repeats every $N = 4$ samples. Figure 11.8 shows the discrete time Fourier transform and the DFT for the signal superimposed. As expected, the DFT magnitudes lie on the DTFT magnitude curve.

EXAMPLE 11.3

The fact that the DFT is a sampled version of the DTFT permits an impulse response $h[n]$ to be computed from a frequency response $H(\Omega)$. Equiripple FIR filters, which were introduced in Section 9.9, require this computation to find a realizable impulse response $h[n]$ from an optimal frequency response $H_1(\Omega)$, expressed as a polynomial in powers of $\cos\Omega$. Suppose that the optimal frequency response for an equiripple filter is given by

$$H_1(\Omega) = 0.3595\cos^4\Omega + 0.8770\cos^3\Omega + 0.3229\cos^2\Omega - 0.1782\cos\Omega - 0.0333$$

This response is purely real, as is always the case for impulse responses $h_1[n]$ that are symmetrical around $n = 0$. Since the frequency response comes from a DTFT, it can be sampled to produce the corresponding DFT. For an Nth order polynomial, $2N + 1$ sample points are needed between $\Omega = 0$ and $\Omega = 2\pi$ to ensure the correct impulse response will eventually be found. Once the DFT is available, an inverse DFT (IDFT) produces the impulse response samples.

For the given polynomial, $N = 4$, so 9 samples are taken in all, in steps of $\Omega = 2\pi/9$ to ensure the range from $\Omega = 0$ to $\Omega = 2\pi$ is covered. The samples are provided in Table 11.3. These samples now serve as the points of the 9-point DFT $H_1[k]$. The time domain samples $h_1[n]$ that produced $H_1[k]$ form the impulse response that generated the frequency response $H_1(\Omega)$ provided above. The inverse DFT of the samples is given in Table 11.4. A DTFT of these samples will verify that they indeed produce the required frequency response. Note that the inverse DFT repeats every $N = 9$ samples. The samples between $n = -4$ and $n = 4$ form a complete picture of the impulse response and may be shifted to the right by four positions to give a causal impulse response $h[n]$.

TABLE 11.3
Samples for Example 11.3

Ω	0	$\dfrac{2\pi}{9}$	$\dfrac{4\pi}{9}$	$\dfrac{6\pi}{9}$	$\dfrac{8\pi}{9}$	$\dfrac{10\pi}{9}$	$\dfrac{12\pi}{9}$	$\dfrac{14\pi}{9}$	$\dfrac{16\pi}{9}$
$H_1(\Omega)$	1.3479	0.5377	−0.0496	0.0494	−0.0281	−0.0281	0.0494	−0.0496	0.5377

TABLE 11.4
Inverse DFT Samples for Example 11.3

n	0	1	2	3	4	5	6	7	8
$h_1[n]$	0.2630	0.2398	0.1706	0.1096	0.0225	0.0225	0.1096	0.1706	0.2398

The easiest way to interpret the DFT equation is to rewrite the complex exponential using Euler's equation from Equation (A.5), $e^{-j\theta} = \cos\theta - j\sin\theta$:

$$X[k] = \sum_{n=0}^{N-1} x[n]\left(\cos\left(\frac{2\pi kn}{N}\right) - j\sin\left(\frac{2\pi kn}{N}\right)\right)$$

The cosine and sine components have digital fundamental frequencies at $\Omega = 2\pi k/N$ radians, where k may take any value from 0 to $N-1$. The more closely the signal $x[n]$ correlates with a sinusoid at one of these frequencies, the larger the spectral component $X[k]$ will be. When $x[n]$ correlates poorly with a sinusoid for a given value of k, cancellations tend to occur to make $X[k]$ small. Example 11.4 will illustrate this idea. Each element $X[k]$ of the DFT can be viewed as the output of a narrow band filter centered at the digital frequency $2\pi k/N$ radians. The DFT as a whole can be seen as a filter bank made up of narrow adjacent filters.

As noted above, N points of the DFT cover the range of frequencies from 0 to f_S Hz, the sampling frequency. The frequency samples are therefore spaced f_S/N Hz apart. This frequency spacing is referred to as the resolution of the DFT because it describes how well the DFT can resolve neighboring signal frequencies. The smaller the frequency spacing, the better the resolution; the wider the spacing, the poorer the resolution of the DFT. Thus,

$$\text{DFT frequency spacing} = \text{DFT resolution} = \frac{f_S}{N}$$

Assuming the sampling frequency remains unchanged, the resolution of a DFT is greatest when a large number of points is used, so that the frequency spacing is small and the fine detail of a spectrum may be captured. In the filter bank view, a DFT with good resolution comprises a large number of extremely narrow band pass filters. For a fixed sampling rate, the resolution is poorest when the number of points used in the DFT is small, and only the coarsest features of a signal's spectrum can be identified.

Elements of the DFT, $X[k]$, are located at the frequencies

$$f = k\frac{f_S}{N} \tag{11.4}$$

Interpretation of DFT plots can often be made easier if they are plotted against these frequencies in Hz instead of indices k. This technique is illustrated in Example 11.5 and is used in many of the other examples in the remainder of this chapter.

Though N points in all are computed for the DFT, there is some replication of information among them. As noted above, N points cover the range of frequencies between 0 and f_S Hz. The points of the DFT occur at the frequencies kf_S/N, so the Nyquist limit of $f_S/2$ Hz is reached when $k = N/2$. Thus, the points of the DFT from $k = 0$ to $k = N/2$ carry all the essential magnitude and phase information of the DFT. The remainder of the points report imaged copies of the important signal frequencies found in the baseband, created as artifacts of sampling. For the magnitude spectrum, these points are symmetrically related to those between $k = 0$ and $k = N/2$, to produce a mirror image across $k = N/2$. This feature is common to all DFT magnitude spectra and explains the "inkblot" nature of the spectral shapes obtained. This symmetry, illustrated in the following examples, is consistent with an even magnitude spectrum that repeats every N samples. For the phase spectrum, the net result is an odd function that repeats every N samples.

EXAMPLE 11.4

To underline the filter bank interpretation of the DFT, consider the 40-second plot of two cosine waves shown in Figure 11.9. These two signals are added together and combined with some random noise to produce a signal $x(t)$ described by the equation

$$x(t) = \cos\left(\frac{2\pi t}{16}\right) + \cos\left(\frac{2\pi 3t}{8}\right) + \text{random noise}$$

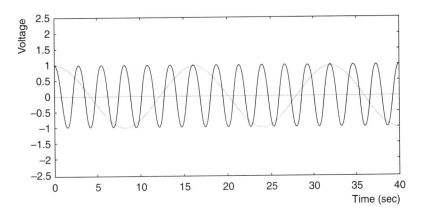

FIGURE 11.9

Two cosine signals for Example 11.4.

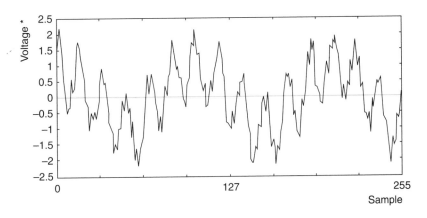

*For clarity, the envelope rather than the individual samples of the signals is plotted.

FIGURE 11.10

Sum of cosines plus random noise for Example 11.4.

This signal contains essentially two main frequency components, 1/16 Hz and 3/8 Hz, and is sampled at 6.4 Hz. The digital frequencies for the two main frequency components in the sampled signal can be found using $\Omega = 2\pi f/f_S$. They are $\Omega_1 = 2\pi/102.4$ and $\Omega_2 = 6\pi/51.2$ radians. The equation for $x[n]$ becomes:

$$x[n] = \cos(n\Omega_1) + \cos(n\Omega_2) + \text{random noise}$$

$$= \cos\left(n\frac{2\pi}{102.4}\right) + \cos\left(n\frac{6\pi}{51.2}\right) + \text{random noise}$$

This digital signal is plotted in Figure 11.10. In 40 seconds, 6.4 Hz sampling collects 256 samples.

FIGURE 11.11

First 16 DFT frequency components for Example 11.4.

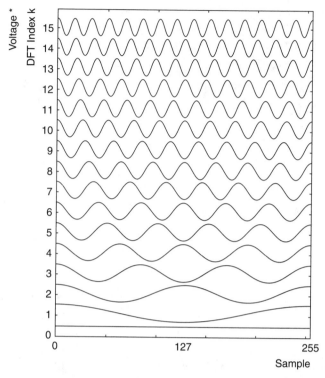

* For clarity, the envelopes of the signals rather than their individual samples are plotted.

The DFT should give a strong response for the two frequencies in $x[n]$ and a weak response for all other frequencies. As indicated by Equation (11.3), a DFT contains only a finite number of frequency elements: For a DFT with $N = 256$ points, each index k corresponds to a digital frequency $\Omega = 2\pi k/N = 2\pi k/256$. The first 16 of the 256 digital frequencies of the DFT are shown in Figure 11.11 for the same number of samples displayed in Figure 11.10. Each trace shows a digital sinusoid $\cos(n2\pi k/256)$ for the given values of k. The frequencies closest to those present in $x[n]$ can be identified in the figure by counting how many peaks occur in the interval. The higher frequency cosine in Figure 11.9 appears to repeat at the same rate as the $k = 15$ trace in Figure 11.11, while the frequency for the slower cosine in Figure 11.9 lies between those of the $k = 2$ and $k = 3$ traces in Figure 11.11. Since the DFT can be seen as a bank of narrow adjacent filters, each centered at a digital frequency $\Omega = 2\pi k/N = 2\pi k/256$ radians, peaks in the spectrum should occur for $2\pi k/256 = 2\pi/102.4$ and $2\pi k/256 = 2\pi/51.2$, or $k = 2.5$ and $k = 15$. Since k must be an integer, the peak at 2.5 will divide into two smaller peaks at $k = 2$ and $k = 3$. Peaks at $k = 2, 3,$ and 15 are confirmed in the DFT magnitude spectrum of Figure 11.12. The first half of the DFT magnitude spectrum, for $k = 0$ to $k = N/2 = 128$, is presented in Figure 11.13. It shows the signal elements below the Nyquist frequency, in the baseband. The peaks in

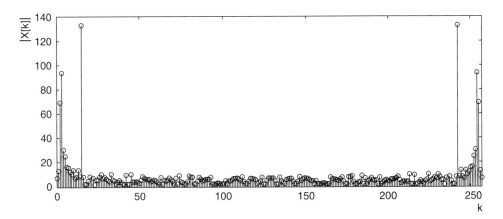

FIGURE 11.12
DFT magnitude spectrum for Example 11.4.

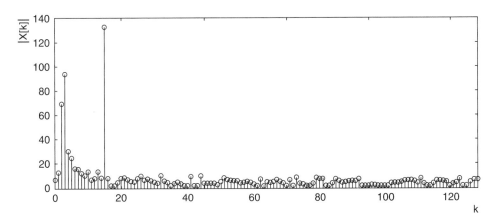

FIGURE 11.13
DFT magnitude spectrum from $k = 0$ to $k = N/2$ for Example 11.4.

the second half of the spectrum are images of the baseband frequencies, at $k = N - 15 = 256 - 15 = 241$, $k = 256 - 3 = 253$, and $k = 256 - 2 = 254$. As expected, repeating the N points of the magnitude spectrum creates an even spectrum.

EXAMPLE 11.5
The DFT magnitude and phase spectra for an 8-point signal sampled at 12 kHz are shown in Figure 11.14. The exact DFT values from which these plots were produced are listed in Table 11.5.

    a. Find the resolution of this DFT.

    b. Find the signal $x[n]$ whose spectrum is shown.

FIGURE 11.14
Spectra for Example 11.5.

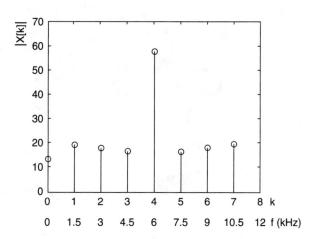

(a) Magnitude Spectrum

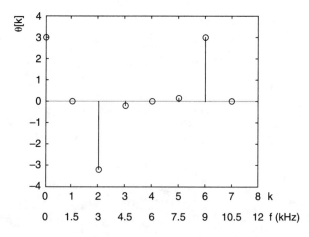

(b) Phase Spectrum

TABLE 11.5
Magnitude and Phase Values for Example 11.5

k	$|X[k]|$	$\theta[k]$
0	14.0000	3.1416
1	19.4230	0.0302
2	18.0000	−3.1416
3	16.9336	−0.2030
4	58.0000	0.0000
5	16.9336	0.2030
6	18.0000	3.1416
7	19.4230	−0.0302

a. The frequency spacing, or resolution, for this DFT is $f_S/N = 12/8 = 1.5$ kHz. This large spacing means poor resolution. With only 8 points to describe the spectrum, very little spectral detail can be observed. Frequencies can be mapped to indices k at 1500 Hz intervals, as marked in Figure 11.14. These frequencies correspond to $k f_S/N$ for $k = 0, ..., 7$.

b. Since $N = 8$ in this case, it is no surprise that the magnitude spectrum is symmetrical about $k = N/2 = 4$. This value of k also corresponds to the highest frequency that can be represented in the system, which is half the sampling frequency, or 6 kHz. The DFT magnitude spectrum shows that the signal's strongest elements occur at this frequency. The high peak at this frequency suggests that the signal whose DFT is given by Figure 11.14 must contain large and rapid changes in level. To find the exact signal, an inverse DFT must be performed using Equation (11.2), with $N = 8$ and the values from Table 11.5. For example, when $n = 0$,

$$x[n] = x[0] = \frac{1}{8}\sum_{k=0}^{7} X[k]e^{j2\pi\frac{k}{8}0} = \frac{1}{8}\sum_{k=0}^{7} X[k]$$

$$= \frac{1}{8}\Big(14.0\,\underline{|3.1416} + 19.423\,\underline{|0.0302} + 18.0\,\underline{|-3.1416} + 16.9336\,\underline{|-0.203}$$

$$+ 58.0\,\underline{|0} + 16.9336\,\underline{|0.203} + 18.0\,\underline{|3.1416} + 19.423\,\underline{|-0.0302}\Big) = 10$$

and, for $n = 1$,

$$x[n] = x[1] = \frac{1}{8}\sum_{k=0}^{7} X[k]e^{j2\pi\frac{k}{8}} = \frac{1}{8}\sum_{k=0}^{7} X[k]e^{j\frac{\pi k}{4}}$$

$$= \frac{1}{8}\Big(14.0\,\underline{|3.1416} + 19.423\,\underline{|0.0302+\pi/4} + 18.0\,\underline{|-3.1416+2\pi/4}$$

$$+ 16.9336\,\underline{|-0.203+3\pi/4} + 58.0\,\underline{|0+4\pi/4} + 16.9336\,\underline{|0.203+5\pi/4}$$

$$+ 18.0\,\underline{|3.1416+6\pi/4} + 19.423\,\underline{|-0.0302+7\pi/4}\Big) = -8$$

The other samples are calculated in a similar way. The 8 unique samples for this inverse DFT are $[10\ -8\ 9\ -9\ -8\ -10\ 11\ -9]$. They are plotted in Figure 11.15. Larger values of n produce additional samples from the same set, repeating with period $N = 8$. As expected, the signal shows dramatic changes from one sample to the next. This characteristic produces the predominantly high frequency content evidenced by the 6 kHz spike in the magnitude spectrum of Figure 11.14(a).

FIGURE 11.15

Inverse DFT for Example 11.5.

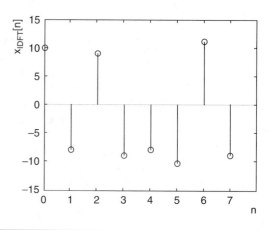

The limited resolution of the DFT has implications for interpreting DFT spectra. Most importantly, a DFT cannot pinpoint a frequency any more exactly than its resolution will allow. This limitation arose in Example 11.4, where a signal containing known frequencies 1/16 Hz and 3/8 Hz was sampled at $f_S = 6.4$ Hz and analyzed by a 256-point DFT. The resolution of this DFT is $f_S/N = 6.4/256 = 0.025$ Hz. Because DFT elements occur only at frequencies that are integer multiples of 0.025 Hz, the 1/16 Hz signal cannot be located exactly, since $(1/16)/0.025 = 2.5$. The 1/16 Hz frequency lies exactly between two DFT frequencies, as noted in the example. Thus, this signal is reported equally at each of the DFT indices on either side of the true signal frequency. The frequency of the 3/8 Hz signal, on the other hand, happens to coincide with a DFT frequency, since $(3/8)/0.025 = 15$, and so can be pinpointed exactly, as observed in Example 11.4. When no frequency in the DFT matches an important frequency in a signal being analyzed, the DFT causes **smearing** of the true spectral results in this way. Example 11.6 further illustrates the point.

EXAMPLE 11.6

The DTFT of a signal sampled at 16 kHz is shown in Figure 11.16(a). A spectral peak occurs at $\Omega = 1.035$ rads. Because the sampling rate is 16 kHz, the digital frequency $\Omega = 1.035$ radians equates to an analog frequency of $f = \Omega f_S/2\pi \approx 2636$ Hz. The signal is also analyzed using 8-point, 16-point, and 32-point DFTs, producing the DFT magnitude spectra in Figure 11.16(b)(c)(d). Each spectrum shows two peaks. The first is close to the frequency 2636 Hz; the second corresponds to an image of this frequency, lying above the Nyquist frequency at $f_S/2 - 2636 = 8000 - 2636 = 5364$ Hz. This imaged frequency is not a true element of the signal but a result of sampling. The frequency spacings for the DFTs are listed in Table 11.6. It is clear from Figure 11.16 that the larger the number of points used, the closer the DFT comes to finding the true spectral peak at 2636 Hz; the largest DFT peak always occurs at the DFT index that is nearest this frequency. All DFT results, however, smear the true peak to some degree. Note that the DFTs in this example are not necessarily perfect sampled versions of the DTFT as suggested in Example 11.2. While all of the transforms here do operate on the same signal, they do not all use the same number of data points.

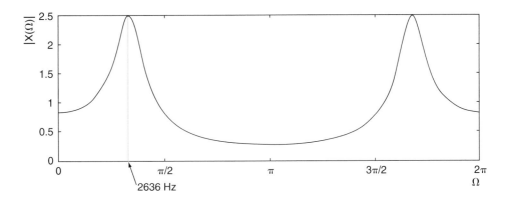

(a) DTFT

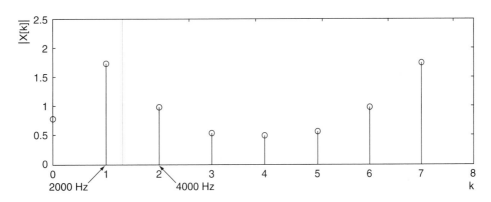

(b) 8-point DFT

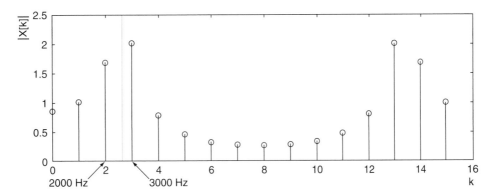

(c) 16-point DFT

FIGURE 11.16
DFT smearing for Example 11.6.

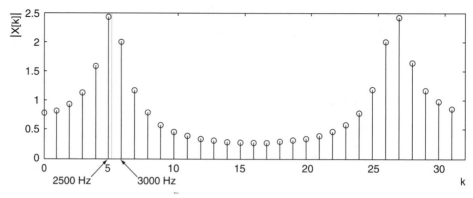

(d) 32-point DFT

FIGURE 11.16
Continued

TABLE 11.6
DFT Spacings for Example 11.6

Number of Points in DFT	DFT Frequency Spacing $\frac{f_S}{N}$ Hz
8	2000
16	1000
32	500

The DFT does not discriminate between periodic and nonperiodic signals, so there is no need to test whether the group of samples selected for analysis repeats or not. The DFT simply operates on a collection of time samples to report a signal's frequency content. Indeed, the DFT can help to uncover whether or not a signal has a periodic nature. The envelopes for the DFT magnitude spectra of nonperiodic signals will show bumps of varying sizes and at varying intervals, but no clear spikes. Periodicity is signaled by the presence of definite narrow spikes at regular intervals in the magnitude spectrum. For a periodic signal, these spikes occur at harmonic frequencies, integer multiples of the signal's fundamental frequency, as established in Section 8.3. Since it may not coincide with the period of the signal, the length of the DFT window affects the shape of the harmonic spikes, as well as the number of points between spikes.

Consider the DFT of a pure sinusoid. Figure 11.17 shows magnitude spectra for a 60 Hz sinusoid sampled at 256 Hz. The digital frequency of the signal is $\Omega = 2\pi f/f_S = 2\pi(60/256)$. Thus, $2\pi/\Omega = 256/60 = 64/15$. This is a ratio of the integers $N_1 = 64$ and $M_1 = 15$, so, from Section 3.3.4, the digital sequence repeats every $N_1 = 64$ samples, covering $M_1 = 15$ cycles of the underlying analog signal. Two perfect spikes appear in the 128-point DFT magnitude spectrum of Figure 11.17(a), the second an image of the first. Thus, when the DFT length

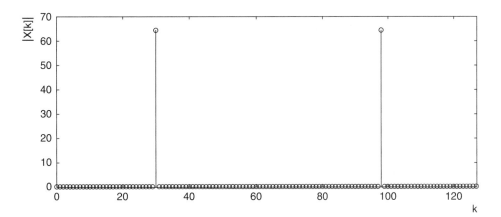

(a) N = 128

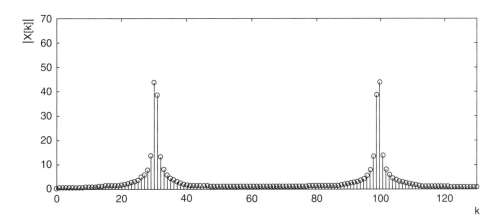

(b) N = 130

FIGURE 11.17
DFT of pure sinusoid.

$N\,(=128)$ happens to be an integer multiple of the number of samples $N_1\,(=64)$ in a digital signal's period, perfect spikes in the spectrum mark the frequency of the sinusoid. In other words, when the length of the DFT is matched to the signal period, the DFT results mimic those of the discrete Fourier series. The same perfect spikes were seen in Example 8.9. In Figure 11.17(b), where N is not a multiple of the digital period 64 of the digital signal, the peaks broaden and shorten. This is a side effect of smearing.

The following examples examine the DFT magnitude spectra for a number of signals. Knowledge of the limitations imposed by the finite resolution of the DFT aids in correct identification of signal properties. In these examples, the envelopes of the DFT spectra are presented rather than their individual samples. Many digital oscilloscopes and spectrum analyzers present DFT spectra in this manner.

EXAMPLE 11.7

An analog signal with a period of 2 seconds, or a frequency of 0.5 Hz, is shown in Figure 11.18. This signal is sampled at a rate of 20 samples per second. The DFT magnitude spectrum for the first 5 seconds of the signal is shown in Figure 11.19. At a sampling rate of 20 samples/sec, 100 DFT points are produced. To highlight the shape of the spectrum, the points are connected with a smooth curve. The spectrum does not point clearly to a periodic function, which would show clearly defined spectral spikes. The problem is the resolution of the DFT, which here equals $f_S/N = 20/100 = 0.2$ Hz. This relatively coarse

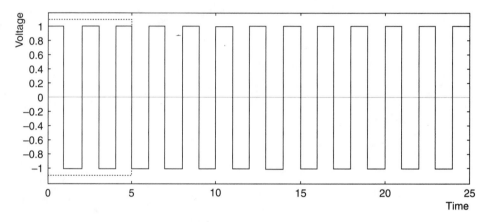

FIGURE 11.18
Analog signal for Example 11.7.

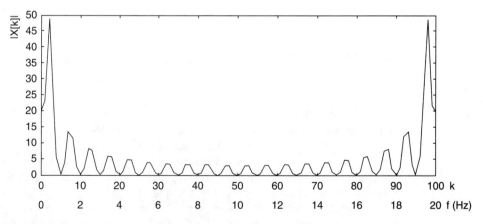

FIGURE 11.19
100-point DFT magnitude spectrum for Example 11.7.

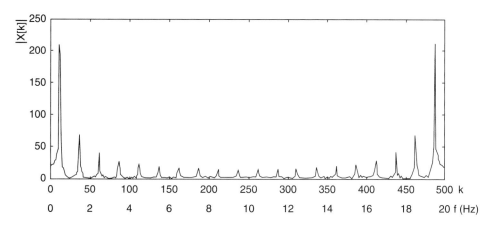

FIGURE 11.20
500-point DFT magnitude spectrum for Example 11.7.

resolution means that peaks that may be present in the spectrum cannot be clearly identified. Figure 11.19 does not show the definitive spikes that would be evidence of a periodic signal.

In this case, the only way to see the true nature of the signal in the DFT magnitude spectrum is to improve the resolution f_S/N. For a given sampling frequency f_S, the only way to do this is to increase the number of samples N being analyzed. This can be accomplished by lengthening the DFT window. A 25-second window is chosen to replace the 5-second window used before. After 25 seconds of sampling at 20 Hz, 500 samples are collected. The 500-point DFT magnitude spectrum is shown in Figure 11.20. The true, periodic nature of the signal is now obvious. The first four spikes occur for $k = 13, 38, 64$, and 89. Since points for the DFT occur only at the frequencies $k f_S/N = k (20/500) = 0.04k$ Hz, these spikes must lie at 0.52, 1.52, 2.56, and 3.56 Hz. Indeed, the first spike marks the fundamental frequency of the square wave signal, ideally 0.5 Hz. The other spikes can be recognized as odd multiples of this fundamental, as Section 8.1 shows is expected for a square wave. The fact that the spike frequencies are not exact is due to the finite length of the window. If an even longer window had been used, the spikes would lie even closer to the theoretical values.

EXAMPLE 11.8

Figure 11.21 shows a portion of the Touch-Tone signal for the number "4." The 1024 samples are collected at 8 kHz. No repeating pattern is obvious. A portion of the DFT magnitude spectrum of the tone appears in Figure 11.22. It is plotted against the frequencies $f = k f_S/N = 7.8125k$ Hz. Tone frequencies were noted in Figure 10.25. The spikes in the DFT magnitude spectrum indicate that the signal has two sinusoidal elements. They occur, as expected, at 770 Hz and 1209 Hz.

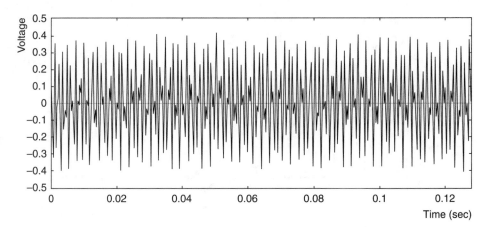

dtmfex.wav

FIGURE 11.21

DTMF signal for Touch-Tone key "4" for Example 11.8.

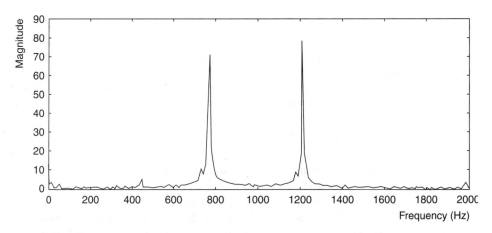

FIGURE 11.22

DFT magnitude spectrum for DTMF tone for Example 11.8.

EXAMPLE 11.9

The ECG (electrocardiogram) is a tool used by physicians to detect potentially deadly heart conditions. A normal heart beats at approximately 72 beats per minute. Figure 11.23(a) shows a time trace of 11 complete beats, each one consisting of a systolic and a diastolic beat. This number of beats occurring within about 9 seconds is consistent with $11/9 = 1.22$ beats per second, or $(1.22)(60) = 73$ beats per minute. Figure 11.23(b) shows an expansion of the first half of the fourth beat.

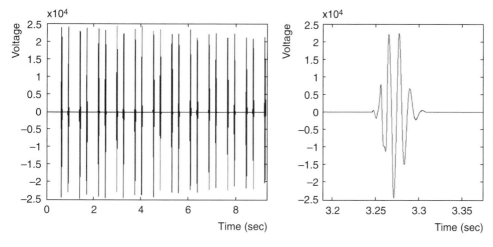

(a) Eleven Heartbeats

(b) Heartbeat Detail

hbeat.wav

Courtesy
Dr. Y.E.
Kocabasoglu,
University of
Groningen

FIGURE 11.23

Heartbeats for Example 11.9.

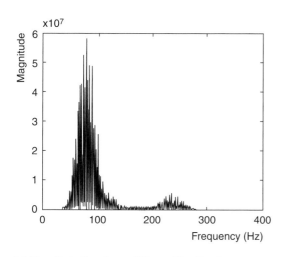

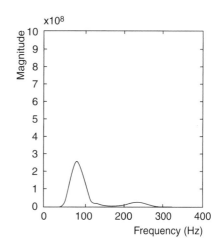

(a) Magnitude Spectrum of Eleven Heartbeats

(b) Magnitude Spectrum of Heartbeat Detail

FIGURE 11.24

DFT magnitude spectrum of heartbeats for Example 11.9.

A portion of the DFT magnitude spectrum obtained after the 11-beat ECG is sampled at 8 kHz is shown in Figure 11.24(a), plotted against frequency. Two major bumps appear in the ECG, but neither occurs near 1.22 Hz, the frequency calculated above. Instead, the bumps occur at 88 Hz and 235 Hz. To investigate the spectrum more carefully, a second DFT is computed, this time for the beat shown in Figure 11.23(b). The magnitude spectrum

FIGURE 11.25

Detail of DFT magnitude spectrum for Example 11.9.

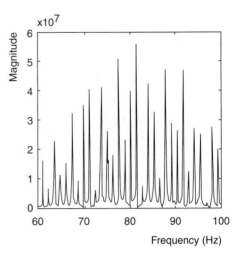

for this DFT is shown in Figure 11.24(b). This spectrum helps to explain the spectrum of Figure 11.24(a). The bumps at 88 and 235 Hz come from the signal variations within each beat. The 1.22 Hz beat repetition frequency, however, seems to be absent from the DFT in Figure 11.24(a). A closer examination of the spiky bumps in this DFT solves this problem, as Figure 11.25 illustrates. This figure shows an expanded version of the peaks in the frequency range between 60 and 100 Hz. While the fundamental at 1.22 Hz is so small as to be negligible, the figure shows the harmonics of this frequency to be very visible indeed. The spikes throughout the spectrum of the heartbeat sequence occur at 1.22 Hz spacing, underlining the periodic nature of the signal.

EXAMPLE 11.10

The DFT magnitude spectrum for the digital white noise signal shown in Figure 11.26 is shown in Figure 11.27. No periodicity is found, since no clear spikes are evident. Since white noise contains equal contributions at all frequencies, the approximate flatness of the magnitude spectrum is not surprising.

11.2 RELATIONSHIP TO FOURIER TRANSFORM

The Fourier transform (not studied in this text) gives spectral information about an analog signal. Because an unadulterated analog signal, rather than a digital signal, is used to compute the Fourier transform, this transform is unaffected by sampling or quantization. Unfortunately, a mathematical function that perfectly describes the analog signal is needed to compute the Fourier transform, which makes it an impractical transform in most cases. The object of Fourier analysis in the digital domain is to approximate as closely as possible the information provided by the Fourier transform. The DTFT, for instance, provides spectral information about a sampled version of the analog signal. As this section will show, the

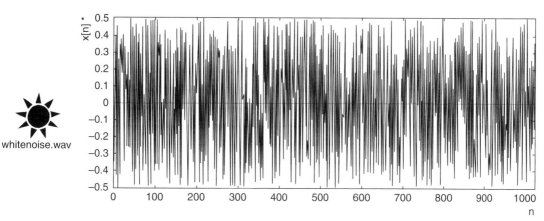

whitenoise.wav

* For clarity, the envelope rather than their individual samples are plotted.

FIGURE 11.26
White noise for Example 11.10.

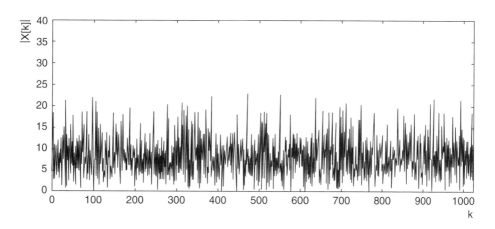

FIGURE 11.27
DFT magnitude spectrum of white noise for Example 11.10.

DTFT approximates the Fourier transform, but is affected by errors due to aliasing and quantization. Both the Fourier transform and the DTFT produce smooth, continuous spectra. The DFT, on the other hand, is simply a processor-friendly means of obtaining a sampled version of the DTFT, as shown in Figure 11.1. It therefore also approximates the original Fourier transform.

The relationships between the Fourier transform, the DTFT, and the DFT can be explained most easily through pictures. In Figure 11.28, the left side corresponds to the time domain and the right side to the frequency domain. For simplicity, only magnitude spectra are shown in the frequency domain; phase spectra are omitted. Figure 11.28(a) shows the analog signal of interest and its spectrum, given by the Fourier transform. The Fourier transform expresses the frequency content of the analog signal. Neither the signal nor its Fourier transform can be handled easily by computers because they are defined at an infinite number of levels and times. The DFT and its inverse, through Equations (11.1) and (11.2), are able to reduce the information in both the time and frequency domains to a finite number of points. The object of this section is to provide meaning behind the DFT and the inverse DFT (IDFT) equations and provide a visual connection between the Fourier transform, the DTFT, and the DFT. Obtaining the DFT in Figure 11.28(g)(ii) from the signal in (a)(i) requires three distinct steps:

1. Sample in time
2. Window in time
3. Sample in frequency

These steps are detailed in the following paragraphs.

Sampling a signal amounts to multiplying it by a train of impulse functions. These impulse functions are shown in Figure 11.28(b)(i), and the sampled signal appears in (c)(i). Furthermore, as indicated in Figure 11.28(b)(ii) and shown in Appendix J, the spectrum of the train of impulse functions is itself also a train of impulse functions. Now, since the sampled signal $x[n]$ in Figure 11.28(c)(i) was obtained by multiplying (a)(i) and (b)(i) in the time domain, its spectrum is found by convolving (a)(ii) and (b)(ii) in the frequency domain. This general relation is developed in Appendix D: Multiplication in the time domain is equivalent to convolution in the frequency domain. It is shown in Appendix K that convolving a spectrum with a train of impulses produces copies of the spectrum at every impulse location. Therefore, convolving the original signal's spectrum in Figure 11.28(a)(ii) with the spectrum of the impulse train in (b)(ii) produces copies of the spectrum as shown in (c)(ii). This is the DTFT. It contains multiple copies of the basic shape of the Fourier transform seen in (a)(ii), but is affected by errors due to quantization and aliasing. These errors are ultimately inherited by the DFT, which is derived from the DTFT.

Because a computer cannot handle an infinite number of samples, the sampled signal in the time domain must be windowed as discussed in Section 11.1 to select the finite group of samples that will be analyzed. A DFT window function is shown in Figure 11.28(d)(i). When the sampled signal $x[n]$ is multiplied by this window $w[n]$, the result is the windowed, sampled signal $x[n]w[n]$ shown in Figure 11.28(e)(i). This signal has exactly N samples. Again, because multiplication occurs in the time domain, convolution occurs in the frequency domain. The spectrum for the window is shown in Figure 11.28(d)(ii). It has the shape of a sinc function, defined in Section 9.4. Its oscillatory **ringing** is characteristic of spectra for rectangular signals, as seen in Section 9.5.1. The result of convolving the DTFT with the window spectrum is shown in Figure 11.28(e)(ii). Irregularities occur because of the ringing in window spectrum. Windowing is therefore another source of error for the DFT. Unlike aliasing and quantization, this error is directly due to DFT processing.

At this point in the process, the time signal in Figure 11.28(e)(i) is both sampled and finite in length, which makes it a signal that is easy for a computer to handle. Its periodic spectrum, however, shown in Figure 11.28(e)(ii), is continuous, which means it is defined

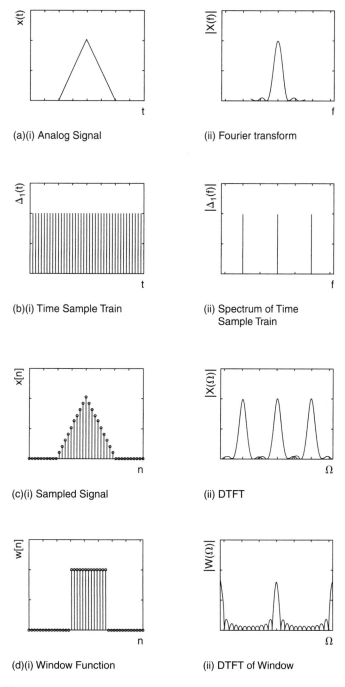

(a)(i) Analog Signal

(ii) Fourier transform

(b)(i) Time Sample Train

(ii) Spectrum of Time
Sample Train

(c)(i) Sampled Signal

(ii) DTFT

(d)(i) Window Function

(ii) DTFT of Window

FIGURE 11.28
Deriving the DFT from the Fourier transform.

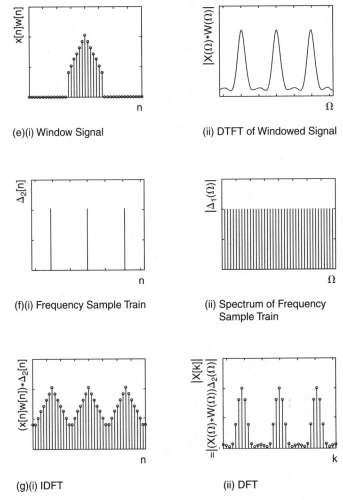

(e)(i) Window Signal

(ii) DTFT of Windowed Signal

(f)(i) Frequency Sample Train

(ii) Spectrum of Frequency Sample Train

(g)(i) IDFT

(ii) DFT

FIGURE 11.28

Continued

at an infinite number of points. To solve this problem, the spectrum must be sampled. Figure 11.28(f)(ii) shows an impulse train in the frequency domain. The spectrum in (e)(ii) is multiplied by this impulse train to produce the sampled version of the spectrum in Figure 11.28(g)(ii), which is the DFT of the digital signal in Figure 11.28(e)(i). The DFT is sampled and periodic, with exactly N distinct values. Just as multiplication in the time domain is equivalent to convolution in the frequency domain, multiplication in the frequency domain is equivalent to convolution in the time domain, as noted in Appendix D. Figure 11.28(f)(i) shows the time signal that produces the impulse train spectrum in (f)(ii), analogous to the relationship between the impulse train and its spectrum in Figure 11.28(b), explained in Appendix K. When the windowed, sampled time domain signal $x[n]w[n]$ is convolved with this impulse train, the result is the sampled, periodic signal of Figure 11.28(g)(i) with exactly N distinct values: Convolution puts copies of the windowed signal at every impulse location in the time domain. This signal is the IDFT. It is faithful to the original signal within the boundaries of the DFT window, but bears no relationship to the original signal otherwise.

Two important points must be noted: First, a single set of DFT samples in Figure 11.28(g)(ii) approximate closely the original spectrum shown in Figure 11.28(a)(ii), though errors do occur due to aliasing, quantization, and windowing. Second, the DFT approximates the original spectrum in exactly N places, where N is the number of time samples collected for the transformation.

EXAMPLE 11.11

A piece of music is sampled at 44.1 kHz. A DFT window 23.22 msec long is used.
  a. How many time samples will be collected within the window?
  b. How many DFT samples do these time samples produce?
  c. What is the resolution of the DFT?

  a. A total of $(44100 \text{ samples/sec})(23.22 \times 10^{-3} \text{ sec}) = 1024$ samples are collected by the window.
  b. When 1024 time samples are collected, 1024 spectrum estimates are produced by the DFT.
  c. The DFT produces an N-point estimate for the spectrum between 0 and f_S Hz. For a sampling rate of 44.1 kHz, the resolution of the DFT is $f_S/N = 44100/1024 = 43.07$ Hz. In other words, each DFT sample covers a range of just over 43 Hz, so the DFT cannot resolve frequencies more finely than this.

11.3 RELATIONSHIP TO FOURIER SERIES

The equations for the discrete Fourier transform (DFT) and the discrete Fourier series (DFS) are identical, as noted in Section 11.1. For convenience, they are repeated here from Sections 8.3 and 11.1:

$$\text{DFS coefficients:} \quad c_k = \sum_{n=0}^{N-1} x[n]e^{-j2\pi\frac{k}{N}n} \tag{8.3}$$

$$\text{DFS expansion:} \quad x[n] = \frac{1}{N}\sum_{k=0}^{N-1} c_k e^{j2\pi\frac{k}{N}n} \tag{8.2}$$

$$\text{DFT:} \quad X[k] = \sum_{n=0}^{N-1} x[n]e^{-j2\pi\frac{k}{N}n} \tag{11.1}$$

$$\text{Inverse DFT:} \quad x[n] = \frac{1}{N} \sum_{k=0}^{N-1} X[k] e^{j2\pi \frac{k}{N} n} \tag{11.2}$$

The DFS operates on N points, where N is the period of a periodic signal. Provided that the DFT operates on the same number of points, the DFS and the DFT are identical. This should come as no surprise. As described in Section 8.3, the DFS coefficients provide a way to find the spectrum of a periodic digital signal. Figure 11.28(g) shows that the DFT coefficients also describe the spectrum of a periodic signal. In this case, the periodic signal is the IDFT, created by repeating windowed samples of a nonperiodic digital signal. Thus, the only real difference between the DFS and the DFT is one of interpretation. The DFS analyzes signals that are by nature periodic. The DFT analyzes a windowed portion of a signal that may be periodic or nonperiodic; its IDFT is always periodic. The magnitude spectrum of a periodic signal may be produced from the DFS coefficients, as $|c_k|/N$ as explained in Section 8.3, or from the DFT magnitude spectrum, as $|X[k]|/N$. The calculations and results are identical.

11.4 DFT WINDOW EFFECTS

The finite set of time samples selected by a DFT is often said to lie within the DFT window, as mentioned early in Section 11.1. In its simplest form, the window simply passes the samples through unmodified. This window, the rectangular window, has been used in all DFT calculations so far in this chapter. As shown in Figure 11.28(d)(i), it gives equal weightings of unity to all samples inside the window and weightings of zero to all samples outside the window. The length of the window is equal to the number of points in the DFT, or N. Example 11.6 showed how window length can influence DFT results. When the signal being studied is time-invariant (that is, its characteristics do not change with time) and when the sampling rate is fixed, a long window reports the signal's frequency spectrum most accurately. In other words, smearing is minimized when the DFT resolution, or frequency spacing, is as small as possible. When the characteristics of the signal change with time, however, a longer window may confound the results sought.

Speech is one type of time-varying signal. To uncover the frequencies present in a single vowel in the middle of a word requires a window no longer than the duration of the vowel, since to use a wider window would capture frequency information not only about the vowel but also about neighboring sounds. At a fixed sampling rate, though, a short window means few captured samples in the time domain, which translates to few samples in the frequency domain as well: The frequency spacing f_S/N is large, so the resolution is poor. A longer window could capture a larger number of signal samples and would therefore provide a larger number of points in the spectrum. With the sampling rate f_S fixed, the larger number of points N for the longer window mean smaller frequency spacing and greater resolution. Unfortunately, a DFT employing the longer window is not capable of furnishing the spectral characteristics of the single vowel, since the extended window includes the vowel and neighboring sounds as well, and will report the frequency content of all.

Choosing window length involves a time-frequency trade-off. At a fixed sampling rate, the DFT can focus on a small detail in time, such as a single vowel, only at the expense of poor resolution in the frequency domain. Improved frequency resolution can be obtained only through widening the view in the time domain, which degrades the DFT's ability to report the frequency content of individual sounds. Nevertheless, variable window length is one important advantage of the DFT, as it permits the DFT to zoom in on detail or zoom out for a big picture.

EXAMPLE 11.12

The DFT magnitude spectrum for 3.5 seconds of a piece of music sampled at 16 kHz is shown in Figure 11.29. This figure plots $|X[k]|$ versus k, but also shows the corresponding frequencies in Hz. The DFT magnitude spectrum for a 0.42-second excerpt of the same music, corresponding to a single note, is shown in Figure 11.30. The two spectra are obviously

pass.wav

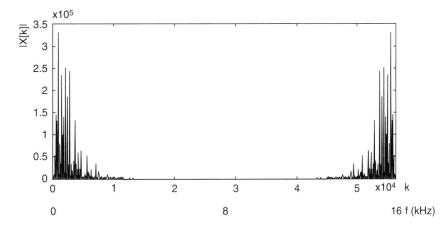

FIGURE 11.29
DFT magnitude spectrum of music for Example 11.12.

pass1.wav

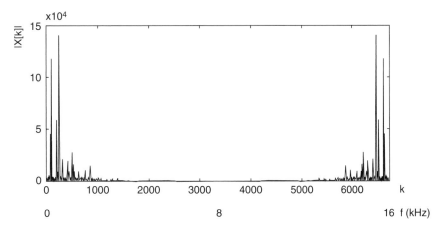

FIGURE 11.30
DFT magnitude spectrum of one note for Example 11.12.

quite different. The one in Figure 11.29 contains the frequencies apparent in Figure 11.30, but includes the frequencies of all other notes as well.

The DFT deals with digital data, so it must contend with whatever aliasing and quantization errors result from sampling. These are two sources of error for the DFT's approximation of the true spectrum of the analog signal being sampled. Windowing is the other major source of error in the DFT. Using a window function $w[n]$ other than the rectangular window used in Figure 11.28(d)(i), for example, has a big impact on the shape of the DFT in Figure 11.28(g)(ii). The effect of window shape is easiest to see when discrete time Fourier transforms, rather than DFTs, are used. As discussed in Section 11.1, the DFT values are identical with the DTFT values at the chosen sample points, but the sampled view makes the window shapes more difficult to discern. The conclusions drawn for DTFTs will apply to DFTs as well.

Because sinusoids consist of a single frequency, they make good test signals for examining the effects of window shape. A perfect analysis should produce a magnitude spectrum with a single spike at the frequency of the sinusoid, as shown in Appendix I. As an example, the ideal magnitude spectrum for the digital sinusoid

$$x[n] = \cos\left(n\frac{3\pi}{7}\right) \tag{11.5}$$

is shown in Figure 11.31. A single spike occurs at $\Omega = 3\pi/7$ radians.

To see the effect of windowing, the DTFT is applied to a windowed portion of the sinusoidal signal $x[n]$. The windowed signal is calculated in the time domain as

$$x_{\text{windowed}}[n] = x[n]w[n] \tag{11.6}$$

where $w[n]$ is the window function. The window function has N nonzero terms and therefore selects the signal samples of interest. The windowed DTFT is then computed as[1]

$$X_{\text{windowed}}(\Omega) = \sum_{n=-\infty}^{\infty} x_{\text{windowed}}[n]e^{-jn\Omega} = \sum_{n=0}^{N-1} x[n]w[n]e^{-jn\Omega}$$

Since the window function $w[n]$ has a finite number of terms, the windowed DTFT spectrum $X_{\text{windowed}}(\Omega)$ is straightforward to calculate. Both linear and logarithmic magnitude spectra are shown in Figure 11.32 for rectangular window lengths of 128 samples, 64 samples, and 32 samples. Since the rectangular window $w[n]$ has equal weightings of one for all samples, it has no effect on the samples $x[n]$ other than to limit how many are used in the computation. The signal windowed in this case is a pure sinusoid, so its true magnitude spectrum should contain only a spike, but the windowed DTFTs in Figure 11.32 can manage

---

[1]   Similarly, a windowed DFT would be computed as

$$X_{\text{windowed}}[k] = \sum_{n=0}^{N-1} x[n]w[n]e^{-j2\pi\frac{k}{N}n}$$

In all previous references to DFTs in this chapter, $w[n]$ has been omitted because it has been assumed to equal one for the duration of the window; that is, a rectangular window has been used.

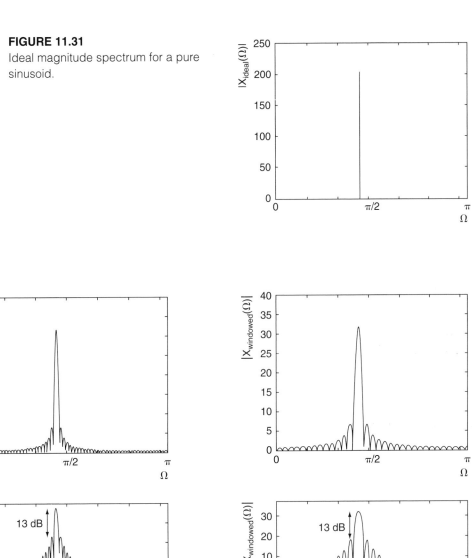

FIGURE 11.31
Ideal magnitude spectrum for a pure sinusoid.

(a) 128 samples

(b) 64 samples

FIGURE 11.32
Spectrum of sinusoid using rectangular windows of different lengths.

FIGURE 11.32
Continued

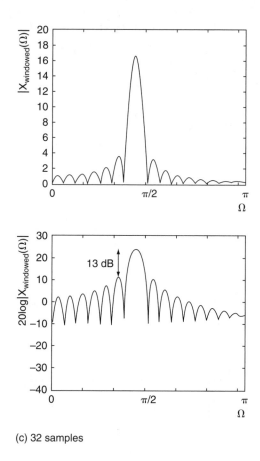

(c) 32 samples

only an approximation to this single spike. For a signal whose characteristics do not change with time, the best approximation to the spectrum is obtained for a long window. For a fixed sampling rate, the resolution of the DFT improves, and the peak that marks the frequency of the sinusoid becomes narrower as the window lengthens; the shorter the window, the broader the peak. The signal $x[n]$ used for this example exists for all n. Unfortunately, real signals often persist for a limited time only, so windows cannot always be lengthened as desired.

All of the logarithmic spectra in Figure 11.32 show evidence of side lobes about 13 dB below the main peak. These side lobes are evidence of what is called **spectral leakage**, which arises as a result of windowing in the time domain. The more abrupt the edges of the window function $w[n]$, the greater the leakage. Spectral leakage is sometimes called ringing. Specifically, it results from the convolution that occurs between the spectrum of the sampled signal and that of the window, as discussed in Section 11.2. The same kind of ringing that occurs in the magnitude spectra of Figure 11.32 occurred also in the FIR windows introduced in Section 9.5. It can be reduced by using windows with smoother boundaries. In fact, the same selection of windows used in FIR design in Section 9.5 is also useful for Fourier calculations. When a rectangular window is used in

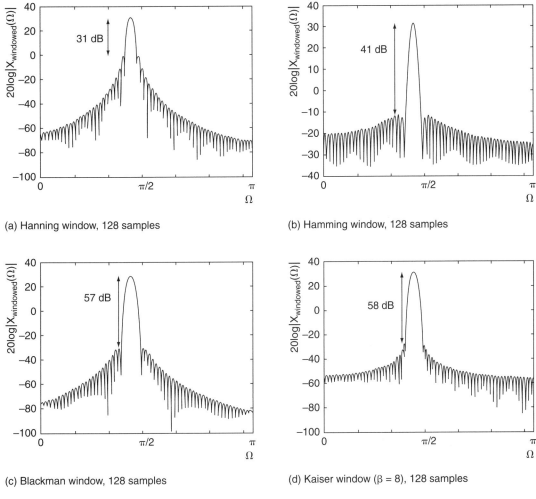

(a) Hanning window, 128 samples

(b) Hamming window, 128 samples

(c) Blackman window, 128 samples

(d) Kaiser window (β = 8), 128 samples

FIGURE 11.33
Spectrum of sinusoid using nonrectangular windows.

Equation (11.6), sample values are multiplied by one inside the window and by zero outside it, and the transform of the windowed samples is taken. Other windows, like Hanning, Hamming, Blackman, and Kaiser windows, weight the samples near the middle of the window more heavily than those at the boundaries, as shown in Figure 9.34.

The magnitude spectrum for the sinusoid of Equation (11.5), calculated using several different nonrectangular windows, is shown in Figure 11.33. While the windows in the figure do not further narrow the width of the spectral peak, but in fact broaden it, they greatly reduce the spectral leakage that occurs on both sides of the peak. To make the peak narrower, a wider window, able to collect a larger number of samples, can be used. The best chance for a good approximation to a clear sinusoidal peak will be obtained when the side

FIGURE 11.34

Approximation to ideal cosine spectrum.

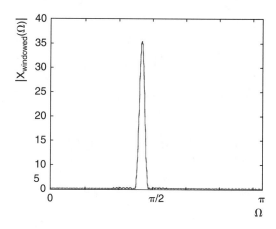

lobes are at least 40 dB below the main lobe. The Hamming, Blackman, and Kaiser windows all meet this objective. When the magnitudes are plotted in dB, the spectra may not seem to resemble a spike, but an attenuation of more than 40 dB indicates that the side lobes are extremely small. Figure 11.34, for example, shows a linear plot of a 128-point spectrum that has a 40 dB difference between the main lobe and the side lobes. The spike in this spectrum approximates the ideal spectrum reasonably well. The advantages of reducing spectral leakage for pure sinusoids carry over to more complex signals, since spectral leakage around spectral peaks can mask legitimate signals at nearby frequencies.

Though discrete time Fourier transforms were used in this section to analyze the effects of windowing, the conclusions drawn hold equally well for DFTs. Windows of the Hanning, Hamming, Blackman, and Kaiser varieties reduce the spectral leakage that occurs for the simple rectangular window; and, where practical, longer windows improve the accuracy of the analysis.

11.5 SPECTROGRAMS

The length of the window N used for DFT calculations has an enormous impact on the kind of information the DFT can provide, as Example 11.12 showed. When sounds are rapidly changing, too long a window gives a murky picture of signal content: All frequencies appearing in the signal are reported, with no indication of their temporal relationships. For example, the DFT for the first three notes of "Mary Had a Little Lamb" would be the same as the DFT of the chord in which the same three notes are played at once, as shown in Figure 11.35.

Since the sampling rate is determined, through Nyquist, by the frequency content of the signal, the key for useful DFT analysis is to choose a window length that suits the signal being analyzed. For music, the length of a note might be a good measure for the choice of window length. For speech, the length of an individual sound in a word might be a good choice. For whale songs, which often progress at much slower rates than human conversa-

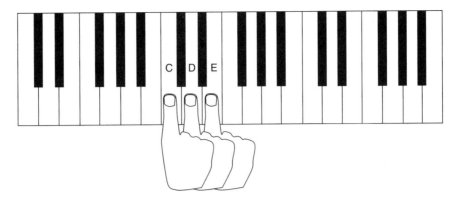

(a) DFT of Three Separate Notes

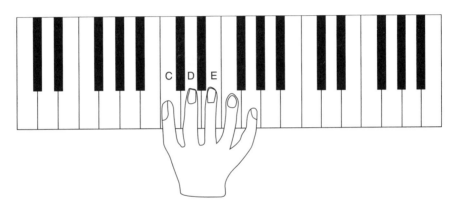

(b) DFT of Chord

FIGURE 11.35
DFT for three single notes and chord.

tion, a longer window might be advocated. Clearly a single DFT can provide only limited information about signals like music or speech. More helpful might be a collection of DFTs, each reporting on a different interval of time. A **spectrogram** is exactly such a collection of DFTs. It plots frequency against time so that each vertical slice of the diagram contains the DFT magnitudes for one window of time. Figure 11.36 contains the spectrogram for the song "Mary Had a Little Lamb," played on a piano. The length of the window used for each slice determines the level of detail. Larger magnitudes appear as darker areas; smaller magnitudes appear lighter. The stripes at the lowest frequencies in each interval are the fundamental frequencies. Their contour across successive intervals follows the tune of the song. The equally spaced stripes above the fundamentals in each interval are the other harmonics, that is, multiples of the fundamental frequencies for each note. The spectrogram is just one means of presenting time-frequency information. A more familiar form of time-

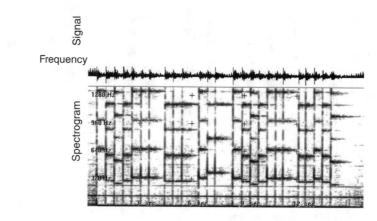

* Spectrograms in this section are produced by the spectrogram program included on the CD courtesy of Richard Horne.

FIGURE 11.36

Spectrogram of "Mary Had a Little Lamb."*

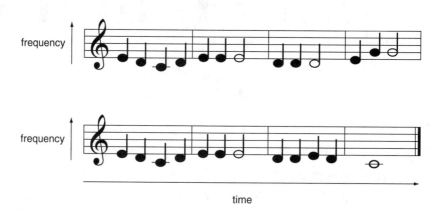

FIGURE 11.37

Musical notation for "Mary Had a Little Lamb."

frequency diagram is musical notation (McClellan, Schafer, & Yoder, 1998). Figure 11.37 shows, in two lines, the same notes as Figure 11.36. Frequency is again plotted vertically and time horizontally.

A variety of spectrograms is presented in Figure 11.38. The spectrograms can provide great insight into the characteristics of a sound, since each vertical slice is essentially a DFT magnitude spectrum. In the spectrogram in Figure 11.38(a), for example, the t's in the phrase can be readily located. T sounds are classed as stop consonants because they include a brief period of silence, obvious in the spectrogram as short gaps. Also, the harmonic content of the vowel sounds shows up clearly as dark horizontal stripes. Away from

alligators.wav

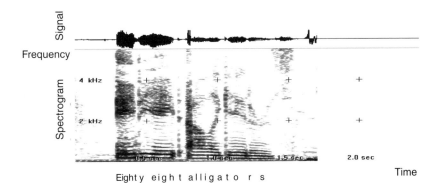

(a) Spectrogram of Phrase

orca.wav

Courtesy
Michael Noonan,
Ph.D., Canisius
College

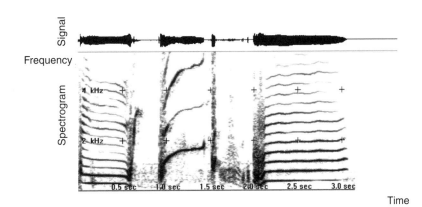

(b) Spectrogram of Orca Sounds

hbsong.wav

Courtesy
Cynthia D'Vin-
cent Intersea
Foundation

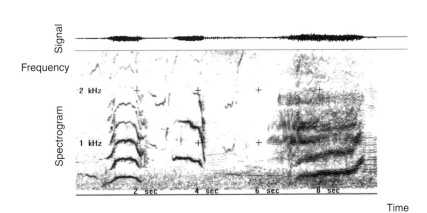

(c) Spectrogram of Humpback Whale Sounds

FIGURE 11.38

Spectrogram examples.

bird.wav

© Kim
Harrington,
WAV place.com

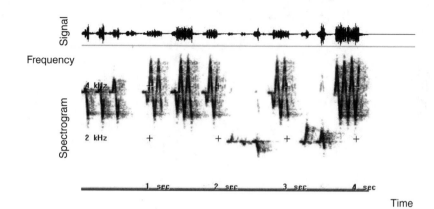

(d) Spectrogram of Bird Song

bat.wav

Courtesy Frank
Hamacher

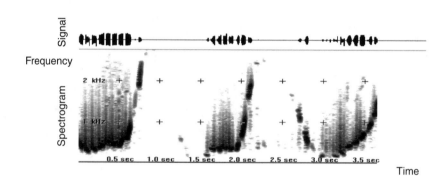

(e) Spectrogram of Bat Echolocation Sounds (Rhinolophus hipposiderus)

did.wav

Courtesy
Jim Hall,
Hall Crystal
Flute and
Didjeridus

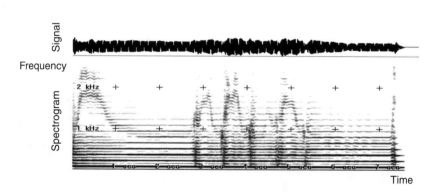

(f) Spectrogram of Didjeridoo Sounds

FIGURE 11.38
Continued

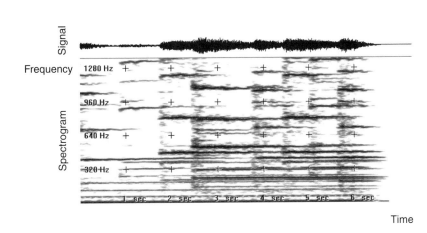

(g) Spectrogram of Big Ben Tolling

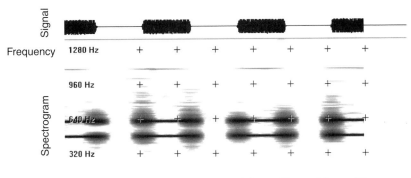

(h) Spectrogram of Telephone Busy Signal

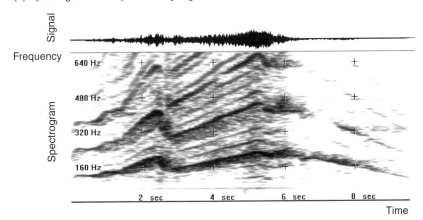

(i) Spectrogram of Passing Motorcycle

FIGURE 11.38
Continued

the fundamental frequencies, these stripes gradually become lighter, reflecting the decreasing magnitudes of the harmonics. Harmonics are also evident in the orca sounds of Figure 11.38(b) and the humpback whale song of Figure 11.38(c), but the wider spacing indicates a higher fundamental frequency. The bird song, analyzed in Figure 11.38(d), shows extremely rapid up and down changes in frequency, whereas each bat echolocation chirp in Figure 11.38(e) contains broad ranges of frequencies for most of its duration. Both the Australian didjeridoo in Figure 11.38(f) and Big Ben in Figure 11.38(g) show strong tonal elements, since each horizontal line corresponds to a sustained sinusoidal tone. Big Ben in particular shows deep, low frequency tones. The telephone busy signal in Figure 11.38(h) is a dual tone multifrequency (DTMF) signal. As expected, the spectrogram for this signal shows a repeating sequence of two pure tones. Finally, the spectrogram for the motorcycle sound in Figure 11.38(i) shows a Doppler shift as the motorcycle passes.

11.6 FFT BASICS

The DFT is an essential digital signal processing tool, yet it is never found in DSP software packages. The reason for the omission is the **fast Fourier transform**, or **FFT**, which gives the same output as the DFT but uses far fewer calculations to do so. The most common FFT is the **radix-2 decimation-in-time FFT**, described in this section. The FFT method relies on breaking an N-point calculation into two $N/2$-point calculations, each of which is further broken down into $N/4$-point calculations, and so on. Beginning with the DFT equation of Equation (11.1), the signal $x[n]$ is partitioned according to its sample numbers, even samples in the first group and odd samples in the second group. The even samples are designated $y[n] = x[2n]$, and the odd samples are designated $z[n] = x[2n+1]$.

$$X[k] = \sum_{n=0}^{N-1} x[n]e^{-j2\pi\frac{k}{N}n} = \sum_{n=0}^{N/2-1} x[2n]e^{-j2\pi\frac{k}{N}(2n)} + \sum_{n=0}^{N/2-1} x[2n+1]e^{-j2\pi\frac{k}{N}(2n+1)}$$

$$= \sum_{n=0}^{N/2-1} y[n]e^{-j\frac{4\pi kn}{N}} + \sum_{n=0}^{N/2-1} z[n]e^{-j\frac{2\pi k(2n+1)}{N}}$$

Factoring out $e^{-j2\pi k/N}$ from the second term gives

$$X[k] = \sum_{n=0}^{N/2-1} y[n]e^{-j\frac{4\pi kn}{N}} + e^{-j\frac{2\pi k}{N}} \sum_{n=0}^{N/2-1} z[n]e^{-j\frac{4\pi kn}{N}}$$

Each of the two terms in this expression contains a DFT of the form of Equation (11.1) for an $N/2$-point signal. Therefore, another way of writing the DFT $X[k]$ is

$$X[k] = Y[k] + e^{-j\frac{2\pi k}{N}}Z[k]$$

for $k = 0, ..., N-1$, where $Y[k]$ is the DFT of the even samples, and $Z[k]$ is the DFT of the odd samples. The $e^{-j2\pi k/N}$ multiplier in front of $Z[k]$ is known in FFT parlance as a **twiddle factor**. The last expression can be further divided to gain additional efficiencies. The first half of the calculations is

$$X[k] = Y[k] + e^{-j\frac{2\pi k}{N}}Z[k] \tag{11.7}$$

for $k = 0, \ldots, (N/2) - 1$. The second half of the calculations is

$$X[k+N/2] = Y[k+N/2] + e^{-j\frac{2\pi(k+N/2)}{N}}Z[k+N/2] \tag{11.8}$$

for $k = 0, \ldots, (N/2) - 1$. Since $Y[k]$ is the DFT of an $N/2$-point signal, it is periodic with period $N/2$, so $Y[k+N/2] = Y[k]$. Similarly, $Z[k+N/2] = Z[k]$. Finally, since $e^{-j\pi} = \cos(\pi)$ $-j\sin(\pi) = -1$, the twiddle factor in Equation (11.8) can be written as

$$e^{-j\frac{2\pi(k+N/2)}{N}} = e^{-j\frac{2\pi k + \pi N}{N}} = e^{-j\frac{2\pi k}{N}}e^{-j\pi} = -e^{-j\frac{2\pi k}{N}}$$

With these simplifications, Equation (11.8) becomes

$$X[k+N/2] = Y[k] - e^{-j\frac{2\pi k}{N}}Z[k] \tag{11.9}$$

for $k = 0, \ldots, (N/2) - 1$. Equations (11.7) and (11.9) together form one step of the FFT, since they convert an N-point DFT into two $N/2$-point DFTs.

Figure 11.39 presents Equations (11.7) and (11.9) in diagram form. It produces an N-point FFT $X[k]$ from two $N/2$-point FFTs, $Y[k]$ and $Z[k]$. Exactly the same idea can be used to produce each of the $N/2$-point FFTs from two $N/4$-point FFTs, following the same format. This process can be continued backward to 2-point FFTs. Figure 11.40 illustrates the process for $N = 8$. The 8-point FFT $X[k]$ of eight time samples in Figure 11.41(a) appears at the far right of the diagram. This FFT is produced from the 4-point FFT $Y_4[k]$ of the even-numbered samples and the 4-point FFT $Z_4[k]$ of the odd-numbered samples. These sets of odd and even time samples are identified in Figure 11.41(b). The 4-point FFT $Y_4[k]$ is in turn produced by the 2-point FFT $Y_2[k]$ of the even-numbered even samples and the 2-point FFT $Z_2[k]$ of the odd-numbered even samples. The FFT $Z_4[k]$ is obtained in a similar manner, from 2-point FFTs for the even-numbered odd samples and the odd-numbered odd samples. For example, $Y_2[k]$ is the 2-point FFT of the samples $x[0]$ and $x[4]$, shown as the even-numbered even samples in Figure 11.41(c). This and the other sample pairs in Figure 11.41(c) appear at the input to Figure 11.40. Note also that, from Equations (11.7) and (11.9), the equations for a 2-point FFT are:

$$X[0] = Y[0] + Z[0]$$
$$X[1] = Y[1] - Z[1]$$

For this reason, $x[0]$ may also be interpreted as $Y_1[0]$, and $x[4]$ as $Z_1[0]$, which provides consistency with Figure 11.39. The same observation holds for the other three 2-point FFTs in stage 1 of Figure 11.40.

In Figure 11.40, each 2-point FFT is called a **butterfly**. Four sets of one butterfly each are required for the first stage. In the second stage, two sets of two butterflies are performed. In the last stage, a group of four butterflies is needed. Even in this small example, the

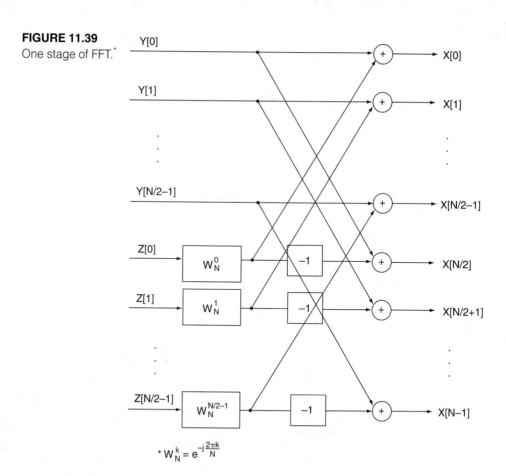

FIGURE 11.39
One stage of FFT.*

$$* W_N^k = e^{-j\frac{2\pi k}{N}}$$

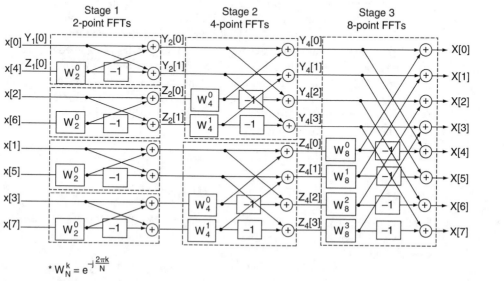

$$* W_N^k = e^{-j\frac{2\pi k}{N}}$$

FIGURE 11.40
All three stages of 8-point FFT.*

FIGURE 11.41

Regrouping into even and odd sequences.

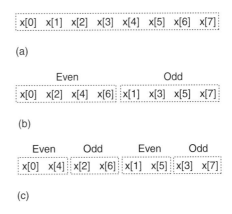

(a)

(b)

(c)

repetition of twiddle factors in different parts of the diagram is evident. For instance, it is easy to see, using the equation for W_N^k provided in the figure, that $W_2^0 = W_4^0 = W_8^0$ and $W_4^1 = W_8^2$. Calculation time may be saved by reusing twiddle factors, instead of recalculating them, wherever possible.

The standard DFT equation requires, for each point, N complex multiplications and $N-1$ complex additions. In all, then, for N points, N^2 complex multiplications and $N(N-1)$ complex additions are needed. The number of operations is frequently used as a measure of the difficulty of the calculation. In the case of the DFT, then, the difficulty is close to being proportional to N^2. Each stage of the FFT requires N complex multiplications and N complex additions. There are $\log_2 N$ stages in all, which means that the difficulty of the FFT is proportional to $N\log_2 N$. This result assumes that N is a perfect power of two. Because of the high efficiency of the FFT when this is so, calculations are almost always based on a power-of-two number of samples. When the number of samples in a signal is not a perfect power of two, zeros are added at the end of the signal to fill it out. This process is called **zero padding**. From Equation (7.1), extra zero samples cannot affect a signal's DTFT. Therefore, they have no influence on the shape of the signal's DFT or FFT either, as these simply sample the DTFT. While zero padding does not add information about a signal and thus cannot improve the accuracy of the picture provided by a DFT, it can reduce the frequency spacing, as a result of an increase in the number of DFT points.

The great advantages of the FFT over the DFT are not seen for simple problems like the 8-point example in Figure 11.40. As N grows, however, the difference between the two different calculation methods becomes dramatic. Figure 11.42 shows a plot of the difficulty of the calculation (as measured by the number of operations) against number of samples N. The figure shows clearly that the FFT is always more efficient than the DFT, especially for $N > 100$. For this reason, the DFT is always computed using an FFT.

FIGURE 11.42
Comparing DFT and FFT
efficiency.

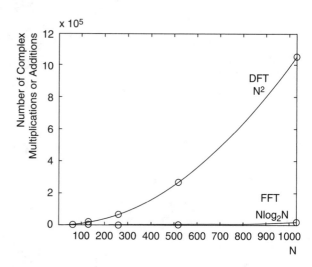

11.7 2D DFT/FFT

Calculations of DFTs and FFTs are not limited to one-dimensional signals like speech and music. Spectral information is also important for two-dimensional signals like digital images or data collected by sonar, radar, or seismic systems. The equations for the 2D DFT are natural extensions of the 1D version. The 2D DFT is

$$X[i,k] = \sum_{m=0}^{M-1} \sum_{n=0}^{N-1} x[m,n] e^{-j2\pi\frac{i}{M}m} e^{-j2\pi\frac{k}{N}n} \qquad (11.10)$$

for $i=0, ..., M-1$, and $k=0, ..., N-1$. The inverse 2D DFT is

$$x[m,n] = \frac{1}{MN} \sum_{i=0}^{M-1} \sum_{k=0}^{N-1} X[i,k] e^{j2\pi\frac{i}{M}m} e^{j2\pi\frac{k}{N}n} \qquad (11.11)$$

where $m=0, ..., M-1$, and $n=0, ..., N-1$. In essence, the 2D DFT first takes 1D DFTs along each row in an image, which produces DFT data in rows, in a matrix having the same shape as the original image. Then 1D DFTs are taken along the columns of this matrix, to give a 2D DFT of the original image. Tricks similar to those discussed in Section 11.6 are used to create the 2D FFT algorithm, which affords huge economies in calculation time compared to the 2D DFT. Two-dimensional transforms are used on digital images in Section 15.7.

CHAPTER SUMMARY

Matlab
Support

1. The discrete Fourier transform (DFT) for a signal $x[n]$ is obtained from the equation

$$X[k] = \sum_{n=0}^{N-1} x[n] e^{-j2\pi\frac{k}{N}n}$$

This transform represents a very practical way to identify the frequency content of the signal. It operates on a signal with N samples and produces a DFT with N points. Each index k marks a point of the DFT.

2. The DFT $X[k]$ provides a DFT magnitude spectrum $|X[k]|$ and a DFT phase spectrum $\theta[k]$.
3. The DFT is periodic with period N.
4. The DFT samples the DTFT at the digital frequencies described by the equation

$$\Omega = 2\pi\frac{k}{N}$$

5. The DFT indices k refer to analog frequencies in Hz given by

$$f = k\frac{f_S}{N}$$

where N is the number of points in the DFT and f_S is the sampling frequency. The spacing between these frequencies determines the resolution of the DFT. Because of its limited resolution, the DFT cannot report all frequencies in a signal perfectly. The DFT tends to smear sharp peaks in a spectrum, although increasing the number of points in the DFT can reduce the amount of smearing.

6. The spectra of both nonperiodic and periodic signals may be obtained using the DFT. The DFT simply operates on a group of signal samples selected by the DFT's window. The type of window used (rectangular, Hanning, Hamming, Blackman, or Kaiser) affects the details of the spectrum reported.

7. Spectrograms combine multiple DFTs together to show how a signal's frequency content changes with time.

8. The FFT is a computationally efficient way of implementing the DFT. It obtains exactly the same results as the DFT.

9. The spectra for digital images are calculated using the 2D DFT (or 2D FFT).

REVIEW QUESTIONS

11.1 A digital signal $x[n]$ is shown in Figure 11.43.
   a. Plot the DFT magnitude and phase spectra for the samples of $x[n]$ for $0 \leq n \leq 7$.
   b. What is the period of the spectra in (a)?

11.2 For the digital signal $x[n] = e^{-0.5n} (u[n] - u[n-4])$, plot:
   a. The DFT magnitude spectrum
   b. The DTFT magnitude spectrum

11.3 The DTFT for a signal is $H(\Omega) = 1 - 0.2e^{-j\Omega} + 0.35e^{-j2\Omega}$. Find the magnitudes and phases for an 8-point DFT of the same signal.

11.4 A digital signal is described as $x[n] = \sin(n\pi/2)$ for $0 \leq n \leq 3$. Plot a 4-point DFT $X[k]$ for this signal.

11.5 The impulse response for a filter is $h[n] = (-0.95)^n$ for $0 \leq n \leq 3$. Plot a 4-point DFT $H[k]$ for this filter.

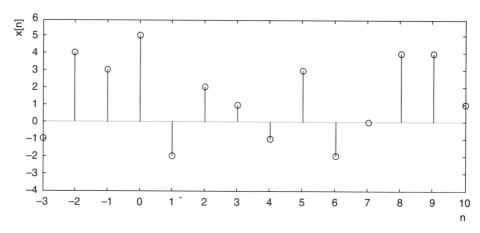

FIGURE 11.43
Signal for Question 11.1.

11.6 Compare the DFT magnitude spectra for the signals:
  a. $x_1[n] = [4\ 3\ 2\ 1]$
  b. $x_2[n] = [4\ 3\ 2\ 1\ 4\ 3\ 2\ 1]$
  c. $x_3[n] = [4\ 3\ 2\ 1\ 4\ 3\ 2\ 1\ 4\ 3\ 2\ 1]$
11.7 Five samples of a digital signal $x[n]$ are $[3\ -1\ 0\ 2\ 1]$.
  a. Find a 5-point DFT magnitude spectrum for these samples.
  b. Zero-pad the signal to 8 points, and find an 8-point DFT magnitude spectrum. Compare the result to the spectrum in (a). To understand the results, it may be necessary to plot the DTFT magnitude spectrum for $x[n]$.
11.8 The DFT magnitudes and phases computed from four samples of a signal are listed in Table 11.7. Find the values of the four samples.

TABLE 11.7
DFT Values for Question 11.8

k	$|X[k]|$	$\theta[k]$
0	10.000	0.0000
1	2.8284	−0.7854
2	2.0000	0.0000
3	2.8284	0.7854

11.9 The 8-point DFT of a signal $x[n]$ defined beginning at $n = 0$ is $X[k]$. The inverse DFT of $X[k]$ is shown in Figure 11.44.
  a. Plot the signal $x[n]$.
  b. How do 4-point and 8-point DFT magnitude spectra for $x[n]$ compare?

FIGURE 11.44

Inverse DFT for
Question 11.9.

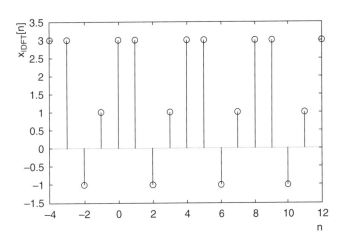

11.10 The frequency response for a (noncausal) equiripple filter is given by

$$H_1(\Omega) = 0.2273(\cos\Omega)^3 + 0.5778(\cos\Omega)^2 + 0.3505\cos\Omega + 0.0311$$

Find the noncausal impulse response $h_1[n]$ that corresponds to this frequency response.

11.11 An equiripple filter is described by the optimal filter shape

$$H_1(\Omega) = 0.6269\cos^3\Omega + 0.5134\cos^2\Omega - 0.1084\cos\Omega - 0.0551$$

Obtain a causal impulse response $h[n]$ for the filter.

11.12 An analog signal is sampled at 16 kHz. A 512-point DFT is computed.
  a. What is the resolution of the DFT?
  b. Find the equivalent frequency in Hz for each of the following points of the DFT:
    (i) $k = 0$
    (ii) $k = 127$
    (iii) $k = 255$
    (iv) $k = 511$

11.13 A 6 kHz analog sine wave is sampled at 40 kHz. Determine where the peaks in its DFT magnitude spectrum will occur for a:
  a. 32-point DFT
  b. 64-point DFT
  c. 128-point DFT

11.14 A 6 kHz analog sine wave is sampled at 7.5 kHz. Determine where the peaks in its DFT magnitude spectrum will occur for a:
  a. 32-point DFT
  b. 64-point DFT
  c. 128-point DFT

11.15 A 16-point DFT is computed for a signal sampled at 12 kHz. Predict the values of k where peaks in the magnitude spectrum will occur if the signal is:

a. $x[n] = \cos(n\pi/7)$
b. $x[n] = \sin(n2\pi/3)$
c. $x[n] = \cos(n3\pi/4)$
d. $x[n] = \cos(n\pi/4) + \cos(n5\pi/9)$

11.16 A DFT is computed for $x[n] = \sin(n4\pi/7)$, sampled at 22 kHz.

a. What is the digital frequency of the sinusoid?
b. What analog frequency does this digital frequency correspond to?
c. What is the smallest number of points N that can be used for the DFT to ensure the spectral peak is located within 10 Hz of the correct location? Consider N values that are powers of two only.
d. For the number of points N identified in (c), where do spectral peaks occur?

11.17 DTMF signals may be decoded using IIR filters, as suggested at the end of Section 10.8, or using a DFT magnitude spectrum. The frequencies for the DTMF tones are shown in Figure 10.25. For an 8 kHz sampling rate, what is the minimum width of the DFT window, in samples and in seconds, to ensure that all keys may be distinguished from one another?

11.18 The 128-point DFT magnitude spectrum in Figure 11.45 was obtained for a signal sampled at 4 kHz. Estimate the major frequencies, in Hz, that are present in the signal.

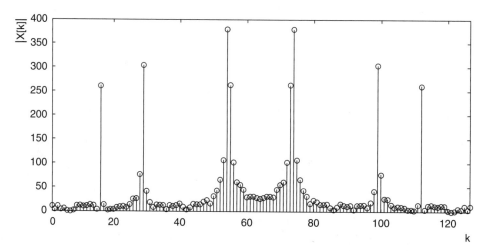

FIGURE 11.45
DFT magnitude spectrum of unknown signal for Question 11.18.

11.19 A signal is sampled at 22.05 kHz. Samples are collected for 2 msec, and a DFT is performed on the data.

a. How many signal samples are collected?
b. How many points does the DFT have?
c. What is the resolution of the DFT?

11.20 Three thousand samples of a signal sampled at 44.1 kHz are collected. How long a segment of the signal in seconds does this represent?

11.21 A digital signal is obtained by sampling the analog signal

$$x(t) = \cos(240\pi t) + \cos(320\pi t) + \cos(420\pi t) + \cos(720\pi t)$$

at the rate of 500 samples per second. A 64-point DFT magnitude of the digital signal is shown in Figure 11.46. With reference to the original signal, explain each peak in the spectrum.

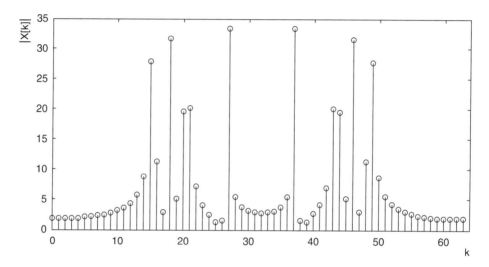

FIGURE 11.46
Magnitude spectrum for Question 11.21.

11.22 A signal containing frequencies up to 500 Hz is sampled, and a DFT is computed. If the frequency spacing of the DFT must be no greater than 0.5 Hz, what is the minimum number of samples needed?

11.23 A magnitude spectrum $|X[k]|$ for a 1.8 kHz sine wave sampled at 8 kHz is computed using a:

    (i) 64-point rectangular window
    (ii) 128-point rectangular window
    (iii) 64-point Hamming window

Without computing a DFT, answer the following questions:

  a. For each window, identify the index k, and its corresponding analog frequency, for the baseband spectral peak.

  b. For each window, what is the magnitude difference, in dB, between the height of the main peak and the biggest side lobe?

  c. Which window provides the best approximation to the single spectral peak expected for the sine wave?

11.24 A spectrogram for a short recording is shown in Figure 11.47.
  a. Describe the frequency content of the sound at:
     (i) 0.4 seconds
     (ii) 1 second
     (iii) 1.8 seconds
  b. Identify the intervals in which clear harmonics occur. Estimate the fundamental frequencies of the sounds in these intervals.

shot.wav

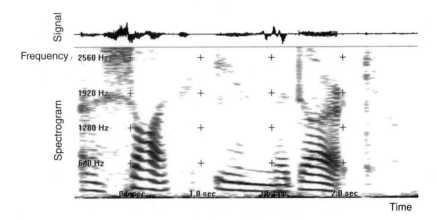

FIGURE 11.47
Spectrogram of unknown recording for Question 11.24.

11.25 For the signal $x[n] = [0\ 1\ 2\ 3\ 4\ 5\ 6\ 7]$:
  a. Find the FFT using the butterfly diagram in Figure 11.40.
  b. Using the results from (a), compute and plot the magnitude and phase spectra for the signal.
11.26 Demonstrate the following for these FFT twiddle factors:
  a. $W_2^0 = W_4^0 = W_8^0$
  b. $W_4^1 = W_8^2$
11.27 For a 512-point signal:
  a. Determine how many complex multiplications (or additions) must be performed to compute a:
     (i) DFT
     (ii) FFT
  b. How many times more computations are needed for the DFT?
11.28 An FFT is performed on a 512-point signal, sampled at 8 kHz.
  a. What is the frequency spacing for the FFT?
  b. The signal is zero-padded to a length of 4096 samples, and a new FFT is computed. What is the frequency spacing for this FFT?

12

HARDWARE FOR DSPs

Though DSP algorithms may be implemented on any processor, specialized digital signal processing hardware enables the greatest speed and efficiency. This chapter surveys elements of DSP hardware and shows why it is so well suited to DSP tasks. The chapter:

➤ highlights the differences in architecture between digital signal processors (DSPs) and standard microprocessors
➤ examines how pipelining, multiple buses, and multiple memories improve throughput
➤ surveys number formats frequently used by DSPs, including two's complement
➤ illustrates how quantization affects numbers stored by DSPs
➤ discusses the structure of multiplier/accumulator, shifter, and address generator units
➤ introduces the use of assembly language as the command language for DSPs
➤ identifies factors for choosing a DSP for an application, including speed, memory, power, convenience, and cost
➤ surveys the products of three major DSP manufacturers

12.1 DIGITAL SIGNAL PROCESSOR BASICS

All processors have the capability to perform digital signal processing tasks. All processors, however, do not perform these tasks equally quickly or efficiently. DSP processors, designed specifically for the operations common in DSP, have special features that permit them to accomplish in real time what other processors cannot. **Real time** means that outputs keep pace with the collection of input samples during actual operation. For some operations, like filtering, this means that a new output sample can be produced as each new input sample is received. For others, like FFTs, output information can be produced only when a block of input samples has been recorded. Some of the features that make a processor suitable for DSP tasks include:

1. Architecture optimized for DSP operations
   - Multiple buses and on-chip memories
   - Multiple memory accesses in a single instruction cycle
2. Specialized hardware units
   - Fast multiply/accumulate operations
   - Fast shifting operations
   - Sufficient register width to handle calculations without overflow
   - Support for circular buffers
   - Address generators
   - Heavy use of parallelism
3. Assembly language program instructions to facilitate repetitive operations
   - Flexible number formats
   - Efficient looping with zero overhead
   - A minimum of one instruction per clock cycle

These features are described in the following sections.

12.2 DSP ARCHITECTURES

A computer's architecture describes how its main hardware components are interconnected. Many standard processors are based on what is called a **Von Neumann architecture**. The simplest Von Neumann processors contain a single, shared memory for programs and data, with only a single bus for memory access. Because only one instruction or piece of data can be retrieved at once, most parts of the computer are inactive at any given time. Programs for these processors proceed in a very serial fashion: An instruction is fetched from memory, the instruction is decoded, and the instruction is executed. When execution requires data to be written to or read from memory, the fetch of the next instruction must wait for execution of the previous instruction to complete. The bottleneck in the system is the single, shared memory, where instructions and data reside. The only way to speed things up is to increase the speed of the bus, the memory, or the computational units. Figure 12.1 shows what a simple Von Neumann architecture looks like.

Most DSP processors are based on a **Harvard architecture** design. This architecture features separate program and data memories. The program memory is designed to contain instructions, while the data memory is designed to hold data. The separation of these two memories means that program code and data items can be fetched simultaneously, one from each memory. Naturally, additional buses are needed in order to realize the advantages of the separate memories. A Harvard architecture is displayed in Figure 12.2. Certainly this architecture is more complex, but the speed advantages are dramatic. In a modified Har-

FIGURE 12.1
Von Neumann
architecture.

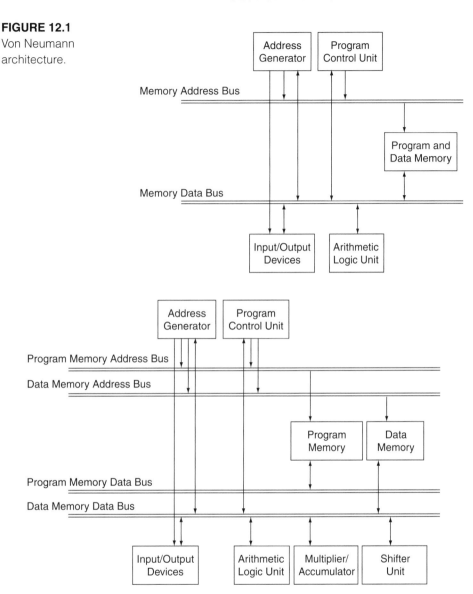

FIGURE 12.2
Harvard architecture.

vard architecture, also very common in DSP devices, separate memories are maintained but program memory may contain a mixture of program instructions and data.

The addition of buses and memories immediately increases the throughput of the processor, even if its absolute speed is unchanged. Harvard architecture permits an instruction to be fetched from program memory while data are being transferred to data memory.

This overlapping of tasks is called **pipelining**. The instructions that perform DSP tasks, called **assembly language** instructions, are listed in a program that a DSP executes. In DSPs, as in other processors, numbers used in calculations are stored in **registers**. The task of retrieving two numbers from memory and adding them together, for example, can be accomplished using three registers, two to store the numbers to be added and one to store the result. The assembly language instructions might be:

(1) Reg1 = GET(Addr0)
(2) Reg2 = GET(Addr1)
(3) Reg3 = Reg1 + Reg2

Each instruction can be broken down into two basic cycles that require bus use: the fetch cycle and the execute cycle. The fetch cycle is responsible for fetching and decoding the instruction, while the execute cycle is responsible for fetching necessary operands and executing the operation. In a simple single-bus Von Neumann architecture, each step that requires the use of the bus must wait for the bus to be free before it can begin. The program fragment above requires several uses of the bus. Initially, the first instruction must be fetched from memory and decoded. When it executes, the value stored at Addr0 in memory is copied to register Reg1. Memory is accessed during both the fetch and the execute portions. The second instruction works the same way, but its instruction fetch must wait until memory use from the first instruction has ceased. When the second instruction executes, the value stored at Addr1 is copied to register Reg2. After the third instruction is fetched and decoded, the addition is performed in the arithmetic logic unit (ALU), and the result is placed in Reg3. Table 12.1 summarizes these steps, and Figure 12.3 illustrates the time line.

When the same three instructions are executed on a Harvard architecture, significant time savings occur. Suppose the two operands are each stored in data memory. Because instructions are stored in a separate program memory, instructions can be fetched at the same time as data memory is accessed. Table 12.2 lists the steps required to perform the operations. After the first fetch is finished, the second fetch can begin immediately, since program memory is not required for the execution of the first instruction. Figure 12.4 illustrates the advantages of pipelining, through overlapping of tasks. The three instructions that once required six clock cycles require only four when parallelism is suitably exploited.

To say that processors fall neatly into either the Von Neumann or the Harvard architecture group would be a gross oversimplification. In addition to multiple buses, many

TABLE 12.1

Fetch and Execute Cycles on Von Neumann Architecture

Step	Action	Bus Status
Fetch instruction (1)	Read instruction from memory.	BUS BUSY.
Execute instruction (1)	Read operand #1 from memory.	BUS BUSY.
Fetch instruction (2)	Read instruction from memory.	BUS BUSY.
Execute instruction (2)	Read operand #2 from memory.	BUS BUSY.
Fetch instruction (3)	Read instruction from memory.	BUS BUSY.
Execute instruction (3)	Add operands #1 and #2 in ALU.	NO BUS ACTIVITY.

FIGURE 12.3

Fetch and execute
time line in Von
Neumann
architecture.

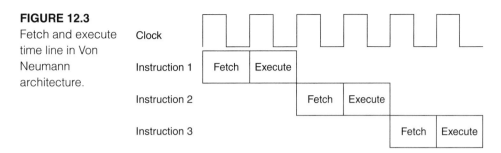

TABLE 12.2
Fetch and Execute Cycles on Harvard Architecture

Step	Action	Bus Status
Fetch instruction (1)	Read instruction from program memory.	PROGRAM BUS BUSY.
Execute instruction (1)	Read operand #1 from data memory.	DATA BUS BUSY.
Fetch instruction (2)	Read instruction from program memory.	PROGRAM BUS BUSY.
Execute instruction (2)	Read operand #2 from data memory.	DATA BUS BUSY.
Fetch instruction (3)	Read instruction from program memory.	PROGRAM BUS BUSY.
Execute instruction (3)	Add operands #1 and #2 in ALU.	NO BUS ACTIVITY.

FIGURE 12.4

Pipelining in
Harvard
architecture.

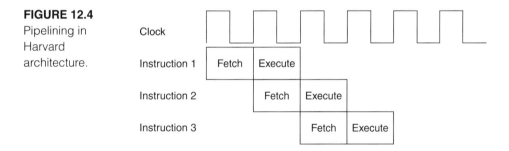

modern processors based on the Harvard scheme now feature multiple cache memories that
increase throughput dramatically. **Cache memories** are memories that make the data that
are most likely to be needed, most rapidly available to the processor. The data to be placed
in the cache are selected according to complex statistical algorithms, so data access times
are not consistent. Also, even though DSP chips are optimized for real-world signal acquisition and processing, high-end general purpose PC and workstation central processing
units (CPUs) are beginning to compete in the same arena. General purpose CPUs can offer
ease of programming, high speed, and fast computation. On the down side, heavy use of
caching means that CPUs cannot offer the predictable execution times that DSPs can, and
power requirements for CPUs are significantly higher than for DSPs. For some applications, hybrid CPU/DSP architectures are being developed that combine the advantages and
functionalities of CPUs and DSPs.

At the same time as conventional DSP and CPU designs begin to blur, new DSP designs are emerging. Some DSPs now offer multiple multiplier/accumulators, or a specialized coprocessor dedicated to FIR filtering. Other DSP architectures obtain performance gains by working with multiple data items or instructions in parallel. Many of these are classed under the heading of **superscalar architectures**, which are capable of executing two or more instructions in the same clock cycle. In SIMD (single instruction on multiple data) architectures, for example, individual program instructions can be decoded to multiple execution units to operate on more than one data item at a time. MIMD (multiple instructions on multiple data) architectures extend this concept by allowing different portions of the same program, or even different programs, to execute simultaneously (Mlynek & Leblebici, 1998). VLIW (very long instruction word) architectures (Berkeley Design Technology, 1999; Weddell, 1999) may fetch groups of eight or more instructions at once and then decode and execute the instructions in parallel. The more complex the architecture, the more important high level programming tools become, as advanced features cannot be exploited effectively using ordinary assembly language tools.

12.3 FIXED POINT AND FLOATING POINT NUMBER FORMATS

In order to operate on real-world data, DSPs must be able to handle calculations with real numbers. The number formats used by DSPs to represent real numbers fall into two major categories: fixed point and floating point. Each of these formats allows a DSP to store and combine real numbers, but each has important advantages and disadvantages. While DSPs operate exclusively on binary data, DSP programs most often rely on binary or hexadecimal formats to represent numbers. Conversions between decimal (base 10), binary (base 2), and hexadecimal (base 16) numbers are explained in Appendix A.11. As indicated in the appendix, numbers can be marked by a subscript according to their bases, and hexadecimal numbers are frequently marked with the annotations "0x" or "h."

To compare fixed point and floating point formats requires the concepts of dynamic range and resolution. **Dynamic range**, reported in dB, refers to the ratio of the maximum possible range of the signal, called the **full scale range**, to the smallest signal difference that can be resolved, the quantization step size.

$$\text{Dynamic range} \ = 20 \log \frac{\text{Full scale range}}{\text{Smallest resolvable difference}} \qquad \textbf{(12.1)}$$

An alternative form of this definition, which gives a slightly larger result, is:

$$\text{Dynamic range} \ = 20 \log \ (\text{Number of possible signal levels})$$

Resolution refers to the size of the step between two adjacent numbers in a number system. The smaller the step size, the better the resolution of the system and the more closely a real number can be approximated. For a given full scale range, the better the resolution, the greater the dynamic range of the system. Resolution may be increased by increasing the number of fractional bits in a number format. In a binary number, **fractional bits** are the bits to the right of the binary point (an analog of the decimal point). Since a limited number of bits is available, quantization errors affect all numbers represented in fixed or floating point formats, no matter how many fractional bits are used. Except for the very largest signals in range, these errors can never be greater than half of the step size between adjacent quantization levels, as explained in Section 2.3.

EXAMPLE 12.1

Find the dynamic range for an 8-bit unsigned integer system.

Eight-bit unsigned integers can represent numbers from 0 to 255. Full scale range for this system is 255, and the smallest resolvable difference is 1. Thus,

$$\text{Dynamic range} = 20 \log \frac{255}{1} = 48.13 \text{ dB}$$

Fixed point DSPs may use signed or unsigned formats, though signed formats are most common. For unsigned formats, all bits may be used to represent magnitude. In the signed formats, the most significant bit is called the **sign bit**. It is zero for positive numbers and one for negative numbers. To implement a signed format requires a method of representing negative numbers. **Two's complement** notation is the most common choice. This binary representation of a negative decimal number is computed by moving from right to left through the binary representation of the positive decimal equivalent and complementing all bits after the first "1" bit encountered, as Example 12.2 demonstrates. The same process applied to a two's complement binary number produces the binary representation of the positive decimal number again, as shown in Example 12.3. As both of the following examples illustrate, the two's complement representation may be expressed in either binary or hexadecimal form.

EXAMPLE 12.2

Find a two's complement hexadecimal representation for the decimal integer -26 in a 16-bit system. Express the result in hexadecimal format.

Since $2^4 + 2^3 + 2^1 = 26$, the binary representation for positive 26 is $0000\ 0000\ 0001\ 1010_2$. For this number, the 2 rightmost bits are unchanged in the two's complement, while all others are complemented. Thus, $1111\ 1111\ 1110\ 0110_2$ is the binary representation for the decimal number -26. The hexadecimal version of this binary number is found by converting each group of 4 bits to a hexadecimal digit, giving 0xFFE6.

EXAMPLE 12.3

Find the decimal value of the two's complement hexadecimal number 0xB781.

The binary form of the number is found by converting each hexadecimal digit into a 4-bit binary value, which gives $1011\ 0111\ 1000\ 0001_2$. The number is negative because the sign bit is 1. Since the rightmost bit is 1, this bit remains unchanged in the two's complement. All others are complemented to give $0100\ 1000\ 0111\ 1111_2$. This number has the decimal value 18559_{10}. The binary number $1011\ 0111\ 1000\ 0001_2$, then, has the decimal value -18559_{10}.

15	14	13	12	11	10	9	8	7	6	5	4	3	2	1	0
$-(2^0)$	2^{-1}	2^{-2}	2^{-3}	2^{-4}	2^{-5}	2^{-6}	2^{-7}	2^{-8}	2^{-9}	2^{-10}	2^{-11}	2^{-12}	2^{-13}	2^{-14}	2^{-15}

FIGURE 12.5
Signed 1.15 number format.

Fixed point formats represent numbers in the form $N.M$, where N is the number of bits to the left of the binary point and M is the number of bits to the right of the binary point. $N + M$ is the total number of bits used to represent the number. Thus, for a 16-bit system, possible fixed point formats include 1.15, 4.12, or 16.0. Signed 16.0 format, for example, is equivalent to a simple signed integer format. Of all the fixed point formats, the signed 1.15 format, illustrated in Figure 12.5, is the most common. This format consists of 1 sign bit and 15 fractional bits. The weightings shown in Figure 12.5 give rise to two's complement notation for negative numbers. Whenever a finite number of bits is used to represent a number, quantization errors occur, calculated as the difference between the quantized and the actual values.

EXAMPLE 12.4

Find the signed 1.15 representation for the decimal number 0.285157 and determine the quantization error.

The largest fractional bit that is less than 0.285157 is $2^{-2} = 0.25$, leaving 0.035157 still to be accounted for. The fractional bits 2^{-3} and 2^{-4} are too large, but 2^{-5} is not, leaving 0.003907. Subtracting 2^{-8} leaves 0.75×10^{-6}. This remainder is smaller than 2^{-15}, so it cannot be represented in the 1.15 format. Thus, the closest approximation to 0.285157 in 1.15 format is $0.010\ 0100\ 1000\ 0000_2$, or 0.28515625_{10}, or 0x2480. The difference of -0.75×10^{-6} is a quantization error. A faster way to convert to 1.15 representation is to truncate the real number $2^{15}(0.285157) + 0.5$ to obtain an integer and then convert to binary or hexadecimal. The 0.5 offset ensures that a value in the top half of a quantization interval will move to the higher binary value after truncation.

EXAMPLE 12.5

Find the 1.15 representation for the decimal number -0.75.

The 1.15 representation for 0.75_{10} is $0.110\ 0000\ 0000\ 0000_2$. To find the representation for -0.75_{10} requires finding the two's complement. From the right end, bit 13 is the first "1" bit. Thus the rightmost 14 bits remain unchanged and the top 2 are complemented. The 1.15 format for -0.75_{10} is, then, $1.010\ 0000\ 0000\ 0000_2$, or 0xA000. Alternatively, the decimal number -0.75 can be written as $-1 + 0.25$ to use the weightings in Figure 12.5. Since $0.25_{10} = 0.010\ 0000\ 0000\ 0000_2$, $-0.75_{10} = 1.010\ 0000\ 0000\ 0000_2$.

EXAMPLE 12.6

Find the decimal equivalent for the signed 1.15 number 1.001 1000 0000 0000$_2$.

This is a negative number. The weightings in Figure 12.5 can be used to find the decimal equivalent as $-2^0 + 2^{-3} + 2^{-4} = -0.8125$.

Naturally the number of bits available in a fixed point format limits the largest numbers that can be represented. In the $N.M$ format, larger N values offer a larger range of values, but larger M values (more fractional bits) offer greater resolution. In the signed 1.15 fixed point format, the sign bit is bit 15, and the binary point is fixed between bits 15 and 14. The largest positive number that can be represented is 0.111 1111 1111 1111$_2$, or $2^{-1} + 2^{-2} + ... + 2^{-15} = 1 - 2^{-15} = 0.999969482421875_{10}$. The most negative number is 1.000 0000 0000 0000$_2$, or -1.0_{10}, for a range of almost 2. Despite these limits, numbers outside the range can still be used in calculations using 1.15 format, provided that they are scaled down before calculations are performed and then scaled back up when calculations are completed. Unfortunately, scaling shifts all binary numbers the same distance, which means that some of the smallest numbers will be lost altogether after scaling down. In effect, scaling down represents a loss in precision.

Since the decimal equivalent of the smallest 1.15 number is -1.0 and of the largest is $1 - 2^{-15}$, the full scale range is nearly 2.0. The smallest difference that can be resolved is 0.000 0000 0000 0001$_2$ = 2^{-15} = 0.000030517578125. The dynamic range is, then, $20 \log[(2 - 2^{-15})/2^{-15}] = 20 \log 65535 = 96.33$ dB. In fact, all 16-bit fixed point formats have the same dynamic range. In the $N.M$ format, increasing N increases the full scale range, but not the dynamic range, of the numbers. For example, in 16.0 format, integers from -32768 to 32767 can be represented, giving a full scale range of 65535; but with a step size of 1, the dynamic range is again 96.33 dB.

When two fixed point numbers are added or subtracted, they must have identical formats. That is, the number of bits to the right of the binary point must be the same for both numbers. If the input format is $N.M$, the format for the result is also $N.M$. There is, however, the possibility that adding two numbers may give a result larger or smaller than can be represented by the number of bits available. These conditions are called **overflow** and **underflow**, respectively, and are signaled by status bits. In the case of signed overflow, the result saturates to 0x7FFF, the most positive value; and in the case of signed underflow, the result becomes 0x8000, the most negative value.

When two fixed point numbers are multiplied, they need not be in the same format. An $A.B$ number multiplied by a $C.D$ number, for example, will give an $(A+C).(B+D)$ result that occupies a larger number of bits than either of the original numbers. It is typical for this "extra wide" result to be rounded to a smaller number of bits when the multiplication is complete. Furthermore, when the numbers being multiplied are signed, the result has 2 identical sign bits, and 1 sign bit can be shifted out before rounding. For example, when a 10.6 signed number is multiplied by a 3.13 signed number, the result is 13.19. After shifting out the redundant sign bit, the format becomes 12.20, which can be rounded to a 16-bit result in 12.4 format. Naturally, some precision is lost in the rounding process. This is one of the penalties of doing arithmetic with a finite number of bits. Examples of multiplication are given in Section 13.3.

Floating point DSPs use formats similar to scientific notation. A real number is written as

$$x = M2^E$$

where M is the mantissa and E is the exponent. Both the mantissa and the exponent are signed. The dynamic range is set by the length of the exponent. The larger the number of bits assigned to the exponent, the larger the dynamic range. The resolution is completely defined by the length of the mantissa. The larger the number of bits assigned to the mantissa, the greater the resolution. Examples 12.7 and 12.8 review these concepts.

EXAMPLE 12.7

A common partition in a 16-bit number format has 12 bits allocated to the mantissa and 4 bits allocated to the exponent. This format is shown in Figure 12.6. The mantissa uses 1.11 format. Find the resolution and the dynamic range for this format.

15	14	13	12	11	10	9	8	7	6	5	4	3	2	1	0
$-(2^3)$	2^2	2^1	2^0	$-(2^0)$	2^{-1}	2^{-2}	2^{-3}	2^{-4}	2^{-5}	2^{-6}	2^{-7}	2^{-8}	2^{-9}	2^{-10}	2^{-11}

FIGURE 12.6
Floating point format for Example 12.7.

Since the mantissa is expressed in 1.11 format, the smallest mantissa is -1 and the largest mantissa is $1 - 2^{-11} = 0.99951171875$. The smallest resolvable step, or resolution, is 2^{-11}, or 0.00048828125. With 4 bits assigned to the signed exponent, the largest multiplier is 2^7 and the smallest multiplier difference possible is 2^{-8}. Therefore, the dynamic range of the multiplier is

$$\text{Dynamic range } = 20 \log\left(\frac{2^7}{2^{-8}}\right) = 90.31 \ \text{dB}$$

This dynamic range is lower by 6.02 dB than that obtained for a signed fixed point format with the same number of bits. As the number of bits available increases, though, the dynamic range of floating point formats eclipses that of fixed point formats.

EXAMPLE 12.8

Use the floating point format of Figure 12.6 to represent 0.0562375 and compute the quantization error.

To give maximum precision, the exponent that makes the mantissa as large as possible must be selected. However, since the mantissa has 1.11 format, the mantissa must be kept between -1.0 and $(1 - 2^{-11}) = 0.99951171875$. The best choice for the exponent is therefore -4, since $0.0562375 = 0.8998 \times 2^{-4}$. Any exponent more negative than

−4 will require a mantissa greater than 1. The mantissa that can be represented in 1.11 format and is closest to 0.8998 is $0.89990234375 = 0.111\ 0011\ 0011_2$. The two's complement representation for −4 is 1100_2. Therefore, the best binary representation for 0.0562375 is $1100\ 0111\ 0011\ 0011_2$, where the first 4 bits correspond to the exponent and the last 12 to the mantissa. The decimal equivalent for this number is $0.89990234375 \times 2^{-4} = 0.056243896484375$. The quantization error is $0.056243896484375 - 0.0562375 = 6.396484374997236 \times 10^{-6}$.

In order to add or subtract two floating point numbers, they must have identical exponents. To arrange this, one of the numbers is shifted until its exponent matches that of the other. The mantissas of the two numbers are then added directly, and the exponent of the result matches the exponents of the operands. The process can be illustrated as follows with decimal numbers:

$$2.34 \times 10^5 + 0.5117 \times 10^3 = 2.34 \times 10^5 + 0.005117 \times 10^5$$
$$= (2.34 + 0.005117) \times 10^5$$
$$= 2.345117 \times 10^5$$

When two floating point numbers are multiplied, their mantissas are multiplied and their exponents are added. To demonstrate the idea using the above decimal operands:

$$(2.34 \times 10^5)(0.5517 \times 10^3) = (2.34)(0.5517) \times 10^{(5+3)} = 1.290978 \times 10^8$$

Multiplying the mantissas follows the same process used for multiplying fixed point numbers and requires the same handling of sign bits and rounding.

Whatever format is chosen, fixed point or floating point, attention must be given to the possibility of overflow or underflow, which can occur with multiplication and division as well as with addition and subtraction. Prevention of these conditions is the reason for the extra register width frequently provided in the computational units of the processor. If the result of a calculation can initially be captured in a high precision register, it can be rounded to normal precision later, giving the most accurate answers possible.

12.4   DSP HARDWARE UNITS

12.4.1   Multiplier/Accumulators

All DSPs have specialized hardware units that allow them to carry out the operations most common in DSP algorithms rapidly and efficiently. One such unit is the MAC, or **multiplier/accumulator**. This unit, as its name suggests, is optimized to perform rapid, repetitive multiplication and addition, operations that are central to DSP filtering. On ordinary processors, multiplications are time-consuming operations that may require multiple clock cycles to complete. On a DSP, multiplications occur within a single clock cycle or better. The basic structure of the MAC is shown in Figure 12.7.

FIGURE 12.7

Structure of multiplier-accumulator (MAC).

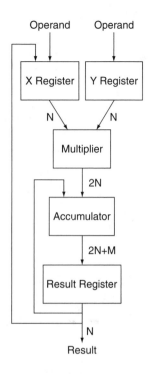

Consider the filtering operation

$$y[n] = x[n] + 0.75x[n-1] + 0.5x[n-2] + 0.25x[n-3]$$

Initially the result register is set equal to the value of $x[n]$. Next, the X register is loaded with the value of $x[n-1]$, and the Y register is loaded with the value 0.75. The two registers are multiplied together and added to the result register to produce the partial sum $x[n] + 0.75x[n-1]$. The X and Y registers are loaded next with the value of $x[n-2]$ and 0.5. The product of these terms is computed and added to the previous partial sum. This process continues until the result register contains the final sum $y[n]$.

Notice in Figure 12.7 that, though the operands have a data width of N bits, the multiplier result has a width of $2N$ bits. Double data width gives the result double precision, which means that errors can be kept as small as possible. After all, two unsigned integers that can be represented in 16 bits can easily multiply to give a number that requires more than 16 bits. For example, squaring $1111\ 1111\ 1111\ 1111_2 = 65535_{10}$ gives 4294836225_{10}, which equals $1111\ 1111\ 1111\ 1110\ 0000\ 0000\ 0000\ 0001_2$: The result occupies 32 bits.

For many DSPs, the accumulator result is widened by M bits beyond the multiplier result. These bits are often called "guard bits" because they guard against overflows and underflows that can occur when numbers of the same sign are added together. For example, the sum of the two 32-bit numbers 4294836225_{10} and 4294836225_{10} equals 8589672450_{10}, or $1\ 1111\ 1111\ 1111\ 1100\ 0000\ 0000\ 0000\ 0010_2$, which occupies 33 bits, a gain of 1 bit. When several MAC operations occur in succession, a number of extra bits can be generated: M guard bits allow a total of 2^M MAC cycles before overflow or underflow will occur. The widening of the data path to $2N$ for the multiplier and $2N + M$ for the accumulator permits more accurate computational results to be carried forward. Naturally, when a sequence of MAC operations terminates, the $2N + M$ result must be rounded before it can be accommodated by the normal N-bit data path of the DSP.

12.4.2 Shifters

All processors have shifting capability. What makes DSP **shifters** special is that they can shift an input as much as the full width of the shift register left or right, in a single instruction cycle. As mentioned in Section 12.3, data used in DSP calculations sometimes need to be scaled to maintain a consistent number format. Scaling requires dividing all incoming data by some scaling factor and then multiplying by the same scaling factor after processing is complete. Fast arithmetic shifting is a convenient way to implement this scaling. When a binary number is shifted one position to the right, it is divided by 2; when a binary number is shifted one position to the left, it is multiplied by 2. Arithmetic left shifts zero-fill the result, which means the new bits at the right are zeros. Arithmetic right shifts **sign-extend** the result, which means the sign bit, or the most significant bit, is repeated at the left for each shift position to the right.

EXAMPLE 12.9
A signed 8-bit integer is given by $0101\ 1001_2$. The decimal value of this number is 89_{10}. If it is shifted arithmetically to the right by 1 bit, it becomes $0010\ 1100_2$, which has the decimal value 44_{10}. Notice that the right shift truncates the rightmost bits and repeats the zero sign bit at the left. The result of the right shift is to halve the original number, but note that the correct result, 44.5, is not obtained. This loss of precision is due to the limited number of bits available.

EXAMPLE 12.10
The signed 8-bit number $1101\ 1100_2$ is negative because its sign bit is 1. It has the decimal value -36_{10}. An arithmetic shift to the right by 2 bits changes it to $1111\ 0111_2$, which has the decimal value -9_{10}. The right shift repeats the sign bit at the left, once for each shift to the right. Two right shifts have the effect of halving the original value twice.

EXAMPLE 12.11
A signed integer is given by $0010\ 0110_2$. The integer has the decimal value 38_{10}. After an arithmetic left shift by 1 bit, it becomes $0100\ 1100_2$, which has the decimal value 76_{10}, twice the original value. Note that one more shift to the left will cause the sign bit to change. This indicates imminent overflow, since shifting a positive number to the left should only make it more positive.

As Example 12.12 will show, shifters can facilitate scaling by providing functions for **normalization** and denormalization. When a word of data contains a number of identical leading bits, these bits can be identified as sign bits and the data can be left-shifted until just 1 sign bit remains. The number of shifts is stored as an exponent. On denormalization, this exponent contains the number of arithmetic right shifts required to restore the data to its original form. Remember that arithmetic right shifts sign-extend the result. Increased calculation precision can sometimes be obtained for fixed point DSPs by using a shifter to normalize an entire block of numbers at once. In this scheme, all fixed point values are shifted by the same amount and the size of the shift is stored as a block exponent. This can be done only when all numbers in the block have room to grow, but with the benefit that the bits gained can allow greater precision for operations with numbers in the block. Example 12.13 demonstrates this strategy.

EXAMPLE 12.12

The sign bits of a number are shown in bold: **1111 111**0 1001 0101. The decimal value of the number is -363_{10}. This number can be normalized by left-shifting six places, shifting in zeros at the right end, to give **1**010 0101 0100 0000. The decimal value of this normalized number is -23232_{10}. An exponent records the number of shifts. In this case, the exponent has the value -6. It is negative because the true number is smaller than the normalized number. In decimal form the normalization process may be summarized by the expression $-363 = -23232 \times 2^{-6}$. Denormalization may be achieved by right-shifting according to the exponent value, with sign extension at the left to give **1111 1110** 1001 0101 again.

EXAMPLE 12.13

The values in a block are given by:

$$0000\ 0001\ 1101\ 1100$$
$$0000\ 0111\ 0100\ 0010$$
$$0000\ 0010\ 0011\ 1001$$

After normalization, the numbers in the block become:

$$0001\ 1101\ 1100\ 0000$$
$$0111\ 0100\ 0010\ 0000$$
$$0010\ 0011\ 1001\ 0000$$

with an exponent of -4. The 4 extra bits gained at the right through normalization give additional precision for operations like multiplication, although they do not improve precision for addition or subtraction.

12.4.3 Address Generators

DSPs are equipped with **address generators** whose job it is to provide fast access to memory locations from an address stored in a register. These special hardware units perform the

FIGURE 12.8

Circular buffer.

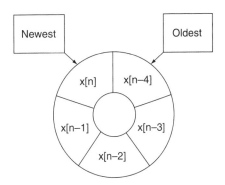

address generation tasks that would otherwise occupy other parts of a busy microprocessor. In fact, address generators contain their own ALUs (arithmetic logic units), which makes them powerful processors. They support address wraparound, which enables them to support **circular buffers**. Since so many DSP algorithms rely on the use of a fixed number of past data values, a sliding window in the form of a circular buffer is a natural choice for access. For example, a nonrecursive filter defined by the equation

$$y[n] = b_0x[n] + b_1x[n-1] + b_2x[n-2] + b_3x[n-3] + b_4x[n-4]$$

requires five input values for each computation. These values can be stored in a circular buffer, as shown in Figure 12.8. When a new input sample is obtained, it replaces the oldest sample in the circle, and the pointers shift by one position. The address generator facilitates these accesses by automatically moving an address pointer from one element to the next in a list, and also from the end of the list back to the beginning again, to complete the circle.

In the circular buffer of Figure 12.8, the address pointer moves by only one position after each access. Address generators need not read from memory in this sequential manner. They can be programmed with an offset that will initiate a read of every second data value, or every fourth, a handy feature for decimation, discussed in Section 14.1.2.

12.5   DSP ASSEMBLY LANGUAGE

Assembly language is the command language for DSPs. While the assembly language specifics may be different for each DSP manufacturer, a number of useful features recur in many families. Beyond the basic computing and control capabilities that assembly languages for all microprocessors must offer, assembly languages for DSPs exploit instructions that can speed up repetitive DSP tasks. Naturally, increasing speed of execution must be a part of this effort. Most DSP instructions now execute within one clock cycle or less. The faster the processor, therefore, the smaller the time to execute a single instruction. A 33 MHz DSP that executes one instruction per clock cycle, for example, completes execution of an instruction in 30.3 nsec, which is equivalent to 33 MIPS (millions of instructions per second).

Another way to reduce execution time for repetitive tasks is to offer efficient looping constructs. Most DSPs provide what is called **zero overhead looping**. All processors offer some form of a do-until loop, which executes a group of instructions until an end condition is reached. DSPs speed the looping process with a counter register whose status is automatically checked and whose value is automatically decremented at the end of each loop. No additional statements are required to test whether the count has expired, nor to decrement the count value.

Because of the high degree of hardware parallelism present in most DSPs, **multifunction instructions** are also common. When multiple buses are available, one read from each memory can occur simultaneously. Similarly, a memory read or memory write can occur together with a computation. Even reads from two memories and a MAC (multiply/accumulate) operation can occur at the same time.

Some DSPs offer a **repeat instruction** that repeats the next instruction the number of times specified. It can be used, for example, with a single move instruction to move a group of data values to memory, or with a single MAC instruction to perform the repeated multiplications and additions required to implement an FIR filter.

When special instructions are not available to perform a task, standard ones must be used in the most efficient possible way. To this end, many tricks exist to perform complex tasks using simpler subtasks. For example, the function

$$f(x) = x^3 + 2x^2 - 4x + 1$$

requires five multiplications and three addition/subtractions. The same function can be re-expressed as:

$$f(x) = x(x(x + 2) - 4) + 1$$

which requires two multiplications and three addition/subtractions. This is an example of **Horner's method**, which states that a polynomial

$$f(x) = a_n x^n + a_{n-1} x^{n-1} + \cdots + a_1 x + a_0$$

can be rewritten as

$$f(x) = x(\cdots(x(x(a_n x + a_{n-1}) + a_{n-2}) + a_{n-3}) + \cdots) + a_0$$

Trigonometric, exponential, and logarithmic functions are not normally available in assembly languages. Often, they are approximated by their **power series expansions**:

$$\sin(x) = x - \frac{x^3}{3!} + \frac{x^5}{5!} - \frac{x^7}{7!} + \frac{x^9}{9!} - \cdots$$

$$\cos(x) = 1 - \frac{x^2}{2!} + \frac{x^4}{4!} - \frac{x^6}{6!} + \frac{x^8}{8!} - \cdots$$

$$\tan(x) = x + \frac{x^3}{3} + \frac{2x^5}{15} + \frac{17x^7}{315} + \cdots$$

$$e^x = 1 + x + \frac{x^2}{2!} + \frac{x^3}{3!} + \cdots$$

$$\ln(1 + x) = x - \frac{x^2}{2!} + \frac{x^3}{3!} - \frac{x^4}{4!} + \cdots$$

where the factorial $n! = n(n-1)(n-2)\ldots(2)(1)$. These series are truncated at the point when additional terms cannot change the result for the number of bits available in the DSP. The coefficients for the expansion are normally precalculated and stored in memory. Frequently used functions like those above can also be stored in **lookup tables**. A lookup table for $\sin(x)$, for example, might contain the values of the function for $0 \le x \le \pi/2$ radians in steps of $\pi/8$ radians. The same list of values can be used in reverse for $\pi/2 \le x \le \pi$. The sine values for $\pi \le x \le 3\pi/2$ can be obtained by simply negating the values in the table, and the values for $3\pi/2 \le x \le 2\pi$ by reversing and negating the values. The function $\cos(x)$ may make use of the same lookup table, by selecting the values in a suitable order.

12.6 HOW TO CHOOSE A DSP

12.6.1 Fixed Point or Floating Point

Choosing between a fixed point DSP and a floating point DSP is one of the most basic decisions a DSP designer must make. With fixed point devices, the programmer must select a number format and be conscious of it throughout all programming. At times, numbers must be scaled down before calculation and then scaled back up afterward. Floating point devices, on the other hand, offer a convenient and natural representation for decimal values. No scaling is needed to perform calculations, so programs for floating point processors are often simpler to write than those for fixed point DSPs.

Floating point devices usually represent numbers with many more bits than do fixed point devices. On the down side for floating point DSPs, performing all operations with this large number of bits takes time. When a job can be done with a fixed point DSP, it nearly always takes less time than it will on a floating point DSP. Furthermore, fixed point DSPs are cheaper than floating point DSPs, though the margin is narrowing. For these reasons, floating point devices are normally selected only for applications where floating point accuracy and number range are a necessity. The cost of programming, however, may influence the choice. For low volume production, the extra expense associated with fixed point programming effort may not be justified, so a floating point device may be chosen instead.

Finally, fixed point devices generally offer smaller chip area and lower power consumption than floating point devices. These considerations may also influence design choices.

12.6.2 Data Width

For both fixed point and floating point devices, the number of bits assigned to a number, called the **data width**, determines the dynamic range that will be available to the programmer. In the case of fixed point, the dynamic range is the same regardless of the position of the binary point. For floating point, the dynamic range is determined by the length in bits of the exponent. Generally, the bigger the data width, the more accurately numbers can be represented, but the slower DSP operations will be. Furthermore, the larger the data width, the larger the size of the chip. Related to data width are the number formats supported by the DSP. Formats that offer suitable levels of resolution for the application intended are essential. Furthermore, flexible choices of number formats make a DSP easier to program.

12.6.3 Hardware and Software Features

DSPs have many special features that allow them to perform well on computationally intensive and repetitive DSP algorithms. For most DSPs, for example, circular buffers are provided in hardware to facilitate filtering. At each step, the oldest datum in the buffer is replaced by a new value. These hardware buffers permit automatic updating of addresses around the circle. Hardware address generators are also commonly featured on DSP chips, to work out addresses at the same time as computations are occurring. Finally, DSPs place a huge emphasis on fast arithmetic units to multiply, add, and shift binary numbers. These units are the workhorses of the DSPs, enabling them to produce results at lightning speeds. For external connections, DSPs offer a variety of input/output ports, like serial ports, direct memory access (DMA) ports, and interface ports. These ports facilitate communication with external memory, host processors, and data converters (A/D or D/A).

On the software side, DSPs offer specialized instructions that their unique hardware structures make possible. For instance, zero overhead looping is handled by a program sequencer that is independent of the computational units. Because counting and comparing are managed by this program sequencer, extra cycles need not be stolen from calculation time. Further speed and efficiency advantages are gained through multifunction instructions. As mentioned, these instructions permit multiple memory reads and writes to occur in the same instruction cycle as a multiplication and addition. They are made possible through the use of multiple memories, accessed by multiple buses. The combinations of instructions that are possible depend on the specific hardware structure of the DSP.

12.6.4 Speed

Instruction cycle time may be the simplest way to assess DSP speed. This is the time for one instruction to execute. Its reciprocals are **MIPS**, millions of instructions per second, for fixed point DSPs, and **MFLOPS**, millions of floating point operations per second, for floating point DSPs. Fast new DSP technology has led to the introduction of BOPS, billions of operations per second, and GFLOPS, billions of floating point operations per second, where G stands for "giga," or 10^9. The trouble with these measures is that the amount of useful work accomplished by an instruction varies from processor to processor. A single instruction might be enough for one DSP, while another might require three or four instructions to do the same job. The specific hardware choices made during the design of a DSP determine how easily and how quickly various tasks can be performed. Sometimes **clock cycle time**, or its reciprocal **clock frequency**, is used as a measure of DSP speed. Unfortunately, these measures may not provide an accurate assessment of a DSP's ability to perform a task either. While many DSP instructions are designed to execute in a single clock cycle, some require two or three.

12.6.5   Memory

Most DSPs offer a combination of on- and off-chip memory options. On-chip memory can be accessed much more quickly than off-chip memory. For this reason, most DSPs offer some quantity of on-chip random access memory (RAM) or read only memory (ROM). A portion of this memory is earmarked as program memory, and the rest is designated as data memory. Not infrequently, memories are capable of holding both program instructions and data, as specified by the programmer. Advantages of multiple memories are realized only when each memory is accessed by a separate bus. This way, memories can be accessed simultaneously without collisions occurring. Harvard architecture describes the structure of the first processors to offer separate program and data spaces. Modified Harvard architecture covers most other combinations of program and data memories.

Often, external memory can also be connected to a DSP, through an input/output port. Naturally, access to external memories is very slow compared to on-chip accesses. The DSP may need to wait additional clock cycles while items are retrieved from off-chip memories. For this reason, internal memories are now much larger than in the past, so that little or no external memory will be required.

12.6.6   Power

When a product must perform for extended periods without service, low power consumption is a must. Most DSP chips offer low voltage operation options, as well as sleep modes that can reduce power use to almost zero when the calculations are not needed. Power savings can also be realized when frequency of operation can be reduced.

12.6.7   Supporting Hardware

Finding a suitable data converter (A/D and D/A) is critical to all DSP designs. After all, a DSP that can handle 8000 MIPS, or 8 billion instructions per second, is probably sitting idle a great deal of the time if the data converter can provide only a million samples per second. Many manufacturers produce data converters and power supplies specially designed to complement their DSPs. Parts that are made to work together usually make interconnection easy. Furthermore, when a wide variety of data converters are available, a good match to analog signal characteristics can be found.

12.6.8   Convenience

Ease of development is an important consideration for DSP programmers. It follows from high quality development tools with both high level and assembly language support. For example, good debugging tools, simulators, and emulators can contribute to short development time. Choice of language is a major consideration as well. Most DSPs can be programmed in both C and assembly languages. Programming in a high level language is generally considered to be easier than programming in assembly language, but there are still costs for this convenience. Assemblers and linkers are used for assembly language programs, while compilers are needed for C programs. These compilers are DSP-dependent, because they must translate C instructions into assembly code that makes good use of the specific hardware of the DSP. In most cases, the assembly code produced in this manner is less than half as efficient as hand-coding in assembly language. The newest Texas Instruments DSPs, however, have been designed specifically for use with C. For these processors, C programs can produce assembly language code that is 80% as good as hand-coding (Texas Instruments, 2000). As the costs of hardware decrease in relation to the costs of software, the trend toward C coding will continue.

In addition to tools that simplify single processor designs, features that facilitate multiprocessor designs may also influence the choice of DSP. For applications that require several DSPs to work in parallel to obtain the necessary computing power, the ease with which processors can be interconnected is a consideration. Finally, good vendor and third party support can help to make a chip popular. For DSPs that are widely used, a large collection of software for common applications is available on the Internet. Reusing or adapting this software is a quick way to develop new code.

12.6.9   Cost

As mentioned, fixed point DSPs tend to cost less than floating point DSPs. Cost goes down when chips are purchased in large batches instead of individually. Even unit prices from most manufacturers are remarkably low considering the massive computing power offered by today's DSPs.

12.6.10   Application

The considerations mentioned in Sections 12.6.1 through 12.6.9 must be judged with a view to the intended application of the DSP. While a fixed point DSP might be a suitable choice for a disk drive controller, the arithmetic precision of a floating point DSP is likely to be necessary for a robotics application. An indication of the processing power the DSP must have can be found by calculating the number of instructions required per sample, as

$$\# \text{ Instructions per sample} = \frac{\text{Sampling period}}{\text{Instruction cycle time}}$$

This measure estimates the number of instructions that can be executed between samples. For example, a processor with a 10 nsec instruction cycle time used with a 16 kHz sampling rate will permit 6250 instructions between samples. A processor with a 5 nsec instruction cycle used with the same sampling rate will allow twice as many, or 12,500, instructions between samples. The complexity of the algorithm that must be applied to each sample determines which processor will meet the needs of the application. Table 12.3 lists a few sample applications of DSPs and their estimated requirements. As a rule, the fastest processor (or processors) that can meet all other demands of an application, such as memory requirements and system cost, will be selected. The more complex the processing required, the

TABLE 12.3
DSP Requirements by Application

Application	Fixed or Floating	Approximate	
		Memory	Cost
Speech recognition	Fixed	Low	Low
Disk drive control	Fixed	Low	Low
Cochlear implants	Fixed	Low	Medium
Data encryption	Fixed	Low	Medium
Digital still camera	Fixed	Medium	Medium
Music synthesis	Fixed	Medium	Medium
High speed fax/modem	Fixed	Medium	Low
Digital answering machine	Fixed	Medium	Low
Audio or image compression	Floating	High	Low
Robotics	Floating	Medium	High
Video recorder	Floating	Medium	Low
3D ultrasound	Floating	High	High
Sonar	Floating	High	High
Radar	Floating	High	High
Electroencephalogram	Floating	High	High

more likely a floating point device will be chosen. The data width for fixed point devices is generally 16 bits, while that for floating point devices is typically 32 bits, offering greater precision for numerically intensive algorithms.

12.7   DSP MANUFACTURERS

12.7.1   Analog Devices

Analog Devices offers a fixed point family, ADSP-21xx, and a floating point family, ADSP-21xxx. A block diagram of the base ADSP-2100 architecture appears in Figure 12.9. This family has two on-chip memories, a program memory and a data memory, in a modified Harvard architecture. The primary function of the program memory is to hold program instructions, but it may store data as well. The data memory contains data alone. The amount of memory varies among members of this DSP family. A total of five buses are available on the chip, four of which connect to memory. Fourteen-bit addresses for memory access are placed on the program or data memory address buses. Data move in and out of memory on the program or data memory data buses. The program memory data bus is 24 bits wide to accommodate 24-bit instructions. The data memory data bus is 16 bits wide for fixed point devices, giving 96.33 dB of dynamic range. For floating point devices, the bus width is 32 bits. A fifth bus is used to carry results of calculations back into the computational units when they are needed for succeeding instructions.

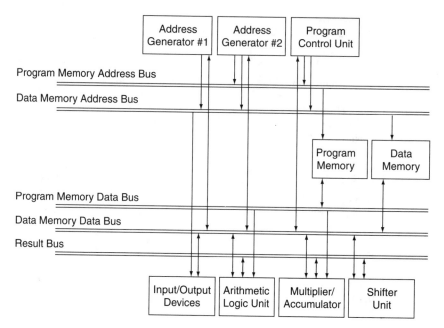

FIGURE 12.9
ADSP family base architecture.

The Analog Devices family uses two address generators, one that provides addresses for data memory and one that provides addresses for both data and program memory. Three computational units are the workhorses of the DSPs: one arithmetic logic unit, one multiplier/accumulator, and one shifter unit. Sixteen-bit addition and 16×16 multiplication are available on fixed point devices.

Multiple memories and multiple buses for access make various multifunction instructions possible. For example, the instruction

$$ar = ax0 + ay0, ax0 = dm(i0, m0);$$

places the sum of the registers ax0 and ay0 in register ar at the same time as a value is collected from data memory (dm) and placed in ax0. The value of ax0 used in the sum is the value present in the register at the beginning of the instruction cycle; at the end of the cycle, ax0 contains the new value obtained from memory. The partner example,

$$dm(i0, m0) = ar, ar = ax0 + ay0;$$

writes the original value of register ar to data memory, while a sum is computed that will replace this value before the end of the cycle. As another example, the following instruction multiply-accumulates into register mr while registers mx0 and my0 are refilled with new values ready for the next step, one from data memory (dm) and one from program memory (pm):

mr = mr + mx0 * my0 (ss), mx0 = dm(i0, m0), my0 = pm(i4, m4);

Devices in the fixed point family perform from 20 MIPS (ADSP-2101) to 75 MIPS (ADSP-2189M), with one instruction occurring in each clock cycle. The ADSP-2181 operates at 33 MIPS. The newest fixed point family, ADSP-219x, is projected to run at 300 MIPS. The ADSP-2101 provides 2k × 16-bit words of program RAM and 1k × 16-bit words of data RAM. The ADSP-2189M offers 32k × 16-bit words of program RAM and 48k × 16-bit words of data RAM.

Devices in the floating point family can perform both fixed and floating point arithmetic. The ADSP-21065L can execute 180 MIPS or 180 MFLOPS, sustained, while the ADSP-21160, operating at 100 MHz, performs 600 MFLOPS. Both are SHARC (Super Harvard Architecture Computer) trademark devices. The newest floating point devices, in the TigerSHARC family, are projected to offer 5 GFLOPS.

12.7.2 Texas Instruments

Texas Instruments manufactures a wide range of both fixed and floating point DSPs. The fixed point families have on-chip program and data memories in varying sizes and configurations, supporting 16-bit data. The TMS320C50, for example, has 2k of program ROM, about 0.5k of data RAM, and 9.5k of RAM that can be used for program code or data. The TMS320LC541, on the other hand, has 20k of program ROM, 8k of ROM that can be used for program instructions or data, and 5k of RAM primarily designed to contain data but also able to hold program code. To support these program and data memories, four data buses and four address buses are available. The basic architecture of the TMS320C54x DSP is illustrated in Figure 12.10.

Computational units include an arithmetic logic unit, a multiplier/accumulator, and a shifter unit. The arithmetic logic unit contains two independent 40-bit accumulators, allowing 40-bit addition. The multiplier can perform 17-bit × 17-bit multiplications. The 40-bit shifter permits all shifts for scaling and fractional mathematics to occur in parallel with arithmetic operations. In addition to the computational units, two address generators, one for data and one for program addresses, facilitate rapid memory access.

The advanced modified Harvard architecture of this device means that several actions can occur within a single instruction cycle. For instance, the multiply-accumulate instruction

MACA *AR5+, B

multiplies two numbers and adds the result to the value in accumulator B within a single cycle. The first number is provided by the top 16 bits of accumulator A. The second number is taken from the data memory address held by register AR5. This address is automatically postincremented as part of the same instruction cycle, ready for the next data memory read.

TMS320C50 DSPs perform between 20 and 40 MIPS. The TMS320C54x group provides 40 MIPS. The newest fixed point DSPs, the TMS320C64x family, can provide a

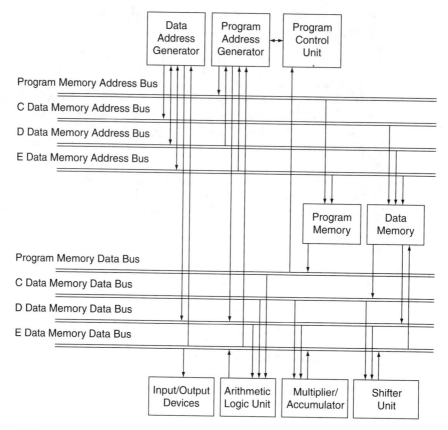

FIGURE 12.10

Basic architecture of TMS320C54x family.

FIGURE 12.11

VLIW instruction.

[B1]	SUB	B1, 1, B1
|| [!B1]	ADD	B8, B5, B5
|| [!B1]	ADD	A6, A0, A0
||	MPY	B9, B0, B8
||	MPYHL	B0, A5, A6
|| [A1]	SUB	A1, 1, A1
||	LDW	*B6++, B0
||	LDH	*+A3(2), A6

staggering 8800 MIPS. This DSP offers advanced VLIW architecture bearing Texas Instruments' velociTI trademark. With this architecture, up to eight instructions can be performed in each clock cycle; so while the DSP runs at 1100 MHz, its instruction rate is much higher. Figure 12.11 shows a sample VLIW instruction that is part of an FIR filter program for this processor. Two subtractions, two additions, two multiplications, and two loads all execute in a single clock cycle. In addition, the 'C64x offers instruction set extensions that simplify programming for digital communication and image processing applications.

Among the floating point devices Texas Instruments offers, the TMS32067x is the most recent. It has a highly parallel velociTI VLIW architecture. One billion floating point operations can occur per second, though the clock rate is only 150 MHz. This DSP contains eight computational units: two floating or fixed point arithmetic logic units, two fixed point arithmetic logic units, two shifters, and two floating or fixed point multipliers. A powerful new multiprocessor DSP, the TMS320C80, contains four advanced DSPs processing in parallel, each of which has its own computational units. A master processor controls the operation of the parallel processing units, called PPs, for two billion operations per second.

12.7.3 Motorola

Motorola offers a range of fixed point DSPs with varying capabilities. The basic architecture of the DSP56000 family is shown in Figure 12.12. It contains a program memory and two separate data memories. The memories are supported by seven buses: three address buses, three data buses, and a global data bus used for data transfer. The data width is 24 bits, one and a half times the width of most other fixed point processors. This large bus width provides 144.49 dB of dynamic range and greater resolution than most other fixed point DSPs.

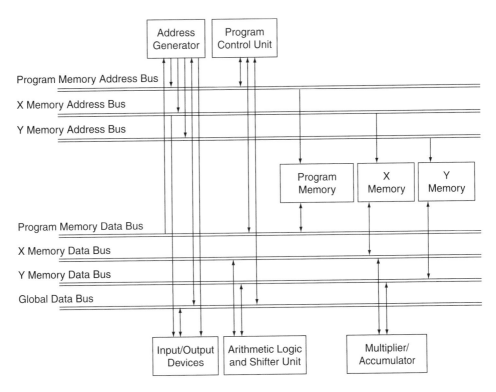

FIGURE 12.12

Base architecture of the DSP56000.

Motorola devices offer a range of different operating speeds and on-chip memory sizes. The DSP56004, for instance, contains 512k of program RAM, 256k each of X and Y data ROM, and 256k each of X and Y data RAM. It runs at 66 MHz and performs 33 MIPS. The DSP56011 has 12.5k of program ROM, 3.5k of X data ROM, 4k of X data RAM, 2k of Y data ROM, and 4.25k of Y data RAM. It offers 40.5 MIPS, at 81 MHz. The DSP56364 runs at 100 MHz and provides 100 MIPS, with yet another combination of program and data ROM and RAM.

Motorola devices combine arithmetic logic and shifter functions into a single computational unit. It contains two 56-bit accumulators and performs 56-bit addition. The multiplier-accumulator unit is separate. It permits 24-bit $\times$ 24-bit multiplications. In addition, an address generator is available to calculate addresses in parallel with other computations.

The modified Harvard architecture of the Motorola devices permits multifunction instructions. For example, the multiply-accumulate instruction

$$\text{MAC X0, Y0, A} \quad \text{X: (R0)+, X0} \quad \text{Y: (R4)+, R0}$$

multiplies X0 by Y0 and adds the result to accumulator A. Registers R0 and R4 contain data memory addresses that are incremented automatically in preparation for the next memory access. These increments occur in the same instruction cycle as the multiply. As another example, the instruction

$$\text{ABS B} \qquad \text{\#\&12, X0} \qquad \text{B, Y0}$$

takes the absolute value of the B register, initializes register X0 with the value 0x12, and saves the value in register B to register Y0. Because the moves occur in parallel to the calculation of the absolute value, the value in B that is moved to register Y0 is the value that was held in B prior to the execution of the absolute value function. Moves may form part of many parallel instructions, but destination and source registers may not be duplicated within the same instruction. In addition to facilitating parallel instructions, hardware features may also help to make program design more convenient. The presence of two data memories, for example, permits the real parts of an FFT to be stored in one memory, while the imaginary parts are stored in the other.

CHAPTER SUMMARY

Matlab
Support

1. Digital signal processors (DSPs) perform DSP tasks more efficiently and more rapidly than standard processors. The architecture for DSP processors is based on the Harvard model, which employs multiple buses and memories to permit several processing activities to occur in parallel.
2. Special hardware units allow extremely fast multiplication, addition, and shifting.
3. DSPs usually execute a minimum of one assembly language instruction per processor clock cycle. Advanced architectures allow eight or more instructions to execute in each cycle, to make the best use of hardware capabilities.
4. DSPs may be fixed point or floating point devices. Each type of device supports particular number formats. Floating point devices support fractional number formats im-

plicitly. For fixed point devices, programmers must manage fractional number formats and scaling explicitly.

5. Multiplier/accumulator hardware performs rapid multiplications, and additions or sub-tractions.

6. Hardware shifters can be used to scale numbers up and down quickly.

7. Address generators support rapid sequential memory accesses using addresses stored in registers. They can be used to create circular buffers, which are useful in filtering and other DSP tasks.

8. The best DSP for a certain application may depend on the choice between fixed point and floating point, the required data width, hardware and software features, speed, available memory, power requirements, supporting hardware (such as A/D and D/A converters), programming tools, and cost.

REVIEW QUESTIONS

12.1 Describe the essential differences between Von Neumann and Harvard architectures.

12.2 What is pipelining and how does Harvard architecture make it possible?

12.3 The decimal numbers 44803 and 299 are multiplied.
   a. Find unsigned 16-bit binary representations for 44803 and 299.
   b. Find an unsigned binary representation for the product. What minimum number of bits are needed to represent the product exactly?
   c. If the result in (b) must be rounded to its 16 most significant bits, what is the product after rounding? What error is committed?
   d. If the result in (b) is written as an unsigned 32-bit number and then rounded to its 16 most significant bits, what is the product after rounding? What error is committed?

12.4 Find the results, in binary and decimal form, when the following shifts are applied to the 8-bit binary number 0010 1001:
   a. No shift
   b. Left shift by 1 bit
   c. Right shift by 1 bit
   d. Right shift by 2 bits

12.5 Find the decimal representation for each of the following signed 16-bit binary numbers:
   a. 0101 1100 1100 0001
   b. 1011 0111 1101 1100
   c. 1111 1111 1111 1111
   d. 1000 0000 0000 0000
   e. 0111 1111 1111 1111

12.6 Find signed 8-bit representations for each of the following decimal values:
   a. 12
   b. -79

    c. -128

    d. -1

    e. 127

12.7 The unsigned 8-bit binary number 0110 1110 is shifted once to the left. Find the decimal value of the number before and after shifting.

12.8 The signed 8-bit binary number 0110 1110 is shifted once to the left. Find the decimal value of the number before and after shifting.

12.9 **a.** Choose an exponent to normalize the following block of binary numbers:

$$
\begin{array}{cccc}
1111 & 1110 & 1011 & 0001 \\
1111 & 1001 & 0000 & 0110 \\
1111 & 1100 & 0111 & 1000 \\
1111 & 1101 & 1111 & 0011
\end{array}
$$

    b. Find the decimal equivalents of the numbers before and after normalization.

12.10 Find the dynamic range in dB for an A/D converter using:

    a. 10 bits

    b. 12 bits

    c. 14 bits

    d. 16 bits

12.11 Write the decimal equivalents for the 1.15 numbers:

    a. 1.000 1011 0100 0000

    b. 0.111 0100 0000 0111

    c. 0.000 0000 0000 0001

    d. 1.111 1111 1111 1111

    e. 1.000 0000 0000 0000

    f. 0.111 1111 1111 1111

12.12 **a.** Find 1.15 representations for the decimal numbers. Express the answers in binary and hexadecimal form.

    (i)   0.875

    (ii) 0.203

    (iii) 0.56879

    (iv) 0.90311

    (v) -0.8222

    (vi) -0.5194

    b. What is the decimal value for each quantized number in (a), and how much quantization error is incurred by using the 1.15 representation?

12.13 **a.** A 16-bit floating point format contains a signed 4-bit exponent followed by a 12-bit mantissa in 1.11 format. Find a representation that uses this format for each of the following numbers:

    (i)   0.0259

    (ii) 1.5712

    (iii) 6.04

(iv) -0.355

(v) -2.111

b. Find the quantization error incurred for each number in (a).

12.14 Find the dynamic range for a 16-bit floating point format with:

a. A signed 3-bit exponent

b. A signed 5-bit exponent

12.15 If the full scale range for an A/D converter is 5 V and the quantization step is 310 mV:

a. How many bits does the converter employ?

b. What is the dynamic range of the converter?

12.16 a. Rewrite the function $f(x) = x^4 - 3x^3 + 2x^2 - 1$ using Horner's method.

b. How many multiplications are needed to compute the original function? How many are needed for the rewritten function?

12.17 The power series expansion for $f(x) = e^{2x}$ is:

$$e^{2x} = 1 + 2x + \frac{(2x)^2}{2!} + \frac{(2x)^3}{3!} + \frac{(2x)^4}{4!} + \frac{(2x)^5}{5!} + \cdots$$

Compare $f(0.5)$ to the approximate value obtained using the expansion with

a. Four terms

b. Six terms

12.18 a. Write a 16-point table for the sine function $y = \sin(x)$ for $0 \le x \le \pi/2$ radians.

b. Suggest how to use the table to generate one complete cycle that begins at $x = 0$ for:

(i) $y = \sin(x)$

(ii) $y = \cos(x)$

(iii) $y = \sin(2x)$

12.19 Write a 16-point lookup table for one full cycle of $f(x) = \cos(x)$ using the first three terms of its power series expansion. Report table entries in hexadecimal 1.15 format.

12.20 A 33 MHz DSP processor executes one instruction per clock cycle. If it is to be used in a 25 kHz sampled system, how many instructions can be executed for each sample?

12.21 A DSP system samples at 44.1 kHz and must execute 10,000 assembly instructions for each sample. A processor must be selected for the task.

a. What number of MIPS is required?

b. If the processor executes one instruction per clock cycle, what is the minimum clock speed for the processor?

c. If the processor executes eight instructions per clock cycle, what is the minimum clock speed for the processor?

13

PROGRAMMING DSPs

This chapter discusses DSP programming issues, with application to the ADSP-2181 EZ-KIT Lite development kit. The chapter:

- ➤ introduces the fixed point ADSP-2181 processor
- ➤ describes the elements on the EZ-KIT Lite development board
- ➤ discusses the number formats available for the ADSP-2181
- ➤ describes how to scale FIR and IIR filters
- ➤ lists the major registers available in the ADSP-2181
- ➤ provides examples of assembly language instructions
- ➤ describes how to set up and initialize the EZ-KIT Lite board
- ➤ supplies assembly language programs for a sine wave generator, an FIR filter, and an IIR filter

13.1 ADSP-2181 PROCESSOR

The ADSP-2181 is a high-performance 16-bit fixed point DSP processor. The architecture for this processor is shown in Figure 13.1. In addition to the essential capabilities offered by the computational units, the processor contains two serial ports, a timer, an internal direct memory access (IDMA) port, and a byte DMA (BDMA) port. The ADSP-2181 features 80 kilobytes of on-chip RAM and a 30.3 nsec instruction cycle time.

The computational units include an ALU for standard arithmetic and logic operations; a MAC for single-cycle multiply, multiply-add, and multiply-subtract operations; and a shifter for logical and arithmetic shifts, and scaling operations. A result bus is connected to the computational units so that the output of any unit can be used as the input to any unit in the next cycle. Two synchronous serial ports support serial communications. The ports are bidirectional and include transmit and receive buffers. The timer unit is accessed by way of registers that control the reloading and decrementing of a counter. Each time a count is

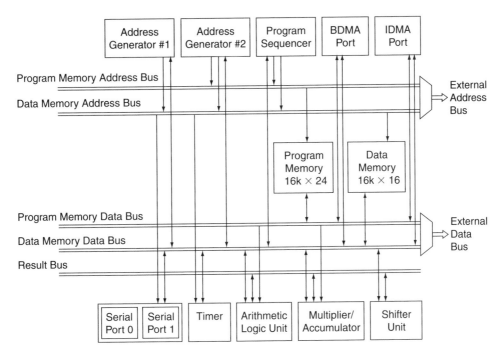

FIGURE 13.1
Architecture of ADSP-2181.

completed, an interrupt occurs. The DMA ports are designed to provide rapid access to and from the outside world. The internal DMA port (IDMA) allows high speed access to on-chip memory by a host system. The byte DMA port (BDMA) permits the DSP rapid access to external RAM or ROM.

On-chip RAM is configured as 16k of 24-bit program memory words and 16k of 16-bit data memory words. The data memory is supported by a data (data memory data) bus and an address (data memory address) bus; and the program memory, which may contain both instructions and data, is supported by its own data and address (program memory data and program memory address) buses. The buses are connected to external address and data buses for access to external memory.

The architecture and instruction set for the ADSP-2181 permit more than one operation to occur at the same time. In one instruction cycle, one or two data moves and a computation can be performed at the same time as a new program address is calculated and the next instruction is fetched. Furthermore, the data moves may include updating of data address pointers as well, also within the same clock cycle. While the ADSP-2181 is occupied with these functions, transfer of data through the serial and DMA ports continues, as do timer decrements. This high degree of parallelism makes the ADSP-2181 an efficient platform for DSP algorithms.

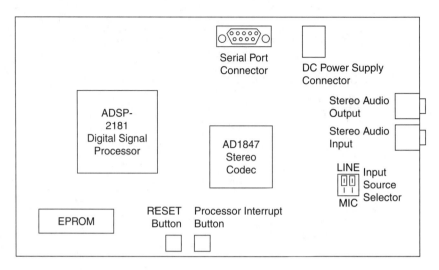

FIGURE 13.2
EZ-KIT Lite development board.

13.2 EZ-KIT LITE DEVELOPMENT BOARD

The EZ-KIT Lite development board provides a convenient platform on which to implement DSP algorithms. The board features the ADSP-2181 digital signal processor, which operates at 33 MIPS (million instructions per second). The A/D and D/A functions are handled by an AD1847 Stereo SoundPort Codec. This chip accepts microphone or line input and provides speaker output, but microphone and speakers are not included with the kit. A power supply and a serial port cable are provided. Figure 13.2 illustrates the layout of the EZ-KIT Lite board.

When the board is powered up or reset, ADSP-2181 code is automatically transferred from an EPROM (erasable programmable ROM) on the board to the internal memory of the processor. This code permits the ADSP-2181 to connect to the serial port of a PC for downloading programs. Host software for the assembly, linking, and downloading of applications is provided with the kit, which means that all phases of DSP development can be investigated. Demonstration programs are included as well.

The audio output from the EZ-KIT Lite board is suitable for driving ordinary amplified speakers. The board is designed to accept either microphone or line input. Stereo line input is attenuated by a factor of 2. By shifting jumpers on the board for stereo microphone input, an amplification factor of 47 is obtained. This large gain is applied because the high quality microphones recommended for use with the kit produce very small voltage signals. For the purposes of lab investigations, the EZ-KIT Lite boards also work well with inexpensive, lower quality electret microphones often used with computer sound cards; but the input gain choices provided on the board may not be able to accommodate the signal level they provide, which is about 1 V peak-to-peak. The simplest solution is to change the feedback resistors in the microphone input stage to provide a gain of 4 or 5. In addition, electret microphones require DC biasing, which must also be added, and which occupies one of the stereo input channels. The EZ-KIT Lite Reference Manual (Analog Devices, 1995a) provides schematics of the input stage to assist with these changes. The size of the input signal to the on-board codec is not absolutely critical, as gains or attenuations may be applied to input signals in software as well as in hardware, to obtain signals of suitable size. The software gains are useful as long as the hardware gains are not so small that the input signal is lost in noise, or so large that it exceeds the limits imposed by the power supplies. For that matter, the software gains themselves can cause register overflows that are audible in the speaker output.

To use the EZ-KIT Lite, the 9 V, 300 mA DC power supply is connected to the board, and a microphone and pair of speakers are connected at the stereo input and output, respectively. To make communication with a host PC possible, the serial cable is connected between the EZ-KIT Lite board and a serial port on the PC. The software that accompanies the kit permits assembly and linking of assembly language programs for the ADSP-2181. Tools for writing C code for the chip are also available, but this chapter will investigate only the use of assembly language. The following sections introduce the aspects of the ADSP-2181 essential for producing assembly language programs.

13.3   NUMBER FORMATS AND SCALING

The ADSP-2181 is a fixed point processor, but can operate using several different number formats in either integer mode or fractional mode. Unsigned and signed formats are available in both modes. In all signed formats, two's complement notation is used to represent negative values. Figure 13.3 shows the hexadecimal equivalents for unsigned and signed

FIGURE 13.3
Unsigned and signed integers, hexadecimal format.

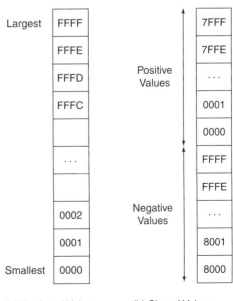

(a) Unsigned Values   (b) Signed Values

15	14	13	12	11	10	9	8	7	6	5	4	3	2	1	0
2^{15}	2^{14}	2^{13}	2^{12}	2^{11}	2^{10}	2^9	2^8	2^7	2^6	2^5	2^4	2^3	2^2	2^1	2^0

(a) Unsigned

15	14	13	12	11	10	9	8	7	6	5	4	3	2	1	0
$-(2^{15})$	2^{14}	2^{13}	2^{12}	2^{11}	2^{10}	2^9	2^8	2^7	2^6	2^5	2^4	2^3	2^2	2^1	2^0

(b) Signed

FIGURE 13.4
Signed and unsigned integer formats.

binary numbers in all 16-bit formats. Conversion between binary and hexadecimal numbers is reviewed in Appendix A.11.

Unsigned integers consist of 16 data bits, while signed integers comprise a sign bit and 15 data bits. Both integer formats are shown in Figure 13.4. Weights are shown for each bit position from 0 to 15. Notice that the weight for bit 15 in the signed format is negative. This ensures that two's complements representations are implemented correctly.

EXAMPLE 13.1

What is the decimal value of the unsigned integer 0xA627?

The hexadecimal digits in the integer may be converted directly to 4-bit binary equivalents. The binary representation for the integer is $1010\,0110\,0010\,0111_2$. The weightings provided in Figure 13.4(a) provide the decimal value, 42535_{10}.

EXAMPLE 13.2

What is the decimal value of the signed integer 0xA627?

As in Example 13.1, the binary representation for the number is $1010\,0110\,0010\,0111_2$. The weightings indicated in Figure 13.4(b) provide the decimal value, -23001_{10}.

Fractional mode for the 2181 permits operations with both high and low resolution numbers. The most common fractional format is 1.15 representation. It was introduced in Section 12.3 but is repeated in Figure 13.5 for convenience.

The largest number that can be represented using this signed format is 0x7FFF = 0.111 $1111\,1111\,1111_2$. Using the weights in Figure 13.5 gives a decimal value for this number of $2^{-1} + 2^{-2} + 2^{-3} + \ldots + 2^{-15} = 0.999969482421875$. The smallest number that can be represented is 0x8000 = $1.000\,0000\,0000\,0000_2$. The decimal value is $-(2^0) = -1.0$. Because of the limited resolution of the 1.15 format, many decimal numbers cannot be represented exactly. In these cases, a quantization error is committed. Errors of this type cannot be avoided when the number of bits used is finite. Examples of 1.15 representation follow.

15	14	13	12	11	10	9	8	7	6	5	4	3	2	1	0
$-(2^0)$	2^{-1}	2^{-2}	2^{-3}	2^{-4}	2^{-5}	2^{-6}	2^{-7}	2^{-8}	2^{-9}	2^{-10}	2^{-11}	2^{-12}	2^{-13}	2^{-14}	2^{-15}

FIGURE 13.5
1.15 Fractional number format.

EXAMPLE 13.3

Use the signed 1.15 format to represent the decimal number 0.9375. State the answer in both binary and hexadecimal forms.

Since $0.9375 = 2^{-1} + 2^{-2} + 2^{-3} + 2^{-4}$, the 1.15 representation for this number is a $0.111\ 1000\ 0000\ 0000_2$. The hexadecimal version of this number is 0x7800.

EXAMPLE 13.4

Find a 1.15 representation of the decimal number -0.9375.

From the previous example, $0.9375 = 0.111\ 1000\ 0000\ 0000_2$. The 1.15 representation of -0.9375 is the two's complement of this number, or $1000\ 1000\ 0000\ 0000_2$. Thus, the decimal number -0.9375 can be represented as $1.000\ 1000\ 0000\ 0000_2$, or 0x8800. To verify this result, the weightings of Figure 13.5 can be used to interpret $1.000\ 1000\ 0000\ 0000_2 = -(2^0) + 2^{-4} = -1.0 + 0.0625 = -0.9375_{10}$.

There are options other than the 1.15 representation. Figure 13.6, for example, shows the weightings for the 12.4 representation. This 16-bit representation contains 12 nonfractional bits and 4 fractional bits. Because a greater number of bits is allocated as nonfractional bits, very large numbers can be represented in this format. With fewer fractional bits, though,

15	14	13	12	11	10	9	8	7	6	5	4	3	2	1	0
$-(2^{11})$	2^{10}	2^9	2^8	2^7	2^6	2^5	2^4	2^3	2^2	2^1	2^0	2^{-1}	2^{-2}	2^{-3}	2^{-4}

FIGURE 13.6
12.4 Fractional number format.

the errors that occur in coding decimal values are, on average, greater. The largest 12.4 value, $0111\ 1111\ 1111.1111$, corresponds to a decimal value of $2^{10} + 2^9 + 2^8 + \ldots + 2^{-3} + 2^{-4} = 2047.9375$. The smallest value, $1000\ 0000\ 0000.0000_2$, corresponds to $-(2^{11}) = -2048.0$. Thus, the full scale range of the 12.4 format is much greater than that of the 1.15 format, while the resolution is much reduced, as the following examples illustrate.

EXAMPLE 13.5

Find a 12.4 representation for the decimal number 0.9375. Express the result in both binary and hexadecimal formats.

In Example 13.3, this decimal number was represented in 1.15 format. Since $0.9375 = 2^{-1} + 2^{-2} + 2^{-3} + 2^{-4}$, Figure 13.6 gives the 12.4 representation as $0000\ 0000\ 0000.1111_2$, or 0x000F.

EXAMPLE 13.6

Compare 1.15 and 12.4 representations for the decimal number 0.5039148.

In the 1.15 format, the decimal that is closest to 0.5039148 is $0.100\ 0000\ 1000\ 0000_2$, or 0x4080, which equals 0.50390625. Thus, an error of $0.50390625 - 0.5039148 = -0.00000855$ is committed. In the 12.4 format, the closest choice is $0000\ 0000\ 0000.1000_2$, or 0x0008, which equals decimal 0.5. The error committed is $0.5 - 0.5039148 = -0.0039148$. The 1.15 format gives a smaller error than the 12.4 format because its number of fractional bits, and therefore its resolution, is better.

The fractional fixed point number formats are listed in Table 13.1. As a group, they offer a continuum of increasing resolution with decreasing full scale range. The best resolution for an application is obtained by choosing the format with the smallest step size that still offers the required range. Management of the number format is left to the programmer. For addition and subtraction, numbers must be in the same format. The same is true for division, since this operation relies on repeated additions or subtractions. For multiplication, however, numbers can have different formats. As indicated in Section 12.3, an $A.B$ number

TABLE 13.1

Ranges for Signed Fractional Fixed Point Number Formats

Signed Number Format	Minimum Value	Maximum Value	Step Size (Resolution)
16.0	−32768.0	32767.0	1.0
15.1	−16384.0	16383.5	0.5
14.2	−8192.0	8191.75	0.25
13.3	−4096.0	4095.875	0.125
12.4	−2048.0	2047.9375	0.0625
11.5	−1024.0	1023.96875	0.03125
10.6	−512.0	511.984375	0.015625
9.7	−256.0	255.9921875	0.0078125
8.8	−128.0	127.99609375	0.00390625
7.9	−64.0	63.998046875	0.001953125
6.10	−32.0	31.9990234375	0.0009765625
5.11	−16.0	15.99951171875	0.00048828125
4.12	−8.0	7.999755859375	0.000244140625
3.13	−4.0	3.9998779296875	0.0001220703125
2.14	−2.0	1.99993896484375	0.00006103515625
1.15	−1.0	0.999969482421875	0.000030517578125

multiplied by a *C.D* number will give an $(A+C).(B+D)$ result, as Example 13.7 illustrates. When two signed numbers are multiplied, the result has 2 sign bits, one of which is redundant. Thus, after any multiplication in fractional mode, the result is always shifted left 1 bit before being stored, as illustrated in Examples 13.8 and 13.9. The automatic left shift after multiplying signed numbers makes the 1.15 format by far the most convenient to use. Two 1.15 numbers multiplied together produce a 2.30 number, which, after shifting, becomes a 1.31 number. This result can easily be converted to a 1.15 result again by rounding off the 16 rightmost bits.

EXAMPLE 13.7

Two decimal numbers, 1.9375 and 3.75, are multiplied. The first number can be represented exactly using 1.4 unsigned fractional number format: 1.1111_2. The second number can be represented without error in unsigned 2.2 format: 11.11_2. To avoid binary multiplication, the binary product of these numbers may be found by multiplying their decimal equivalents and then converting the result to binary:

$$1.1111_2 \times 11.11_2 = 1.9375_{10} \times 3.75_{10} = 7.265625_{10} = 111.010001_2$$

As expected, the product of a 1.4 number with a 2.2 number requires a minimum $(1+2).(4+2) = 3.6$ format to avoid errors. Note that a DSP would perform this multiplication in hardware in binary form, but would obtain the same result.

EXAMPLE 13.8

Two decimal numbers, -0.0625 and -0.25, are multiplied. The first number uses a signed 1.4 fractional number format, and the second uses a signed 2.2 format. Multiplying the two signed fractional numbers gives

$$1.1111_2 \times 11.11_2 = -0.0625_{10} \times -0.25_{10} = 0.015625_{10} = 000.000001_2$$

The product initially contains 6 fractional bits and 1 redundant sign bit. When this sign bit is stripped off by shifting the number 1 bit to the left, the result is 00.0000010_2, in 2.7 format. The value of the number is unchanged by this shift, since the distance of the content bit to the binary point is the same in both cases.

EXAMPLE 13.9

Two decimal numbers, -0.0625 and 3.75, are multiplied. Using signed fractional number formats gives the 1.4 representation 1.1111_2 for the first number and the 3.2 representation 011.11_2 for the second number. Multiplying the two numbers gives a 4.6 representation result

$$1.1111_2 \times 011.11_2 = -0.0625_{10} \times 3.75_{10} = -0.234375_{10} = 1111.110001_2$$

The result is shifted left by one position to give 111.1100010_2, in 3.7 format.

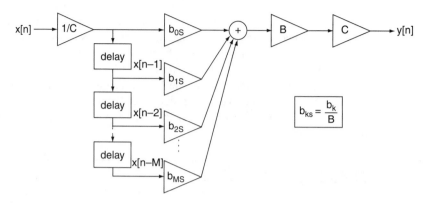

FIGURE 13.7
Scaling for FIR filters.

Numbers of any size can be combined in the ADSP-2181, but programming is most convenient using 1.15 representation because minimal format management is required. To use the 1.15 format, input data and coefficients are scaled to fit into the -1 to $0.999969\ldots$ range. When filter coefficients fall outside this range, they are scaled down on input. When the sizes of input signals are too large, they are scaled down also, to ensure that the digitized signal samples fall within the correct range. At the output, the results are scaled back up to account for the scaling factors used on the input side. Because scaling can be done most efficiently by using a shifter, scale factors are normally powers of 2. Dividing a data value by 2 is the same as shifting its binary value to the right by 1 bit. Multiplying by 2 is the same as shifting to the left by 1 bit. Scaling by 4 means shifting by 2 bits, scaling by 8 means shifting by 3 bits, and so on.

For FIR filters, scaling is necessary when the coefficients or the inputs fall outside the permissible 1.15 range. Figure 13.7 illustrates the scaling mechanisms for this type of filter. B is the scaling factor for the coefficients, and C is the scaling factor for the inputs. Both B and C are normally selected to be powers of 2 to speed up scaling. The scaled coefficients are denoted by b_{kS} and are calculated by dividing the original filter coefficients b_k by B.

EXAMPLE 13.10
The FIR filter

$$y[n] = 2x[n] + x[n-1]$$

must be implemented on a 1.15 fixed point DSP for samples $x[n]$ that lie between -1 V and 1 V. Choose a suitable scaling strategy.

The input samples fall into the correct range and therefore do not need to be scaled. The filter coefficients, however, must be scaled. Dividing the b_k coefficients by 2 is not enough, since the number 1.0 cannot be represented in 1.15 format. Therefore, a scaling factor of $B = 4$ must be chosen, which may be implemented by shifting the binary representations of the coefficients to the right by 2 bits. The equations become

$$y_S[n] = 0.5x[n] + 0.25x[n-1]$$

$$y[n] = 4y_S[n]$$

The first equation can now be evaluated directly, since the coefficients satisfy the 1.15 format. The output is multiplied by 4, by shifting left by 2 bits, to balance the effect of dividing the coefficients by 4.

Scaling is frequently necessary for IIR filters as well. Figure 13.8 illustrates how suitable scaling of inputs and coefficients may be accomplished for a direct form 2 difference equation. After scaling, all coefficients lie within the required range. Again, the original filter coefficients a_k and b_k are divided by their respective scale factors A and B to produce the scaled coefficients a_{kS} and b_{kS}. The inputs are unavoidably scaled down by A as part of the process of scaling the a_k coefficients. If additional scaling is necessary, the scaling factor C can be used.

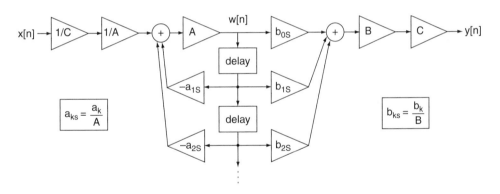

FIGURE 13.8
Scaling for IIR filters, direct form 2.

EXAMPLE 13.11
A microphone signal providing samples up to 1 V in size is to be filtered using a fourth order Chebyshev Type I filter. The filter coefficients are as follows:

$$b = [0.0109 \ \ 0.0435 \ \ 0.0653 \ \ 0.0435 \ \ 0.0109]$$

$$a = [1.0000 \ -2.1861 \ \ 2.3734 \ -1.3301 \ \ 0.3273]$$

Find a scaling strategy that will permit this filter to be implemented on a 1.15 fixed point DSP.

The coefficients given correspond to the transfer function

$$H(z) = \frac{0.0109 + 0.0435z^{-1} + 0.0653z^{-2} + 0.0435z^{-3} + 0.0109z^{-4}}{1 - 2.1861z^{-1} + 2.3734z^{-2} - 1.3301z^{-3} + 0.3273z^{-4}} \quad \textbf{(13.1)}$$

with the difference equation

$$y[n] = 2.1861y[n-1] - 2.3734y[n-2] + 1.3301y[n-3]$$
$$- 0.3273y[n-4] + 0.0109\,x[n] + 0.0435x[n-1]$$
$$+ 0.0653x[n-2] + 0.0435x[n-3] + 0.0109x[n-4]$$

From Section 4.6.2.2, this difference equation gives the direct form 2 representation

$$w[n] = x[n] + 2.1861w[n-1] - 2.3734w[n-2] + 1.3301w[n-3] - 0.3273w[n-4]$$
$$y[n] = 0.0109w[n] + 0.0435w[n-1] + 0.0653w[n-2]$$
$$+ 0.0435w[n-3] + 0.0109w[n-4]$$

In order to use a 1.15 representation, the a_k coefficients must be scaled down, since they do not fall within the range specified in Table 13.1. A suitable power-of-two scaling factor would be $A = 4$. The b_k coefficients need not be scaled down, but can benefit by scaling up to keep computation errors to a minimum. To keep the b_k coefficients within the 1.15 range, they can be scaled up by as much as a factor of 8, that is $B = 1/8$. After scaling, the coefficients become

$$b_S = [0.0872 \quad 0.3480 \quad 0.5224 \quad 0.3480 \quad 0.0872]$$
$$a_S = [0.2500 \quad -0.5465 \quad 0.5934 \quad -0.3325 \quad 0.0818]$$

The new coefficients are used to construct the equations:

$$w_S[n] = 0.25x[n] + 0.546525w[n-1] - 0.59335w[n-2]$$
$$+ 0.332525w[n-3] - 0.081825w[n-4]$$
$$w[n] = 4w_S[n]$$
$$y_S[n] = 0.0872w[n] + 0.3480w[n-1] + 0.5224w[n-2]$$
$$+ 0.3480w[n-3] + 0.0872w[n-4]$$
$$y[n] = \frac{y_S[n]}{8}$$

Notice that, although the inputs $x[n]$ do not themselves require scaling, they are, in the process of scaling, divided by the same scaling factor as the a_k coefficients. If this scaling reduces the size of the inputs unduly, the scaling factor C can be used to keep the inputs at a reasonable level during the calculations.

In practice, much of this scaling can be avoided by expressing the fourth order Chebyshev filter as a cascade of two second order Chebyshev filters. Using methods that are outside the scope of this text, the transfer function $H(z)$ in Equation (13.1) can be decomposed into two (rounded) second order factors:

$$H_{21}(z) = \frac{0.0491 + 0.1186z^{-1} + 0.0754z^{-2}}{1 - 1.1908z^{-1} + 0.4339z^{-2}} \tag{13.2a}$$

and

$$H_{22}(z) = \frac{0.2221 + 0.3495z^{-1} + 0.1446z^{-2}}{1 - 0.9953z^{-1} + 0.7543z^{-2}} \tag{13.2b}$$

As discussed in Section 4.6.2.2, the fourth order IIR filter can be implemented by cascading together two second order blocks with transfer functions as described in Equation (13.2). In this implementation, the b_k coefficients are no longer in great need of scaling, and the a_k coefficients need be scaled down only by a factor 2.

Because the ADSP-2181 is a fixed point processor, scaling is often needed for filter implementations. However, the programmer must be cognizant of the loss of precision it represents. Arithmetic right shifts, for example, frequently used to divide by 2, can cause a loss of precision. For instance, the number 0.1111_2, in 1.4 format, is equivalent to the decimal number 0.9375. An arithmetic right shift produces the 1.4 number 0.0111_2, equivalent to the decimal value 0.4375. Unfortunately, this result is not equal to half of the original number due to the finite word length of the processor. Furthermore, arithmetic left shifts, used to multiply by 2, can cause a register to overflow. When this happens, the register takes its maximum value, which is not an accurate representation of the shifted result. Finite word length and overflow errors accumulate as numbers are manipulated in a DSP program. Moreover, these errors add to the errors of aliasing and quantization already committed during A/D conversion. To minimize signal degradation, all unnecessary computations must be eliminated.

Coefficients may not be supplied to the ADSP-2181 in ordinary decimal notation. They must instead be supplied as 16-bit binary numbers in 1.15 notation. For convenience, these 16-bit binary values are presented as 4-digit hexadecimal values. As Example 13.3 showed, for instance, the decimal value 0.9375 has the 1.15 format 0.111 1000 0000 0000_2, which converts to 0x7800. It is hexadecimal numbers like this one that are transferred to the ADSP-2181. A list of 4-digit hex coefficients may be supplied as an initializing list within an assembly language program or may be read from a list in a file. When values are stored in program memory, which is 24 bits wide, only the most significant 16 bits are extracted as data. To be sure these values can be accessed correctly, initializing hexadecimal values for program memory contain 6 digits: the top 4 express the data value; the bottom 2 may be set to zero. For example, the coefficient 0.75_{10} is supplied to data memory as the hexadecimal number 6000, and to program memory as the hexadecimal number 600000.

13.4 REGISTERS

The ADSP-2181 has many registers. They may contain data, addresses, or status information. Some of the more frequently used registers are listed in Table 13.2 with their descriptions and lengths. Registers cannot be matched freely to the assembly language instructions described in the next section. The ADSP-2100 Family User's Manual (Analog Devices, 1995b) describes each instruction and its legal operands. For example, the instruction that adds two numbers has the form

$$\text{sum} = \text{xop} + \text{yop};$$

where the sum register must be register ar or register af; the xop operand must be one of the registers ax0, ax1, ar, mr2, mr1, mr0, sr1, or sr0; and the yop operand must be one of the registers ay0, ay1, or af. Other instructions have similar restrictions.

TABLE 13.2
Registers of the ADSP-2181

Register	Type	Length
ax0, ax1	ALU X input	16 bits
ay0, ay1	ALU Y input	16 bits
ar	ALU result	16 bits
af	ALU feedback	16 bits
mx0, mx1	Multiplier input	16 bits
my0, my1	Multiplier input	16 bits
mr0 + mr1 + mr2	Multiplier result	40 bits
si	Shifter input	16 bits
sr0 + sr1	Shifter result	32 bits
i0 . . . i7	DAG index registers	14 bits
m0 . . . m7	DAG modify registers	14 bits
l0 . . . l7	DAG length registers	14 bits

The data address generator, or DAG, registers are used for fast, convenient access to either data or program memory. For example, the command

$$\text{mx0} = \text{dm}(\text{i0,m0});$$

reads from data memory, dm. The address of the item to be read is contained in the index register i0. After the read is complete, the value of the modify register, m0 in this case, is added to the index register to update it to point to the next item to be read. If m0 = 1, for example, then the index register is incremented by 1 following each read. When index register i0 is used, length register l0 determines the style of access. For linear memory access, l0 is set to zero. For circular buffer memory access, l0 is set to the length of the buffer. When this style of addressing is used, items are read from memory, in order, until the end of the buffer is reached, at which point the index register is reset to the beginning again. Reading from program memory follows the same formula, as does writing to both data and program memory. For example,

$$my0 = pm(i7,m4);$$

$$pm(i5,m5) = ar;$$

$$dm(i2,m1) = sr1;$$

are instructions for reading from and writing to program memory, and writing to data memory. Modify and index registers need not be matched: Index registers i0, i1, i2, and i3 may make use of only modify registers m0, m1, m2, and m3; while index registers i4, i5, i6, and i7 may use only modify registers m4, m5, m6, and m7. Within these groups, modify registers may be freely reused. Finally, index registers i0, i1, i2, and i3 can provide only data memory addresses.

13.5 ASSEMBLY LANGUAGE INSTRUCTIONS

13.5.1 Instructions for ADSP-2100 Family

With few exceptions, the assembly language instructions for the ADSP-2181 are used for all members of the ADSP-2100 family. In fact, some version of the basic instructions is common to nearly all DSPs. The instructions are divided into categories, according to function:

- ALU (arithmetic logic unit)
- MAC (multiplier/accumulator)
- Shifter
- Move
- Program flow control
- Multifunction
- Miscellaneous

The instructions for each category are listed in Tables 13.3 through 13.9. Details of usage may be found in the ADSP-2100 Family User's Manual (Analog Devices, 1995b).

TABLE 13.3
ALU Instructions

• Add	• XOR	• Not
• Add with Carry	• Test Bit	• Absolute Value
• Subtract X–Y	• Set Bit	• Increment
• Subtract X–Y with Borrow	• Clear Bit	• Decrement
• Subtract Y–X	• Toggle Bit	• Divide
• Subtract Y–X with Borrow	• Pass	• Generate ALU Status
• AND	• Clear	
• OR	• Negate	

TABLE 13.4
MAC Instructions

• Multiply	• Clear
• Multiply/Accumulate	• Transfer MR
• Multiply/Subtract	• Conditional MR Saturation

TABLE 13.5
Shifter Instructions

• Arithmetic Shift	• Block Exponent Adjust
• Logical Shift	• Arithmetic Shift Immediate
• Normalize	• Logical Shift Immediate
• Derive Exponent	

TABLE 13.6
Move Instructions

• Register Move	• Data Memory Write (Direct Address)
• Load Register Immediate	• Data Memory Write (Indirect Address)
• Data Memory Read (Direct Address)	• Program Memory Write (Indirect Address)
• Data Memory Read (Indirect Address)	• I/O Space Read
• Program Memory Read (Indirect Address)	• I/O Space Write

TABLE 13.7
Program Flow Control Instructions

• Jump	• Return from Subroutine
• Call	• Return from Interrupt Service Routine
• Jump on Flag In Pin	• Do Until
• Call on Flag In Pin	• Idle
• Modify Flag Out Pin	

TABLE 13.8
Multifunction Instructions

• Computation with Memory Read	• Data and Program Memory Read
• Computation with Register-to-Register Move	• ALU with Data and Program Memory Read
• Computation with Memory Write	• MAC with Data and Program Memory Read

TABLE 13.9
Miscellaneous Instructions

• Stack Control	• No Operation
• Mode Control	• Interrupt Enable
• Modify Address Register	• Interrupt Disable

13.5.2 Assembly Language Examples

This section provides illustrations of some of the instructions mentioned in Section 13.5.1. Details for the usage of all ADSP-2181 instructions appear in the ADSP-2100 Family User's Manual (Analog Devices, 1995b).

Figure 13.9 shows a schematic of the essential registers in the arithmetic logic unit (ALU) of the ADSP-2181, and Table 13.10 shows some examples of ALU instructions. The ALU is responsible primarily for adding and also for subtracting, though subtracting a number is handled by adding its two's complement. ALU instructions are relatively straightforward, but there are restrictions on which registers may be combined. Table 13.10 shows just a few of the legal combinations.

The essential registers in the MAC (multiply/accumulate) unit of the ADSP-2181 are shown in Figure 13.10. MAC instructions, like the ALU instructions, have restrictions about the pairs of registers that may be combined. The lists of possibilities are contained in the ADSP-2100 Family User's Manual. A number of examples are provided in Table 13.11. Many of the instructions require a designation of (uu), (us), (su), (ss), or (rnd). The first four

FIGURE 13.9

Basic elements of ADSP-2181 ALU.

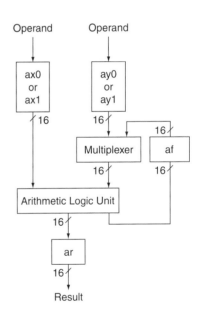

TABLE 13.10

Examples of ALU Instructions

Instruction	Explanation
ar = ax0 + ay0;	Adds two 16-bit values and places 16-bit result in ar.
ar = mr1 − af;	Subtracts one 16-bit result from another and places the 16-bit result in ar.
af = ax1 − ay1;	Subtracts one 16-bit result from another and places the 16-bit result in af.

FIGURE 13.10

Basic elements of ADSP-2181 MAC.

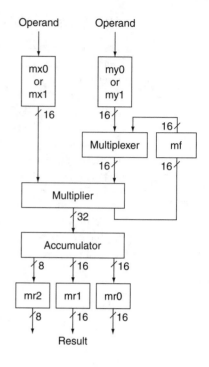

TABLE 13.11

Examples of MAC Instructions

Instruction	Explanation
mr = mx0 * my0 (ss);	Multiplies two signed 16-bit values and places the 40-bit result in mr.
mr = mx1 * my1 (rnd);	Multiplies two signed 16-bit values and rounds the 40-bit result to a 24-bit result in mr2 mr1.
mr = mr + mx1 * my1 (ss);	Multiplies two signed 16-bit values and adds the 40-bit result to a 40-bit value in mr.
mr = mr + sr1 * my0 (rnd);	Multiplies two signed 16-bit values, adds the 40-bit result to a 40-bit value in mr, and rounds to a 24-bit value in mr2 mr1.
mr = mr − ar * my0 (ss);	Multiplies two signed 16-bit values and subtracts the 40-bit result from a 40-bit value in mr.
mr = mr − mx0 * my0 (rnd);	Multiplies two signed 16-bit values, subtracts the 40-bit result from a 40-bit value in mr, and rounds to a 24-bit value in mr2 mr1.
if mv sat mr;	Saturates mr if the multiply overflow status bit (mv) is set. This bit is set when nonsign bits appear in mr2. When mr is positive, the saturated value is 00 7FFF FFFF; when mr is negative, the saturated value is FF 8000 0000.

TABLE 13.12

Examples of Shifter Instructions

Instruction	Explanation
sr = ashift mr1 by 2 (hi);	Copies 16-bit number in mr1 into sr1 and shifts to the left by 2 bits to give a 32-bit result in sr. Two zeros are shifted in at the right.
sr = ashift ar by −3 (lo);	Copies 16-bit number in ar into sr0 and shifts to the right by 3 bits to give a 32-bit result in sr. Three extra sign bits are shifted in at the left.
sr = lshift ar by −3 (lo);	Copies 16-bit number in ar into sr0 and shifts to the right by 3 bits to give a 32-bit result in sr. Three zeros are shifted in at the left.

of these modes refer to the types of numbers being multiplied, where u stands for an unsigned format and s stands for a signed format that assumes two's complement representations. The designation (ss), for example, is used when two signed numbers are multiplied, while (us) might be used when the most significant half of a double precision number is multiplied by the least significant half of another double precision number. The last designation, (rnd), is a rounding mode that rounds the 40-bit product of two signed numbers in mr2 mr1 mr0 to the most significant 24 bits in mr2 mr1.

Table 13.12 shows examples of shifter instructions. Most common is the arithmetic shift ashift, which can be used for scaling numbers. The shift distance may be between −128 and 127 bits. For every left shift, the size of a number doubles; for every right shift, the size of a number halves. The sr register is the 32-bit combination of the sr1 and sr0 registers that holds the result of the shift. The shift is designated (hi) if the 16-bit value to be shifted is placed in the most significant half of the shifter register sr1, or (lo) if it is placed in the least significant half of the shifter register sr0. Left shifts truncate bits from the most significant end of a binary number and add zeros at the least significant end. Right shifts truncate bits from the least significant end of a binary number and add sign bits at the most significant end, zero for positive numbers and one for negative numbers. The logical shift lshift differs from an arithmetic shift in only one way: On right shift, zeros are shifted in at the left instead of sign bits.

Moving data from a memory location to a register, from a register to a memory location, or from a register to another register is accomplished using an assignment-style instruction, as the examples in Table 13.13 illustrate. Again, these examples do not provide an exhaustive collection of all legal register combinations.

Loops, jumps, subroutine calls, and conditional statements are provided by program flow control instructions. Some examples of these instructions are presented in Table 13.14.

Multifunction instructions are made possible by the parallel architecture of the ADSP-2181, particularly the multiple buses and on-chip memories. A "read first, write last" rule applies to all multifunction instructions. This means that the values needed for the computation are collected at the beginning of an instruction cycle and the results are written at the end of the cycle. Table 13.15 gives several examples of multifunction instructions. Parts of the instructions are separated by commas.

At the beginning of each assembly language program, variables used in the program must be declared and perhaps initialized. Table 13.16 illustrates several declarations and initializations. The items in the table are called assembler directives rather than program instructions.

TABLE 13.13
Examples of Move Instructions

Instruction	Explanation
ar = my1;	Assigns the 16-bit value in my1 to ar.
i0 = ^rx_buf;	Stores a pointer to an array "rx_buf" in the index register i0.
l0 = %rx_buf;	Stores the length of an array "rx_buf" in the length register l0.
ax0 = dm(rx_buf);	Copies the 16-bit first element of an array "rx_buf" into ax0.
mr1 = dm(i2,m3);	Copies the 16-bit value in the data memory location pointed to by index register i2 into mr1 and postmodifies the pointer i2 by adding the value in the modify register m3 to it.
sr1 = pm(i5,m4);	Copies the 16-bit value in the program memory location pointed to by the index register i5 into sr1 and postmodifies the pointer i5 by adding the value in the modify register m4 to it.
dm(tx_buf) = ax0;	Copies the 16-bit value in ax0 to the first element of the array "tx_buf" in data memory.
dm(i0, m0) = ax0;	Copies the 16-bit value in ax0 to the data memory location pointed to by the index register i0 and postmodifies the pointer i0 by adding the value in the modify register m0 to it.
pm(i7, m7) = mr0;	Copies the 16-bit value in mr0 to the program memory location pointed to by the index register i7 and postmodifies the pointer i7 by adding the value in the modify register m7 to it.

TABLE 13.14
Examples of Program Flow Control Instructions

Instruction	Explanation
if not ce jump top_of_loop;	Jump back to the top of the loop unless the counter expired status bit (ce) is set.
if lt ar = −ar;	If the result of the last ALU operation is less than zero, the negative of ar replaces ar.
if av call reduce_coeffs;	If the ALU overflow status bit (av) is set, execute the instructions in the subroutine "reduce_coeffs," then return to point of call.
if ne ar = −ax0;	If the result of the last ALU operation is not zero, the negative of ax0 is copied into ar.

cntr = 40;
i0 = ^data;
m0 = 1;
l0 = 125;
do loop1 until ce;
loop1: dm(i0, m0) = 0
	The instructions within the "do" loop execute until the count (recorded in the register cntr) expires. In this case, the first 40 elements of a circular array containing 125 elements are set to zero. The index register i0 is initially set to point to the start of the array "data" using the ^ operator. Each time a write to data memory occurs in the loop, the index register is postmodified by the modify register m0 to point to the next array element.

TABLE 13.15
Examples of Multifunction Instructions

Instruction	Explanation
ar = ax0 + ay0, ax0 = pm(i5, m6);	An addition and a read from program memory occur at the same time. Register ax0 provides a value for the add operation at the start of the cycle and is written with a new value at the end of the cycle.
ar = ay0 − ax0, ay0 = dm(i1, m1);	A subtraction and a read from data memory occur at the same time. Register ay0 provides a value for the subtract operation at the start of the cycle and is written with a new value at the end of the cycle.
ar = ax0 + ay0, ax0 = mr1;	An addition and a read from a register occur at the same time. Register ax0 provides a value for the add operation at the start of the cycle, and is written with a new value at the end of the cycle.
dm(i0, m0) = ar, ar = ax0 + ay0;	A write to data memory and an addition occur at the same time. Register ar is written to data memory at the start of the cycle and acquires a new value at the end of the cycle.
mr = mr + mx0 * my0 (ss), mx0 = dm(i0, m0), my0 = pm(i5, m5);	A multiply-accumulate operation occurs at the same time as reads from data and program memory. The values of registers mx0 and my0 at the start of the cycle are multiplied together and added to mr. At the end of the cycle, a value from data memory is written to mx0 and a value from program memory is written to my0.

TABLE 13.16
Examples of Declarations and Initializations

Instruction	Explanation
.const TSCALE = 0x3FFB;	Declares a constant TSCALE and initializes it with the hexadecimal value 3FFB.
.const A = 40; .var/pm x[A+2];	Declares a constant A and initializes it with the value 40. Also declares an array "x" of length $40 + 2 = 42$ in program memory.
.var/dm g_ind; .var/dm/circ tx_buf[3]; .init tx_buf: H#C000, H#0000, H#0000;	Declares a variable "g_ind" in data memory. Declares a circular buffer of length 3 in data memory, "tx_buf," and initializes the array with the hexadecimal values C0000, 0000, and 0000.
.const TAPS = 125; .var/pm/circ a_coeffs[TAPS]; .init a_coeffs: <alist.dat>;	Declares a constant TAPS and initializes it with the value 125. Declares a circular buffer of length 125 in program memory, "a_coeffs," and initializes the array with values from the file "alist.dat." "Alist.dat" resides in program memory, so the top 16 bits of each "alist.dat" value contain coefficient values, while the bottom 8 bits are set to zero.

```
        .const          N = 8;
        .var/dm/circ    x_data[N];              /* declares a circular buffer of length 8    */
                                                /* in data memory                            */

        i0 = ^x_data;                           /* pointer to buffer                         */
        l0 = %x_data;                           /* length of buffer                          */
        m0 = 1;
        ar = 0;                                 /* set sum register to zero                  */
        ay0 = dm(i0,m0);                        /* copy first value from buffer into ax0     */

        cntr = N–1;                             /* initialize counter to loop 7 times        */

        do sum until ce;
sum:    ar = ar + ay0, ay0 = dm(i0, m0);        /* for each loop, add to sum,                */
                                                /* read new element from buffer              */

        ar = ar + ay0;                          /* at conclusion of loop, add final value    */
                                                /* from buffer to sum                        */

        sr = ashift ar by –3 (hi);              /* divide resulting sum by 8                 */
```

FIGURE 13.11
Example of simple filter.

The instructions to implement a simple eight-element moving average filter are given in Figure 13.11. The difference equation for the filter is

$$y[n] = \frac{1}{8}(x[n] + x[n-1] + x[n-2] + x[n-3]$$
$$+ x[n-4] + x[n-5] + x[n-6] + x[n-7])$$

The program declares a constant N with the value 8 and also a circular buffer of length N in data memory, called "x_data." The index register i0 points to the array, and the length register l0 is set to the length of the circular buffer through the use of the % operator. Prior to the start of the loop, register ar, which will contain the cumulative sum, is given the initial value zero. Also, the first element of the "x_data" array is written to ax0. After the write occurs, the index register i0 is postmodified by the value of m0 to point to the next element of "x_data." Finally, the cntr register is set to 7. This value determines the number of times the loop will execute. Within the loop, a multifunction instruction adds the value of ax0 to the value in register ar and reads a new value from data memory into register ax0. Because the modify register m0 has the value 1, the array pointer is updated to point at the next element each time through the loop. By the time the counter has expired, ar contains the sum of the first seven elements of "x_data," and ax0 contains the final element of the array. The first instruction after the loop terminates adds this last element to the sum. The last instruction in the program shifts the sum to the right by 3 bits, which has the effect of dividing it by 8. The 16-bit result of filtering is held in register sr1, the most significant 16 bits of 32-bit register sr, as a result of the (hi) designation. Note that a single execution of the instructions in Figure 13.11 produces a single filter output. Once a new input value is slotted into the array "x_data," a new output can be found by executing the program segment again.

13.6 SETUP AND INITIALIZATION OF THE EZ-KIT LITE BOARD

To implement a DSP filter or any other algorithm on the EZ-KIT Lite requires a DSP program, or set of instructions. Unfortunately, the program must contain a great deal more than just the instructions related to the algorithm. To behave correctly, all the hardware on the EZ-KIT Lite board must be initialized properly through software. This includes the codec, which performs A/D and D/A conversions; the serial ports, which handle communication between the ADSP-2181 and the codec; and the interrupts, which signal important events such as data received or data ready to transmit. Fortunately, a template assembly language program "mic2out.dsp" is provided with the EZ-KIT Lite that includes all the required steps. Apart from the necessary initializations, this program does nothing useful at all; it simply passes input from the microphone straight to the speaker output. For this reason it is called a "talk-through" program. This template program can be used as the basis for other programs.

The assembly language listing for the talk-through program, including all set-up and initialization steps for the EZ-KIT Lite board, is shown Figure 13.12. It is informative to get a general sense of the six important elements in the program:

- Description of constants
- Declarations and initializations
- Interrupt vector table
- Initialization of ADSP-2181 DSP
- Initialization of ADSP1847 codec
- Interrupt service routines

The constants used in the program consist of names assigned to registers that control the behavior of the DSP's serial ports and timer. These constants are defined when the program is assembled. The list of constants is followed by the declarations and initializations of a receive buffer, a transmit buffer, and a set of initialization codes.

The interrupt vector table, the second element of the program, provides addresses for the routines to be executed when the DSP is reset or receives data. On reset, the DSP jumps to the "start" interrupt service routine. Every time an input data sample is received, the DSP executes the "input_samples" interrupt service routine. The "start" routine begins with an initialization of the ADSP-2181. This initialization sets pointers to the receive buffer, transmit buffer, and initialization commands, and sets up the serial ports of the DSP. Initialization of the ADSP-1847 codec, which performs A/D and D/A conversion on the EZ-KIT Lite board, follows. This step relies on the input service routine "next_cmd," whose sole purpose is to send initialization codes to the codec. Codes are sent to set the gain on each input channel, to choose a conversion rate, and to enable receive and transmit autobuffering through the buffers "rx_buf" and "tx_buf." The "start" interrupt service routine finishes

```
/*********************************************************************************
 * EZ-KIT LITE TALK-THROUGH DEMO
 * This demo program performs a talk-through, loopback function. The DSP reads the stereo
 * output of the AD1847 and immediately writes the stereo data back to the AD1847.
 *
 * This sample program is organized into the following sections:
 *   Definition of assemble-time constants
 *   Interrupt vector table
 *   ADSP 2181 initialization
 *   ADSP 1847 codec initialization
 *   Interrupt service routines
 *********************************************************************************/

.module/RAM/ABS=0 loopback;


/*================== ASSEMBLE TIME CONSTANTS ==========================*/

.const IDMA=                    0x3fe0;
.const BDMA_BIAD=               0x3fe1;
.const BDMA_BEAD=               0x3fe2;
.const BDMA_BDMA_Ctrl=          0x3fe3;
.const BDMA_BWCOUNT=            0x3fe4;
.const PFDATA=                  0x3fe5;
.const PFTYPE=                  0x3fe6;
.const SPORT1_Autobuf=          0x3fef;
.const SPORT1_RFSDIV=           0x3ff0;
.const SPORT1_SCLKDIV=          0x3ff1;
.const SPORT1_Control_Reg=      0x3ff2;
.const SPORT0_Autobuf=          0x3ff3;
.const SPORT0_RFSDIV=           0x3ff4;
.const SPORT0_SCLKDIV=          0x3ff5;
.const SPORT0_Control_Reg=      0x3ff6;
.const SPORT0_TX_Channels0=     0x3ff7;
.const SPORT0_TX_Channels1=     0x3ff8;
.const SPORT0_RX_Channels0=     0x3ff9;
.const SPORT0_RX_Channels1=     0x3ffa;
.const TSCALE=                  0x3ffb;
.const TCOUNT=                  0x3ffc;
.const TPERIOD=                 0x3ffd;
.const DM_Wait_Reg=             0x3ffe;
.const System_Control_Reg=      0x3fff;
```

FIGURE 13.12

Listing of talk-through program.

```
/*=============== DATA BUFFER DECLARATIONS ============================*/

.var/dm/ram/circ    rx_buf[3];          /* declare receive buffer, Status + L data + R data   */
.var/dm/ram/circ    tx_buf[3];          /* declare transmit buffer, Cmd + L data + R data     */
.var/dm/ram/circ    init_cmds[13];      /* declare buffer with AD1847 initialization codes    */
.var/dm             stat_flag;          /* declare a variable to be used as a status flag      */


/*================ DATA BUFFER INITIALIZATIONS ==========================*/


/* The AD1847 CODEC communicates with the DSP serially. The AD1847 transmits and receives three 16
 * bit words per sample. In this program we declare two, three-location circular buffers, tx_buf for the transmit
 * data buffer and rx_buf for the receive data buffer. The AD1847 receives from the DSP's tx_buf, a
 * command word, left channel data, and right channel data each sample. The AD1847 transmits to the DSP
 * rx_buf a status word, left channel data and right channel data each sample. The DSP's serial port
 * communicating with the AD1847 is configured to use autobuffering and multi-channel mode. This
 * allows the DSP's serial port to automatically access memory and automatically transmit/receive three
 * words at a time. Once the DSP's serial port is configured and running, it will run forever interrupting the
 * DSP when new data has been received. This means the DSP reads the AD1847's data by reading the
 * rx_buf buffer and writes to the AD1847 by writing to the tx_buf buffer. The DSP knows when it can
 * read/write the AD1847 via the serial port interrupts.                                      */

.init tx_buf:        0xc000, 0x0000, 0x0000;     /* Initialize the tx_buf with these values.
                                                  * Command word = 0xc0000
                                                  * Left channel data = 0x0000
                                                  * Right channel data = 0x0000.
                                                  * This command word will place AD1847 in command
                                                  * mode by setting the MCE bit             */

/* Initialize the init_cmds buffer with the following values. These are the initial configuration values for
 * the AD1847 CODEC. Bits 0-7 are the 8 bits of data for the index registers ( data0-7 ). Bits 8-11 are the
 * address for the index registers ( IA0-3 ). Bit 13 sets the part into readback mode ( RREQ ). Bit 14 is the
 * Mode Change Enable bit ( MCE ). This must be set when changing the index registers. Bit 15 is the Clear
 * Overrange bit ( CLOR ). This bit determines the over-range status mode. The program is set up to automatically
 * load this buffer into the AD1847.                                                          */

.init init_cmds:        0xc002,              /* CLOR set, MCE set, Index reg address=0, data=0x2
                                             * Left input control reg
                                             * b7-6: 0=left line 1
                                             *   1=left aux 1
                                             *   2=left line 2
                                             *   3=left line 1 post-mixed loopback
                                             * b5-4: res
                                             * b3-0: left input gain x 1.5 dB                */
```

FIGURE 13.12
Continued

539

```
0xc102,         /* CLOR set, MCE set, Index reg address=1, data=0x2
                * Right input control reg
                * b7-6: 0=right line 1
                *       1=right aux 1
                *       2=right line 2
                *       3=right line 1 post-mixed loopback
                * b5-4: res
                * b3-0: right input gain x 1.5 db                              */

0xc288,         /* CLOR set, MCE set, Index reg address=2, data=0x88
                * Left aux 1 control reg
                * b7 : 1=left aux 1 mute
                * b6-5: res
                * b4-0: gain/atten x 1.5, 08= 0dB, 00= 12dB                    */

0xc388,         /* CLOR set, MCE set, Index reg address=3, data=0x88
                * Right aux 1 control reg
                * b7 : 1=right aux 1 mute
                * b6-5: res
                * b4-0: gain/atten x 1.5, 08= 0dB, 00= 12dB                    */

0xc488,         /* CLOR set, MCE set, Index reg address=4, data=0x88
                * left aux 2 control reg
                * b7 : 1=left aux 2 mute
                * b6-5: res
                * b4-0: gain/atten x 1.5, 08= 0dB, 00= 12dB                    */

0xc588,         /* CLOR set, MCE set, Index reg address=5, data=0x88
                * right aux 2 control reg
                * b7 : 1=right aux 2 mute
                * b6-5: res
                * b4-0: gain/atten x 1.5, 08= 0dB, 00= 12dB                    */

0xc680,         /* CLOR set, MCE set, Index reg address=6, data=0x80
                * left DAC control reg
                * b7 : 1=left DAC mute
                * b6 : res
                * b5-0: attenuation x 1.5 dB                                   */

0xc780,         /* CLOR set, MCE set, Index reg address=7, data=0x80
                * right DAC control reg
                * b7 : 1 = right DAC mute
                * b6 : res
                * b5-0: attenuation x 1.5 dB                                   */
```

FIGURE 13.12

Continued

```
0xc850,         /* CLOR set, MCE set, Index reg address=8, data=0x5c
                 * data format register
                 * b7 : res
                 * b5-6: 0 = 8-bit unsigned linear PCM
                 *    1 = 8-bit u-law companded
                 *    2 = 16-bit signed linear PCM
                 *    3 = 8-bit A-law companded
                 * b4 : 0 = mono, 1 = stereo
                 * b0-3: 0 =  8.00000 Khz
                 *    1 =  5.51250 Khz
                 *    2 = 16.00000 Khz
                 *    3 = 11.02500 Khz
                 *    4 = 27.42857 Khz
                 *    5 = 18.90000 Khz
                 *    6 = 32.00000 Khz
                 *    7 = 22.05000 Khz
                 *    8 = .
                 *    9 = 37.80000 Khz
                 *    a = .
                 *    b = 44.10000 Khz
                 *    c = 48.00000 Khz
                 *    d = 33.07500 Khz
                 *    e =  9.60000 Khz
                 *    f =  6.61500 Khz
                 * (b0) : 0 = XTAL1 24.576 Mhz; 1 = XTAL2 16.9344 Mhz          */


0xc909,         /* CLOR set, MCE set, Index reg address=9, data=0x09
                 * interface configuration reg
                 * b7-4: res
                 * b3 : 1 = autocalibrate
                 * b2-1: res
                 * b0 : 1 = playback enabled                                   */


0xca00,         /* CLOR set, MCE set, Index reg address=0xa, data=0
                 * pin control reg
                 * b7 : logic state of pin XCTL1
                 * b6 : logic state of pin XCTL0
                 * b5 : master - 1 = tri-state CLKOUT
                 *     slave - x = tri-state CLKOUT
                 * b4-0: res                                                   */
```

FIGURE 13.12

Continued

```
          0xcc40,              /* CLOR set, MCE set, Index reg address=0xc, data=0x40
                               * miscellaneous information reg
                               * b7 : 1 = 16 slots per frame
                               *     0 = 32 slots per frame
                               * b6 : 1 = 2-wire system
                               *     0 = 1-wire system
                               * b5-0: res                                              */


          0xcd00;              /* CLOR set, MCE set, Index reg address=0xd, data=0
                               * digital mix control reg
                               * b7-2: attenuation x 1.5 dB
                               * b1 : res
                               * b0 : 1 = digital mix enabled                           */
/*================= INTERRUPT VECTOR TABLE ===============================

/* Each interrupt vector has 4 PM memory locations. When an interrupt occurs the
 * DSP will jump to the location of the interrupt vector, providing the interrupt
 * is unmasked.                                                                        */

          jump start;          /* address = 0x00: reset interrupt vector                */
          rti;
          rti;
          rti;


          rti;                 /* address = 0x04: IRQ2 interrupt vector                 */
          rti;
          rti;
          rti;


          rti;                 /* address = 0x08: IRQL1 interrupt vector                */
          rti;
          rti;
          rti;


          rti;                 /* address = 0x0c: IRQL0 interrupt vector                */
          rti;
          rti;
          rti;
```

FIGURE 13.12

Continued

```
ar = dm(stat_flag);  /* address = 0x10: SPORT0 tx interrupt vector          */
ar = pass ar;
if eq rti;
jump next_cmd;

jump input_samples; /* address = 0x14: SPORT1 rx interrupt vector           */
rti;
rti;
rti;

rti;                 /* address = 0x18: IRQE interrupt vector               */
rti;
rti;
rti;

rti;                 /* address = 0x1c: BDMA interrupt vector               */
rti;
rti;
rti;

rti;                 /* address = 0x20: SPORT1 tx or IRQ1 interrupt vector  */
rti;
rti;
rti;

rti;                 /* address = 0x24: SPORT1 rx or IRQ0 interrupt vector  */
rti;
rti;
rti;

rti;                 /* address = 0x28: timer interrupt vector              */
rti;
rti;
rti;

rti;                 /* address = 0x2c: power down interrupt vector         */
rti;
rti;
rti;
```

FIGURE 13.12

Continued

```
/*==================== ADSP 2181 INITIALIZATION =========================*/

/*---------------- Data Address Generator Initialization -------------------------------------*/

start:
        i0 = ^rx_buf;           /* set i0 = the starting address of rx_buf        */
        l0 = %rx_buf;           /* set l0 = the length of rx_buf                  */
        i1 = ^tx_buf;           /* set i1 = the starting address of tx_buf        */
        l1 = %tx_buf;           /* set l1 = the length of tx_buf                  */
        i3 = ^init_cmds;        /* set i3 = the starting address of init_cmds     */
        l3 = %init_cmds;        /* set l3 = the length of init_cmds               */
        l5 = 0;                 /* set l5=0 because fb_gain is not a circular buffer */
        m1 = 1;                 /* modify register m1 can be used with i0, i1, i2 or i3 */

/*---------------------- S E R I A L   P O R T   #0   S T U F F ------------------------------------- */

/* This code configures the serial ports on the DSP. The serial ports are configured by writing
 * to memory mapped control registers. SPORT0 is configured to use autobuffering
 * and multi-channel mode with an external serial clock and external frame sync.
 * Autobuffering can be thought of as serial port DMA. The serial port uses the address
 * generators to access data memory automatically. In this program autobuffering is configured
 * to use i0, i1, and m1. These are data address generator registers. i0 points to the rx_buf buffer,
 * i1 points to the tx_buf buffer and m1 = 1.
 * Multi-channel mode is enabled and configured for 32 channels. The AD1847 is also configured for 32
 * channels. In multi-channel mode each channel is a word and associated with a time slot. In each of
 * the 32 time slots the serial port can transmit and/or receive data or do nothing, tristating the
 * data for that time slot. One frame consists of all 32 channels. One frame sync will start a frame
 * transfer ( all 32 words ). This program configures the serial port 0 to transmit on channels
 * 0, 1, 2, 16, 17 and 18 for transmit, and channels 0, 1, 2, 16, 17, and 18 for receive. It
 * will ignore all other time slots.
 * External serial clock and external frame-syncs mean they are inputs to the DSP.        */

        ax0 =0x287;
        dm (SPORT0_Autobuf) = ax0;      /* enable receive and transmit autobuffering,
                                         * RMREG=m1, RIREG=i0, TMREG=m1, TIREG=i1,
                                         * CLKOUT enabled, BIASRND disabled         */

        ax0 = 0;
        dm (SPORT0_RFSDIV) = ax0;       /* RFSDIV = SCLK Hz/RFS Hz - 1
                                         * using external  receive frame sync ( RFS=input )  */

        ax0 = 0;
        dm (SPORT0_SCLKDIV) = ax0;      /* SCLK = CLKOUT / (2 (SCLKDIV + 1)
                                         * using external serial clock ( SCLK=input )  */
```

FIGURE 13.12

Continued

544

```
                ax0 =0x860f;
                dm (SPORT0_Control_Reg) = ax0;      /* SLEN= 16 bits, right justify, zero-fill unused MSBs,
                                                     * INVRFS=0, INVTFS=0, IRFS=0, ITFS=1,
                                                     * MFD=1, ISCLK=0, MCE=1                              */


                ax0 =0x7;
                dm (SPORT0_TX_Channels0) = ax0;     /* enable channels 0, 1 and 2 for transmit           */

                ax0 =0x7;
                dm (SPORT0_TX_Channels1) = ax0;     /* enable channels 16, 17 and 18 for transmit        */

                ax0 =0x7;
                dm (SPORT0_RX_Channels0) = ax0;     /* enable channels 0, 1 and 2 for receive            */

                ax0 =0x7;
                dm (SPORT0_RX_Channels1) = ax0;     /* enable channels 16, 17 and 18 for receive         */


/*----------------- S Y S T E M  A N D  M E M O R Y  S T U F F ------------------------------------------ */

                ax0 = 0xfff;
                dm (DM_Wait_Reg) = ax0;             /* Set all IOWAIT ranges to 7 wait states,
                                                     * set DWAIT to 0 wait states                         */

                ax0 = 0x1000;                       /* PWAIT = 0, enable SPORT0                           */
                dm (System_Control_Reg) = ax0;

                ifc = 0xff;                         /* clear pending interrupt                            */
                nop;
                icntl = 0;                          /* external interrupts set to level sensitivity, disable
                                                     * nested interrupts                                 */
                mstat = 0x40;                       /* enable go mode                                     */
```

FIGURE 13.12

Continued

```
/*============== AD1847 CODEC INITIALIZATION=============================== */

                    ax0 = 1;
                    dm(stat_flag) = ax0;            /* initialize stat_flag to 1                           */
                    imask = b#0001000000;          /* unmask SPORT0s transmit interrupt                   */
                    ax0 = dm (i1, m1);             /* set ax0 = the first value of the tx_buf             */
                    tx0 = ax0;                     /* begin autobuffer transmit by writing first value to tx0 reg
                                                    * (when autobuffer transmit completes a transmit interrupt
                                                    * will be generated)                                   */

/* Test for entire init_cmds buffer to be sent to the codec                                               */

check_init:
                    ax0 = dm (stat_flag);          /* set ax0 = the value at stat_flag. Flag set in SPORT0 tx interrupt
                                                    * routine                                              */
                    af = pass ax0;                 /* pass the value of ax0 thru the ALU setting status flags  */
                    if ne jump check_init;         /* if result of pass not equal to 0 jump check_init    */

/* Once initialized, wait for codec to come out of autocalibration by testing the ACI bit
 * of the AD1847s status reg. This is done twice, first to ensure the codec is in autocalibration
 * mode, then to determine when autocalibration is complete.                                              */

                    ay0 = 2;
check_aci1:                                        /* loop to test ACI bit = 1 ( in autocalibration )     */
                    ax0 = dm (rx_buf);             /* read status word of AD1847                           */
                    ar = ax0 and ay0;              /* AND 2 with status word                               */
                    if eq jump check_aci1;         /* if result of AND is not = 0 leave loop               */

check_aci2:                                        /* loop to test ACI bit =0 ( autocalibration complete ) */
                    ax0 = dm (rx_buf);             /* wait for bit clear                                   */
                    ar = ax0 and ay0;
                    if ne jump check_aci2;
                    idle;

/* Once autocalibration is complete unmute left and right DAC channels by writing the appropriate
 * index register                                                                                         */

                    ay0 = 0xbf3f;                  /* unmute left DAC                                      */
                    ax0 = dm (init_cmds + 6);
                    ar = ax0 AND ay0;
                    dm (tx_buf) = ar;
                    idle;                          /* wait for transmit to complete                       */
                                                   /* unmute right DAC                                    */
                    ax0 = dm (init_cmds + 7);
                    ar = ax0 AND ay0;
```

FIGURE 13.12

Continued

546

```
                dm (tx_buf) = ar;
                idle;                        /* wait for transmit to complete              */
                ifc = 0xff;                  /* clear any pending interrupt                */
                nop;
                imask = 0x30;                /* unmask rx0 and IRQE interrupts             */


  /* Wait in idle mode for an interrupt. When the program returns from the interrupt jump to
   * idle mode.                                                                             */

talkthru:  idle;
                jump talkthru;

/*================= INTERRUPT SERVICE ROUTINES =========================*/

/*----------------------- Receive Interrupt Service Routine -----------------------------------------------*/

/* NOTE: Serial port data is being accessed using autobuffering. When using autobuffering the
 * DSP uses the data address generators to automatically transfer serial port received data to
 * data memory and transfer transmit data from data memory to the serial port. The serial
 * port autobuffering is configured to access received data in the rx_buf buffer and transmit
 * data in the tx_buf buffer.                                                                */

input_samples:
                ena sec_reg;                 /* use shadow register bank                   */
                ax0 = dm(rx_buf + 1);        /* get left channel data                      */
                ay0 = dm(rx_buf + 2);        /* get right channel data                     */
                                             /* your processing here                       */
                dm (tx_buf + 1) = ax0;       /* send left channel data                     */
                dm (tx_buf + 2) = ay0;       /* send right channel data                    */
                rti;                         /* Return from interrupt, will pop the status register which consists
                                              * MSTAT, ASTAT and IMASK ( returns to primary registers because of
                                              * pop of MSTAT                               */

/*-------------------- Transmit Interrupt Service Routine-------------------------------------------------*/

/* The transmit interrupt service routine is used to initializes the codec (AD1847).
 * i3 points to the init_cmds buffer. The AD1847 is expecting a control word and two data words from the
 * DSP. Here we load the next command word into the first location of the tx_buf buffer, the two data words
 * have been initialized to zero in the beginning of the program. This routine also checks to see if the last data
 * word has been transmitted. If the last data word has been transmitted take the AD1847 out of command
 * mode and return from interrupt.                                                          */
```

FIGURE 13.12

Continued

```
next_cmd:
                    ena sec_reg;
                    ax0 = dm (i3, m1);        /* get command word from init_cmds buffer            */
                    dm (tx_buf) = ax0;        /* place command word in first location of tx_buf
                                              * ( transmit slot 0)                                 */

                    ax0 = i3;
                    ay0 = ^init_cmds;
                    ar = ax0 - ay0;           /* test for additional command words                 */
                    if gt rti;                /* rti if more control words still waiting            */
                    ax0 = 0xaf00;             /* else set done flag and remove MCE if done initialization */
                    dm (tx_buf) = ax0;
                    ax0 = 0;
                    dm (stat_flag) = ax0;     /* reset status flag                                 */
                    rti;

.endmod;
```

FIGURE 13.12   © Analog Devices, Inc.
Continued

with an endless loop during which the DSP waits for an interrupt to signal that a new input sample has been received.

The receive interrupt service routine, at "input_samples," is the core of the program. Through serial port autobuffering, microphone input data sampled by the A/D converter appear in the data memory locations "rx_buf + 1" and "rx_buf + 2." In the receive interrupt service routine, the contents of these locations are read into registers ax0 and ay0. For this talk-through program, these values are copied directly to the data memory locations "tx_buf + 1" and "tx_buf + 2" and then, through autobuffering, transmitted via a serial port to the D/A converter and finally the speakers. In other words, inputs are copied directly to outputs, so no actual digital signal processing occurs. Filtering or other modification of incoming data would occur where the comment "your processing here" appears in bold in the "input_samples" routine.

13.7 RUNNING PROGRAMS ON THE EZ-KIT LITE

As with other processors, the ADSP-2181 requires that instructions be assembled to obtain an object file in machine code and then linked with other object files to create an executable program. These steps are performed using an assembler and a linker that run on a host PC. To produce an object file, the assembler requires a description of the hardware it will target. This description is provided to the assembler in an architecture file, called "adsp2181.ach." This file, among other things, describes the range of memory locations that are assigned as program memory and those that are assigned as data memory. The assembler is invoked with the filename of the assembly language program "user_prog.dsp" and an indication of the processor to be used, as asm21 user_prog − 2181

The linker links the object files produced by the assembler with the command

1d21 user_prog −a adsp2181 −e demo_prog

which produces an executable file, named after the −e flag. Once again, the hardware architecture must be specified. In this case the executable program is named "demo_prog.exe."

Executable programs are downloaded from the PC where they are created to the ADSP-2181 using EZ-KIT Lite host software, included with the kit. Program operation can then be tested by providing input to the microphone and listening to the results from the speakers. Replacing the microphone with a function generator and the speakers with an FFT-capable oscilloscope makes more detailed tests possible. The function generator sets the exact frequency content of the input, and the oscilloscope reports the exact frequency content of the output.

When the program does not work as desired, the most effective way to begin debugging is to follow the program execution using the simulator. This simulator runs on the PC and displays the contents of the registers and the memory locations at each step as the program executes. The simulator is invoked as

sim2181 −a adsp2181 −e demo_prog

In order for the simulator to operate correctly, the PC running the simulator must be booted in MS-DOS mode, as the simulator will not run under Windows. Furthermore, a flag must be added to the link step prior to running the simulator. That is,

1d21 user_prog −a adsp2181 −e demo_prog −g

The −g flag generates a table of program symbols that allows the simulator to reference variables and labels from the program by name. In the simulator, the program can be executed one instruction at a time, and memory and register values can be both examined and modified at will. Even inputs to the serial ports can be simulated, by placing their values in a text file. Serial outputs can be written to a separate text file.

13.8 DSP APPLICATIONS

13.8.1 Sine Wave Generator

A simple but important task for a DSP is to produce a sine wave of a specified frequency. Unlike in high level languages, sine functions are not available as assembly language instructions, so they must be approximated using numerical tricks. In Section 12.5, the power series expansion

$$\sin(x) = x - \frac{x^3}{3!} + \frac{x^5}{5!} - \frac{x^7}{7!} + \frac{x^9}{9!} - \cdots \qquad (13.3)$$

was given for a sine wave. The advantage of such an expansion is that it provides a means to calculate the sine function using only multiplication, addition, and subtraction. To produce an entire cycle of the sine wave, x must take values between $-\pi$ radians and π radians. To permit the use of the 1.15 fractional fixed number format, it is more convenient to rewrite the expansion using $x = \pi y$. Reasonable accuracy may be obtained by taking the first three terms only:

$$\sin(\pi y) \approx \pi y - \frac{(\pi y)^3}{3!} + \frac{(\pi y)^5}{5!} = 3.14159y - 5.16671y^3 + 2.55016y^5 \quad \textbf{(13.4)}$$

In this form, y ranges from -1 to nearly 1 and can be written in 1.15 format. The coefficients, however, must be scaled down by a factor of 8 before the 1.15 format can be used. After scaling, the coefficients have the values 0.39270, -0.64596, and 0.31877, or 0x3244, 0xAD51, and 0x28CD, in 1.15 format.

For additional economy of calculation, the symmetrical nature of the sine wave can be exploited. For example, an angle in the second quadrant can always be written as $\pi - x$, where x is an angle in the first quadrant, and it happens that $\sin(\pi-x)$ has exactly the same value as $\sin(x)$, according to a mathematical identity. An angle in the third quadrant can be written in terms of first quadrant angle x as $\pi + x$, and $\sin(\pi+x)$ equals $-\sin(x)$. Finally, an angle in the fourth quadrant can be written as $-x$, where x is again a first quadrant angle, and $\sin(-x)$ equals $-\sin(x)$. Thus, sines of angles in quadrants two, three, and four can all be calculated using sines for first quadrant angles. Since only the sine values for angles in the first quadrant need to be calculated, only x values between 0 and $\pi/2$ radians are needed in Equation (13.3), and y values in Equation (13.4) may be restricted to the range 0 to 0.5. Using these small values improves the accuracy of the approximation. Figure 13.13 shows the listing of a function "sin.dsp" that computes the sine value of an angle using these ideas. It uses the Boolean operator AND, which sets output bits to one where both inputs contain one bits, and to zero otherwise; and also the Boolean operator OR, which sets output bits to zero where both inputs contain zero bits, and to one otherwise. The pass instruction is a way to set processor flags according to the value of a particular register.

Calculation of sine values is only one part of a sine generation program. Figure 13.14 contains the highlights of a program "sinsnd.dsp" that generates a sine wave using the sine subroutine. Where lines have been skipped, the notation ". . . . ." has been used. The program begins by defining 16 angles. These numbers, in 1.15 format, correspond to decimal values between -1.0 and 0.875, which in turn represent angles between $-\pi$ and 0.875π radians.

Because the timer will be used to determine the rate at which sine values are presented, the interrupt vector table includes an address for the timer interrupt service routine that was not present in the talk-through program. The timer is initialized to count 4096 instruction cycles between interrupts. Since each interrupt produces a new sine value at the output, and because 16 values make up a full cycle, this count corresponds, for a 30.3 nsec instruction cycle time, to a sine wave period of 16(4096 instruction cycles)(30.3 nsec/instruction cycle) = 1.99 msec. Before enabling timer interrupts and starting the timer, sine values are computed for each angle in the table. When a timer interrupt is received, the interrupt service routine simply sends a single sine value in the table to the output via the

```
.module sin_val;
/*
                    Sine Approximation

                    On Input:           ax0 = angle/pi, in 1.15 format

                    On Output:          ar = sin(angle) in 1.15 format

*/

.var/dm             sin_coeff[3];
.init sin_coeff :   H#3244, H#AD51, H#28CD;  /* coefficients divided by eight, in 1.15 format          */

.entry  sin;

sin:                i3 = ^sin_coeff;
                    m3 = 1;
                    l3 = 0;
                    ay0 = H#4000;
                    ar = ax0,  af = ax0 AND ay0;          /* check bit 14 –                              */
                                                          /* if set, angle is quadrant two or four        */
                    if ne ar = -ax0;                      /* if quadrant two or four, take 2's complement  */
                                                          /* converts quadrant two angle to quadrant three */
                                                          /* converts quadrant four angle to quadrant one  */

                    ay0 = H#7fff;
                    ar = ar AND ay0;                      /* removes most significant bit                 */
                                                          /* no effect on quadrant one angles             */
                                                          /* moves quadrant three angles to quadrant one  */
                    my1 = ar;                                 /* my1 = ar = x, in quadrant one          */
                    mf = ar * my1 (rnd), mx1 = dm(i3, m3);    /* mf = x^2                               */
                    mr = mx1 * my1 (ss);                      /* mx1 = c1                               */
                    mf = ar * mf (rnd), mx1 = dm(i3, m3);     /* mf = x^3                               */
                    mr = mr + mx1 * mf (ss);                  /* mx1 = c2                               */
                    mf = ar * mf (rnd);
                    mf = ar * mf (rnd), mx1 = dm(i3, m3);     /* mf = x^5                               */
                    mr = mr + mx1 * mf (ss);                  /* mx1 = c3                               */
                    sr = ashift mr1 by 3 (hi);                /* shift left by 3 (multiply by 8) to     */
                    sr = sr OR lshift mr0 by 3 (lo);          /* undo effect of coefficient scaling     */
                    ar = pass sr1;
                    if lt ar = pass ay0;                  /* negative result indicates overflow, set ar to 7fff  */
                    af = pass ax0;
                    if lt ar = -ar;                       /* if original angle was in quadrant three or four  */
                                                          /* negate sine value                            */
                    rts;

.endmod;
```

FIGURE 13.13

Listing of sine subroutine.

```
             .....
.external       sin;                            /* declare sin subroutine                  */
.const          num_vals = 16;
.var/dm         angles[num_vals];

                                                /* initialize table of 16 angles           */
.init angles :  H#8000, H#9000, H#A000, H#B000,
                H#C000, H#D000, H#E000, H#F000,
                H#0000, H#1000, H#2000, H#3000,
                H#4000, H#5000, H#6000, H#7000;


.var/dm/circ    sin_vals[num_vals];


                .....
                                                /* interrupt vector table                  */
                jump sin_samp;                  /* 28: timer interrupt                     */
                rti;
                rti;
                rti;


                .....

                ax0 = H#2000;                   /* set up timer  to count to 0x2000 = 4096 */
                dm(TPERIOD) = ax0;              /* between interrupts                       */
                dm(TCOUNT) = ax0;
                ax0 = 0;
                dm(TSCALE) = ax0;


                .....

                i0 = ^angles;                   /* index register i0 points to buffer of angles       */
                m0 = 1;
                l0 = 0;
                i4 = ^sin_vals;                 /* index register i4 points to circular buffer of sines */
                m4 = 1;
                l4 = num_vals;

                cntr = num_vals;
                do fill_tbl until ce;
                    ax0 = dm(i0, m0);           /* read an angle into register ax0         */
                    call sin;                   /* calculate the sine of the angle         */
fill_tbl:           dm(i4, m4) = ar;            /* write the sine value to the table of sines */

                imask = b#0000000101;           /* enable timer interrupt and rx0 interrupt */
                mstat = b#1100000;              /* enable timer and go mode                 */
                .....
```

FIGURE 13.14

Listing of sine generation program highlights.

```
sin_samp:                                  /* timer interrupt handler                 */
            mx0 = dm(i4,m4);               /* read a sine value from the circular buffer  */
                                           /* post-modify i4 to point to next sine value  */
            dm (tx_buf + 1) = mx0;         /* send sine value to both speakers         */
            dm (tx_buf + 2) = mx0;
            rti;


            .....
```

FIGURE 13.14

Continued

transmit buffer "tx_buf." Postmodification of the index register i4 by modify register m4 ensures that the next sine value from the table is selected at the time of the next interrupt. The DSP continues to send out sine values from its table "sin_vals," one after another, in an infinite loop.

13.8.2 FIR Filter

FIR filters, defined by the equation

$$y[n] = b_0 x[n] + b_1 x[n-1] + b_2 x[n-2] + \cdots b_M x[n-M]$$

are one of the essential building blocks of DSP. Specialized hardware and multifunction instructions mean that the ADSP-2181 can perform FIR filtering tasks quickly and efficiently. Figure 13.15 highlights the differences between an FIR filtering program and the talk-through program of Figure 13.12. At the beginning of the program, circular buffers for both the filter coefficients b_k and the input data $x[k]$ are declared, in this case with 102 elements each. The coefficients are placed in program memory and the input data in data memory to permit the use of multifunction instructions in the main body of the program. The coefficients are initialized from a file of six-digit hexadecimal values in 1.15 format, "b.dat." To enable the FIR instructions to be as compact as possible, the coefficients are listed in reverse order in the coefficient file, from b_M to b_0, and the index register i4 is initialized to point to b_M. The data buffer contains input data in order from oldest to newest. The index register i2 initially points to the oldest datum in the input data buffer.

Each time a receive interrupt occurs, one new data value is read into the "data" buffer. This newest input replaces the oldest input already in the buffer. When the program begins to execute, it takes as many sampling cycles as there are slots in the buffer to fill the buffer with input values. Since the calculation for each filter output requires all inputs, the input data buffer is zeroed as part of the start procedure, leaving just one spot for the new input datum.

Every time a new sample is received from the microphone, a receive interrupt is raised and the "input_samples" interrupt service routine executes. In the talk-through program, this routine simply copied the inputs to the outputs. In this program, the routine performs an FIR filtering function. In the first segment of the routine, the new input sample replaces the oldest datum in the circular buffer, and the pointer i2 is updated to point to the

```
.const            taps = 102;

.var/pm/circ      fir_coefs[taps];
.init             fir_coefs: <b.dat>;          /* coeffs in reverse order from bM to b0    */
.var/dm/circ      data[taps];                  /* data stored from oldest to newest        */

                  …..


start:

                  …..

                  i2 = ^data;                  /* i2 points to oldest element  in input buffer x[n-M]    */
                  l2 = taps;
                  i4 = ^fir_coefs;             /* i4 points to filter coefficient bM       */
                  l4 = taps;
                  m1 = 1;                       /* set modify registers                     */
                  m4 = 1;

                  cntr = taps - 1;
                  do zero until ce;
zero:             dm(i2,  m1) = 0;             /* zero circular input data buffer          */

                  …..


input_samples:
                  ena sec_reg;                 /* use shadow register bank                 */
                  mr1 = dm (rx_buf + 2);       /* get new codec sample from serial port, right side   */

                  dm(i2,m1) = mr1;             /* read sample into input data buffer       */

                  cntr = taps-1;
                                               /* mr = 0,  read data and coeff             */
                  mr = 0,  mx0 = dm(i2, m1), my0 = pm(i4, m4);

                  do fir1loop until ce;
                                               /* mac, read next data and coeff            */
fir1loop:         mr = mr + mx0 * my0 (ss), mx0 = dm(i2, m1), my0 = pm(i4, m4);
                  mr = mr + mx0 * my0 (rnd);   /* one last mac to give y[n]                 */
                  if mv sat mr;                /* if overflow, saturate result register    */

                  dm (tx_buf + 1) = mr1;       /* copy filter output to serial port for speakers  via codec   */
                  dm (tx_buf + 2) = mr1;
                  rti;
```

FIGURE 13.15

Listing of FIR filter program highlights.

next element, which now becomes the oldest sample in the buffer, $x[n-M]$. The main feature of the receive interrupt service routine is a loop that calculates a single output value

$$y[n] = \sum_{k=0}^{M} b_k x[n - k]$$

at each receive interrupt that occurs. Before this loop begins, the register mr that will hold $y[n]$ is set to zero, $x[n-M]$ is read into mx0, and b_M is read into my0. As soon as the reads are complete, index registers i2 and i4 are updated to point at the next input value and coefficient. The FIR loop multiplies mx0 by my0 and adds the result to the sum in register mr. Also during the loop, the next input data value, $x[n-(M-1)]$, is read into mx0, and the next coefficient, b_{M-1}, is read into my0. The loop continues until the product $b_1x[n-1]$ has been added to the sum, $x[n]$ has been read into mx0, and b_0 has been read into my0. The first instruction after the loop adds the product $b_0x[n]$ to the sum to obtain the final value for $y[n]$, as specified by the FIR difference equation. This instruction is the reason the counter begins at M instead of $M+1$. Because the data buffer is circular, the data buffer pointer i2 now points at the oldest element in the input buffer, ready to accept a new input sample at the next interrupt. The last steps in the interrupt service routine serve to copy the rounded value of $y[n]$, stored in mr1, to the transmit buffer, which passes the output via the serial port to the speakers. Both speakers receive the same signal.

13.8.3 IIR Filter

Like FIR filters, IIR filters are used in many DSP tasks. The algorithm is very similar for both filters, but because IIR filters rely on both inputs and outputs, two loops are required instead of just one. Figure 13.16 shows a listing of the main differences between an IIR filter program and the talk-through program shown in Figure 13.12. The method presented relies on a direct form 2 realization:

$$w[n] = x[n] - \sum_{k=1}^{N} a_k w[n - k]$$

$$y[n] = \sum_{k=0}^{M} b_k w[n - k]$$

where, for simplicity, $N = M$.

In the declaration section of the program, both the a_k and b_k coefficients are listed, initialized from files of four-digit hex values in 1.15 format, "ai.dat" and "bi.dat." IIR filter coefficients frequently exceed unity, so the code in Figure 13.16 assumes all coefficients have been scaled down by an appropriate scaling factor, as described in Section 13.3. Otherwise, the code for the IIR filter is nearly identical with that for the FIR filter described in the previous section. The main difference is the presence of two loops, one to calculate $\sum_{k=1}^{N} a_k w[n - k]$ and one to calculate $\sum_{k=0}^{M} b_k w[n - k]$. Each time a receive interrupt occurs,

```
.const              taps = 5;                       /* the number of b coefficients          */
.var/pm/circ        a_coefs[taps-1];
.init a_coefs:      <ai.dat>;                       /* coeffs in order from a1 to aN         */
.var/pm/circ        b_coefs[taps];
.init b_coefs:      <bi.dat>;                       /* coeffs in order from b0 to bN         */
.var/circ           data[taps];

.const              a_scale = 0;                    /* assume all a coefficients have been divided by 2^a_scale */
.const              b_scale = 0;                    /* assume all b coefficients have been divided by 2^b_scale */

                    .....

                    i2 = ^data;                     /* i2 points to the element w[n-1]       */
                    l2 = taps;
                    i4 = ^a_coefs;                  /* i4 points to filter coefficient a1    */
                    l4 = taps-1;
                    i7 = ^b_coefs;                  /* i7 points to filter coefficient b0    */
                    l7 = taps;
                    m0 = 0;                          /* initialize modify registers          */
                    m1 = 1;                          /* m0, m1, m2 used with i0, i1, i2 and i3 */
                    m2 = 2;
                    m4 = 1;                          /* m4 and m7 used with i4, i5, i6 and i7 */
                    m7 = 1;

                    cntr = taps;
                    do zero until ce;
zero:               dm(i2, m1) = 0;                 /* zero circular delay line buffer       */

                    .....

input_samples:
                    ena sec_reg;                    /* use shadow register bank              */

                    mr1 = dm (rx_buf + 2);          /* get new codec sample from serial port, right side */
                    ax1 = mr1;

                    cntr = taps-2;
                                                    /* mr = 0, read delay line element and ak coeff */
                    mr = 0, mx0 = dm(i2, m1), my0 = pm(i4, m4);

                    do iir1loop until ce;
                                                    /* mac, read next delay line element and ak coeff */
iir1loop:           mr = mr + mx0 * my0 (ss), mx0 = dm(i2, m1), my0 = pm(i4, m4);
                    mr = mr + mx0 * my0 (rnd);      /* one last mac                          */
                    if mv sat mr;                   /* if overflow, saturate result register */

                    sr = ashift mr1 by a_scale (hi); /* undo effects of scaling ak coeffs    */
                    af = -sr1;                      /* negative sum of akw[n-k]              */
                    ar = ax1 + af;                  /* add x[n] to get w[n]                  */
                    dm(i2, m0) = ar;                /* write wn to delay line buffer         */
                    cntr = taps - 1;

                                                    /* mr = 0, read delay line element and bk coeff */
```

FIGURE 13.16

Listing of IIR filter program highlights.

```
              mr = 0, mx0 = dm(i2, m1), my0 = pm(i7, m7);

              do iir2loop until ce;
                                          /* mac, read next delay line element and bk coeff      */
iir2loop:     mr = mr + mx0 * my0 (ss), mx0 = dm(i2, m1), my0 = pm(i7, m7);
              mr = mr + mx0 * my0 (rnd);   /* one last mac                                       */
              if mv sat mr;                /* if overflow, saturate result register              */

              sr = ashift mr1 by b_scale (hi);/* undo effects of scaling bk coeffs              */

              dm(tx_buf + 1) = sr1;        /* copy filter output y[n] to serial port for speakers via codec  */
              dm(tx_buf + 2) = sr1;
              rti;
```

FIGURE 13.16

Continued

a new input sample $x[n]$ is used to compute a value for the next delay line element $w[n]$ and hence a value for one new output $y[n]$, according to the equations above. The index registers are used with appropriate modify registers throughout to ensure that the program is faithful to the direct form 2 equations and that a newly computed delay line element $w[n]$ always replaces the stalest datum in the delay line.

CHAPTER SUMMARY

Matlab
Support

1. The ADSP-2181 is a 16-bit fixed point processor.
2. The ADSP-2181 can provide integer or fractional number formats. The fractional number formats range from 15.1 to 1.15. The 1.15 format is the most convenient and popular. It can represent numbers between -1.0 and nearly 1.0.
3. When a value such as a coefficient or input that must be used by the processor lies outside the legal range for the number format chosen, the value must be scaled. Scaling and all other numerical manipulations reduce the precision of the results.
4. Registers are used as temporary storage for values being processed by a DSP.
5. The basic instructions for most DSPs are very similar, though the language may differ.
6. Assembly language programs for DSPs involve a large number of set-up and initialization instructions in addition to the main code for an algorithm.

REVIEW QUESTIONS

13.1 How does the ADSP-2181 represent the integer 2483 as:
   a. An unsigned integer?
   b. A signed integer?

13.2 a. How is the decimal value 0.93207 represented if an ADSP-2181 programmer wishes to use:
   (i) 1.15 representation?
   (ii) 6.10 representation?
   (iii) 12.4 representation?
   b. What error is committed for each case in (a)?

13.3 Compute the dynamic range for each of the following fixed point configurations using Equation (12.1):
   a. 6.10
   b. 12.4
   c. 16.0

13.4 The decimal number 0.613, quantized in unsigned 1.3 format, is multiplied by the decimal number 1.85, quantized in unsigned 2.2 format.
   a. Find the quantized value for each of the decimal numbers.
   b. What is the product of the two quantized numbers, and what minimum $N.M$ format is required to represent it?
   c. How great is the difference between the product calculated in (b) and the true product of 0.613 and 1.85?

13.5 The FIR filter

$$y[n] = 5x[n] + x[n-1] + 5x[n-2]$$

is to be implemented on the ADSP-2181 using a 1.15 fractional fixed point format. Input samples are expected to lie between -2 and 2 V. Draw a difference equation diagram for the scaled filter.

13.6 The IIR filter

$$y[n] = 2.0651y[n-1] - 1.5200y[n-2] + 0.3861y[n-3]$$
$$+ 0.0086x[n] + 0.0258x[n-1] + 0.0258x[n-2] + 0.0086x[n-3]$$

is to be implemented on the ADSP-2181 using a 1.15 fractional fixed point format. Input samples are expected to lie between -1 and 1 V.
   a. Write the direct form 2 equations for the filter.
   b. Write the scaled direct form 2 equations for the filter.
   c. Draw a direct form 2 difference equation diagram for the scaled filter.

13.7 The coefficients for an FIR filter are:

$$b = [0.0059 -0.0489\ 0.1716\ 0.7426\ 0.1716 -0.0489\ 0.0059]$$

Find a 1.15 representation for each of the coefficients. Write the result in binary and hexadecimal format.

13.8 A buffer is initialized to contain the integer values from 1 to 13. The index register i1 points to the first element in the buffer. The length register l1 is assigned the value 0. The modify register m0 has the value 1. After the instruction

$$my1 = dm(i1, m0);$$

has been executed 10 times, what does the register my1 contain?

13.9 A buffer is initialized to contain the integer values from 1 to 13. The index register i1 points to the first element in the buffer. The length register l1 is assigned the value 8. The modify register m0 has the value 1. After the instruction

$$my1 = dm(i1, m0);$$

has been executed 10 times, what does the register my1 contain?

13.10 The register ax0 contains 0x114A, and the register ay0 contains 0x2002. Determine the contents of the register ar after the execution of the instruction:
   a. ar = ax0 + ay0;
   b. ar = ax0 − ay0;

13.11 The register mx0 contains 0x2000. The register my0 contains 0xB557. Both numbers are stored in 1.15 format. Determine what the 40-bit register mr contains after the execution of the instruction:
   a. mr = mx0 * my0 (ss);
   b. mr = mx0 * my0 (rnd);
   c. mr = mx0 * my0 (uu);

13.12 The register ar contains 0x2E96. What does the 32-bit register sr contain after the execution of the instruction:
   a. sr = ashift ar by −3 (lo);
   b. sr = ashift ar by 1 (lo);
   c. sr = ashift ar by −2 (hi);
   d. sr = ashift ar by 3 (hi);

13.13 The first instruction in the program segment in Figure 13.17 forces the processor to use integer instead of fractional mode. After the rest of the instructions in the segment execute, what do the registers my1, ar, cntr, and mr0 contain?

FIGURE 13.17
Program segment for Question 13.13.

```
mstat = 0x10;
my1 = 1;
ar = 6;

cntr = 6;
do factrl until ce;
mr = ar * my1 (uu);
my1 = mr0;
factrl:  ar = ar − 1;
```

13.14 Explain what the program segment in Figure 13.18 does. All coefficients and data are expressed in 1.15 format.

```
.const        N = 4;
.var/pm/circ  coeff[N];
.init coeff:      0x2000, 0x4000, 0x2000, 0x1000;
.var/dm/circ  data_in[N];
.init data_in:    0x4800, 0x2000, 0x6000, 0x5000;

i0 = ^data_in;
m0 = 1;
l0 = N;
i5 = ^coeff;
m5 = 1;
l5 = N;

cntr = N–1;
mr = 0, mx0 = dm(i0, m0), my0 = pm(i5, m5);
do firloop until ce;
firloop:    mr = mr + mx0 * my0 (ss), mx0 = dm(i0, m0), my0 = pm(i5, m5);
            mr = mr + mx0 * my0 (rnd);
            if mv sat mr;
```

FIGURE 13.18
Program segment for Question 13.14.

13.15 Determine the contents of the registers ar, ax0, and ay0 after each instruction in the program segment of Figure 13.19 executes.

FIGURE 13.19
Program segment for Question 13.15.

```
ar = H#43B2;
ax0 = H#101A;
ay0 = H#0030;
i1 = 0;
m2 = 0;
l1 = 0;
dm(i1, m2) = ar, ar = ax0 + ay0;
ar = ax0 – ay0, ay0 = dm(i1, m2);
```

14

SIGNAL PROCESSING

This chapter examines a number of in-depth applications of digital processing. The applications include:

> ➤ digital audio
> ➤ compact discs
> ➤ MP3
> ➤ speech recognition
> ➤ music synthesis
> ➤ speech synthesis
> ➤ geophysical processing
> ➤ encryption
> ➤ motor control

14.1 DIGITAL AUDIO

14.1.1 Digital Audio Basics

The components of a basic digital audio system mirror the essential components of all DSP systems. As shown in Figure 14.1, an antialiasing filter must first remove all frequencies outside the range of interest from a musical signal to prepare for sampling. The sampled analog signal is then converted to a set of digital codes. For digital recording, this marks

FIGURE 14.1
Digital recording.

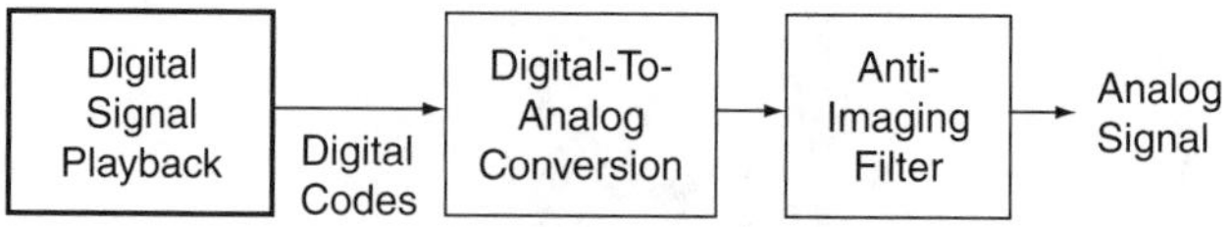

FIGURE 14.2

Digital playback.

the end of the process. Digital playback, on the other hand, illustrated in Figure 14.2, begins with the digital codes, converts them back to an analog signal, and finishes by removing the artifacts introduced by D/A conversion. The result is an audible analog signal.

The more demanding the audio requirements, the more advanced the features that improve the performance of each element. High fidelity audio means that aliasing, quantization noise, phase distortion, and other factors that reduce audio quality are all carefully minimized. The following sections describe ways to accomplish these goals.

14.1.2 Oversampling and Decimation

The antialiasing filter, introduced in Section 2.2.2, is designed to band-limit the analog signal to ensure that aliasing does not occur during sampling. A suitable cut-off frequency for this low pass filter must be determined once the maximum desired signal frequency W Hz has been identified. Ideally, the roll-off for the filter should be very steep and the cut-off frequency as close as possible to W samples per second to keep the sampling rate, and therefore the amount of processing, as small as possible. Unfortunately, it is very difficult and expensive to construct an analog filter with a sufficiently steep roll-off.

Oversampling, briefly introduced in Section 2.2.2, provides a means to obtain a tightly filtered signal with only a very simple analog antialiasing filter. The process relies on sampling a signal more often than necessary. Nyquist theory requires a sampling rate twice the maximum signal frequency as a minimum. Oversampling means sampling at a rate of perhaps 4×, 8×, 16×, or even more times the Nyquist requirement. Eight or sixteen times oversampling, for example, is common for CD players. In Figure 14.3, a 4× oversampling rate is used for illustration. The signal in (a) contains important information between 0 and W Hz only, and noise above W Hz. When important signal information lies below W Hz, the minimum Nyquist sampling rate is $2W$. Therefore, a 4× oversampling rate will be $8W$, four times the Nyquist minimum.

Prior to sampling at $8W$ samples per second, the signal is filtered by an analog antialiasing filter. In this case, because the important signal information lies below W Hz, it is not necessary for this filter to satisfy the usual condition, that no frequency above half the sampling rate be permitted to enter the system. As Figure 14.3(c) shows, even an antialiasing filter with a gradual roll-off like the one in (b) is adequate to leave the important signal information intact after $8W$ sampling. At worst, this filter produces an output that, once sampled at $8W$ samples per second, has the spectrum described by the dotted lines in Figure 14.3(c). Spectral images lie at every multiple of the sampling frequency, and aliasing occurs between W and $7W$ Hz, as the crossover in the figure indicates. When the signal shown in Figure 14.3(a) is sampled, aliasing is caused by the noise in the range from W to $7W$ Hz, resulting in the altered spectral shape shown in Figure 14.3(c). Though aliasing does occur, signal integrity is maintained because aliasing has no impact on the important 0 to W Hz range, as the figure suggests.

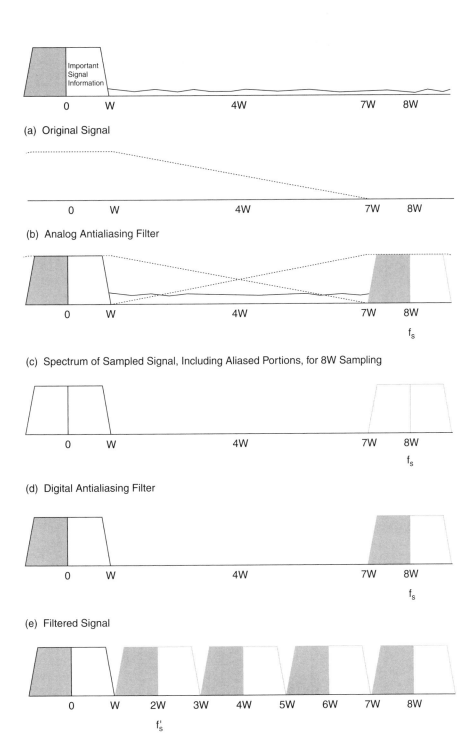

(a) Original Signal

(b) Analog Antialiasing Filter

(c) Spectrum of Sampled Signal, Including Aliased Portions, for 8W Sampling

(d) Digital Antialiasing Filter

(e) Filtered Signal

(f) Spectrum of Sampled Signal After 4x Decimation, 2W Sampling Rate

FIGURE 14.3

Oversampling and decimation.

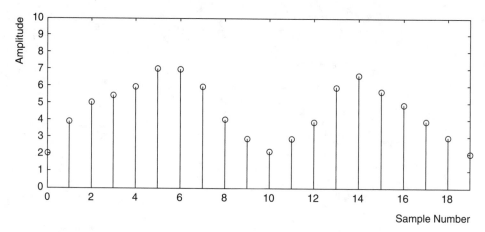

(a) Original Signal Samples

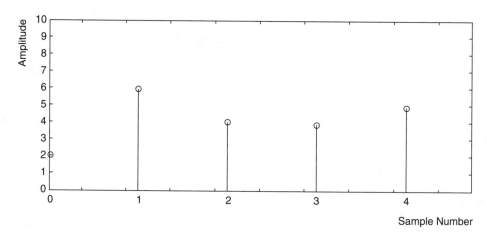

(b) After 4x Decimation

FIGURE 14.4
4× decimation.

After rough analog filtering and sampling, a digital antialiasing filter with a steep roll-off, illustrated in Figure 14.3(d), can be used on the sampled signal to make a sharp cut at W Hz. Digital filters can approach an ideal filter shape more successfully and more cheaply than their analog counterparts. A high order digital antialiasing filter can now extract exactly the portion of the spectrum that contains the audio signal of interest, as seen in Figure 14.3(e). This is the result sought, but maintaining the higher sampling rate means that 4× as many calculations will be necessary in the processing phase. For this reason, rough analog filtering, sampling, and digital filtering are followed by **decimation**, which drops digital samples from the signal in the time domain. In this example, 4× decimation throws away three out of every four samples, as illustrated in Figure 14.4. The effect of decimation is to reduce the sampling rate back to its minimum level, since the sampling rate reverts from $8W$ back to $2W$ samples per second in this case. The result is shown in Figure 14.3(f). Oversampling followed by decimation gives a spectrum that could otherwise be obtained by $2W$ sampling, but without the need for a high order analog filter. The flexibility offered by this digital technology, known as **multirate DSP**, means that the effects of aliasing can be nearly eliminated by using a high quality digital filter. The cost is the requirement for a faster, more expensive DSP.

As evidenced by Figure 14.3, oversampling and tight digital filtering generate space in the spectrum of a sampled signal that is available until decimation removes it. The space created by oversampling simplifies antialiasing filter design, as described above, but it has several incidental benefits as well. For one, the effect of quantization noise is reduced, because it is spread over a wider range of frequency, from 0 to $4W$ instead of 0 to W Hz, for example. This weakens the effect of the noise on the signal in the range from 0 to W Hz. Second, some analog-to-digital converters change the spectrum of the quantization noise and move the bulk of it to the spaces in the spectrum produced by oversampling, where it will be eliminated by filtering through a process called **noise shaping**. With noise shaping, less quantization noise remains in the frequency band of interest and the signal-to-noise ratio increases. When noise reduction is the goal, as it is for many instrumentation applications, oversampling ratios as high as 256:1 may be desirable (Analog Devices, 1995).

14.1.3 Zero Insertion and Interpolation

Once processing is complete, the digital codes must be prepared for conversion back to the analog domain. The codes are first converted to analog using a zero order hold, which produces a signal with a staircase characteristic. The steps in the staircase contain high frequency elements that are introduced by digital processing and must be removed. A Nyquist anti-imaging filter, with cut-off at W Hz, does the trick. The sharper the filter, the more successfully extraneous signal elements will be removed. Building an analog filter to do this job is difficult, but a suitable digital filter is quite easy to design. Unfortunately, analog filtering cannot be avoided. Once digital-to-analog conversion has occurred, a final analog filtering step to remove high frequency images from the signal is always necessary. The object in D/A conversion is similar to that in A/D conversion, that is, to accomplish the required filtering without the need for a complex, high order analog filter. For D/A conversion, the first step is to increase the sampling rate of the digital signal, through a process called **zero insertion**. As its name suggests, zero insertion means adding zeros between existing samples in the time domain. In Figure 14.5(b), for example, three zeros are added between each pair of samples in (a). This has the effect of multiplying the sampling rate by a factor of 4, and therefore increasing the Nyquist frequency from W to $4W$. In the frequency domain, the shape of the digital signal's spectrum does not change before and after zero insertion, as seen in Figure 14.6(a) and (b), but the change in sampling rate is very significant at the next step, interpolation.

Following zero insertion, the signal is filtered by a high order digital anti-imaging filter. This filter has a low pass nature and has the effect of cleanly removing the spectral elements between W and $7W$ Hz. Figure 14.6(c) shows the filter and (d) shows the filtered signal. A low order analog filter, shown in Figure 14.6(e), is now capable of recovering the

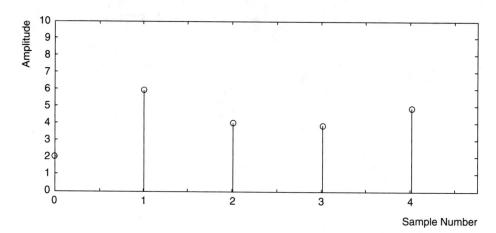

(a) Original Signal Samples

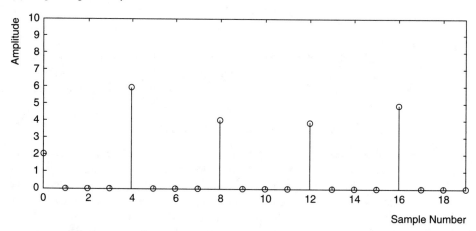

(b) After 4x Zero Insertion

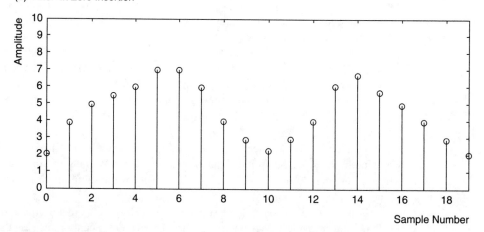

(c) After Interpolation

FIGURE 14.5

4× zero insertion and interpolation.

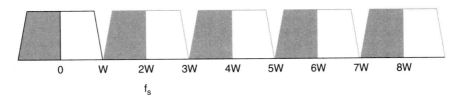

f_s

(a) Digital Signal, 2W Sampling Rate

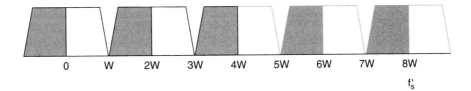

f'_s

(b) 4x Zero Insertion Raises Sampling Rate to 8W

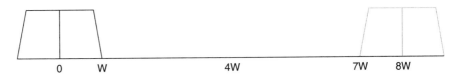

(c) Digital Anti-Imaging Filter

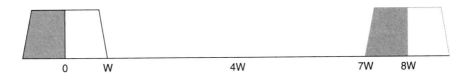

(d) Spectrum of Digital Signal After Digital Anti-Imaging Filter is Applied

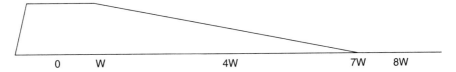

(e) Analog Anti-Imaging Filter

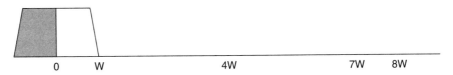

(f) Recovered Signal

FIGURE 14.6

4× zero insertion and interpolation.

band-limited analog signal in Figure 14.6(f). This analog anti-imaging filter is also known as an **interpolation filter**. In the time domain, it forces inserted zeros to take values that are interpolated between known samples, as shown in Figure 14.5(c). The amplitudes of the interpolated samples reflect the fact that the signal can contain elements no greater than W Hz in frequency. The sharp transitions between the existing samples and the inserted zeros disappear, resulting in a smooth signal with no spurious elements above the Nyquist frequency.

Had the sampling rate not been increased through zero insertion, the digital anti-imaging filter would not have been able to change the spectrum in Figure 14.6(a) because the spectral image of the filter shape in (c) would occur at $2W$, not $8W$. Zero insertion creates space in the spectrum that makes the task of the analog anti-imaging filter straightforward. After digital anti-imaging, a low order analog filter can easily recover the required signal spectrum. This result can be achieved with $2W$ sampling as well, but only with a high order analog filter. Instead, arbitrary filter sharpness can be obtained through the use of digital technology, improving the fidelity of the output signal. As for oversampling, the increase in effective sampling rate achieved through zero insertion spreads quantization noise more thinly across the spectrum, further improving the output signal-to-noise ratio. Zero insertion and interpolation, another example of multirate DSP, can have beneficial visual as well as auditory effects. Thirty-five mm movie projectors project each frame of a film three times. If the film was shot at 24 frames per second, the resulting sequences play at 72 frames per second. In this case, the human eye interpolates between the frames, giving the effect of a higher apparent frame rate.

Decimation and interpolation are used in digital audio because the need for faithful reproduction of music and voices is so great. The same techniques are applicable to high precision instrumentation, sonar processing, and other applications where signal integrity is important.

14.1.4   Dithering and Companding

High quality digital audio systems must have large dynamic range. This means that they must be capable of faithfully reproducing both very quiet and very loud sounds, as required by the music. In digital audio applications, dynamic range is defined as

$$\text{Dynamic range} = 20 \log\left(\frac{\text{Maximum signal}}{\text{Minimum signal discernible from noise}}\right)$$

This definition follows that given in Equation (12.1). Dynamic range is determined largely by the number of bits used in the digital system. Additional tricks are possible, however, that can further extend the ability of a digital audio system to respond. These include dithering and companding.

During quantization, each analog sample must be mapped to the nearest available digital level. Figure 14.7 shows that signals of adequate size can be captured with reasonable accuracy, but small signals may be lost entirely. The figure shows two signals, with quantization levels marked at 0, 1, and 2. For each sample taken, the nearest quantization level is selected. In (a), some sense of the signal can be recovered, but in (b) all signal variation is lost in quantization errors. **Dither** improves this situation through the unlikely strategy of

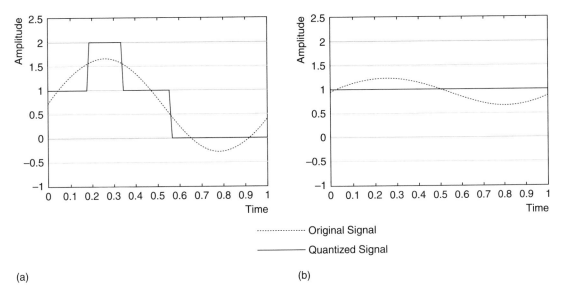

(a)

(b)

········· Original Signal

———— Quantized Signal

FIGURE 14.7

Quantization of small signals without dither.

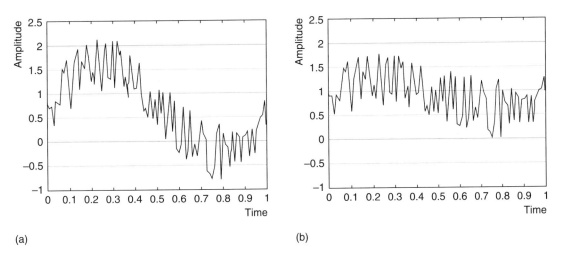

(a)

(b)

FIGURE 14.8

Signals with dither added.

adding a small amount of white noise to the analog signal before sampling. The spectrum of this noise is flat. Figure 14.8 shows what the signals look like with dither noise added. After quantization, the shape of the signals has a strange new nature, as shown in Figure 14.9. The original shapes of the signals may be recovered by finding running averages through the points, marked by dotted lines in the figure. This way, the variations of the signals around

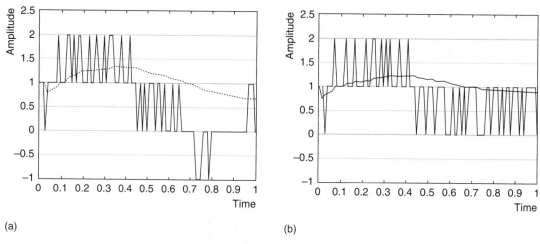

(a)

(b)

FIGURE 14.9

Quantization of small signals with dither.

the quantization levels become more obvious, and more of the original character of the signal is preserved. The effect is especially beneficial for small signals, less than one quantization step in amplitude, which can otherwise be lost completely.

Dither has another advantage as well. For most signals, quantization errors are assumed to be random and independent of the signal being coded. This assumption leads to an rms[1] quantization noise amplitude of $Q/\sqrt{12}$, where Q is the quantization step size, as shown in Appendix B. Unfortunately, quantization errors for sinusoidal inputs are not independent of the signal and, in fact, vary with the same period. As a result, distortion that is harmonically related to the input appears in the quantized signal. When dither is added to a sine wave, quantization errors are randomized, but dither noise must have an rms amplitude of at least $Q/\sqrt{12}$ to have this effect. The addition of dither therefore at least doubles the noise, which means the overall signal-to-noise ratio is changed by $10 \log(1/2) = -3$ dB. The signal-to-noise ratio for a quantized sine wave, for example, is $6.02N + 1.76$ dB, as shown in Appendix B. With dither added, the signal-to-noise ratio reduces to $6.02N - 1.24$ dB. The advantage, though, is that the noise spectrum is flattened throughout the Nyquist band, on average interfering less with important signal harmonics.

When low amplitude signals represent a large proportion of the signals that must be handled by a digital system, these small signals can be boosted relative to their larger cousins with a process called **companding**. Because quantization noise contributes equally at all signal levels, the smallest signals are the most affected. To avoid losing them in noise, companding increases the level of quiet sounds relative to loud sounds prior to quantization. The curve in Figure 14.10(b) shows how small inputs are boosted compared to the usual situation shown in Figure 14.10(a). This is accomplished by compressing the input

[1]   The root-mean-square (rms) amplitude of a signal refers to the square root of the mean, or average, squared signal amplitude.

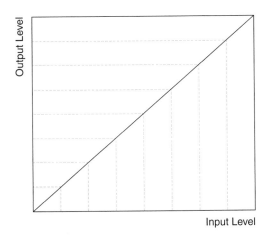

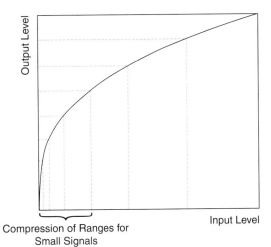

(a) Without Compression

(b) With Compression

FIGURE 14.10

Compression in companding.

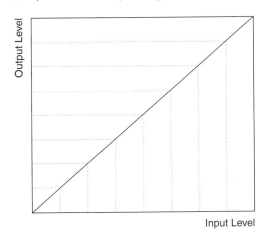

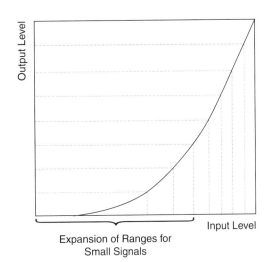

(a) Without Expansion

(b) With Expansion

FIGURE 14.11

Expansion in companding.

ranges for quantization, as the horizontal axes in the figure show. This compression at small amplitudes explains the first half of the term "compand." On playback, the process is reversed, as illustrated in Figure 14.11. Here, the smallest signals are attenuated to restore the original signal balance. As shown along the horizontal axis of part (b), input ranges previously squashed together are expanded, which accounts for the second half of the term "compand." The consequence of companding is that the effects of quantization noise are evened out over all signal levels.

Companding may be achieved using analog circuitry or DSP. In the analog solution, operational amplifier circuits are used to compress analog signals prior to A/D conversion and then to expand them after D/A conversion. In the DSP solution, sound samples are first quantized to 13 or 14 bits, which the compander requantizes to 8 bits using a compression function. This function is an approximation to the one shown in Figure 14.10(b), made up of straight line segments to make implementation on a DSP easier. The expansion function uses a similar set of straight line segments to approximate the curve in Figure 14.11(b). Expansion begins with 8-bit samples and produces the original 13- or 14-bit samples again. In North America and Japan, μ-law companding, which begins with 13-bit sound samples, is most common. Europe favors A-law companding, which uses 14 bits.

By reducing the effect of quantization noise on small signals, companding can improve the signal-to-noise ratio for the signal being processed. This in turn means that the number of quantization bits can be reduced and still yield uncompanded performance. Companding does have a downside: Noise levels are modulated along with signal levels, which can mean that noise will become more audible in the reconstructed signal. Also, care must be taken to ensure that the degree of compression is not too extreme, or important variations in the loudest sounds may be lost. Companding is commonly used in telephony applications like cordless telephones and modems, for which 8 kHz sampling is common. With companding, reasonable voice quality can be obtained at the low bit rate of (8000 samples/sec)(8 bits/sample) = 64 kbps.

14.1.5 Audio Processing

14.1.5.1 Compact Discs Compact discs (CDs) are a popular storage medium for audio signals. They record sound samples as 16-bit values, one for the left channel and one for the right channel. This number of bits provides a dynamic range of $20 \log(2^{16}) = 96.3$ dB, a reasonably good match for the 120 dB of dynamic range human hearing possesses. Because humans can hear up to 22.05 kHz, the samples are collected at the rate of 44,100 samples per second. As shown in Figure 14.12, the left and right stereo channels are multiplexed into one data stream, which is combined with error correction codes that can be checked when the CD is played.

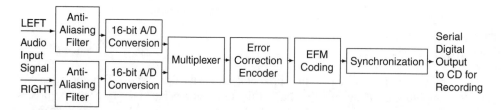

FIGURE 14.12
Recording a compact disc.

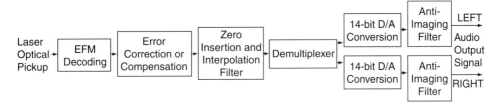

FIGURE 14.13
Playing a compact disc.

Most production-quality audio is recorded using 24 bits to ensure that the dynamic range is large enough to record both very quiet and very loud sounds accurately. To record this audio in a 16-bit format suitable for a CD, the 8 least significant bits of the original 24 must be truncated. To prevent the information in these 8 bits from being lost entirely, 8 bits of dither are added to each 24-bit sample before truncation. As a result of dithering, the least significant bit after truncation bears some sense of the low level variations in level contained in the truncated bits. This effect of dithering was illustrated in the last section.

Data are recorded onto CDs using a laser, which etches pits 0.5 μm wide along a track on the plastic surface of the CD. The beginning and end of each pit mark the presence of a one in the binary data. All bits between the ones are assumed to be zeros, where each bit occupies a 0.23 μm distance along a track. To minimize the number of pits that need to be inscribed, each 16-bit sample is coded using an EFM (eight-to-fourteen) code. This code maps every 8 bits of data to a 14-bit code that is specially designed to minimize the number of transitions between zeros and ones, thus minimizing the number of pits that must be etched.

As Figure 14.13 shows, the pits are detected by a laser when the CD is played. The error codes for each group, or frame, of data bits are checked for correctness before a segment of music is played. If an error occurs—perhaps because of disc defects, dust, or scratches—an interpolated sample is produced or the datum is simply set to zero. To reduce the possibility that several samples in a row will be lost, an error that would be audible, adjacent samples in the music are interleaved across several different frames. The digital data from the CD are then subjected to zero insertion and interpolation, as described in Section 14.1.3. This step permits a low order anti-imaging filter to serve in the last step of the D/A process. Moreover, the fact that it reduces the effect of quantization noise on the signal means that a 14-bit D/A converter suffices. The recovered left and right stereo channels are sent to speakers at the output.

Apart from the overhead created by the use of the EFM code, a great deal of other control, error correction, and synchronization information must be appended to each piece of sound data. In fact, out of every 49 bits stored on the CD, only 16 bits are audio sound (Pohlmann, 1994)! To produce two 16-bit stereo samples at the rate of 44.1 ksamples/sec requires the enormous audio bit rate of $2(16)(44100) = 1.41$ Mbps. Including the overhead, the total bit rate becomes $49(1.41)/16 = 4.32$ Mbps. A 5-minute tune, then, must occupy 1296 megabits, or 162 megabytes, which explains the need for massive storage capability on the CD.

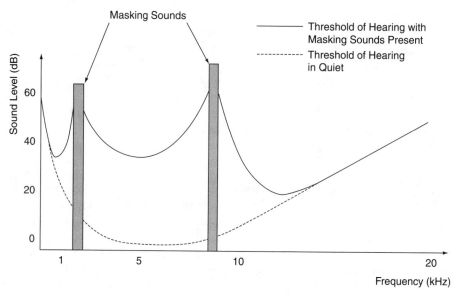

FIGURE 14.14
Auditory masking effects.

14.1.5.2 MPEG Audio Layer 3 (MP3) MP3 is a coding format that permits nearly perfect CD quality to be reproduced with an audio bit rate of only 128 kbps instead of the 1.41 Mbps, mentioned in the last section as being required for normal CD coding. MP3 is able to achieve this remarkable 11:1 reduction in bit rate by exploiting the characteristics of the human auditory system. In Figure 14.14, the dashed line shows the normal threshold of hearing. The sound level shows how loud a sound has to be at a particular frequency to be detected by a human listener. As the figure indicates, midrange frequencies are detected much more easily than either low or high frequencies.

The presence of loud sounds influences the shape of the hearing threshold. Figure 14.14, for example, shows two strong sounds marked as "masking sounds." When a loud sound is present in a particular frequency range, quieter sounds in the same range become completely inaudible. In fact, quiet sounds in nearby frequency ranges may also be rendered inaudible. This explains the "curtains" on either side of the masking sounds. In these cases, the loud sounds are said to **mask** the quieter sounds present at the same time. The new threshold of hearing rises from the dashed line to the solid line in the figure. In addition, loud sounds can mask quieter sounds that occur a few milliseconds before to a few hundred milliseconds after. Because sounds masked by other, louder sounds cannot be perceived by human listeners, they can be removed from coding altogether without degrading sound quality. Note that compression achieved in this manner does remove elements present in the original signal, but ideally does not affect a listener's experience of the signal. This is the strategy used in MP3.

The compression scheme used by MP3 is called **perceptual coding** because it achieves its results by modeling the behavior of the human ear. Like the ear, MP3 uses a bank of filters to identify the frequency content of each short interval of sound. By registering the strength of the sounds at its output, each filter determines dynamically what hearing threshold should be used in its frequency range. Where quiet sounds are masked and need not be coded, fewer bits can serve for quantization. Though reducing the number of quantization bits increases the quantization noise, this noise has no effect as long as it lies below the threshold of hearing, in a particular frequency range and at a particular time. In the end, each frequency range will use the smallest number of bits possible for coding, and different ranges may use different numbers of bits. Thus, the MP3 coder produces a data stream whose bit rate varies with the complexity of the sound being coded. In addition, redundancy between stereo channels, which carry very similar information, can be exploited to reduce the bit rate further. Finally, after quantization, both run-length encoding and Huffman encoding, to be discussed in Section 15.8, are used for additional compression. Despite very low bit rates, audio signals coded with MP3 are nearly indistinguishable from CD signals. MP3 stereo encoders and decoders have been implemented on floating point devices of all major DSP manufacturers (Fraunhofer Institut Integnerte Schaltungen, 2000).

14.2   SPEECH RECOGNITION

Speech sounds are produced when air passes through the human vocal tract, shown in Figure 14.15. The overall size of the channel determines the basic frequency of the voice; while the specific configuration of the throat, glottis (or vocal cords), tongue, teeth, and jaw determines the nature of individual sounds. Humans recognize speech without thinking about it, but the problem of automatic speech recognition is difficult. For one, each person's voice is as unique as a fingerprint. Even within a single language, dialects and regional differences make word sounds highly variable. Furthermore, a sound has a different character when quiet instead of loud. **Isolated word recognition** refers to the recognition of words spoken on their own, with pauses in between. **Continuous speech recognition** refers to the recognition of natural-sounding speech. The latter type of recognition is the most challenging. In continuous speech, a signal is present at all times, so even determining where one word ends and the next begins is a difficult task. Think of identifying word boundaries in Mandarin Chinese when you speak only English! Moreover, in continuous speech, word sounds tend to blend together across adjacent words. Said rapidly, "how to wreck a nice beach" sounds much like "how to recognize speech."

FIGURE 14.15
Human vocal tract.

vocal
cords

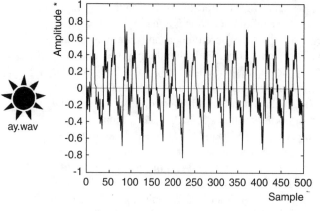

ay.wav

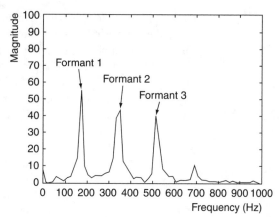

* For clarity, the envelope of the signal rather than its
samples is plotted.

(a) Signal

(b) Spectrum

FIGURE 14.16
The vowel sound "ay."

When utterances are presented to a speech recognizer, they are compared to **templates** that are already stored in the memory of the recognizer. The template that is closest to the new utterance is selected as the recognized word. Speech recognizers may be characterized according to whether they are speaker-dependent or speaker-independent. With **speaker-dependent** recognizers, users must provide training samples of each vocabulary item. **Speaker-independent** recognizers, on the other hand, aim to recognize words produced by users with widely varying speech characteristics. These recognizers typically offer more limited vocabulary sizes.

Phonemes are the individual speech sounds that make up a language. The "l" sound in "lap," the "oo" sound in "boot," and the "k" sound in "poke" are all examples of phonemes. In normal speech, a phoneme might last a hundred milliseconds. Vowels are unique among phonemes because they have a highly periodic nature. This periodicity results from the vibration of the vocal cords and produces a relatively small number of bumps in the frequency domain, called **formants**. Sounds like this, produced when the vocal cords are tight and vibrating, are called **voiced** sounds. Figure 14.16(b), for example, shows the frequency spectrum for the "ay" vowel sound. Other sounds, like "sss" and "fff," have a wide range of frequency content, similar to white noise. Sounds like these, created when air flows turbulently through slack vocal cords, are called **voiceless** sounds.

Many speech recognition algorithms rely on analyzing phonemes for their frequency content, because signal variations are too inconsistent over time to make reliable comparisons of entire words. When a short interval of sound is passed through a bank of band pass filters, the response from the filters will reveal the main frequency components present. For example, a band pass filter centered at 300 Hz will resonate best, and produce the strongest response, for an input signal with a strong 300 Hz component. With a bank of filters, the strengths of the output of each filter for each interval of time form a pattern that allows a

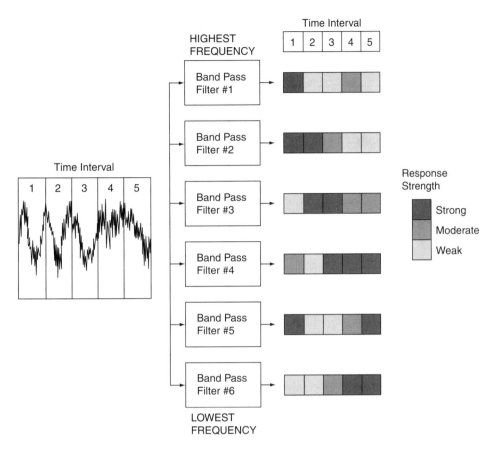

FIGURE 14.17
Analyzing a sound using band pass filters.

sound to be recognized. In Figure 14.17, the speech sound is divided into five time intervals. For each filter, the filter output strength for each interval of the input speech sequence is recorded. The output pattern registers a signal that contains mostly high frequencies at the beginning and mostly low frequencies toward the end.

The kind of output display shown in Figure 14.17, with frequencies changing vertically and time changing horizontally, has much in common with a low resolution spectrogram or FFT. The pattern of output strengths is useful in speech recognition for exactly this reason. It does not have the overwhelming detail of a true spectrogram, discussed in Section 11.5, but it does capture the essence of the changing frequencies in a phoneme or word. For N time intervals and M filters, an analysis like this produces an $N \times M$ matrix of data to represent an utterance. This matrix can be compared to $N \times M$ template matrices already stored in memory to identify which word was spoken. This simple recognition strategy leaves out some of the necessary detail. For example, no mention is made of how the strength of a filter's output can be assessed, nor are the mechanisms for comparing new utterances to templates explained. These issues are discussed in the following paragraphs.

The strength of a filter's response may be measured conveniently using the power of the filtered output signal. The **power** of a digital signal, defined as the average of the sum of the squares of the signal samples, is a good indicator of the signal's strength. Example 14.1 shows how power is calculated.

EXAMPLE 14.1

The difference equation for a band pass filter centered at 300 Hz, with a sampling frequency of 2 kHz, is

$$y[n] = 1.0275\, y[n-1] - 0.725\, y[n-2] + 0.1367\, x[n] - 0.1367 x[n-2]$$

In this example, three signals are considered:

x1.wav

(i) $x_1[n] = \sin\left(n2\pi\dfrac{100}{2000}\right)$

x2.wav

(ii) $x_2[n] = \sin\left(n2\pi\dfrac{300}{2000}\right)$

x3.wav

(iii) $x_3[n] = \sin\left(n2\pi\dfrac{400}{2000}\right)$

y1.wav

(i) The signal $x_1[n]$ is processed through the difference equation to produce $y_1[n]$. Figure 14.18 shows 500 samples of the signal $x_1[n]$, the output $y_1[n]$, and the square of the output $y_1[n]^2$. The power of the filter output, equal to the average value of $y_1[n]^2$, is indicated on the figure. It equals 0.0093.

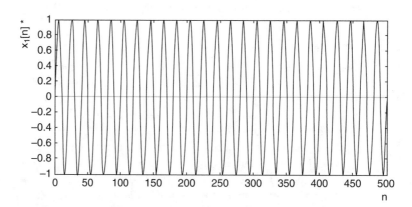

(a) Input Signal

FIGURE 14.18

Calculation of signal power for Example 14.1(a).

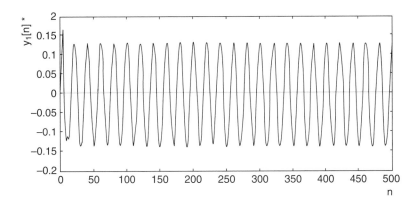

(b) Output Signal

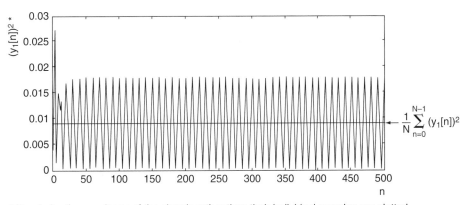

* For clarity, the envelopes of the signals rather than their individual samples are plotted.

(c) Output Signal Power

FIGURE 14.18

Continued

y2.wav

(ii)   When the procedure in (a) is repeated for $x_2[n]$, the power is found to be 0.4833.

(iii)   The power of the filter output is 0.1080 for the input $x_3[n]$.

y3.wav

In each case, the digital input has the form $x[n] = \sin(n\Omega) = \sin(n2\pi f/f_S)$. The signal frequencies f of the three signals are 100, 300, and 400 Hz, while the sampling frequency remains constant at 2 kHz. Since the filter is designed to be most sensitive to 300 Hz, it is not surprising that the power of the filter output is highest in (ii), when the input signal is 300 Hz also. The next highest power occurs for (iii), for an input frequency of 400 Hz; and the smallest output power occurs in (i), when the input frequency is farthest from 300 Hz, at 100 Hz.

$$\text{Utterance matrix} \begin{bmatrix} A & B & C \\ D & E & F \\ G & H & I \end{bmatrix} \qquad \text{Template matrix} \begin{bmatrix} J & K & L \\ M & N & O \\ P & Q & R \end{bmatrix}$$

Distance between Utterance and Template

$$D = \sqrt{(A-J)^2 + (B-K)^2 + (C-L)^2 + (D-M)^2 + (E-N)^2 + (F-O)^2 + (G-P)^2 + (H-Q)^2 + (I-R)^2}$$

FIGURE 14.19
Calculating distance between utterance and template.

When the strengths of filter responses have been compiled, a matrix of data about an utterance can be created. During training, the same word may be repeated several times and several such matrices produced. The average of these matrices might serve as a template, or representative, for the word in question. A template is formed for each word in the speech recognizer's vocabulary. When a new utterance is to be recognized, a matrix of data is produced for it as well. One simple way to compare this matrix to a template matrix is to find the "distance" between the two. Figure 14.19 explains how this can be done. The method is similar to finding the distance between two sets of (x, y) coordinates. In the example shown in Figure 14.19, each word is split into three time intervals and the filter bank contains three filters, so the word is represented by a 3×3 matrix. The distance between the matrices is calculated, as shown in the figure, as the square root of the sum of the squares of the differences between all corresponding elements of the matrices. For practical isolated word recognition, 3×3 matrices are not adequate. To recognize a small vocabulary with reasonable accuracy, 16 time intervals and 16 filters suffice.

Finally, for a reasonable expectation of success, normalization of amplitude, duration, and perhaps pitch is needed. First, a loud sound looks different from a quiet sound, so the amplitude of all utterances and templates should be scaled prior to processing. For example, to scale signals to the range $-R$ to R, every input utterance can be transformed according to the equation

$$x_{scaled} = \left(\frac{x - x_{min}}{x_{max} - x_{min}} \right) 2R - R$$

where x is a sample value in the original utterance, x_{scaled} is the scaled value of this sample, x_{max} is the largest sample value in the utterance, and x_{min} is the smallest. Second, each time a word is spoken, the length of the utterance varies. A method frequently used to address this problem is called **dynamic time warping**. This process operates on the utterance matrix, stretching or compressing it along its time axis to achieve the best possible alignment with a template matrix before distance is calculated. In choosing the best alignment, similar features in the utterance and template matrices are matched as closely as possible. An illustration is provided in Figure 14.20. A direct comparison without dynamic time warping, shown by the dashed line, would compare DIIGITAAL against DIGITAL and incorrectly report a poor match. The solid line shows the best alignment of the two utterances, exposing the fact that the utterance matches the template. Third, when speaker independence is sought, utterances must be normalized for voice pitch as well.

FIGURE 14.20

Dynamic time warping.

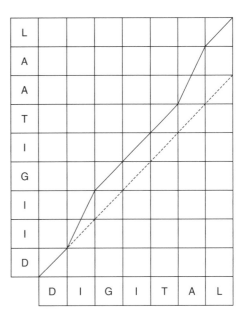

For continuous speech recognition, when word boundaries are not easily identified, the filter bank concept may be applied to overlapping segments of speech. Pauses, or quiet intervals, are not good indicators of word boundaries because these can occur within some sounds and words, as seen in Figure 14.21. Both the signal trace and the spectrogram in the figure show instants of quiet, despite the fact that the single word "bottleneck" is spoken. In Figure 14.22, three vocabulary templates — RED, WHITE, and BLUE — are compared at regular intervals to an input sequence in search of a match. At each juncture, a filter bank analysis is applied to a short segment of the input sequence for the purpose of computing a distance from each of the templates. A match with a vocabulary item is obtained when a sufficiently small distance is found. In the lower half of the figure, the sentence "THE QUICK RED FOX" is transcribed. A match is found to this utterance only in one place: at the word RED in the input sequence. No other templates are found.

The task of recognizing one of thousands of possible words is a difficult one. To improve the chance of recognizing the correct word, many recognizers use some form of grammar model that helps to narrow down the choices. For example, in the sentence shown in Figure 14.23, the missing words are relatively easy to guess. Similarly, when the words "The," "quick," and "red" have already been recognized successfully, the type of word most likely to follow would be a noun: "The quick red smiled" or "the quick red and" are not likely to occur in normal speech. The total number of words that a recognizer must consider at any point in time is called its **perplexity**. Knowledge of language structure can help to reduce the perplexity. Finally, some recognizers provide, along with recognized words, probabilities that the selected words are correct. When the nearest template to an utterance is still fairly distant from it, the chance that the utterance has been correctly identified is small: The problem may be that the word is missing from the template vocabulary.

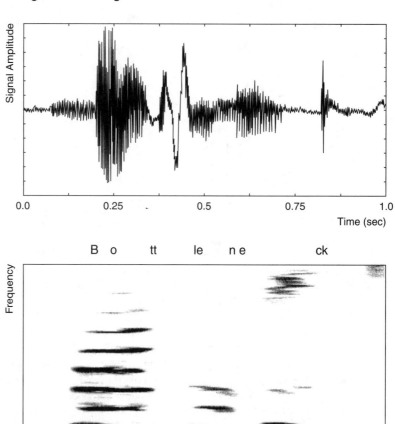

FIGURE 14.21

Quiet intervals within a single word.

Use of DSP technology has speeded progress in continuous speech recognition. The state of the art provides recognition speeds of real time or nearly real time, which means speech is recognized about as fast as it is produced. Error rates are typically between 5% and 15%. While vocabularies may reach 70,000 words, perplexity does not normally exceed 250. **Keyword spotting**, where a small number of words are sought in streams of ordinary speech, is being used in telephone operator service to help improve service. AT&T's "How May I Help You?" service (AT&T, 2000) goes further: It uses a sophisticated language understanding system that permits callers to ask the same questions in a dozen dif-

FIGURE 14.22
Finding words in continuous speech.

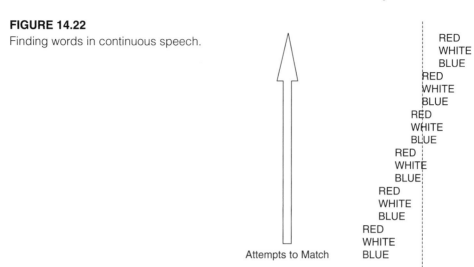

Attempts to Match

Input Speech Sequence: THE QUICK **RED** FOX

The dog buried ___ bone ___ the ground.

FIGURE 14.23
Using knowledge of language to deduce words.

ferent ways and still receive the correct information. Many operator functions are being automated in this way. Naturally, voice recognition systems for public telephone networks must be both reliable and speaker-independent. Other applications include voice car rental, airline reservations, and medical or legal transcription.

Voice-to-text systems are now widely available for home use. To enable continuous speech recognition of large vocabularies, most transcription systems require some kind of training phase to improve the quality of the templates and hence the recognition rate. Speech recognition systems can help prevent fraud with calling cards, cellular phones, or banking services. Voice-activated controls are available in some aircraft, automobiles, and boats, though the most critical operations are left to manual or computer control. In moments of stress, the characteristics of a pilot's voice will change, reducing recognition accuracy at what may be a critical time.

14.3   VOICE AND MUSIC SYNTHESIS

Electronic synthesizers have become standard equipment for musicians. They are capable of mimicking many traditional instruments—like pianos, guitars, violins, and flutes—with increasing integrity. Furthermore, the sounds that synthesizers produce are not limited to existing instruments. They are hugely flexible in their ability to store, mix, join, and transform

t	sin(t)
0.0000	0.0000
0.5236	0.5000
1.0472	0.8660
1.5708	1.0000
2.0944	0.8660
2.6180	0.5000
3.1416	0.0000
3.6652	−0.5000
4.1888	−0.8660
4.7124	−1.0000
5.2360	−0.8660
5.7596	−0.5000

steps.wav

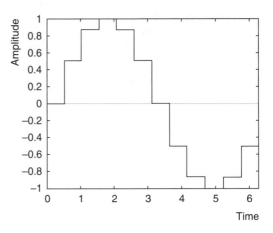

FIGURE 14.24
Twelve-point table lookup for sine wave.

sounds. Synthesis techniques can be used to produce human voices as well. It would be impractical, for example, to record every possible bank balance for automated banking. Instead, simple synthesis systems string together prestored segments to produce the responses required.

The simplest kind of synthesis reproduces audible pure tones, originating from sine and cosine signals. While all processors offer basic operations like addition, subtraction, multiplication, and division, many small, fast processors do not offer sine, cosine, or other mathematical functions. These functions can be produced using lookup tables or power series expansions, as discussed in Section 12.5. The lookup table in Figure 14.24, for example, generates a 12-point sine wave with amplitude 1 and period 2π. The table is stored in memory, and once every processor cycle, a sine value is taken from the table and sent to the output. Notice that, for the sine wave, only the first group of points (first 4 of 12 in this case) are absolutely needed. All the others can be deduced from the regular shape of the wave, as observed in Section 12.5 and Section 13.8.1. As the graph in Figure 14.24 shows, 12 points produce a rather poor sine wave shape. Improvements can be made by using **linear interpolation** between the table values. The result is shown in Figure 14.25. Linear interpolation joins pairs of points by straight lines, where points along the lines are generated between the values in the table in a manner to be detailed shortly. Lookup table methods can be applied to all signals, even when no functional description is available.

When a signal can be described neatly by a mathematical function, it can be expanded into a power series using a tool like the Maclaurin series, described in Appendix A.16. The Maclaurin series expansion of a sine wave, also presented in Section 12.5, is:

$$f(t) = \sin(t) = t - \frac{t^3}{3!} + \frac{t^5}{5!} - \cdots \tag{14.1}$$

where $n! = n(n-1)(n-2)...(2)(1)$. The value of $\sin(t)$ for any value of t can be approximated by using this power series expansion, using multiplication, addition, and subtraction

FIGURE 14.25

Table lookup plus linear interpolation.

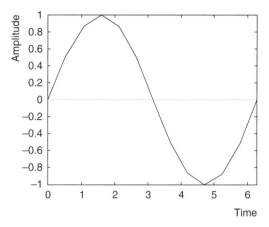

lininterp.wav

only. The approximation can be improved by including more terms. Even the three terms shown in Equation (14.1) are sufficient to ensure an error in the value of sin(t) of less than 0.005 from $t = 0$ up to $t = \pi/4$. As indicated above, values for this first quarter-period are enough to deduce the rest of the signal. These tricks were used in the sine generation program of Section 13.8.1.

Synthesizing good sine waves seems an insignificant beginning when the goal is the reproduction of full symphony orchestras, yet analogous techniques are used. **Wavetable synthesis** (Heckroth, 1995) is one of the most widely used strategies for powerful and flexible music synthesis. Like table lookup for the sine wave, wavetable synthesis builds sounds from high quality digital sound samples stored in a synthesizer's memory. The sounds can be played from memory on demand. Storing segments of sound for synthesis must be distinguished from the simple digital recording of an entire piece of music. The main difference is that the stored segments can be used efficiently to create a multitude of compositions. Simple compositions require only a single note at a time, while more ambitious pieces require multinote combinations from one or more instruments. Music synthesizers that can play more than one note at a time are said to be **polyphonic**. The number of **voices** describes the number of sounds that can be produced at once. Symphonic sound can be created through this kind of synthesis only if the number of voices is sufficient.

Wavetable synthesis seems simple enough, but implementation is not so straightforward. Consider, for example, the task of storing in memory digital samples that represent all possible sounds that a flute can produce. As a starting point, each note could be digitally recorded. However, the frequency content of each note a flutist plays may change as the note progresses through the phases of **attack**, **decay**, **sustain**, and **release**, illustrated in Figure 14.26. Storing all of these portions of each note becomes very memory-intensive. To make the task even more difficult, the loudness with which a note is played also affects its spectral content. If each note must be recorded at different loudness levels, memory requirements are multiplied again.

Clearly, the most simplistic use of wavetable synthesis demands prohibitively massive memories. The problem is solved with a battery of clever tricks that use a relatively

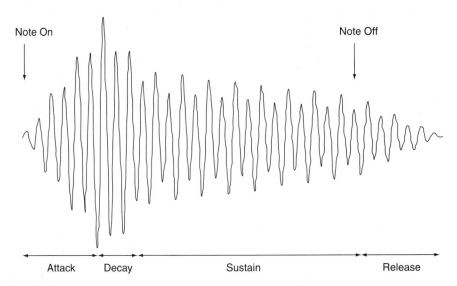

FIGURE 14.26
Features of musical note (adapted from Heckroth, 1995).

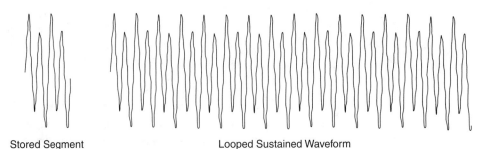

Stored Segment   Looped Sustained Waveform

FIGURE 14.27
Looping (adapted from Heckroth, 1995).

small set of digital samples to create rich and realistic sounds (Heckroth, 1995). One of these tricks is called **looping**. During the attack and decay portions of a note, sound characteristics are changing relatively rapidly. During the sustain portion, on the other hand, sound characteristics are more constant. This can be modeled by repeating a short segment of stored samples over and over again, as shown in Figure 14.27. Some care must be taken to ensure that the loop length is an integer multiple of the fundamental frequency of the sound. If not, the loop segments will fail to join smoothly at loop boundaries.

The sounds produced by looping can be improved by applying an **amplitude envelope** to the sound. This approach assumes that the frequency content of a sound is constant throughout, while the amplitude of the sound changes. Figure 14.28 contains a typical amplitude envelope. Linear line segments describe the overall shape of a complete musical sound. The attack portion shows a steep slope, to reflect a rapid note onset; the sustain por-

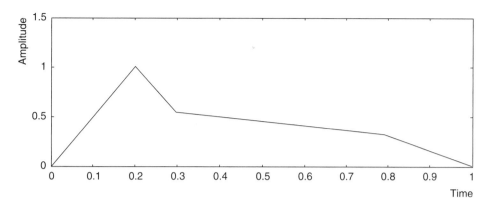

FIGURE 14.28
Amplitude envelope (adapted from Heckroth, 1995).

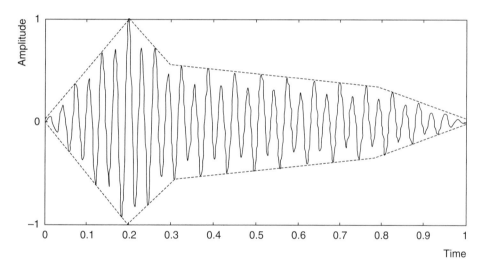

FIGURE 14.29
Note produced with amplitude envelope (adapted from Heckroth, 1995).

tion is nearly flat, showing that the sound is nearly steady in this region. This envelope multiplies with the characteristic sound variation for an instrument. For example, if it is applied to the sustained signal in Figure 14.27, a waveform like the one shown in Figure 14.29 is produced. When the same note must be played more loudly or more softly, a new envelope can be applied to the same sustain waveform to produce a new sound. Although the model of Figure 14.26 can be applied to many musical sounds, such as those produced by horns and stringed instruments, it does not work well for short, impulse-like sounds such as those produced by drums. For these sounds no sustain segment exists, and there is no alternative but to store complete sampled segments.

The use of amplitude envelopes dramatically reduces storage requirements when a single note of a single instrument must be played at different loudnesses. Variations of the note played more loudly or more quietly use the same sampled sound segment. **Pitch shifting** can help to produce notes at other frequencies for the same instrument by reusing the same stored sound segment. For example, if sound segments are stored for a piano's middle C, other notes can be produced by accessing the stored sound segment at different rates. When every other sample of the middle C segment is accessed in successive processor cycles, for instance, the frequency of the new sound doubles, producing the C note of the next higher octave. Interpolation of the kind shown in Figure 14.25 permits values between stored samples to be computed. This means that notes at frequencies below the original can also be produced. For example, by accessing the samples stored for middle C as well as values halfway between them in successive processor cycles, a note at half the original frequency, one octave below the original, is produced.

Linear interpolation is the simplest way to find signal values between stored samples. Figure 14.30 shows two samples, $x[n]$ and $x[n + 1]$. A sample value between them, $x[n + m]$, may be found using similar triangles, as illustrated in the figure. The ratios of height to width for the largest and smallest triangles give

$$\frac{x[n + 1] - x[n]}{n + 1 - n} = \frac{x[n + m] - x[n]}{n + m - n}$$

which simplifies to

$$x[n + 1] - x[n] = \frac{x[n + m] - x[n]}{m}$$

Solving for $x[n + m]$:

$$x[n + m] = m\,x[n + 1] + (1 - m)\,x[n] \qquad \textbf{(14.2)}$$

Figure 14.31 shows a sequence of sample values with amplitudes 0.5, 0.23, 0.7, 0.8, and 0.65 for sample numbers 0, 1, 2, 3, and 4, stored in memory. These samples belong to the piano note middle C, at frequency 262 Hz. Suppose the note B (below middle C) must be

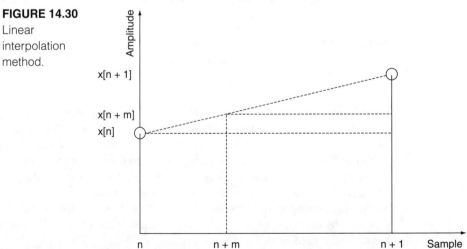

FIGURE 14.30
Linear interpolation method.

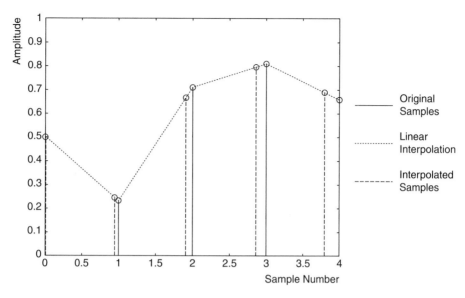

FIGURE 14.31

Example of linear interpolation.

synthesized from the middle C samples. B has a frequency $17/18^{th}$ that of C, or about 247 Hz. Thus, a new sequence of samples must be found at locations with sample numbers 0, 17/18, 34/18, 51/18, 68/18, and so on, in place of the original samples at sample numbers 0, 1, 2, 3, 4, and so on. Figure 14.31 shows how the interpolated samples are related to the original ones. As an example, the third interpolated value would be calculated using Equation (14.2), with $n = 1$ and $m = 16/18$:

$$x\left[\frac{34}{18}\right] = x\left[1 + \frac{16}{18}\right] = \frac{16}{18}x[2] + \left(1 - \frac{16}{18}\right)x[1]$$

$$= \frac{16}{18}(0.7) + \left(1 - \frac{16}{18}\right)(0.23) = 0.6478$$

EXAMPLE 14.2

Compute the exact value of $x[51/18]$ for the samples in Figure 14.31.

Equation (14.2) gives

$$x\left[\frac{51}{18}\right] = x\left[2 + \frac{15}{18}\right]$$

$$= \frac{15}{18}x[3] + \left(1 - \frac{15}{18}\right)x[2] = \frac{15}{18}(0.8) + \left(1 - \frac{15}{18}\right)(0.7) = 0.7833$$

Timbre refers to the character of a note played by a particular instrument. When two instruments like a flute and a violin play the same note at the same loudness, the differences between them are due to their differing timbres. When interpolation is used to shift the pitch of a note, the timbre also changes. Thus, interpolation is normally used only to create notes close in frequency to the original. To cover a complete musical scale for an instrument, several sampled segments must be stored, each of which provides a small range of notes by interpolation. The technique of using a different stored sound sample for each range within an instrument's sweep is referred to as **key splitting**. Care must be taken to ensure that interpolation does not cause notes at the high end of one range to differ in timbre from notes at the low end of the next range.

Other tricks for improving the quality of music synthesis pertain in particular to sampling issues. First, sampling rates must be adequate to capture all audible frequencies. Since the human ear can respond to frequencies up to 22 kHz, 44 kHz is normally used for sampling. Second, musical sounds will be reproduced most faithfully if their amplitudes are matched to the dynamic range of the A/D converter when they are sampled. In this way, the audible effects of quantization noise are minimized. Small analog signals may be amplified and large analog signals may be attenuated prior to sampling, to suit the range of the data converter. Quantization errors affect quiet sounds the most, since the errors can have amplitudes of the same order of magnitude as the sounds themselves. Companding, discussed in Section 14.1.4, can be used to reduce the impact of quantization on these sounds.

Many advanced techniques are used in commercial music synthesizers to model subtle yet important characteristics of musical sounds that cannot be captured by the simple methods introduced in this section. They include oversampling (to improve the results of interpolation), velocity splits (to more accurately model the differences between loud and soft versions of the same note), low frequency oscillators (to add vibrato or tremolo effects), layering (to create a richer sound by combining several constituent sounds), digital filtering (for a wide variety of synthesis improvements), and waveguides (to model instrument sounds mathematically). Analog Devices markets a "real-time music engine" (Analog Devices, 1996), based on its SHARC ADSP-2106x that can synthesize musical sounds using all of these methods concurrently, and that provides 64 voices of polyphony.

The synthesis of a human voice is really just another music synthesis task. Many of the same techniques are used. **Speech synthesis** systems are sometimes referred to as **text-to-speech** systems. Speech may be produced from printed words or from phonetic markings that take the place of words. Other markers may indicate stressed syllables, speech tempo, and speaker accent. When synthesis is intended for a large public audience, to provide airline schedule information by phone, for example, natural-sounding speech is of the utmost importance. When synthesis is used by a highly motivated listener, such as a visually challenged individual, intelligibility and speed are more important than other speech qualities.

Speech synthesis schemes vary according to the requirements of the application. The simplest method is to record and replay whole phrases. This method works well when a limited number of distinct phrases are necessary. For example, interactive banking requires a few phrases, along with numbers. "Savings," "checking," "transfer," "account balance," "bill payment," "to," and "from" are examples of the phrases that might be used.

Number phrases like "thousand," "hundred," "seventy," and "four" are also used. By stringing together recordings of these words, all messages can be constructed. For instance, the message "(You have requested a) (bill payment) (in the amount of) (four) (hundred) (and) (eighty) (five) (dollars) (and) (thirty) (five) (cents) (to be forwarded) (from your) (checking) (account to) (XYZ)" might be constructed from 18 separate phrases as indicated by the parentheses. This simple system produces speech that is very easy to understand but lacks realistic sound. The recorded phrases are, of course, high quality reproductions of natural speech, but the transitions between the recordings are not. Sounds in natural speech tend to blend together at word boundaries, something the simplest speech synthesizers cannot capture.

When a large variety of words or phrases must be synthesized, a more flexible approach is required. Some synthesizers use only the phonemes, the smallest units of a language, to construct words and phrases. Languages like English, French, Spanish, German, and Dutch contain only about 40 or 50 phonemes in all, making the phoneme approach highly efficient and flexible. Unfortunately, abrupt transitions between phonemes make the resulting speech stilted and unnatural. For example, in natural usage, the mouth forms the first "c" in "cucumber" very differently from the second. In each case, while speaking the /k/ sound, the mouth prepares for the vowel to follow. For this reason, the quality of synthesis can be improved if pairs of phonemes, called **diphones**, are stored instead. The segments start in the middle of one phoneme unit and end in the middle of the next. Diphone synthesis creates more natural sounding speech, but a larger number of basic elements must be used: For N phonemes, N^2 diphones can be formed. Partial syllables, syllables, or words can also be used as the basis for synthesis. As the units become longer, the quality improves, but the number of elements stored in the synthesizer memory must increase. For example, English contains over 10,000 possible syllables. Apart from the content of the speech elements themselves, other factors that contribute to the quality of synthesis include features like the rhythm, intonation, and accent patterns of the speech. These characteristics fall under the general heading of **prosody**. Often, synthesizers rely on predeveloped rules to apply prosodic features to synthesized speech segments, with the goal of producing more natural sounding output.

The input to a synthesizer, combined with the attendant prosodic information, must, at the end, be synthesized to produce audible output. There are three main categories of speech generators: articulatory synthesizers, formant synthesizers, and concatenative synthesizers. **Articulatory synthesizers** use a physical model of the human vocal apparatus to simulate speech sounds. Inputs for this type of synthesizer include the positions and speeds of the tongue, lips, vocal cords, and jaw. **Formant synthesizers** concentrate on modeling the acoustical properties of speech sounds. A set of filters is used to model articulation resonances, and outputs are generated by applying appropriate excitations to these filters. For example, a noise source provides the needed excitation for the voiceless sound "f," but a periodic source is better for a vowel sound like "oo" or "ah." **Concatenative synthesizers** link together speech recordings that may be phonemes, diphones, or words, and smooth the transitions from one segment to the next. Formant and concatenative synthesizers have had the most commercial success.

One unusual application of voice synthesis (Depalle, Garcia, & Rodet, 1994; Scammell, 1999) was motivated by castration or, rather, the lack thereof. Castrati, it seems, were blessed with unusually wide vocal range and extremely large lung capacity that resulted from "the surgical intervention they had undergone." Some were able to sustain a note for over a minute. Castrati singers gained in popularity during the 16th century, favored by Pope Clement VIII over young boy singers who tended to behave poorly. Women, who might naturally have been able to sing the soprano roles, were, at that time, flatly forbidden to sing. Offering sons for castration was a path out of poverty for poor families, though many lost their boys to medical complications. Those that survived were intensively trained. Musical scores were written specifically to take advantage of special castrati singing abilities. Sadly, performance of these scores has been limited by the lack of available singers, as castration was outlawed at the end of the 19th century. Depalle and colleagues (1994) undertook to re-create a castrati voice for the soundtrack of a movie (*Farinelli, il castrato*) by transforming the voices of a countertenor and a soprano singer in a process called **voice morphing**. For example, the countertenor voice was attenuated at the higher frequencies, and the timbre of the soprano voice was altered to match the timbre of the countertenor. Notes neither singer could reproduce were generated independently. The work has made new recording of castrati scores possible.

14.4 GEOPHYSICAL PROCESSING

Offshore oil wells are expensive items, perhaps $10 million in the Gulf of Mexico to hundreds of millions of dollars in Arctic waters. In the past, geologists had to work with only the sparsest information about the subsurface to make educated guesses about where to drill. Modern geophysical techniques and instrumentation, combined with digital signal processing, have taken some of the guesswork out of the process. When potentially successful sites can be identified, or poor sites disqualified, a great deal of money can be saved.

Over time, oil deposits form where organic material and heat are present. The oil collects when a reservoir for the oil exists and when a fine shale, or similar material, provides a natural cap for the reservoir. Since oil is lighter than water, the best places to look are often the highest places in the subsurface layers, called anticlines, marked by arrows in Figure 14.32. This figure shows several anticlines, formed by stratified layers of rocks below the surface of the ground or below the sea bottom.

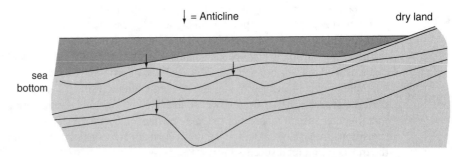

FIGURE 14.32
Anticlines.

In the absence of other information about a potential drill site, anticlines represent the best bets for finding oil or gas deposits. When an anticline occurs in shallow water, relatively low drilling costs provide a further incentive. The goal of an exploration well is to determine if enough oil or natural gas can be found in the rocks to justify drilling production wells. In the process of drilling, a great deal of geological and geophysical information can be gleaned. Perhaps most obviously, the fluids ejected from the hole can be analyzed for oil content. Second, while a normal drilling bit grinds up rock material, a special coring bit can be used to produce rock cores for analysis by a geologist. Third, well logging tools periodically replace the drilling bit and are lowered into the hole to measure a variety of parameters that describe the rocks surrounding the hole, including porosity, fluid permeability, and seismic velocity. The velocity probe is one of these logging tools. It sends a stream of acoustic pulses toward the rock and sediment around the drilled hole and records the acoustic energy that returns, as shown in Figure 14.33.

Because the distance between the transmitter and receiver is known, measuring the time it takes for a sound pulse to move from one to the other allows the velocity with which the sound travels in the material to be calculated. Each type of rock has a characteristic velocity, so this information can be helpful in identifying the type of rock. Table 14.1 lists some common rock types and their associated sound velocities. Because there exist two different types of seismic waves that can propagate in rocks, two velocity ranges are provided in the table. *P* **waves,** or pressure waves, are longitudinal. In these waves, the vibrational motion of the rock particles is in the same direction as the direction of propagation of the acoustic wave. *S* **waves,** or shear waves, are transverse. For these waves, the vibrational motion of the rock particles is perpendicular to the direction of propagation. A velocity probe can be designed to measure both *P* waves and *S* waves. Figure 14.34 provides an example of the output of a velocity meter. The vertical axis is the depth in meters below the sea bottom, known because the exact depth of the probe is known at all times. The horizontal axis is the velocity of sound in the rock at each depth. This particular graph shows the velocity profile for a well drilled off the coast of British Columbia.

FIGURE 14.33

Section of well bore.

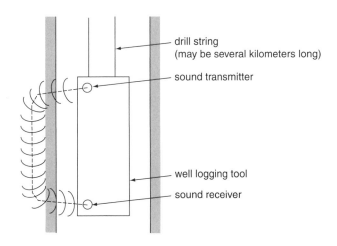

TABLE 14.1
Sound Velocity in Rock

	P Wave Velocities	*S* Wave Velocities
Sandstone	1400 to 3500 m/sec	2000 to 3500 m/sec
Limestone	1700 to 7100 m/sec	2800 to 3500 m/sec
Granite	5000 to 6000 m/sec	2900 to 3200 m/sec
Basalt	5000 to 6500 m/sec	2700 to 3200 m/sec

FIGURE 14.34

Raw velocity data.

(Courtesy Stewart Langton)

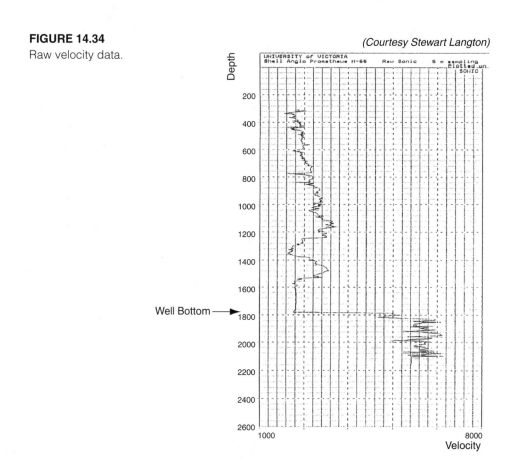

While the velocity data from the well can be used directly for information about the well bore, it is more often used to provide reference data for comparison with seismic data collected by ship at locations away from the well. To make this possible, the velocity data must first be processed to provide a synthetic seismogram that can be compared directly to seismic data. The velocity data in Figure 14.34 is first "blocked" by averaging the velocities over small spans of depth. Figure 14.35 suggests how blocking affects the data at a

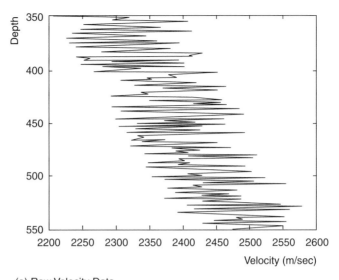

(a) Raw Velocity Data

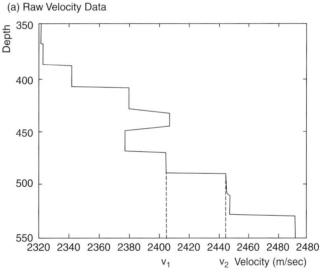

(b) Blocked Velocity Data

FIGURE 14.35
Processing of velocity data.

small scale. The jumps in velocity present in the blocked data, marked for one transition in Figure 14.35(b), are recorded as reflection coefficients using the formula:[2]

$$R = \frac{v_2 - v_1}{v_2 + v_1} \tag{14.3}$$

[2]   This formula assumes the densities of rock in neighboring layers are nearly uniform.

A reflection coefficient is calculated at each transition. For the marked transition, the velocity is $v_1 = 2404.2$ m/sec in one block and $v_2 = 2444.8$ m/sec in the next. The reflection coefficient is 0.0084. When the velocity drops from one block to the next, the reflection coefficient is negative.

EXAMPLE 14.3

The velocities on either side of a rock boundary are $v_1 = 2500$ m/sec and $v_2 = 1600$ m/sec. Find the reflection coefficient for the boundary.

From Equation (14.3), the reflection coefficient is

$$R = \frac{v_2 - v_1}{v_2 + v_1} = \frac{1600 - 2500}{1600 + 2500} = -0.2195$$

The reflection coefficients provide a new view of the data, shown in Figure 14.36. Each spike on this graph represents the boundary between two types of rocks with differ-

FIGURE 14.36

Reflection coefficients.

(Courtesy Stewart Langton)

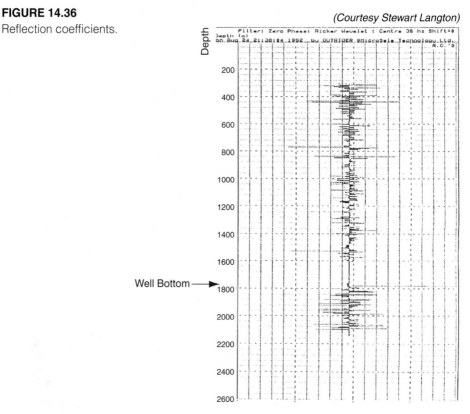

Reflection Coefficients

ent velocities, for example, between shale and sandstone, or between granite and sandstone. The length of the spike indicates the value of the reflection coefficient, which may be positive or negative. The reflection coefficient determines the amount of seismic energy that is reflected from a boundary compared to the amount of energy that is transmitted through the boundary. The same principles apply in optics, where the reflection coefficient of a mirror is nearly 1.0, while the reflection coefficient of a piece of glass is close to 0.0.

A drilled hole in the ocean floor can provide a great deal of information about the oil-producing potential of the rock at the bore site. Gathering information by drilling, however, quickly becomes an expensive proposition. Ideally, it would be possible to extrapolate from the data collected at a drill site to deduce information about the surrounding area. A seismic survey can provide the data that support exactly this kind of extrapolation. These surveys are relatively cheap, costing only a few million dollars for a week or two of data collection. Data describing the sea floor and the sediments below it are gathered by firing an air gun and recording pressure changes with pressure transducers called geophones. At each new ship position, up to 20 seconds of reflection data samples are collected. After processing, the digital data are presented in a diagram like the one in Figure 14.37. In this figure, ship position is shown along the horizontal and time is shown along the vertical. All data are collected from the surface of the sea along the line of the ship's path, marked on the map in Figure 14.38.

To understand how Figure 14.37 should be interpreted, note first that in a seismic cross-section, each vertical trace represents energy that is reflected at the boundaries between different types of rocks. If the earth were made of a single homogeneous type of rock, then energy from the air gun would travel downward without ever returning to the ship. Where a boundary occurs between two types of rocks, some energy is reflected back to the ship, the same way an echo from a shout returns from a distant rock face. The vertical axis in Figure 14.37 records the time taken for the reflection of sound from the air gun to reach the ship, and the dark lines in the figure mark the boundaries where reflections of significant energy occur. Since the subsurface of the earth is made of stratified layers of different

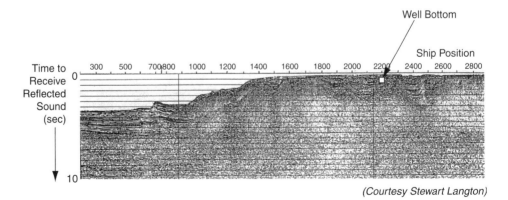

(Courtesy Stewart Langton)

FIGURE 14.37
Seismic cross-section.

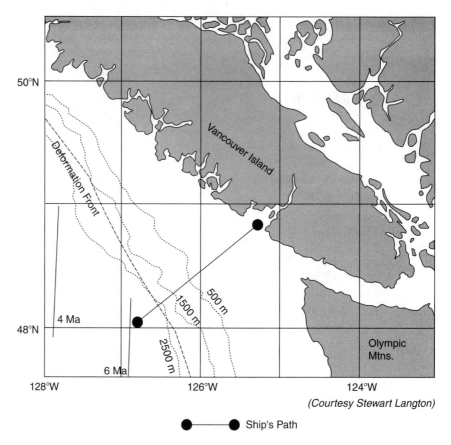

(Courtesy Stewart Langton)

FIGURE 14.38
Ship's path during collection of seismic data.

types of rocks deposited at different times, there are many boundaries from which reflections occur. Even within layers of sandstone, basalt, or other rock, however, there are many additional minor boundaries. Thus, thousands of reflected waves are received at the ship, from both major and minor boundaries. The size of each received pulse is proportional to the value of the reflection coefficient. Figure 14.39 shows the details of the reflected pulses. The sound wave produced by the air gun is not a single spike but a real signal with nonzero width and finite height, so the pulses in the figure are not perfect horizontal spikes like those in Figure 14.36. Since the boundary between different types of rocks produces the largest reflection coefficients, it is the largest pulses on the seismic profile that are of interest. As seen in Figure 14.39, the positive pulses are shaded black to make the data easier to interpret. This shading causes important boundaries to appear as dark bands, which allows important features in the seismic cross-section to be identified. This explains the multitude of fine horizontal dark lines in Figure 14.37. Figure 14.40 presents an expert's interpretation of these seismic data.

FIGURE 14.39
Detail of seismic pulses.

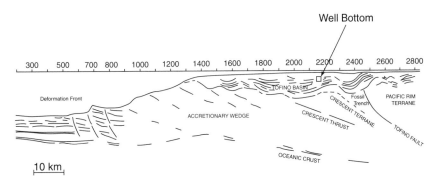

(Courtesy Stewart Langton)

FIGURE 14.40
Interpreted seismic data.

The presentation of the seismic data in Figure 14.37 entails a great deal of background processing. The ship performing the survey drags dozens of geophones behind it and continues to move during the survey. Therefore, a feature at a particular location will be registered by many geophones at many different times, as suggested in Figure 14.41. Also, the longer the path, the greater the attenuation. The pulses in Figure 14.41 all correspond to reflections from the same feature. They are realigned, boosted as necessary, and added together to give a vertical trace in a seismic cross-section.

The shipboard survey can provide some useful indications about the composition of the sediment below the seafloor. Unfortunately, the vertical axis is time, not depth, so major features cannot be easily mapped to absolute locations below the seafloor. The depths of important features can be found only by comparing seismic data to data from a drilled well. Reflection coefficients like those in Figure 14.36, obtained from a well, cannot be

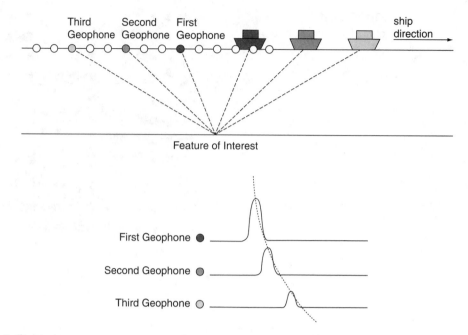

FIGURE 14.41
Correcting geophone data for seismic cross-section.

compared directly to the seismic data, however. To make matching easier, the reflection co-efficients are convolved with the impulse response of an air gun to create a synthetic seis-mogram. Because both depths d and velocities v for the well data are known, the depths along the vertical axis of the reflection coefficient data in Figure 14.36 can be converted to times using $t = d/v$. The relationship between depth and time is nonlinear. The synthetic seismogram for the reflection coefficients of Figure 14.36 is shown in Figure 14.42. It is plotted with time on the vertical axis.

Features in the seismic cross-section of Figure 14.37 and a synthetic seismogram con-structed from traces like the one in Figure 14.42 may now be matched directly. From ve-locity information and an analysis of rocks and fluid from the drilled hole, the depths of lay-ers containing promising rocks can be identified exactly. Layers identified in the well can be followed in the seismic cross-section. The darkest lines in Figure 14.37 show how the contours of the layers change as the distance from the drilled hole increases. The bottom of the well is marked on both the synthetic and the measured seismogram profiles, and also on the graph of reflection coefficients. If oil or natural gas is found in a particular layer of rock in a well, then that layer can be traced in the seismic profile to locations away from the well that may also be promising drill sites.

Choosing a successful drill site is a task driven partly by skill and partly by luck. A great deal of effort is devoted to the task because of the extreme expense associated with drilling into the ocean floor. The costs are especially high when drilling must occur in deep

FIGURE 14.42

One trace of a synthetic seismogram.

(Courtesy Stewart Langton)

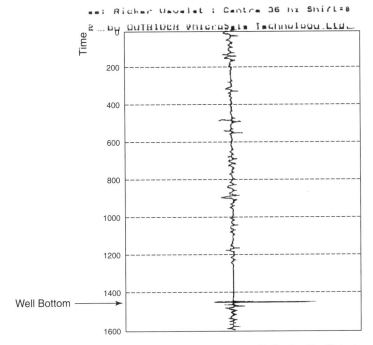

Reflection Coefficients

water or hostile Arctic environments. Heavy use of digital signal processing can provide the means to extrapolate from drilled well data and improve the chances of finding new oil or natural gas deposits in the same region.

14.5 ENCRYPTION

14.5.1 Encryption Basics

When the postal system was the primary means of sending messages from one place to another, normal security was taken care of with the sealing of each envelope. Amateur tampering could easily be detected, and the sheer cost and trouble of trying to find one worthwhile letter among thousands was an unattractive proposition. In this age of electronic communication, huge numbers of messages are passed in the form of binary bitstreams from originators to recipients. The bitstreams move along cables that connect the communicating computers, or pass through the air as infrared or radio signals. Listeners that can access the medium of communication can in principle also eavesdrop on the messages being passed. This includes all the intermediate nodes that conduct a digital message from a sender to a receiver in a computer network. Potential problems exist for electronic mail, automated teller machines, cell phones, and many other systems.

1101001111011000011111111101001011101001011110001010111011111010111110

FIGURE 14.43
Binary message passing between computers.

Translating a string of ones and zeros like the one in Figure 14.43 into a comprehensible message may seem an impossible task. For computer communication to succeed, however, the rules for conversation are carefully structured. Messages are passed in the form of frames that contain known segments. A frame might begin with a flag made up of eight ones, followed by an 8-bit originator number, an 8-bit recipient number, 16 bits of data, and an 8-bit code for error checking. Someone watching a bitstream could easily search for sequences of eight ones and thereby be able to deduce the location, if not the meaning, of the data. A relatively small number of standard formats for computer communication exist, and of these, some are far more common than others.

Suppose Johnny e-mails Grandma a letter thanking her for his birthday money. The characters constituting the letter would most likely be coded using the popular ASCII character set, which assigns a 7-bit code to each letter. If the communication code described above were used to transmit the letter, each frame would contain two ASCII characters. A watcher knowing the format of the frames could quickly reconstruct the message. Where electronic message interception differs from the interception of post is that electronic message theft can proceed without any alteration of the message, and without the recipient having any awareness of the crime. The eavesdropper, however, must know how the data are coded to be able to interpret the data, as suggested by the above example. **Encryption**, or scrambling, can help avert the possibility of unauthorized access to data. It recodes numbers or text to ensure secure communication, so that even if the message is intercepted, it will be unreadable. Of course, the intended receiver must have a way to decrypt the messages he or she receives. This is normally accomplished using some kind of key, as illustrated in Figure 14.44.

The goal of encryption is to destroy or disguise any and all patterns that exist in the data being encrypted. In normal usage, some characters occur more frequently than others; in most encrypted messages, all characters occur with equal frequency. For the sake of privacy, electronic mail messages can be encrypted before being sent through the Internet, and the same goes for credit card numbers and other sensitive personal information. In the case of voice communication over cell phones, digital voice signals can be scrambled as they are

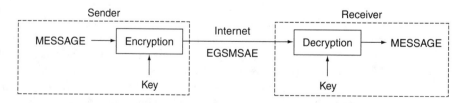

FIGURE 14.44
Symmetric key encryption.

produced by the speaker and be deciphered by the receiver at the other end. For security reasons, many countries have or have had restrictions on importing and exporting encryption algorithms. In the United States, for example, export limitations on encryption were lifted as recently as 1999 (Mills, 1999). The Russian government still prohibits development, production, implementation, and operation of cryptography without a license, something limited primarily to state organizations (Koops, 1999).

Word scrambles are a simple example of an encryption scheme called transposition. **Transposition** requires that the elements of a message be rearranged so that the result is garbled and unreadable. On decryption, the elements resume their original places. Unfortunately, this system does not stand up to attack. Knowledge of word and letter frequency, along with exhaustive word comparisons, could be used to decipher a scrambled message like "EC BRODETOR."[3] For guaranteed security of electronic mail, Internet credit card submissions, as well as voice communication, more advanced encryption methods are needed.

14.5.2 Pretty Good Privacy

Pretty Good Privacy (**PGP**) is one encryption scheme used for electronic mail. Only an overview of this method is included here, with a view to introducing important encryption techniques and terms. A detailed description of a second encryption scheme follows in the next section. PGP is designed to ensure that no one but the intended reader can access a message, and it has proved itself to be resistant to attack by intruders. Simple **symmetric key encryption** systems employ a single key, which both sender and receiver use to encrypt and decrypt as shown in Figure 14.44. Passing the key from one party to the other in a secure way is one problem with this system: The key cannot be encrypted until it is shared by both ends. Couriers or other trusted intermediaries must be used. PGP uses the RSA algorithm, which is a **public key encryption** method. In this scheme encryption and decryption keys are different, but they are mathematically related through the product of two prime numbers (numbers that can be divided only by one and themselves). The keys are generated as a pair. The encrypt key is made public and is called the **public key**. The decrypt key is kept private and is called the **private key**. A message encrypted with the public key can be decrypted only with the private key. Security of the public key is not an issue, and safeguarding the private key is comparatively easy, since it does not need to be shared. The essential elements of public key encryption are illustrated in Figure 14.45.

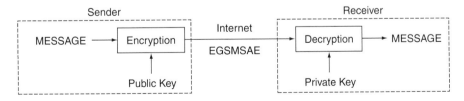

FIGURE 14.45
Public key encryption.

[3] RED OCTOBER

When keys are short, one key can be calculated from the other, meaning that security does not exist. The longer the key, the more difficult it is to compute one key given the other. Successful encryption relies, of course, on the assumption that the private key will remain private. Because key construction is based on long primes, security in turn depends on the ability to find larger and larger prime numbers. Naturally, a key must always be judged against current technology. A fast computer can use the "brute force" approach of trying all possible factors to find the prime numbers that make a product. As computer power increases, the number of bits needed for the public and private keys must increase, to ensure that the brute force approach must still take months at least to succeed. It is believed that 2048-bit public and private keys will guarantee security from compromise for at least the next 10 or 15 years. The drawback of long keys, of course, is that the encryption process takes longer.

When a file is prepared for secure transfer, PGP first chooses a random 128-bit symmetric key to encrypt the file. The public key is then used to encrypt this symmetric key. The encrypted symmetric key and the symmetrically encrypted file are sent. The recipient uses his or her private key to decrypt the symmetric key and then uses this key to decrypt the message. The same file is never encrypted the same way twice because the symmetric key is always chosen randomly.

PGP is also capable of authenticating **digital signatures**, a means of verifying that e-mail is coming from the person who claims to send it. The "signature" is created using a **hash function**, which operates on a message of any length and produces a fixed-length value. Thus, the signature is a garbled and condensed version of the message itself. The digital signature is normally placed at the beginning of a message and is encrypted using the private key before sending. The recipient receives the message and computes his or her own hash result for the message received. This result can be compared to the signature received using the public key that corresponds to the sender's private key. If the two match, the message can be confirmed as coming from the person whose public key the recipient is using. If the recipient can be sure to whom a public key belongs, perhaps by checking a public key server, then a digital signature provides full authentication. Thus, a compromised private key not only permits unauthorized decryption of messages but also creates the potential for fraudulent use of a digital signature.

MD5 and SHA1 (Secure Hash Algorithm) are two examples of hashing algorithms that can be used for digital signatures. MD5 is presently used by PGP, but may be replaced by SHA1 in future versions. A good hash must be easy to calculate, even for large message lengths. Further, it must be very unlikely that two different messages should ever produce the same hash results. Finally, given the hash alone, it must be extremely difficult to deduce the message.

14.5.3 Data Encryption Standard

The Data Encryption Standard (**DES**) is an encryption scheme that has been in use since the late seventies. It is a symmetric key algorithm, with the same key used for both encryption and decryption, as suggested in Figure 14.46. The DES key contains 64 bits, but every eighth bit is a parity bit, used for verifying that the key is legitimate. Thus, the key contains 56 content bits, which means that 2^{56} possible keys exist. Since 2^{56} works out to

FIGURE 14.46

Data Encryption Standard
encryption and decryption.

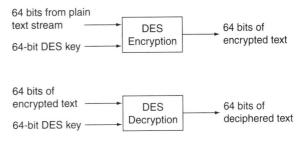

over 72 million billion possibilities, it is no mean feat to try them all, but today's computers have made exactly that a reality. DES first succumbed to brute force attack in July of
1998 (Mills, 1999), after 56 hours of processing. This record has since been broken.

Unlike most encryption algorithms, the algorithm for DES was always public
knowledge. It was issued by the U.S. National Bureau of Standards in 1977 and heavily
tested by the National Security Agency. The success of encryption lay completely in the
secrecy of the key. When compromising the key in a relatively short amount of time became possible, the DES algorithm became useless for applications requiring true secrecy.
Using credit cards on the Internet, accessing ATMs through access cards, and banking via
computer, for instance, all demand the highest possible security. **Triple DES** is one attempt to remedy the situation. It uses the DES algorithm three times and uses either two
or three 56-bit keys. This amounts to using a single key of length 112 or 168 bits. Since
each additional bit doubles the number of possible keys that can be created, triple DES
represents a dramatic increase in the level of security that can be obtained. Nevertheless,
the National Institute of Standards and Technology has invited proposals for an AES, or
Advanced Encryption Standard, to address growing security needs for this millennium.

The DES encryption algorithm is pictured in Figure 14.47. A 64-bit message to be
coded is first rearranged according to sequence A in Table 14.2(a) and then split into left
and right halves consisting of 32 bits each. The right half is coded using a subkey coding
process to be described, producing a 32-bit result. This result is combined with the left half
in an exclusive-OR (XOR) operation: In bit positions where both 32-bit numbers have the
same value (0 and 0, or 1 and 1), the result is 0; where they have differing values (0 and 1),
the result is 1. The number resulting from the XOR operation is coded using subkey 2 and
combined with the right half of the original data. The coded left and right halves are combined and cycled for a total of 16 stages, as depicted in Figure 14.47. After one final re-
arrangement, this time following sequence B of Table 14.2(b), the final 64-bit encrypted result is obtained. Sequences A and B are inverses of one another in the sense that, at the start,
bit 58 of the input stream becomes bit 1 of the output stream and, at the end, bit 1 of the input stream becomes bit 58 of the final output. The subkey coding process remains to be explained.

Figure 14.48 gives details of the subkey coding process required in Figure 14.47.
Each 32-bit sequence is reordered and lengthened (by repeating some bits) using sequence
C in Table 14.3(a). The 48-bit result is XORed with a 48-bit subkey derived from the primary 64-bit DES encryption key. The new sequence is then divided into eight chunks of 6
bits each. Each group of 6 bits is recoded to give 4 bits, using a recoding process described

FIGURE 14.47

DES encryption.

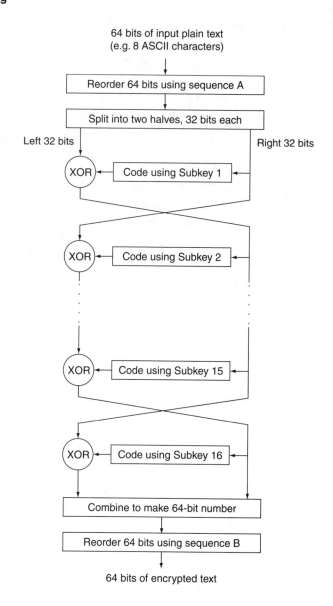

64 bits of input plain text
(e.g. 8 ASCII characters)

Reorder 64 bits using sequence A

Split into two halves, 32 bits each

Left 32 bits   Right 32 bits

XOR ← Code using Subkey 1

XOR ← Code using Subkey 2

XOR ← Code using Subkey 15

XOR ← Code using Subkey 16

Combine to make 64-bit number

Reorder 64 bits using sequence B

64 bits of encrypted text

below. The 32-bit number that results when the eight 4-bit pieces are joined is rearranged using sequence D defined in Table 14.3(b).

Figure 14.49 explains how the 6-bit groups are recoded to produce 4 bits. The first and last bits give a row number between 0 and 3. The other 4 bits give a column number between 0 and 15. These indices are used to select a value from an S-matrix in Table 14.4. All the numbers in the S-matrices lie between 0 and 15, and so they can be coded using 4 bits, as required in Figure 14.48. Suppose, for example, that the 6-bit binary input to the left-most recoding block in Figure 14.48 is 101000. According to Figure 14.49, the first and

TABLE 14.2
Sequences A and B

Output bit #:	1	2	3	4	5	6	7	8	9	10	11	12	13	14	15	16
Input bit #:	58	50	42	34	26	18	10	2	60	52	44	36	28	20	12	4
Output bit #:	17	18	19	20	21	22	23	24	25	26	27	28	29	30	31	32
Input bit #:	62	54	46	38	30	22	14	6	64	56	48	40	32	24	16	8
Output bit #:	33	34	35	36	37	38	39	40	41	42	43	44	45	46	47	48
Input bit #:	57	49	41	33	25	17	9	1	59	51	43	35	27	19	11	3
Output bit #:	49	50	51	52	53	54	55	56	57	58	59	60	61	62	63	64
Input bit #:	61	53	45	37	29	21	13	5	63	55	47	39	31	23	15	7

(a) Sequence A (Rearranges 64-bit Sequence)

Output bit #:	1	2	3	4	5	6	7	8	9	10	11	12	13	14	15	16
Input bit #:	40	8	48	16	56	24	64	32	39	7	47	15	55	23	63	31
Output bit #:	17	18	19	20	21	22	23	24	25	26	27	28	29	30	31	32
Input bit #:	38	6	46	14	54	22	62	30	37	5	45	13	53	21	61	29
Output bit #:	33	34	35	36	37	38	39	40	41	42	43	44	45	46	47	48
Input bit #:	36	4	44	12	52	20	60	28	35	3	43	11	51	19	59	27
Output bit #:	49	50	51	52	53	54	55	56	57	58	59	60	61	62	63	64
Input bit #:	34	2	42	10	50	18	58	26	33	1	41	9	49	17	57	25

(b) Sequence B (Rearranges 64-bit Sequence)

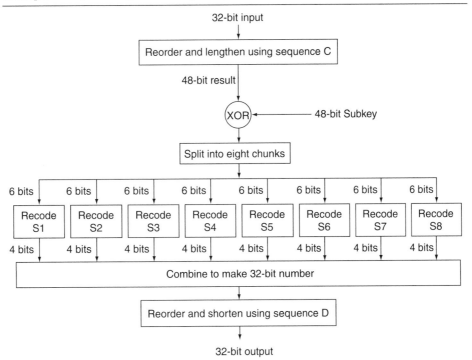

FIGURE 14.48
Code using subkey process.

TABLE 14.3
Sequences C and D

Output bit #:	1	2	3	4	5	6	7	8	9	10	11	12	13	14	15	16
Input bit #:	32	1	2	3	4	5	4	5	6	7	8	9	8	9	10	11
Output bit #:	17	18	19	20	21	22	23	24	25	26	27	28	29	30	31	32
Input bit #:	12	13	12	13	14	15	16	17	16	17	18	19	20	21	20	21
Output bit #:	33	34	35	36	37	38	39	40	41	42	43	44	45	46	47	48
Input bit #:	22	23	24	25	24	25	26	27	28	29	28	29	30	31	32	1

(a) Sequence C (Converts 32-bit Sequence to 48-Bit Sequence)

Output bit #:	1	2	3	4	5	6	7	8	9	10	11	12	13	14	15	16
Input bit #:	16	7	20	21	29	12	28	17	1	15	23	26	5	18	31	10
Output bit #:	17	18	19	20	21	22	23	24	25	26	27	28	29	30	31	32
Input bit #:	2	8	24	14	32	27	3	9	19	13	30	6	22	11	4	25

(b) Sequence D (Rearranges 32-bit Sequence)

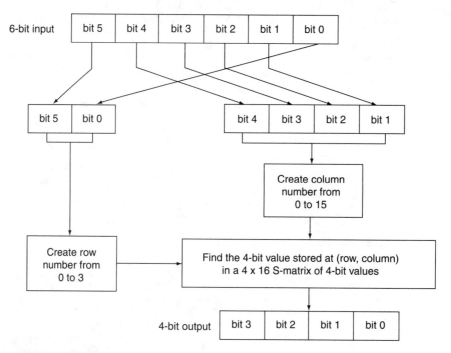

FIGURE 14.49
Recode process.

TABLE 14.4

S-Matrices

*S*1	14	4	13	1	2	15	11	8	3	10	6	12	5	9	0	7
	0	15	7	4	14	2	13	1	10	6	12	11	9	5	3	8
	4	1	14	8	13	6	2	11	15	12	9	7	3	10	5	0
	15	12	8	2	4	9	1	7	5	11	3	14	10	0	6	13
*S*2	15	1	8	14	6	11	3	4	9	7	2	13	12	0	5	10
	3	13	4	7	15	2	8	14	12	0	1	10	6	9	11	5
	0	14	7	11	10	4	13	1	5	8	12	6	9	3	2	15
	13	8	10	1	3	15	4	2	11	6	7	12	0	5	14	9
*S*3	10	0	9	14	6	3	15	5	1	13	12	7	11	4	2	8
	13	7	0	9	3	4	6	10	2	8	5	14	12	11	15	1
	13	6	4	9	8	15	3	0	11	1	2	12	5	10	14	7
	1	10	13	0	6	9	8	7	4	15	14	3	11	5	2	12
*S*4	7	13	14	3	0	6	9	10	1	2	8	5	11	12	4	15
	13	8	11	5	6	15	0	3	4	7	2	12	1	10	14	9
	10	6	9	0	12	11	7	13	15	1	3	14	5	2	8	4
	3	15	0	6	10	1	13	8	9	4	5	11	12	7	2	14
*S*5	2	12	4	1	7	10	11	6	8	5	3	15	13	0	14	9
	14	11	2	12	4	7	13	1	5	0	15	10	3	9	8	6
	4	2	1	11	10	13	7	8	15	9	12	5	6	3	0	14
	11	8	12	7	1	14	2	13	6	15	0	9	10	4	5	3
*S*6	12	1	10	15	9	2	6	8	0	13	3	4	14	7	5	11
	10	15	4	2	7	12	9	5	6	1	13	14	0	11	3	8
	9	14	15	5	2	8	12	3	7	0	4	10	1	13	11	6
	4	3	2	12	9	5	15	10	11	14	1	7	6	0	8	13
*S*7	4	11	2	14	15	0	8	13	3	12	9	7	5	10	6	1
	13	0	11	7	4	9	1	10	14	3	5	12	2	15	8	6
	1	4	11	13	12	3	7	14	10	15	6	8	0	5	9	2
	6	11	13	8	1	4	10	7	9	5	0	15	14	2	3	12
*S*8	13	2	8	4	6	15	11	1	10	9	3	14	5	0	12	7
	1	15	13	8	10	3	7	4	12	5	6	11	0	14	9	2
	7	11	4	1	9	12	14	2	0	6	10	13	15	3	5	8
	2	1	14	7	4	10	8	13	15	12	9	0	3	5	6	11

last bits identify a row number $10_2 = 2_{10}$, and the other 4 bits identify a column number $0100_2 = 4_{10}$. The recoding block in question makes use of the *S*-matrix *S*1, which is included at the top of Table 14.4. At row 2, column 4 is found the decimal number 13, which has the binary value 1101. This is the 4-bit output of this recoding block.

The last piece of the puzzle lies in the generation of the 48-bit subkeys, referred to in Figure 14.48, from a single 64-bit key that defines the entire DES process. Figure 14.50 supplies an explanation of the procedure. Sequence E in Table 14.5(a) reorders the 64-bit primary key and drops the parity bits. The left and right halves of the 56-bit sequence that remains are subjected to circular left shifts, illustrated in Example 14.4, before being

FIGURE 14.50

Calculation of 48-bit subkeys from 64-bit encryption key.

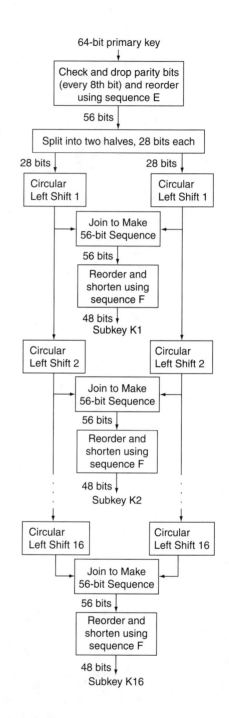

TABLE 14.5

Sequences E and F

Output bit #:	1	2	3	4	5	6	7	8	9	10	11	12	13	14		
Input bit #:	57	49	41	33	25	17	9	1	58	50	42	34	26	18		
Output bit #:	15	16	17	18	19	20	21	22	23	24	25	26	27	28		
Input bit #:	10	2	59	51	43	35	27	19	11	3	60	52	44	36		
Output bit #:	29	30	31	32	33	34	35	36	37	38	39	40	41	42		
Input bit #:	63	55	47	39	31	23	15	7	62	54	46	38	30	22		
Output bit #:	43	44	45	46	47	48	49	50	51	52	53	54	55	56		
Input bit #:	14	6	61	53	45	37	29	21	13	5	28	20	12	4		

(a) Sequence E (Converts 64-bit Sequence to 56-bit Sequence)

Output bit #:	1	2	3	4	5	6	7	8	9	10	11	12	13	14	15	16
Input bit #:	14	17	11	24	1	5	3	28	15	6	21	10	23	19	12	4
Output bit #:	17	18	19	20	21	22	23	24	25	26	27	28	29	30	31	32
Input bit #:	26	8	16	7	27	20	13	2	41	52	31	37	47	55	30	40
Output bit #:	33	34	35	36	37	38	39	40	41	42	43	44	45	46	47	48
Input bit #:	51	45	33	48	44	49	39	56	34	53	46	42	50	36	29	32

(b) Sequence F (Converts 56-bit Sequence to 48-bit Sequence)

TABLE 14.6

Table of Circular Left Shifts

Circular Left Shift #	Number of Left Shifts
1	1
2	1
3	2
4	2
5	2
6	2
7	2
8	2
9	1
10	2
11	2
12	2
13	2
14	2
15	2
16	1

rejoined and then reordered and shortened using sequence F of Table 14.5(b). The number of positions to shift in each circular shift is listed in Table 14.6. In the end, the sixteen 48-bit subkeys that are needed in Figure 14.47 are produced from the single 64-bit primary key.

EXAMPLE 14.4

A binary sequence 1110110100100011 is subjected to a circular left shift. Identify the sequence that results after shifting:

    a. One position

    b. Two positions

    a. All bits in the sequence move one position to the left. The most significant bit moves to the least significant bit position. After one circular left shift the sequence becomes 1101101001000111.

    b. After two circular left shifts the sequence becomes 1011010010001111.

The encrypted data sequence can be decrypted by a receiver as outlined in Figure 14.51, providing the primary key is known. The receiver uses the primary key to generate subkeys that are identical with the set of subkeys used by the sender. Using a process that mirrors the encryption algorithm, the received bits are decrypted to re-create the original message. In essence, the multiple XOR steps act like tumblers in a lock, which finally "unlock" the coded message.

14.5.4   DSP for Security

In the U.S. government, encryption of the most sensitive classified information is always handled with hardware rather than with software. Software, because it must be created and maintained by a group of programmers, and because it can be accessed via a computer network, is, by its nature, too prone to security breaches. Hardware, on the other hand, can be safeguarded against attack by securing its physical location and is, therefore, the method of choice for processing Confidential, Secret, and Top Secret data.

Symmetric encryption algorithms like DES perform much faster than public key algorithms like PGP. The drawback of symmetric algorithms is that security is destroyed when the single shared key is lost. Public key encryption is more secure because only the private key, which need not be shared, has to be kept safe. DSPs are well suited to implementing encryption algorithms, since the algorithms are well defined, rule-oriented number manipulations. PGP, which is very computation-intensive, tends not to be implemented on DSP platforms. DSP hardware is a natural choice for DES and triple DES, however, as these algorithms use smaller keys and smaller chunks of data throughout than PGP. Hash algorithms are also very suited to implementation on DSPs. Analog Devices' ADSP-2141L SafeNet DSP, for example, offers DES at 640 Mbps, triple DES at 214 Mbps, and the hash algorithms MD5 and SHA1 at 315 and 253 Mbps, respectively.

14.6   MOTOR CONTROL

Motors are ubiquitous in modern homes. They appear in CD players, cassette players, VCRs, satellite dishes, bread makers, answering machines, electronic typewriters, microwaves, and many toys, to name just a few uses. In applications like cassette players and

FIGURE 14.51
DES decryption.

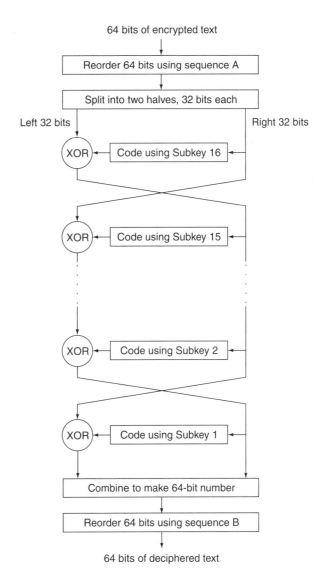

64 bits of encrypted text

Reorder 64 bits using sequence A

Split into two halves, 32 bits each

Left 32 bits   Right 32 bits

XOR ← Code using Subkey 16 ←

XOR ← Code using Subkey 15 ←

XOR ← Code using Subkey 2 ←

XOR ← Code using Subkey 1 ←

Combine to make 64-bit number

Reorder 64 bits using sequence B

64 bits of deciphered text

VCRs, the motor must spin at a constant speed. This requires what is called **speed control**. For applications like satellite dishes, the motor must turn to a particular position and stop. This is referred to as **position control**. The designer must specify the behavior of the motor in both types of applications. Is it most important to reach the final speed or position quickly, or is it best to avoid oscillations? Figure 14.52 presents an underdamped and an overdamped motor response to a request for a set speed or position. The **underdamped** response approaches the desired level rapidly but oscillates around this level in the process. The **overdamped** response is more gradual but does not overshoot the desired level. The response that is best depends on the needs of the application.

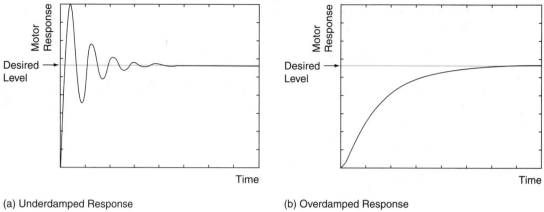

(a) Underdamped Response   (b) Overdamped Response

FIGURE 14.52
Underdamped and overdamped motor step responses.

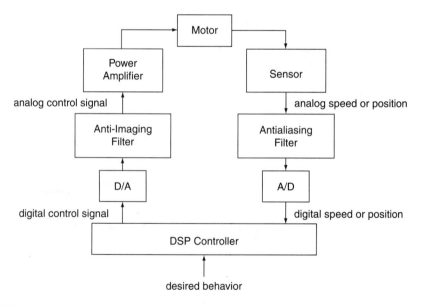

FIGURE 14.53
Closed loop motor control system.

A motor on its own may not display the behavior a designer chooses. When this is the case, a **controller** must be added to form a **closed loop system** that will modify the system's response. A closed loop motor control system is shown in Figure 14.53. All the basic elements in Figure 14.53 are familiar from Section 2.1. Here, the DSP implements the controller that will modify motor behavior. The **sensor** is responsible for measuring the signal being controlled, either the speed or the position of the motor. This sensor may be a tachometer (or tachogenerator), if speed is being measured, or a potentiometer if position is being measured. If necessary, the sensor signal is amplified or otherwise conditioned to permit A/D conversion. The digitized sensor input is used by the DSP controller to compute a suitable digital control signal, which is converted to analog to drive the motor. A power amplifier raises the power of the analog control signal to the level required by the motor. If the control strategy is successful, the motor will approach final speed or position with as little or as much oscillation as the designer is willing to tolerate.

The most common control strategy in use is called PID (proportional integral derivative) control. A PID controller acts on the error in the system, which is the difference between the desired and the actual behavior of the motor shaft. For a speed control system,

$$\text{Error} = \text{Desired shaft speed} - \text{Actual shaft speed}$$

and for a position control system,

$$\text{Error} = \text{Desired shaft position} - \text{Actual shaft position}$$

The control signal at each step consists of three elements:

$$\text{Control} = \text{Proportional} + \text{Integral} + \text{Derivative}$$

Specifically, at each step n, the control signal $m[n]$ is:

$$m[n] = p[n] + i[n] + d[n]$$

In this equation, the term

$$p[n] = K_p e[n] \tag{14.4}$$

is proportional to error $e[n]$. The multiplier K_p is called the proportional gain. The integral control term $i[n]$ is computed from

$$i[n] = i[n-1] + K_i \Delta t\, e[n] \tag{14.5}$$

where Δt is the time between samples and K_i is the integral gain. This term monitors the total error. Finally, the derivative control term $d[n]$, computed as

$$d[n] = K_d \left(\frac{e[n] - e[n-1]}{\Delta t} \right)$$

reflects changes in error. The multiplier K_d is called the derivative gain. The high frequency behavior of the controller improves when the high frequency gains are limited. This can be accomplished by modifying the derivative component of the control signal to:

$$d[n] = \frac{K_d}{K_d + N\Delta t} d[n-1] + \frac{K_d N}{K_d + N\Delta t} (e[n] - e[n-1]) \tag{14.6}$$

where N is the limit on high frequency gain.

Taking the z transforms of Equations (14.4), (14.5), and (14.6) gives the transfer functions for each of the three controller terms, which may be combined to produce the overall transfer function for the controller:

$$\frac{M(z)}{E(z)} = K_p + \frac{K_i \Delta t}{1 - z^{-1}} + \frac{N(1 - z^{-1})}{\left(1 + \frac{N\Delta t}{K_d}\right) - z^{-1}} \qquad \textbf{(14.7)}$$

where $E(z)$ is the z transform of the error $e[n]$ and $M(z)$ is the transform of the control signal. Calculation of the control signals according to the foregoing equations can conveniently be implemented on a DSP platform.

EXAMPLE 14.5

Write the transfer function for a PID controller with $K_p = 2$, $K_i = 1$, $K_d = 0.4$, $\Delta t = 0.2$ sec, and $N = 0.5$.

Using Equation (14.7), the transfer function is

$$\frac{M(z)}{E(z)} = 2 + \frac{0.2}{1 - z^{-1}} + \frac{0.5(1 - z^{-1})}{\left(1 + \frac{0.1}{0.4}\right) - z^{-1}} = \frac{2.6(1 - 1.7538z^{-1} + 0.7692z^{-2})}{1 - 1.8z^{-1} + 0.8z^{-2}}$$

Embedded DSP motor control will definitely feature in the next generations of appliances like washing machines (Lemaire, 1999). The challenge for designers of washing machines is to reduce power, water consumption, and cycle time, while delivering cleaner clothes with less residual moisture. Reducing water content is important also because dryers use so much more power than washers. Several of these objectives dictate that higher spin speeds will be critical. As washer spin speed increases, vibration increases also. Translational vibrations can cause washers to "walk" and can also increase the probability of unbalancing the wash load. At high spin speeds, excessive side-to-side motion can cause severe mechanical problems. The simplest and lowest cost machines available today use mechanical switches to signal out-of-balance conditions. These switches are not numerous or accurate enough to serve high performance machines. Advanced machines not only detect unbalanced loads but also reduce spin speed to accommodate the problem, and even vary the spin pattern to reposition the load.

Figure 14.54 suggests a more sophisticated closed loop scheme. A DSP generates a speed profile that the washer will follow. The rotor speed of the washer is monitored by a digital tachometer, and this speed is constantly compared to the speed profile. When too great a difference exists, the DSP computes a slip speed that will correct the speed of the motor. The slip speed and rotor speed are added together to produce a stator speed. The DSP uses the magnitude and phase of this stator speed to calculate a set of three-phase winding voltages that are applied to the motor through a power inverter, which converts a DC signal into an AC signal that drives the motor. The DC current at the power inverter is used as an estimate of motor current; undue ripple in the current indicates an overload condition. When this occurs, the current signal is used to limit the stator speed. In Figure 14.54, the

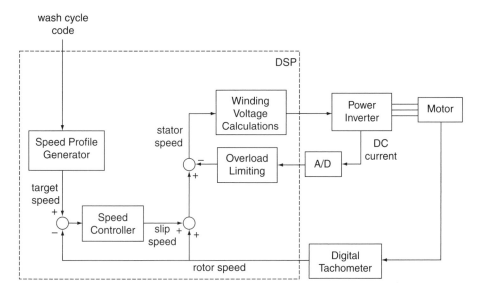

FIGURE 14.54
DSP for washing machine control (adapted from Lemaire, 1999).

DSP governs all of the elements enclosed in the dashed box. It performs all control calculations needed to maintain the correct motor speed profile. A DSP used this way can adequately meet out-of-balance detection requirements for machines whose spin speeds average about 1000 rpm. Higher performance machines must rely on direct measurements provided by acceleration sensors, still too expensive for most economy appliance designs.

CHAPTER SUMMARY

Matlab
Support

1. Oversampling refers to sampling at a rate that exceeds the Nyquist rate. If the Nyquist rate is $2W$, the $4\times$ oversampling rate is $8W$.
2. Decimation drops samples from a signal. Decimation by a factor of 4 drops three out of every four samples.
3. Noise shaping pushes quantization noise out of the frequency band containing the important signal frequencies.
4. Zero insertion adds zeros between existing samples in a signal. In $4\times$ zero insertion, three zeros are added between each pair of samples.
5. An interpolation filter smooths sharp transitions in a signal.
6. Dithering refers to the addition of small amounts of white noise to a signal, which has the effect of capturing small signals that would otherwise be lost in quantization noise.
7. Companding boosts small signals relative to large signals prior to quantization, reducing the effect of quantization noise on these signals.

8. MP3 exploits the limitations of human hearing to reduce the number of bits needed to store audio signals, compared to compact discs.

9. Continuous speech recognizers can recognize natural speech, while isolated word recognizers can recognize only words separated by pauses. Speaker-independent recognizers work with all speakers of a language, while speaker-dependent recognizers must be trained by the individuals who will use them.

10. Wavetable synthesis is a common music synthesis strategy. It uses a small collection of short recordings from an instrument and a large group of clever tricks to create all the sounds the instrument is capable of making.

11. The number of voices a music synthesizer has refers to the number of sounds that can be produced at the same time. A synthesizer that can produce more than one note at a time is polyphonic.

12. A text-to-speech system is a voice synthesizer.

13. Most encryption schemes use a key for encryption and decryption.

14. Digital signatures allow the source of a message to be verified by e-mail.

15. Data Encryption Standard (DES) is a popular encryption scheme, but has been broken. Triple DES, which uses longer keys, may take its place.

15

IMAGE PROCESSING

> This chapter introduces image processing concepts. The chapter:
>
> - ➤ defines pixel, gray scale level, and resolution
> - ➤ introduces histograms
> - ➤ describes how histograms may be equalized to improve image contrast
> - ➤ presents methods for adding, subtracting, and scaling images
> - ➤ illustrates digital image warping and morphing
> - ➤ provides examples of image filtering
> - ➤ shows how to find and refine the edges of objects in a digital image
> - ➤ uses shape and color to identify objects
> - ➤ shows how an object's features may be used for classification
> - ➤ presents magnitude and phase spectra for a selection of digital images
> - ➤ introduces basic concepts of tomography
> - ➤ discusses image compression techniques, including the JPEG standard

15.1 IMAGE PROCESSING BASICS

In the past, digital images were created by dividing images on photographic film into tiny picture elements called pixels and recording the color for each. Special film scanners, capable of focusing on one pixel at a time, were used for recording colors. Today scanners and digital cameras are common and inexpensive enough that digital images are easy to create and obtain, either from live subjects or from printed pictures or photos. The growth of the Internet is one reason for the explosion of digital imagery, since images cannot be transferred through a computer network without a suitable digital format. Furthermore, for network transfers of images to be rapid and efficient, it is prudent to compress image information prior to transmission. This compression is possible only because the images are in digital form. Other reasons for digitizing images exist as well. In digital form, images may be filtered, transformed, or combined with other images. They may also be analyzed for edges, shapes, or other important features. Image processing applications tend to require a great deal of repetitive processing, which makes them well suited for implementation on DSP hardware.

Color digital images require three values to be recorded for each pixel: one red component, one green component, and one blue component. These components are combined with different weightings to produce a range of colors. To avoid the complexities that this tricolor system introduces, this chapter will focus on gray scale images exclusively. Black and white photographic images, for example, contain an infinite number of shades of gray from pure white to pure black. They can be converted into gray scale digital images using a scanner. The scanner records, for each pixel, a single number corresponding to the shade of that pixel. The number matches one of the finite number of levels that are possible in the digital image and is called the gray scale level of that pixel. The number of gray scale levels that can be used to record a digital image depends on the number of bits used. With N bits, 2^N different levels can be represented. That means that a 16-bit system supports 65,536 gray scale levels. While this is a large number, the number of gray levels present in a black and white photograph is much larger. The gray levels from the photograph, then, cannot be matched exactly, so quantization errors arise.

Once an image has been divided into pixels, it becomes a digital image comprising a defined number of rows and columns, as Figure 15.1 illustrates. The number of rows and columns an image comprises in part determines how much detail can be discerned in the image. A $3'' \times 5''$ photo digitized as 1024×1708 pixels, for example, can retain more detail than a version digitized as 256×427 pixels. The other feature that determines how much detail can be displayed is the number of gray scale levels available. The higher the number of pixels in an image, and the higher the number of gray scale levels, the higher the

FIGURE 15.1

Digital image.

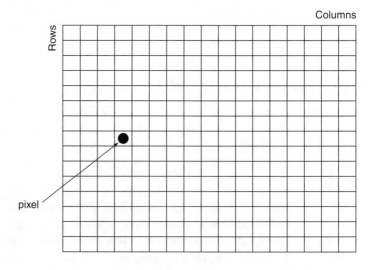

FIGURE 15.2
Decreasing rows and columns.

(a) 808 rows × 562 columns × 256 colors

(b) 202 rows × 141 columns × 256 colors

resolution of the digital image. In Figure 15.2, resolution is decreased by decreasing the number of pixels used to represent the image of a lighthouse. In Figure 15.3, resolution is decreased by decreasing the number of gray scale levels available to code the shade of each pixel. Since dividing an image into pixels amounts to sampling it, aliasing can occur in a digital image if pixels are too large or too far apart for the subject being recorded. When an image contains fine lines or details that are closer together than the pixel spacing, moiré patterns appear that show sampling to be inadequate.

15.2 HISTOGRAMS AND HISTOGRAM EQUALIZATION

An 8-bit image supports 2^8, or 256, possible gray scale levels, where 0 corresponds to pure black and 255 corresponds to pure white. Figure 15.4 shows all 256 colors. A **histogram** for a gray scale image shows how many pixels occur for each gray scale level. An image

FIGURE 15.2
Continued

(c) 101 rows × 70 columns × 256 colors

(d) 51 rows × 35 columns × 256 colors

and its histogram are given in Figure 15.5. The histogram is actually a bar graph, where the height of each bar is the number of pixels with a particular gray scale level. The middle bump in this histogram occurs because the image of the elephant contains a significant number of medium gray elements, while the spike at the right registers the presence of very light-colored elements. A histogram is a record of the proportions of light and dark pixels in an image, but it can provide only general information about the subject of the image. For example, the histogram in Figure 15.6(a) suggests an image that is primarily dark gray and light gray, but there is no way to guess whether it might belong to Figure 15.6(b) or Figure 15.6(c).

Though a histogram cannot give detailed information about the objects in an image, it can determine some of their characteristics. When the area covered by a photograph is known, the area of an object in the photo can sometimes be found by determining how the number of pixels in the object compares to the total number of pixels in the image:

FIGURE 15.3
Decreasing number of
gray scale levels.

(a) Eight-Bit Quantization (808 rows × 562 columns × 256 colors)

(b) Four-Bit Quantization (808 row × 562 columns × 16 colors)

(c) One-Bit Quantization (808 rows × 562 columns × 2 colors)

FIGURE 15.4
256 gray levels.

FIGURE 15.5
Digital image plus
its histogram.

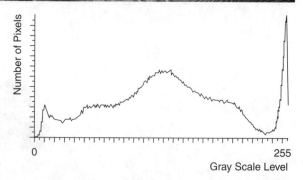

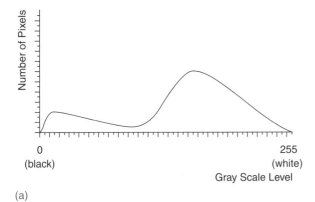

(a)

(b)

(c)

FIGURE 15.6
Histogram plus two possible source images.

$$\frac{\text{Area of object}}{\text{Area of whole picture}} = \frac{\text{Number of pixels in object}}{\text{Total pixels}}$$

Furthermore, if something is known about which gray scale values belong to a background and which belong to an object, a **threshold** can be chosen that marks the boundary between the two. This information can assist in identifying the edges of the object and hence its shape. Example 15.1 illustrates these concepts.

EXAMPLE 15.1

A photograph of a light-colored ball lying on the grass is digitized using 4 bits. The total area of the scene shown in the photo is 600 cm^2. The number of pixels for each gray level is given in Table 15.1, and the histogram produced from this table is shown in Figure 15.7. Is the ball a golf ball, a softball, or a volleyball?

TABLE 15.1
Histogram Data for Example 15.1

Gray Scale Level	Number of Pixels
0	10
1	120
2	577
3	976
4	1287
5	861
6	98
7	21
8	11
9	7
10	30
11	31
12	196
13	123
14	50
15	39

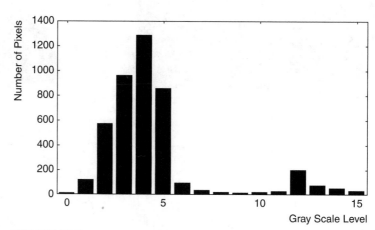

FIGURE 15.7
Histogram for Example 15.1.

The larger bars at the lower gray scale levels make it obvious that the dark background makes up the majority of the photo. The presence of the light-colored ball can be seen in the contributions at the higher gray scale levels. Gray scale level 9 might be chosen as a reasonable threshold between the object and background, since it is a minimum point between the two peaks in the histogram. Since the object is known to be lighter than the background, the total number of pixels associated with the object can be found by summing all the pixels for gray scale levels 10 to 15. This sum is 469. The total number of pixels in the image is 4437. Thus, the ball in the picture takes up 469/4437 of the total area in the picture. Since the photo covers 600 cm^2, the ball covers 469(600)/4437 = 63.4 cm^2. The area of a picture of a ball is πr^2 cm^2, where r is the radius of the ball. Therefore, $\pi r^2 = 63.4$, $r^2 = 20.2$, and $r = 4.5$ cm. The ball has a radius of 4.5 cm, or a diameter of 9 cm, too big for a golf ball and too small for a volleyball. The ball in the photo is a softball.

Histogram information can also be used to increase the contrast in a digital image. Since a histogram shows the number of pixels for each gray scale level, it also indicates the usage of the range of gray scale levels available. The features in the background of the image in Figure 15.8 are difficult to see. The histogram for the image shows why this is the case. The pixels in the image do not use a very wide range of gray scale levels. As a result, the contrast in the image is poor, making automatic feature identification difficult. The contrast can be improved if the use of gray scale levels can be expanded to include all possible gray scale levels.

FIGURE 15.8

Before equalization.

(a) Original Image (807 pixels × 842 pixels)

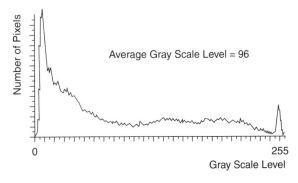

(b) Histogram

Histogram equalization is a process that stretches the range of gray scale levels present in an image and attempts to make more even use of the full range of gray scale levels available. Histogram equalization redefines each gray scale level G according to the equation

$$G_{new} = \frac{\text{\#Pixels with gray scale level} \leq G}{\text{Total pixels}} (\text{Max gray scale level}) \quad \textbf{(15.1)}$$

A total of 807×842, or 679,494, pixels are represented in the histogram of the 8-bit image in Figure 15.8. According to the histogram, half of the pixels, 339,747 in all, are accounted for at or before the average gray scale level of 96. This average lies toward the dark end of the 0 to 255 scale. In order to flatten the histogram as much as possible, and therefore to use the available shades as evenly as possible, the gray scale level 96 should map to a new gray scale level of approximately $(339{,}747/679{,}494)(255) = 127.5 \approx 128$, halfway through the range. The maximum gray scale level in Figure 15.8 is 255, so, according to Equation (15.1), the gray scale level 255 maps to $(679{,}494/679{,}494)(255) = 255$ again. In this example, histogram equalization tends to increase all gray scale levels, thereby increasing the average. Figure 15.9 shows the results of equalization for the image in Figure 15.8. The average gray scale level in the equalized image is 128, which means that half of the image is made up of darker pixels and the other half of lighter pixels. While the aesthetic qualities of the equalized photo are less pleasing, the contrast is increased so dramatically that two new people are evident in the background!

Histogram equalization is not the only means of changing a histogram to manipulate the appearance of an image. Histograms can be used to make color substitutions, for example, to identify interesting parts of an image. This technique is used in Section 15.7.2, to highlight brain tissue in a magnetic resonance image (MRI).

15.3 COMBINING IMAGES

Once images are digitized, they are easy to manipulate. When two digital images have the same size, they can be added or subtracted point by point. Figure 15.10 shows the result of adding two $4 \times 4 \times 8$-bit images. The resultant image no longer conforms to the proper range of gray scale values, from 0 to 255 for an 8-bit image; several gray scale values in the image fall outside this range. This problem frequently arises in image processing. It is solved by scaling the resultant image to ensure the gray scale values are legal for the data width available. Equation (15.2) can be used to perform the required scaling:

$$\text{Scaled output}[m, n] = \frac{\text{Output}[m, n] - \text{Min}}{\text{Max} - \text{Min}} \times \text{Max gray scale level} \quad \textbf{(15.2)}$$

where Min is the smallest value in the unscaled output and Max is the largest value in the unscaled output. If the maximum gray scale value is 255, the results of scaling will be values between 0 and 255. After scaling, each output value is rounded to the nearest integer.

For the sum image in Figure 15.10, the minimum value in the unscaled output is 112 and the maximum value is 355. The scaling equation becomes

FIGURE 15.9

After equalization.

(a) Equalized Image (807 pixels × 842 pixels)

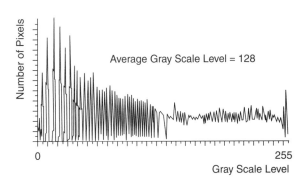

(b) Histogram

$$\begin{bmatrix} 165 & 134 & 98 & 205 \\ 225 & 89 & 100 & 189 \\ 83 & 91 & 120 & 213 \\ 102 & 115 & 198 & 232 \end{bmatrix} + \begin{bmatrix} 67 & 102 & 111 & 128 \\ 130 & 84 & 33 & 10 \\ 44 & 21 & 7 & 3 \\ 19 & 5 & 13 & 40 \end{bmatrix} = \begin{bmatrix} 232 & 236 & 209 & 333 \\ 355 & 173 & 133 & 199 \\ 127 & 112 & 127 & 216 \\ 121 & 120 & 211 & 272 \end{bmatrix}$$

FIGURE 15.10

Adding digital images.

$$\text{Scaled output}[m, n] = \frac{\text{Output}[m, n] - 112}{355 - 112} \times 255 = \frac{\text{Output}[m, n] - 112}{243} \times 255$$

Pixel [0, 3], for example, has the unscaled value 333. The scaled value is

$$\frac{333 - 112}{243} \times 255 = 231.9$$

which rounds to 232. All other gray scale values can be scaled in the same way. The unscaled and scaled images are shown in Figure 15.11.

When two 8-bit images are added, the sum image may contain numbers as low as 0 and as high as 2(255) = 510, which must be scaled back to the 0 to 255 range. Figure 15.12 shows the sum of an image of a die and an image of giraffe skin, after scaling. The smallest sum prior to scaling, 67, occurs near the bottom of the bottom dot on the three-dot side of the die. In the scaled image, this number maps to a gray scale level of 0, or black. The largest sum prior to scaling, 497, occurs near the top left corner of the same side of the die. According to Equation (15.2), this point maps to a gray scale level of 255, or white, in the scaled image. The gray scale levels of all other pixels in the sum image scale to grays on the continuum between black and white. The only way to get a clear combination of two images is to add an image over a black area in another, as shown in Figure 15.13. A pixel summed with a black pixel remains unchanged, since the gray scale level of the black pixel is zero.

FIGURE 15.11
Scaling a digital image.

$$\begin{bmatrix} 232 & 236 & 209 & 333 \\ 355 & 173 & 133 & 199 \\ 127 & 112 & 127 & 216 \\ 121 & 120 & 211 & 272 \end{bmatrix} \qquad \begin{bmatrix} 126 & 130 & 102 & 232 \\ 255 & 64 & 22 & 91 \\ 16 & 0 & 16 & 109 \\ 9 & 8 & 104 & 168 \end{bmatrix}$$

(a) Before Scaling   (b) After Scaling

FIGURE 15.12
Sum of images by addition.

 +  =

© Snap-Shot.com

FIGURE 15.13
Sum of images by masking.

 +  =

Addition of digital images can be useful when several noisy pictures of the same scene are averaged together to reduce noise. Figure 15.14(a) shows a field of noise simulating a grainy photograph. An average can be obtained by adding several similar images together point by point and dividing the result by the number of images. The features that are common to all pictures are reinforced in the summing process, while the noise spots are not. In general, the gray scale levels associated with the noise spots tend to decrease. In other words, the sum image computed from two images of a similar scene de-emphasizes the elements that are different in the two images and highlights the elements that are the same. Figure 15.14 shows the averages of two, four, and eight images of photographic film noise, after scaling. The larger the number of pictures averaged, the greater the degree of noise reduction obtained.

Images can be subtracted pixel by pixel just as easily as they can be added. The results of a subtraction can be negative, so scaling using Equation (15.2) is required in this case as well. Subtraction is especially useful in removing the background from an image. This is possible when a picture of an object in front of a background and also a picture of the background alone are available, as shown in Figure 15.15. The difference image is found by subtracting the background plus subject from the background alone. Its negative is shown in Figure 15.16. Where pixels in the same locations in the two images in Figure 15.15 have identical gray scale values, the difference image records a gray scale value of zero. This produces black pixels in the difference image, which become white pixels in the negative.

FIGURE 15.14
Averaging images.

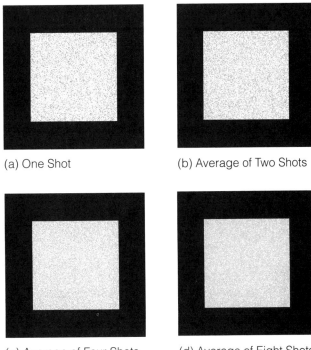

(a) One Shot

(b) Average of Two Shots

(c) Average of Four Shots

(d) Average of Eight Shots

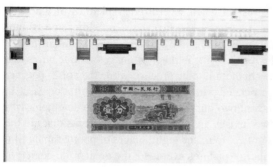
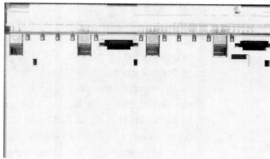

(a) Subject on Scanner Background

(b) Scanner Background Alone

FIGURE 15.15

Background and background with subject.

FIGURE 15.16

Negative of difference image.

Motion detection is another application of image subtraction. The difference image computed from two images of a similar scene de-emphasizes the elements that are common to both images and highlights the elements that move. Figure 15.17 demonstrates motion detection. The first two images record a parking lot at two separate instants of time. A car is evident in the foreground of the picture in (a) that is absent in the other one, but other, subtler differences exist as well. The negative of the difference image in Figure 15.17(c) clearly shows where significant motions of vehicles or people occurred between photos. The fainter lines in other areas of the difference image do not indicate relative motion. They simply show that the two photos are not identically aligned.

15.4 WARPING AND MORPHING

The pixel layout of digital images permits objects in them to change shape. As an example, the 8×10 image in Figure 15.18(a) contains a 2×2 white square that must be transformed. The location of each pixel may be described as $[m, n]$ where m is the row number and n is the column number. With the top left pixel defined as $[0, 0]$, the square occupies the pixels at locations $[0, 0]$, $[0, 1]$, $[1, 0]$, and $[1, 1]$. The color of all pixels in the square is 255, or white. The square may be moved to a new location by adding, say, three to all the row numbers and two to all the column numbers. The result of this translation is shown in Figure 15.18(b). The original square may be magnified by multiplying all corner pixel coordinates by a factor of 4. Figure 15.18(c) shows the result of this expansion.

FIGURE 15.17
Motion detection in images.

(a) Image One

(b) Image Two

FIGURE 15.17
Continued

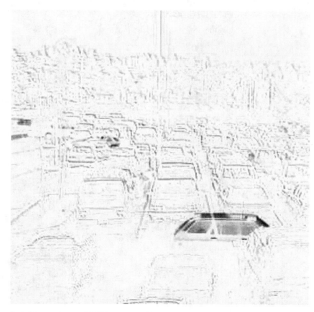

(c) Negative of Difference Image

FIGURE 15.18

Simple transformations of a
white square.

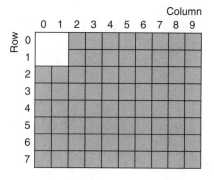

(a) Original Square

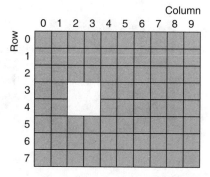

(b) After Translation

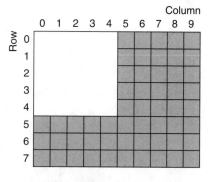

(c) After Magnification

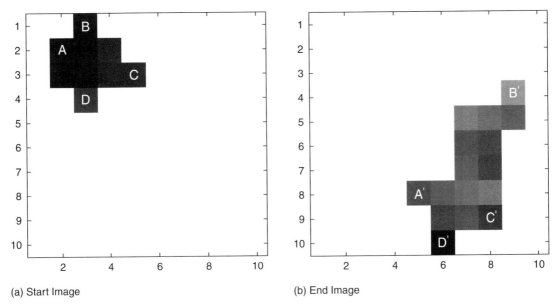

(a) Start Image   (b) End Image

FIGURE 15.19
Transformation of images using control points.

TABLE 15.2
Gray Scale Levels of Control Points in Figure 15.19

Point	Gray Scale Level	Point	Gray Scale Level
A	50	A′	110
B	60	B′	180
C	70	C′	90
D	100	D′	10

Not all transformations can be conveniently expressed by a simple rule. In Figure 15.19, the shape in (a) must be transformed into the shape in (b). All the important features in the start image are marked by control points *ABCD*. The same control points are mapped to their new locations *A′B′C′D′* in the end image. To map the control points correctly from the start to the end image, both their locations and their colors must be transformed. The locations of the control points are indicated in Figure 15.19, and their gray scale values are listed in Table 15.2. The transition between the two images may be smoothed by inserting a number of intermediate images. A set of shifting control points forms the skeleton for each such intermediate image, as shown in Figure 15.20.

The simplest way to transform the control point locations is to use linear interpolation. The two intermediate images identified in Figure 15.20 lie one third and two thirds of the way between the start and end images. The intermediate locations for the first control

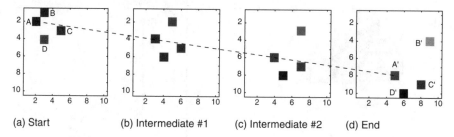

(a) Start   (b) Intermediate #1   (c) Intermediate #2   (d) End

FIGURE 15.20
Intermediate control point locations.

point, therefore, are one third and two thirds of the distance along a straight line joining A and A'. The simplest way to transform the control point colors is to use weighted averages. For example, using the gray scale levels for A and A' from Table 15.2, the colors of the pixels that mark the transition from A to A' through two intermediate images are:

$$A: \qquad \frac{3}{3}(50) + \frac{0}{3}(110) = 50$$

$$\frac{2}{3}(50) + \frac{1}{3}(110) = 70$$

$$\frac{1}{3}(50) + \frac{2}{3}(110) = 90$$

$$A': \qquad \frac{0}{3}(50) + \frac{3}{3}(110) = 110$$

This color blending scheme is often called **cross dissolving**.

Linear interpolation of location, and weighted averaging of gray scale levels, produces the progression of control points shown in Figure 15.20. The problem of determining the gray scale levels of the pixels in the intermediate images that are not control points, however, has not yet been addressed. These pixels are moved according to their proximity to neighboring control points. Pixels that are about the same distance from two control points A and B, for example, are affected by each control point equally and move in a manner that is a compromise between the transformations AA' and BB'. Other schemes use lines or triangular patches instead of points to specify how shape and color are transformed. The details can be quite complex, and the above illustration is intended only to hint at the essential principles.

Warping refers to stretching or reshaping a single image or part of an image. **Morphing**, on the other hand, refers to a transformation of one image into another. Marker points like those in Figure 15.19 can be used for both warping and morphing. Figure 15.21 illustrates a warp of a puppet's head using a small set of 13 control points. As another example, Figure 15.22 shows the sets of source and destination points that morph a pussycat

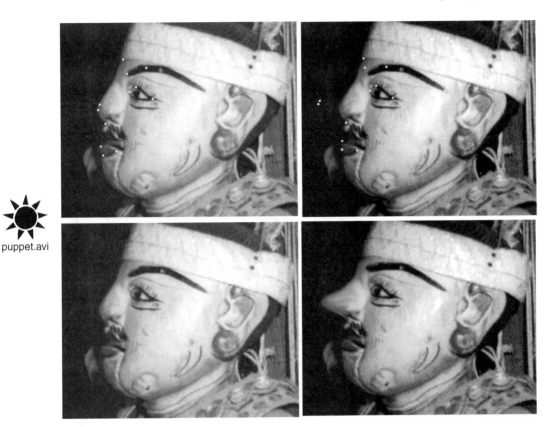

FIGURE 15.21
Warping Pinnochio.

puppet.avi

into a lion. For instance, the points surrounding the cat's nose in the source image encircle
the lion's nose in the destination image. The start and end images, along with four intermediate images, are shown in Figure 15.23. Warping and morphing are widely used in
movies and television. The most compelling sequences often begin with morphing software
but end with hand-tuning. Warping is also finding uses in medical imaging and other areas.
In one application, called "Digital Personnel" (Braukus & Lovato, 2000), facial features in
a still image of a person's face warp in real time, in accordance with a live or recorded voice
signal, to give the illusion of live speech.

15.5 FILTERING IMAGES

Digital image filters were introduced in Section 5.4, where a low pass filter was used to
blur edges in a black and white image. Filtering an image is a way of changing its frequency properties. In the case of the low pass filter, sharp, high frequency edges in an image are removed. Figure 15.24 shows progressively more heavily low pass filtered images

FIGURE 15.22
Source and destination points for morphing.

© A. Anwar Sacca

© Snap-Shot.com

of jellyfish. Figure 15.24(a) shows the original image. In Figure 15.24(b), (c), and (d), gray scale levels are averaged across neighborhoods 3×3, 5×5, and 7×7 pixels in size. For each case, a convolution kernel, the 2D equivalent of an impulse response, is given. The larger the neighborhood of averaging, the stronger the blurring. Because of boundary effects, the filtered images are always smaller than the original. For an $M \times M$ filter, $(M - 1)/2$ rows at the top and bottom and $(M - 1)/2$ columns at the left and right edges of the image are influenced by boundary effects. Thus, when an $M \times M$ filter operates on an $N \times N$ digital image, only $(N-(M-1)) \times (N-(M-1))$ pixels in the filtered digital image are unaffected by boundary effects, as discussed in Section 5.4. In Figure 15.24(d), a 7×7 filter operates on a 186×248 image. The filtered image is 180×242 pixels in size.

FIGURE 15.23
Six images from a morphing sequence.

cattolion.avi

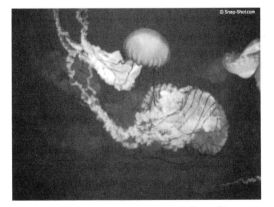

© Snap-Shot.com

(a) Original

(b) $\frac{1}{9} \begin{bmatrix} 1 & 1 & 1 \\ 1 & 1 & 1 \\ 1 & 1 & 1 \end{bmatrix}$

FIGURE 15.24
Low pass filtering.

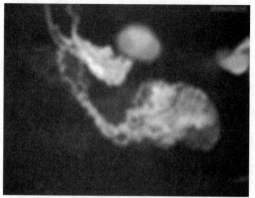

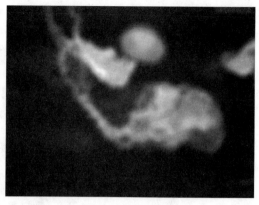

(c)

$$\frac{1}{25}\begin{bmatrix} 1 & 1 & 1 & 1 & 1 \\ 1 & 1 & 1 & 1 & 1 \\ 1 & 1 & 1 & 1 & 1 \\ 1 & 1 & 1 & 1 & 1 \\ 1 & 1 & 1 & 1 & 1 \end{bmatrix}$$

(d)

$$\frac{1}{49}\begin{bmatrix} 1 & 1 & 1 & 1 & 1 & 1 & 1 \\ 1 & 1 & 1 & 1 & 1 & 1 & 1 \\ 1 & 1 & 1 & 1 & 1 & 1 & 1 \\ 1 & 1 & 1 & 1 & 1 & 1 & 1 \\ 1 & 1 & 1 & 1 & 1 & 1 & 1 \\ 1 & 1 & 1 & 1 & 1 & 1 & 1 \\ 1 & 1 & 1 & 1 & 1 & 1 & 1 \end{bmatrix}$$

FIGURE 15.24
Continued

FIGURE 15.25
Convolution kernel for high pass filter.

$$\begin{bmatrix} 0 & -1 & 0 \\ -1 & 4 & -1 \\ 0 & -1 & 0 \end{bmatrix}$$

Naturally, high pass digital image filters can also be constructed. The convolution kernel for one such filter is shown in Figure 15.25. As described in Section 5.4, a convolution kernel provides filtering action as it moves over all regions of the image. At each pixel location, a new gray scale value is calculated by summing the products of the kernel with gray scale levels in a neighborhood of the pixel. In the case of a high pass filter, the result of filtering is close to zero when the kernel aligns with a region of similar gray scale levels. When the center pixel has a gray scale level quite different from its neighbors, the result is large and positive, or large and negative. High pass filter results are subject to scaling described by Equation (15.2). After scaling, the largest positive values map to white and the largest negative values to black, while the zeros map to medium gray.

EXAMPLE 15.2

When the 8-bit image of Figure 15.26(a) is filtered with the high pass filter of Figure 15.25, the result is as shown in Figure 15.26(b). To illustrate the filtering process, the gray scale levels for a portion of the original image are provided in Figure 15.27(a), along with pixel row and column numbers. From this 6×5 area of the original image, the 4×3 area of the filtered image within the dashed box can be computed. The calculation of the filtered value for the top left element in the dotted box is illustrated by superimposing the kernel elements of Figure 15.25 as superscripts and results in an unscaled output of $(0)(0) + (0)(-1) + (0)(0) + (0)(-1) + (0)(4) + (0)(-1) + (0)(0) + (0)(-1) + (0)(0) = 0$. This result is not unexpected, since the pixels that contribute to the output have identical gray scale values, and the high pass filter responds only to changes in gray scale level. Repeating the calculation for every element of the dotted box gives the unscaled filtered image values in Figure 15.27(b). As with addition and subtraction of images, the filtered matrix must be scaled to ensure that the final gray scale levels lie in the range 0 to 255. The scaled versions of the filtered values in Figure 15.27(b) are obtained using Equation (15.2) with maximum 4(255) and minimum 4(−255), and are shown in Figure 15.27(c). The gray scale variations uncovered in this manner may be confirmed by examining the dotted box in Figure 15.26(b).

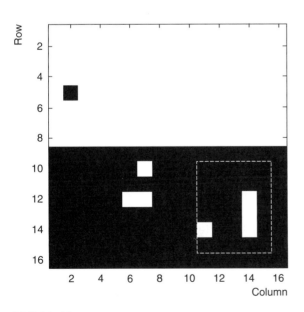

(a) Original Image

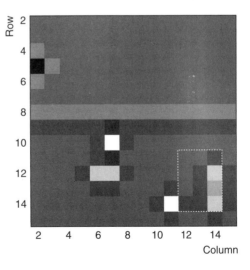

(b) After High Pass Filtering

FIGURE 15.26
High pass filtering a simple digital image for Example 15.2.

Row \ Column	11	12	13	14	15
10	0^0	0^{-1}	0^0	0	0
11	0^{-1}	0^4	0^{-1}	0	0
12	0^0	0^{-1}	0^0	255	0
13	0	0	0	255	0
14	255	0	0	255	0
15	0	0	0	0	0

(a) Portion of Original Image

Row \ Column	12	13	14
11	0	0	−255
12	0	−255	765
13	0	−255	510
14	−255	−255	765

(b) Portion of Filtered Image, Unscaled

Row \ Column	12	13	14
11	128	128	96
12	128	96	223
13	128	96	191
14	96	96	223

(c) Portion of Filtered Image, Scaled

FIGURE 15.27
High pass filtering a portion of an image for Example 15.2.

Applying the above procedure to the entire original image yields the filtered image in Figure 15.26(b). The large medium-gray areas correspond to the regions of low filter response, where the gray scale levels in the original image are more or less constant. The very dark and very light areas in Figure 15.26(b) correspond to the regions of high filter response, where the gray scale levels in the original image are changing sharply. Notice that the strongest responses of the filter, evidenced by pure black and pure white colors, are obtained where a single pixel stands out dramatically from its neighbors in the original image.

The Chinese street market depicted in Figure 15.28(a) is high pass filtered twice over in Figure 15.28(b) by the convolution kernel in Figure 15.25. The second picture seems sharper because the areas where gray scale levels are changing, such as edges, are emphasized.

A variation on the high pass filter in Figure 15.25 is the Sobel edge filter, which goes one step beyond looking for sharp random changes in gray scale level and instead looks for regular changes that may be the edges of an object in an image. The convolution kernels for two Sobel edge filters are shown in Figure 15.29. When the vertical edge kernel in Figure 15.29(a) lies over a region of an image of constant gray level, the output is zero. However, when the kernel straddles the sharp vertical edge of an object, the output is either strongly positive or strongly negative, depending on whether the colors change from dark to light or from light to dark across the edge. When it is the presence of an edge that is important and not the direction of the change, the absolute value of the output image is most useful. The horizontal edge kernel in Figure 15.29(b) works in a similar fashion. When both horizontal and vertical edges must be identified, the absolute values of the output images for both convolution kernels are calculated, and the maximum gray scale value at each pixel location is taken. Scaling, if necessary, is the last step in the edge detection process. Figure 15.30 provides examples of vertical, horizontal, and all edge detection. Edge detection capability is required in many pattern recognition tasks, including industrial robot vision and military target identification.

(a) Original Image

(b) Filtered Image

FIGURE 15.28
High pass filtering.

FIGURE 15.29
Sobel edge filters.

$$\begin{bmatrix} -1 & 0 & 1 \\ -2 & 0 & 2 \\ -1 & 0 & 1 \end{bmatrix}$$

$$\begin{bmatrix} -1 & -2 & -1 \\ 0 & 0 & 0 \\ 1 & 2 & 1 \end{bmatrix}$$

(a) Vertical Sobel Edge
Convolution Kernel

(b) Horizontal Sobel Edge
Convolution Kernel

15.6   PATTERN RECOGNITION

15.6.1   Identifying Features

Finding the edges of objects is often the first step toward automatic recognition of the objects. The phenomenal recognition abilities of the human eye and brain permit many objects in the X-ray scan of Figure 15.31(a) to be identified easily. These same abilities must be programmed into a computer to allow patterns to be recognized automatically. Simple

FIGURE 15.30
Edge detection.

(a) Original Image

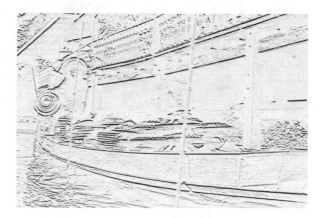

(b) Detection of Horizontal Edges

edge detection filtering is usually not enough: While the edges traced in Figure 15.31(b) may seem to delineate the objects rather well, a closer look, provided in Figure 15.32(b), uncovers many gaps and discontinuities along the border of the handgun. These gaps prevent a computer from recognizing that the edges all belong to a single object.

Dilation and erosion are two tricks of image processing that assist a computer in making sense of the object boundaries present in a digital image. Figure 15.33(a) shows a portion of an object boundary, defined through edge detection filtering. Because of the gaps, it in effect consists of a number of objects. **Dilation** adds a layer of pixels to all objects in a digital image, as shown in Figure 15.33(b); it tends to thicken boundaries by adding a layer of pixels, as seen in Figure 15.33(c) and in Figure 15.32(c) for the handgun. **Erosion** removes a layer of pixels from all objects in an image. It removes the outer layer of pixels in Figure 15.33(d), yielding the eroded boundary in Figure 15.33(e). The boundary is now much smoother than the original, though two small gaps remain. Figure 15.32(d) shows that

FIGURE 15.30
Continued

(c) Detection of Vertical Edges

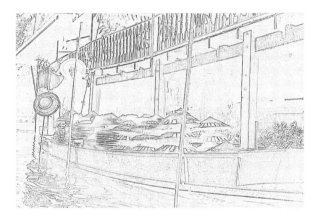

(d) Detection of All Edges

the boundaries of the handgun can be more clearly defined by applying dilation and erosion to the original image. The reverse process, erosion followed by dilation, is also useful. It has the effect of removing isolated specks of noise from an image.

Once the boundaries of an object are clearly established, identification or classification of the object becomes possible. Several features can help in this goal. In the gray scale image of Figure 15.34, for example, shading can be used to identify how many snooker balls of each color lie on the billiard table. Each ball has a characteristic shade of gray, and a computer can search for these shades in the image. Texture provides another means of identification. In satellite photos of the earth's surface, for example, the colors produced by different textures can be used to distinguish between cultivated and uncultivated regions of land. Textures in three dimensions can be identified as well. A set of altitude measurements made at 10 m intervals on a 100 m^2 grid, for instance, is presented in Figure 15.35, with the surface fitted to these points depicted in Figure 15.36. This figure shows what may be a narrow river gorge between two sharp inclines.

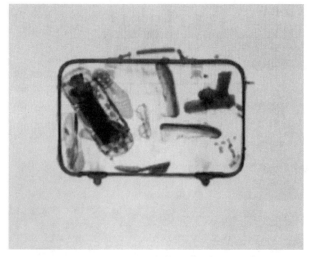

Courtesy Analogic Corporation

(a) Original Image

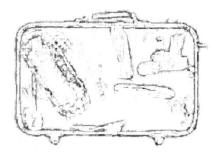

(b) Edge Detection

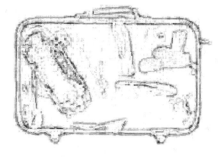

(c) After Dilation

(d) After Dilation and Erosion

FIGURE 15.31
X-ray image of airport baggage.

FIGURE 15.32
X-ray detail.

(a) Original

(b) Edge Detection

FIGURE 15.32

Continued

(c) After Dilation (d) After Dilation and Erosion

FIGURE 15.33

Dilation and erosion.

(a) Boundary of Object Detected by Edge Filtering

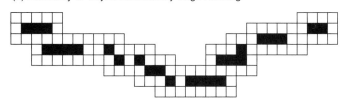

(b) Pixels Added by Dilation

(c) Dilated Boundary

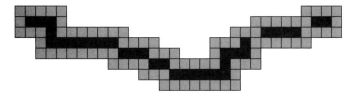

(d) Pixels to be Removed by Erosion

(e) Eroded Boundary

FIGURE 15.34
Snooker balls on
billiard table.

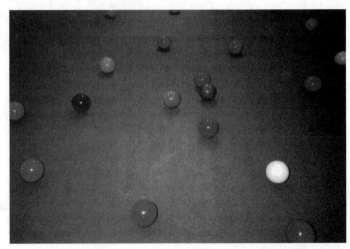

FIGURE 15.35
Altitude
measurements.

160	475	634	790	791	793	792	950	946	1104
157	318	790	790	946	1106	1106	1261	1341	1421
159	240	202	396	474	553	792	1107	1105	1106
476	551	197	554	318	238	315	792	790	790
709	753	318	476	239	239	790	790	791	791
791	714	631	789	315	789	790	397	473	478
792	791	792	792	788	792	316	399	632	790
871	867	947	792	867	950	988	1027	1106	1105
792	792	791	870	1026	1027	1107	1261	1261	1262
752	791	827	870	868	1024	1065	1183	1342	1498

FIGURE 15.36
Surface fit to
altitude
measurements.

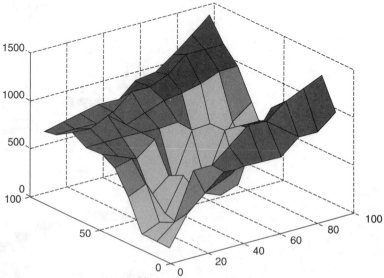

Object size is another important feature for automatic object recognition. An estimate of size may be made by computing the area of an object in a digital image. This may be done in various ways, though the value of the results is limited by the fact that a digital image can record only a two-dimensional projection of a three-dimensional object. An image of a coin lying flat, for instance, presents a larger object than that of a coin balanced on its edge. In Example 15.1, the boundary of an object was identified by setting a suitable histogram threshold, and the object's size was calculated by counting pixels. This method works well for simple images containing a small number of objects. For more complicated images, area may be calculated by using the boundary vertex coordinates. The pixels in the boundary of an object are shown in Figure 15.37(a) as dark squares. An interior point is chosen for reference, and each vertex, where the boundary changes direction, is connected to it to form a large number of small slices. One such slice is shaded, and two others are indicated by dashed lines. As Figure 15.37(b) suggests, the area of each slice can be calculated from the vertex coordinates — (x_0, y_0), (x_1, y_1), and (x_2, y_2) — by subtracting the areas of the three shaded triangles from the area of the square, after Castleman (1996).

$$\text{Area} = (x_0 - x_1)(y_2 - y_0) - \frac{1}{2}(x_0 - x_1)(y_1 - y_0)$$

$$- \frac{1}{2}(x_0 - x_2)(y_2 - y_0) - \frac{1}{2}(x_2 - x_1)(y_2 - y_1)$$

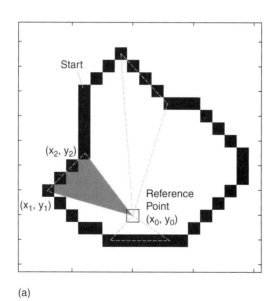

(a)

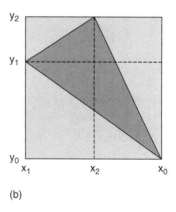

(b)

FIGURE 15.37
Calculating area and perimeter from boundary points.

FIGURE 15.38
Boundary chain codes.

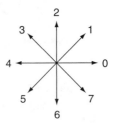

The total area of an object is computed by summing the absolute values of the areas of the slices. Perimeter is another possible measure of object size. It may be calculated from the boundary vertex coordinates as well. For an object like the one in Figure 15.37(a), the distance between each pair of vertices (x_1, y_1) and (x_2, y_2) can be found as $\sqrt{(x_2 - x_1)^2 + (y_2 - y_1)^2}$. The total perimeter of the object will be the sum of these distances for all adjacent pairs of vertices on the boundary.

Both area and perimeter can be used to help determine object shape. Shape is perhaps the most problematic object feature to identify. Very rough shape information can be obtained using some simple checks. For instance, if the width and length of an object are approximately equal, then the object is square or circular in shape. If the width and length are very different, the object is long and thin. The circularity measure (Castleman, 1996)

$$\text{Circularity} = \frac{(\text{Perimeter})^2}{\text{Area}} \tag{15.3}$$

can also give an indication of object shape. The ratio takes its minimum value of 4π when the object in question is a circle. A square gives a circularity measure of 16. The more complex the object, the higher the measure becomes. Direction codes or boundary chain codes (Castleman, 1996; Gonzalez & Woods, 1992) can help to record the boundary shapes for more complicated objects. These codes, as indicated in Figure 15.38, tabulate the approximate directions traced out by an object's boundary. The boundary of the object in Figure 15.37(a), for example, might code, clockwise from "Start," as 1-1-1-7-7-7-0-0-7-7-7-7-6-6-5-5-5-5-5-4-4-4-4-4-4-3-4-3-3-3-1-1-1-2-2-2-2-2. A code repeated several times indicates a straight-line edge of the object. The number of repetitions suggests the edge's length.

15.6.2 Object Classification

Automatic classification of an object in a digital image normally consists of selecting the most likely candidate from a list of objects that are possible. To do this, a number of distinguishing features for the set must be identified. For example, color and shape might be chosen to identify fruits, with shape evaluated using Equation (15.3). Color and shape easily distinguish between oranges, red apples, bananas, grapes, and lemons, but classification becomes more difficult for yellow apples, apricots, and yellow plums. Size can be added to the list of features to continue to make classification possible. The larger the group of possible objects, and the more similar the objects in the group, the larger the number of features that will be necessary.

FIGURE 15.39

Classification of fruit images by color and shape.

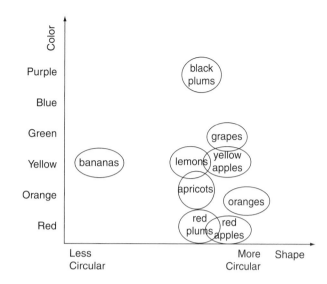

FIGURE 15.40

Distributions of shapes for yellow fruit.

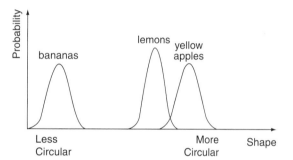

In Figure 15.39, color and shape are used to classify images of fruit. The ovals represent the likely range of possibilities for each fruit. Problems arise wherever the ovals overlap. In these areas, confusions between different kinds of fruit can occur. For example, a fairly spherical red fruit, nearly circular in an image, might be identified as either a red plum or a red apple. These overlaps motivate the addition of extra features to improve classification accuracy. Figure 15.40, for instance, shows the distributions for the yellow fruits. The most likely fruit shapes occur where the distributions peak. The overlapped regions of the distributions show where fruits of more than one type may occur.

Feature selection is one of the most difficult tasks in designing object classification systems. Walker (1998), for example, selected from over 123 features to classify six species of algae, illustrated in Figure 15.41. The algae, belonging to the cyanobacteria genera Microcystis and Anabaena, cause blooms in freshwater systems that can create health risks if they infect drinking water supplies. To ensure supplies are safe, large numbers of water samples must be collected. Automatic classification of the algae is anticipated to reduce the time needed for testing.

FIGURE 15.41
Six cyanobacteria samples.

Courtesy Dr. Ross F. Walker, Lake Biwa Research Institute, Japan

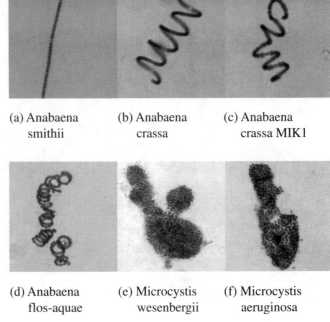

(a) Anabaena smithii (b) Anabaena crassa (c) Anabaena crassa MIK1

(d) Anabaena flos-aquae (e) Microcystis wesenbergii (f) Microcystis aeruginosa

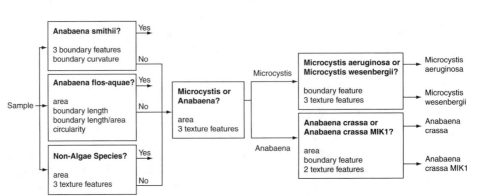

FIGURE 15.42
Bacteria classification.

The shapes of several of the organisms in Figure 15.41 are extremely similar. For best results, the classification is carried out in several stages, as illustrated in Figure 15.42. At each stage, a different subset of four features is selected, to focus on the distinctive characteristics of the bacteria being classified in that stage. For example, Anabaena flos-aquae is identified by a set of features including area, boundary length, boundary length divided by area, and circularity. Using this classification scheme, laboratory algae specimens were classified with only 3% error.

FIGURE 15.43

Neural network.

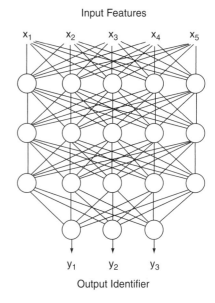

Input Features

x_1 x_2 x_3 x_4 x_5

y_1 y_2 y_3

Output Identifier

TABLE 15.3

Classification of Marine Animals

| | | Features (1 = Yes; 0 = No) | | | | |
Marine Animal	Output Identifier	Scales	Legs	Swimming	Crustacean	Tail
Salmon	000	1	0	1	0	1
Octopus	001	0	1	1	0	0
Dolphin	010	0	0	1	0	1
Sea star	011	0	0	0	0	0
Crab	100	0	1	0	1	0

Statistical methods exist that can improve the chances that an object described by a set of features will be classified correctly (Castleman, 1996; Gonzalez & Woods, 1992). Some of these methods rely on a set of known probabilities that a given object will be present in the image. Others use sets of training images to estimate the distributions of features that objects will have. These more advanced techniques may improve recognition accuracy, but they cannot overcome the inherent problems of overlapping classes. Another popular approach to pattern recognition uses **neural networks**, which attempt to model the way the human brain learns. Figure 15.43 shows a diagram of such a network. The network is a web of nodes whose inputs are the known features of an object to be classified, and whose outputs are patterns that identify an object. Table 15.3 shows an example. The input features 01100, for example, for a swimming noncrustacean with legs but no scales or tail, produce the output identifier 001, identifying an octopus pattern. A network is "trained" by presenting sets of features of known objects on the input side and providing the correct identifying patterns at the output. Each link in the neural network web is assigned a weighting factor, and after each training item, the weights in the network are adjusted. Once a great deal of training has occurred, the weights in the network guarantee that input features like those in the training set will produce the correct output patterns. The network in Figure 15.43 expects five features at the input and produces a 3-bit identification code at the output, like the example in Table 15.3. Details of neural network pattern recognition methods may be found in Castleman (1996) and Gonzalez and Woods (1992).

15.7 IMAGE SPECTRA

15.7.1 Image Spectra Basics

Images have spectra just as signals do, though they are interpreted a little differently. For signals, low frequency components are those that change slowly. For images, low frequency regions are those where gray scale levels change little. High frequency elements for signals mean rapid changes. For images, high frequency elements are edges. The spectrum for a two-dimensional image is computed using the 2D DFT introduced in Section 11.7 or the 2D FFT, which computes the spectrum more efficiently than the 2D DFT. For convenience, the equation for the 2D DFT is repeated here:

$$X[i, k] = \sum_{m=0}^{M-1} \sum_{n=0}^{N-1} x[m, n] e^{-j2\pi \frac{i}{M} m} e^{-j2\pi \frac{k}{N} n} \tag{11.10}$$

As usual the spectrum comprises a magnitude spectrum $|X[i, k]|$ and a phase spectrum $\theta[i, k]$. Because it is obtained from a two-dimensional object, the spectrum contains frequency data in two directions, one along the rows of the image and one along the columns. The magnitudes and phases, therefore, must be presented in a third dimension. Normally different color intensities are used to show the relative sizes of these quantities on a two-dimensional graph. Heights are used on a three-dimensional graph. Recall from Section 11.7 that the 2D DFT first takes 1D DFTs along the rows of an image and then takes 1D DFTs along the columns of data that result. The inverse DFT, defined in Equation (11.11), reproduces the original image $x[m, n]$. The examples that follow show the magnitude spectra for three stark images: a set of vertical stripes, a set of horizontal stripes, and a checkerboard. In each, the magnitude spectrum reflects the frequency content of the image.

EXAMPLE 15.3

The magnitude spectrum of the 50×50 image of vertical stripes in Figure 15.44 is given in Figure 15.45. This magnitude spectrum is obtained from the 2D DFT of the image, which, as mentioned above, may be viewed as a two-step process. In the first step, DFTs are taken along the rows of the image. In any given row, the pixels alternate white and black, so the DFT for the row produces the spectrum of a square wave like the one in Example 8.4, with a sinc-shaped envelope. The magnitude spectrum for each row m might be designated as $|X_m[k]|$ versus k; it reports the frequencies present in row m of the image.

Together, the DFTs $X_m[k]$ for $m = 0, \ldots, 49$ form a 50×50 matrix. In the second step toward computing the 2D DFT, DFTs are taken along the columns of this matrix to produce the final result $|X[i, k]|$. Since the spectra $|X_m[k]|$ are identical for every

FIGURE 15.44

Vertical stripes $x[m, n]$ for Example 15.3.

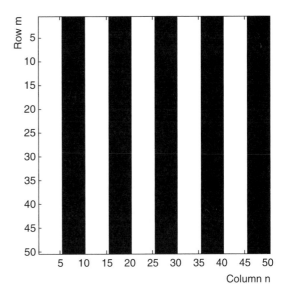

FIGURE 15.45

Magnitude spectrum of vertical stripes for Example 15.3.

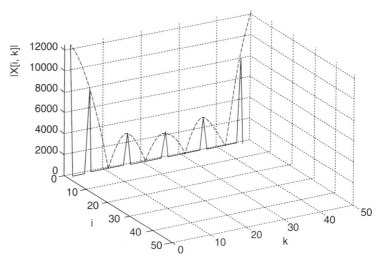

row m, every column n of the 50×50 matrix contains a (different) constant value. The DFT for a constant signal reports only a DC component, which explains why only one nonzero spectral element appears in each column k, at index $i = 0$, which corresponds to DC.

EXAMPLE 15.4

The magnitude spectrum of the 50×50 image of the horizontal stripes in Figure 15.46 appears in Figure 15.47. Each row in the image has a constant, DC, value, either 0 for black or 255 for white. Thus, following the discussion in the previous example, the 1D DFT

FIGURE 15.46

Horizontal stripes $x[m, n]$ for Example 15.4.

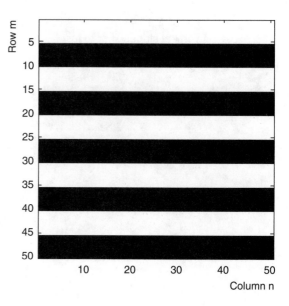

FIGURE 15.47

Magnitude spectrum of horizontal stripes for Example 15.4.

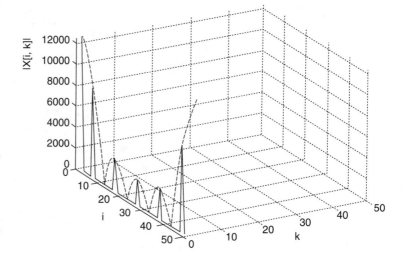

$X_m[k]$ for every row gives a single nonzero spectral component, at k = 0. The size of this component varies from one row to the next, alternating between high (for white) and low (for black). As in the previous example, the DFT results $X_m[k]$ for the rows are combined into a 50 × 50 matrix. The first, or $k = 0$, column contains the high-low alternations already described; the rest of the columns are zero. Thus, when the 2D DFT is completed, by taking a 1D DFT of these columns, the result is a sinc-shaped envelope along the i axis, where $k = 0$, and zeros elsewhere, as Figure 15.47 shows.

EXAMPLE 15.5

A checkerboard pattern is shown in Figure 15.48. The envelope of the magnitude spectrum along any single row or column has the sinc variation shown in Figure 15.49. The image as a whole, however, has the magnitude spectrum shown in Figure 15.50. The sinc shape shows up in all directions, reflecting the square wave variation in both the vertical and horizontal directions in the original image. Along the diagonal directions in the original image, however, the strong square wave variation is lost, which accounts for the smaller peaks in the center of the magnitude spectrum in Figure 15.50. A two-dimensional version of Figure 15.50 is shown in Figure 15.51, where the lightest areas mark the highest peaks.

FIGURE 15.48

Checkerboard $x[m, n]$ for Example 15.5.

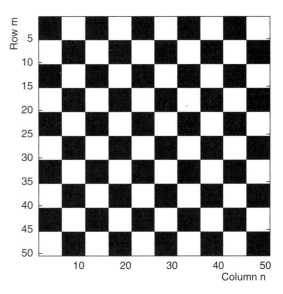

FIGURE 15.49

Magnitude spectrum of one row or column of checkerboard for Example 15.5.

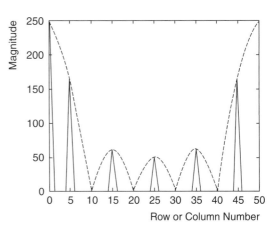

FIGURE 15.50

Magnitude spectrum of checkerboard (shown in 3D) for Example 15.5.

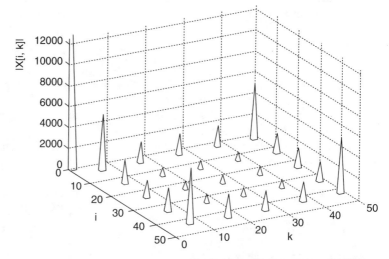

FIGURE 15.51

Magnitude spectrum of checkerboard (shown in 2D) for Example 15.5.

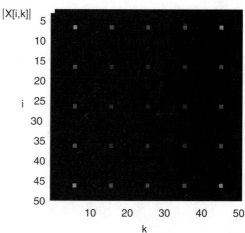

The spectrum of an image comprises two parts: a magnitude spectrum $|X[i, k]|$ and a phase spectrum $\theta[i, k]$. As the preceding examples show, the magnitude spectrum of an image can give clues about the structure of an image. However, the magnitudes $|X[i, k]|$ carry only part of the information needed to completely specify an image $x[m, n]$. As with any other complex quantity, the magnitudes $|X[i, k]|$ and the phases $\theta[i, k]$ together are needed to give $X[i, k]$:

$$X[i, k] = |X[i, k]|e^{j\theta[i, k]}$$

Given both of these elements, the inverse 2D DFT of $X[i, k]$, described in Equation (11.11), reproduces the original image $x[m, n]$ exactly. Example 15.6 illustrates the somewhat surprising result that the phases of an image alone carry enough information to create a facsimile of that image, while the magnitudes do not.

EXAMPLE 15.6

Figure 15.52(b) and (c) show the 2D magnitude and phase spectra for the image in (a). When an inverse 2D DFT is performed using these magnitudes and phases, the result is, naturally, the original image. In Figure 15.53, the image is reconstructed instead by performing an inverse 2D DFT on the magnitudes of Figure 15.52(b) with all phases set to zero. Clearly, no trace of the original image is evident when phase information is lost. In Figure 15.54, the image is reconstructed from the phases in Figure 15.52(c) with all magnitudes set to unity. As the figure shows, even when magnitude information is lost, some sense of the original image remains.

FIGURE 15.52

Spectrum of image for Example 15.6.

(a) Original Image

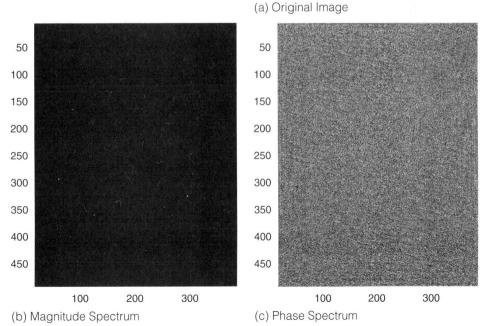

(b) Magnitude Spectrum

(c) Phase Spectrum

FIGURE 15.53
Reconstruction from magnitudes
(phases = 0) for Example 15.6.

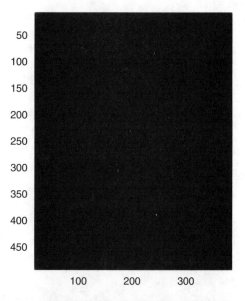

FIGURE 15.54
Reconstruction from phases
(magnitudes = 1) for
Example 15.6.

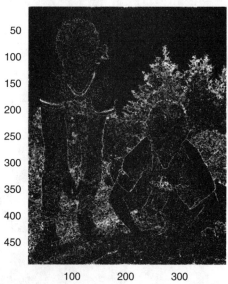

15.7.2 Tomography

Tomography studies the details in a two-dimensional slice of a three-dimensional object. In medicine, **computed tomography** (CT) images are reconstructions of cross-sections of the human body. They are produced by combining the information from rotating X-rays of the subject. Because bone and tissue have different densities, the X-ray beams are attenuated differently in each region. Thus, the received signal can be used to delineate the structures being scanned. CT scans provide excellent detail for bone structures and are of diagnostic use in the identification of tumors and other abnormalities.

FIGURE 15.55
Density function for object to
be scanned.

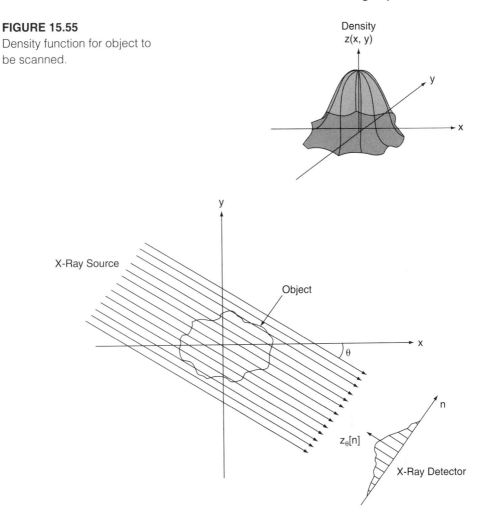

FIGURE 15.56
Measuring X-ray attenuation through scanned object.

The 3D function in Figure 15.55 records the densities $z(x, y)$ of an object at all (x, y) positions. In Figure 15.56, an X-ray beam passes through the object in the direction u and reaches a row of detectors that record X-ray attenuation. The received signal $z_\theta[n]$ can be thought of as a kind of shadow: The denser the object, the darker the shadow. The X-ray beams that cross through the edges of the object experience little attenuation, while those that pass through the object's center are greatly attenuated. The total attenuation along any single line through the object is proportional to the total area under the $z(x, y)$ function along that line. In Figure 15.56, the graph $z_\theta[n]$ versus n records the total attenuation at each position along the row of detectors. This graph does identify the places where the object is densest, but only for a single direction. Also, there is no way to determine exactly where the

FIGURE 15.57
DTFT of attenuation function.

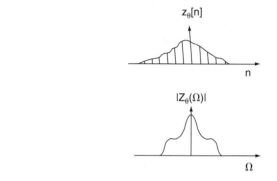

FIGURE 15.58
Computed tomography scanner.

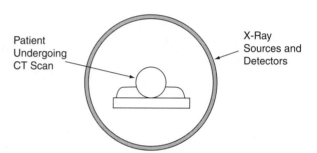

object lies. It must lie between the X-ray source and detector, but its position cannot be specified more closely than this. For both of these reasons, X-rays of the same slice must be taken from a number of different directions to form a complete picture of the slice being scanned. To obtain pictures of other slices, either the CT scanner or the object being scanned must be moved.

The goal of **projection-slice tomography** is to construct the density function $Z(x, y)$ of Figure 15.55 using the attenuation functions $Z_\theta[n]$ for a number of directions θ. A one-dimensional discrete time Fourier transform (DTFT) may be taken of each attenuation function $z_\theta[n]$ to give $Z_\theta(\Omega)$, as illustrated in Figure 15.57. This transform reflects the information collected for the direction θ. When a patient is scanned, data are collected from several different directions within the CT scanner, as shown in Figure 15.58. The DTFTs computed from all the slices taken are combined to create an approximation to the DTFT $Z(\Omega_x, \Omega_y)$ of the density function $z(x, y)$. Figure 15.59 suggests how this is done. Because not every direction can be tested, the regions between tested angles are interpolated. Once the DTFT $Z(\Omega_x, \Omega_y)$ has been estimated, it can be sampled and subjected to a two-dimensional inverse discrete Fourier transform (IDFT), in a manner similar to that used in Example 11.3, to produce $z[m, n]$, a sampled version of the density function $z(x, y)$. An image of $|z[m, n]|$ that uses different colors for different densities will show where hard and soft tissues are located in the slice being studied. The projection-slice approach is not normally used in practice. An equivalent but computationally more efficient reconstruction method, known as **filtered back projection**, is more often used instead (Kak, 1979).

FIGURE 15.59
Construction of two-dimensional
DTFT magnitude spectrum.

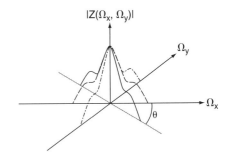

FIGURE 15.60
CT reconstruction of
thrinaxodon skull.

img4.mpg

© Dr. Timothy Rowe, University of Texas

parasaur.wav

When multiple CT slices are combined, accurate three-dimensional presentations of objects whose interiors are not normally visible can be constructed. Figure 15.31(a), for example, showed a scan produced by Analogic's EXACT AN6000, a multislice CT system that scans baggage moving along airport conveyor belts. The system is designed to detect explosives, drugs, and other contraband, and obtains images from all angles via a rotating X-ray detector array. A three-dimensional color-coded picture is created from these images for presentation to security personnel. CT can also be a tool for archaeologists analyzing fragile fossil remains. The image in Figure 15.60, for example, is a digital reconstruction of a thrinaxodon skull of length 3.8 cm. In the past, meticulous studies of such a skull might include physically cutting the skull into slices using a saw (West, 1996). The disadvantages of this approach are that the specimen is destroyed and that only one slice direction can be examined. Using the CT scan, the skull can be studied inside and out from any angle without damaging the fossil specimen.

Scientists at Sandia National Laboratories (1996) used CT technology to create, from fossil evidence, a three-dimensional model of the crest of the rare Parasaurolophus dinosaur, whose rendering appears in Figure 15.61. The model revealed a number of air tubes running back and forth along the length of the crest, as shown in Figure 15.62. Though the fossil evidence available could not provide information about the soft tissues that would contribute to vocalizations, knowledge of the bone structure permitted scientists to make educated guesses about the types of sounds that the dinosaur could produce.

FIGURE 15.61
Parasaurolophus dinosaur.

FIGURE 15.62
Model of parasaurolophus crest.

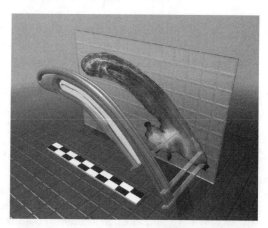

Magnetic resonance imaging (MRI) relies on technology very different from that of CT scans. MRIs do not use X-rays but depend instead on the fact that protons (principally the protons in hydrogen atoms) emit signals whose frequencies change as the strength of surrounding magnetic fields is varied. To produce a scan, an MRI makes use of several smooth changes in magnetic field strength, called magnetic field gradients. Figure 15.63, for example, shows a magnetic field gradient along the x axis, between the north and south poles of a magnet. Further to the right, the poles of the magnets are closer together and the field strength is greater. In all, an MRI uses three magnetic field gradients. As illustrated in Figure 15.64, the first is the slice selection gradient. This gradient is applied in a direction perpendicular to the slice being studied and determines the plane of the image to be recorded. The other gradients are the phase encoding gradient and the frequency encoding gradient. These are applied in the plane of the slice of interest, at right angles to one another.

Within the slice being imaged, a matrix of magnetization values is collected. The matrix is usually 256×256 pixels in size, with each pixel recording a 16-bit digital amplitude. A two-dimensional DTFT is taken of this matrix of data to give a matrix of complex values. The MRI image that physicians use to help with their diagnoses corresponds to a plot of the DTFT magnitudes. In other words, the MRI is the magnitude spectrum of the raw magnetization data. In some cases, smoothing is done in the frequency domain to improve the quality of the image. An example of an MRI scan is provided in Figure 15.65(a). The largest magnitudes in the image are typically represented with the lightest colors.

FIGURE 15.63

Magnetic field gradient.

FIGURE 15.64

Slice selection, phase encoding, and frequency encoding gradients.

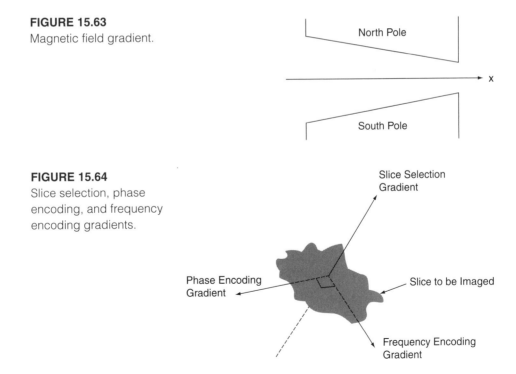

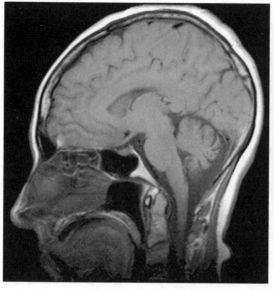
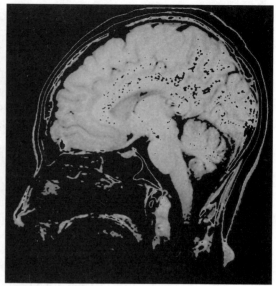

With permission, The Basics of MRI
Joseph P. Hornak, © 2000,
http://www.cis.rit.edu/htbooks/mri/

(a) Original MRI

(b) Processed MRI

FIGURE 15.65
Magnetic resonance images.

FIGURE 15.66
Histogram of original MRI.

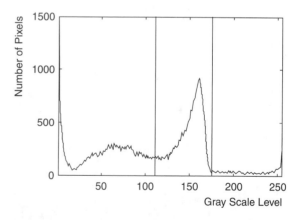

Tissues of different types show different magnetic properties, so their images in MRI scans have different gray scale levels. The histograms of MRI images can help distinguish among various tissue types. Figure 15.66, for example, presents the histogram for the MRI in Figure 15.65(a). To show detail in the shadings of the different tissues, the count for the black pixels (gray scale value zero) has been set to zero. The dips in the histogram can be used to distinguish between different elements in the MRI. For example, all pixels having gray scale values below 115 or above 175 might be set to black to remove the lightest and darkest elements and focus on the vital brain tissue, as shown in Figure 15.65(b). In fact assigning different colors to different ranges of gray scale levels can sometimes help to make interpretation of MRI scans easier (Hornak, 1999).

MRI scans have several advantages over CT scans. MRI images provide excellent detail for soft tissues, and medical images in any plane can be investigated without moving the patient or the scanner. MRI scans also remove the risk of X-ray exposure presented by CT scans. CT scans remain useful, though. They provide the best imaging for hard structures like bones and are normally more accessible than their MRI counterparts because the cost is far less.

15.8 IMAGE COMPRESSION

The amount of storage needed for an image is determined primarily by the number of pixels and the number of bits dedicated to the color coding of each pixel. Even a relatively small image with decent resolution takes up a lot of space. For example, a $1024 \times 1024 \times 8$-bit image requires 8,388,608 bits, or 8 Mbytes, of storage. Memory space is not the only concern: With the rise of the Internet, the desire to transmit images and other signals electronically has increased dramatically. Access to the Internet is most often achieved by using modems over shared lines. Most users do not want to waste available services by paying for lengthy downloads, nor do they want to invest unnecessary time. The intention of signal or image **compression** is to reduce the size of a file to be transferred without fundamentally affecting the content. This reduction is made possible by the use of formats that capitalize on patterns that repeat in the data being compressed. **Lossless compression** means that the data file is compacted without losing any information; that is, the file re-created from the compressed form is identical with the original. **Lossy compression** indicates that some, hopefully small, amount of information is lost in the compression process. The re-created file is very similar to, but not identical with, the original. The **compression ratio** for a compressed file measures its degree of compaction. For example, a compression ratio of 16:1 means the compressed file is 1/16 the size of the original, where size is normally measured in bytes.

EXAMPLE 15.7

A 3 megabyte image file is compressed to 93 kilobytes. Find the compression ratio.

A megabyte contains $1024 \times 1024 \times 8 \approx 8.39$ million bits. Thus, the original image contains about 25.17 million bits. A kilobyte contains $1024 \times 8 = 8192$ bits, so the compressed image contains 0.76 million bits. The compression ratio is (25.17/0.76):1, or 33.1:1. In other words, the compressed file is $1/33.1 = 0.0302$ times the size of the original.

Run-length encoding is a simple compression scheme that gains compression advantages where a signal repeats a sample value more than three times. For each such instance, three copies of the sample are followed by a count of the additional copies needed. For instance, the signal

$$12 \quad 7 \quad 0 \quad 0 \quad 0 \quad 0 \quad 0 \quad 4 \quad 2 \quad 1 \quad 1 \quad 1 \quad 1 \quad 1 \quad 1$$

might be recoded as

$$12 \quad 7 \quad 0 \quad 0 \quad 0 \quad \mathbf{2} \quad 4 \quad 2 \quad 1 \quad 1 \quad 1 \quad \mathbf{3}$$

for a saving of three samples for this 15-sample example. The values shown in bold are counts. **Huffman encoding** recodes a signal using the shortest codes for the most commonly occurring signal segments. Consider, for example, a system that uses standard ASCII codes for alphabetic characters. Each ASCII code contains 7 bits. The code for the letter "e," for example, is 1100101, and the code for the letter "t" is 1110100. A message containing 100 characters contains 700 bits. If the message is Huffman-encoded, the most common characters are assigned the shortest codes. The shortest Huffman codes are 4 bits long to permit error detection; so the letter "e," the most common character, might be coded as "0000" and the letter "t," the second most common character, as "0001." When the 4-bit codes are exhausted, 5-bit codes are used, as suggested by Table 15.4, which contains a sample set of Huffman codes for letters only.

A passage from Thoreau's *Walden* (Krutch, 1962, p. 343) reads:

> *I left the woods for as good a reason as I went there. Perhaps it seemed to me that I had several more lives to live, and could not spare any more time for that one. It is remarkable how easily and insensibly we fall into a particular route, and make a beaten track for ourselves. I had not lived there a week before my feet wore a path from my door to the pond-side; and though it is five or six years since I trod it, it is still quite distinct. It is true, I fear that others may have fallen into it, and so helped to keep it open. The surface of the earth is soft and impressible by the feet of men; and so with the paths which the mind travels. How worn and dusty, then, must be the highways of the world, how deep the ruts of tradition and conformity! I did not wish to take a cabin passage, but rather to go before the mast and on the deck of the world, for there I could best see the moonlight amid the mountains. I do not wish to go below now.*
>
> (Courtesy Bantam Books, Inc.)

For simplicity, the message is coded using lowercase letters only. The 735 letters in the excerpt require $735 \times 7 = 5145$ bits if encoded directly using ASCII codes. Using the Huffman codes in Table 15.4, the same set of letters can be coded using just 2998 bits, achieving a 1.716:1 compression ratio. Huffman encoding can be applied to words or patterns of data as well as to individual letters. The assignment of codes may be based on a survey of the actual message to be compressed, as in the case of Table 15.4, or on known statistics of

TABLE 15.4
Huffman Codes

Letter	Percentage of Total Letters in Thoreau Passage	ASCII Code Lowercase	ASCII Code Uppercase	Huffman Code
e	12.79%	1100101	1000101	0000
t	11.16	1110100	1010100	0001
o	9.12	1101111	1001111	0010
a	7.76	1100001	1000001	0011
i	7.62	1101001	1001001	0100
s	6.26	1110011	1010011	0101
h	6.12	1101000	1001000	0110
r	5.85	1110010	1010010	0111
n	5.44	1101110	1001110	1000
d	4.76	1100100	1000100	1001
l	3.40	1101100	1001100	1010
f	2.99	1100110	1000110	1011
m	2.72	1101101	1001101	1100
w	2.45	1110111	1010111	1101
u	1.90	1110101	1010101	1110
p	1.77	1110000	1010000	1111
b	1.63	1100010	1000010	00000
c	1.50	1100011	1000011	00001
y	1.50	1111001	1011001	00010
v	1.09	1110110	1010110	00011
g	0.95	1100111	1000111	00100
k	0.95	1101011	1001011	00101
q	0.14	1110001	1010001	00110
x	0.14	1111000	1011000	00111
j	0.00	1101010	1001010	01000
z	0.00	1111010	1011010	01001

character frequency. For instance, the English letters occurring most frequently are, in order, e t a o n i h s r d l u m w c f g y p b v k q x j z.

Both run-length and Huffman encoding schemes can be used for compression of images as well as one-dimensional signals. They are, for example, both applied in **JPEG** (Joint Photographic Experts Group), the most common image compression standard in use. JPEG is based on the discrete cosine transform (DCT), which, in a manner similar to the discrete Fourier transform (DFT), responds to signal details of different frequencies. The equation governing the 2D DCT for a square $N \times N$ image is:

$$C[i, k] = 4 \sum_{m=0}^{N-1} \sum_{n=0}^{N-1} x[m, n] \cos\left(\frac{(2m + 1)i\pi}{2N}\right) \cos\left(\frac{(2n + 1)k\pi}{2N}\right) \quad \textbf{(15.4)}$$

where $i, k = 0, \ldots, N-1$. The inverse 2D DCT is

$$x[m, n] = \frac{1}{N^2} \sum_{i=0}^{N-1} \sum_{k=0}^{N-1} \beta[i]\beta[k]C[i, k] \cos\left(\frac{(2m + 1)\pi}{2N}\right) \cos\left(\frac{(2n + 1)k\pi}{2N}\right) \quad (15.5)$$

where $m, n = 0, \ldots, N-1$, and

$$\beta[p] = \begin{cases} \frac{1}{2} & p = 0 \\ 1 & p = 1, \cdots, N - 1 \end{cases}$$

Of note is the fact that the discrete cosine transform does not involve complex arithmetic. For this reason, it can be calculated very rapidly. The DCT has the property that it tends to concentrate information from an image into the lower coefficient numbers, that is, small values of i and k. This property, to be illustrated shortly, is shared by the discrete wavelet transform, to be investigated in Chapter 16, which is the reason that wavelet transforms form the basis for JPEG 2000, the newest JPEG compression standard. Wavelet compression schemes are examined in Section 16.7. The DFT, on the other hand, is not a good transform for compression, because it does not concentrate information efficiently.

The JPEG standard operates on 8 × 8 subblocks of an original image. A sample block is shown in Figure 15.67(a). JPEG's first step is to shift the gray scale levels until they lie symmetrically around zero. For an 8-bit image, for example, which may contain gray scale levels from 0 to 255, this is achieved by subtracting 128 from each level, producing a matrix like the one in Figure 15.67(b). The 2D DCT transforms this matrix, with the rounded result shown in (c). JPEG takes advantage of the DCT's ability to concentrate information by weighting the matrix of DCT results by an 8 × 8 normalization matrix, an example of which appears in Figure 15.67(d). Each element in the DCT is divided by the corresponding element in the normalization matrix and rounded, which causes in particular the small, high frequency coefficients to be zeroed, as shown in Figure 15.67(e). The matrix is then unraveled using the zigzag pattern in Figure 15.67(f), which tends to place the important coefficients first and the zeros last. The trailing zeros are removed and are replaced by an end of block (EOB) symbol signaling that all remaining coefficients are zero. Since JPEG always acts on 8 × 8 blocks, the number of trailing zeros is always known. The resulting vector, shown in Figure 15.67(g), codes the original subimage completely, though with some small errors introduced through rounding. For this reason, JPEG is a lossy compression scheme. Note that the vector contains only 36 numbers, many fewer than the original 64. The more similar the gray scale levels in the original subblock, the fewer the elements in the compressed vector. In the extreme case that the subblock has the same gray scale level for all 64 pixels, for example, the compressed vector contains only one number.

The vector is subjected to further compression through Huffman encoding. The first element is actually encoded as a difference from the previous subblock's first element. Since the value of this difference is normally smaller than the element itself, some bits are saved in this manner. The other elements are encoded according to their size and the number of preceding zeros, a form of run-length encoding. This means that zeros that are embedded in the vector are not coded on their own, but as part of the next element in the list, providing further compaction. The use of Huffman codes for JPEG is described in detail in Gonzalez and Woods (1992).

FIGURE 15.67
DCT compression in
JPEG.

50	52	57	59	61	83	79	75
51	49	56	60	68	80	75	70
67	75	90	102	119	11	88	72
58	59	75	88	113	104	91	75
67	64	69	93	99	85	84	62
69	70	73	65	80	71	73	76
78	74	71	70	66	69	72	78
70	68	72	80	81	68	66	64

(a) Subblock of Original Image

−78	−76	−71	−69	−67	−45	−49	−53
−77	−79	−72	−68	−60	−48	−53	−58
−61	−53	−38	−26	−9	−17	−40	−56
−70	−69	−53	−40	−15	−24	−37	−53
−61	−64	−59	−35	−29	−43	−44	−66
−59	−58	−55	−63	−48	−57	−55	−52
−50	−54	−57	−58	−62	−59	−56	−50
−58	−60	−56	−48	−47	−60	−62	−64

(b) After Gray Scale Level Shift

−13732	−927	−966	464	37	4	11	127
−80	−552	−205	178	−189	20	101	−65
−1070	110	554	−208	−80	31	95	−132
−788	−21	502	−108	−60	37	−9	−26
−48	−127	−225	89	50	93	−78	−9
401	−85	73	−18	−63	−56	75	−14
483	88	−205	−13	−21	−102	14	62
444	176	−153	21	23	89	−69	−93

(c) Discrete Cosine Transform

128	88	80	128	192	320	408	488
96	96	112	152	208	464	480	440
112	104	128	192	320	456	552	448
112	136	176	232	408	696	640	496
144	176	296	448	544	872	824	616
192	280	440	512	648	832	904	736
392	512	624	696	824	968	960	808
576	736	760	784	896	800	824	792

(d) A Normalization Matrix

FIGURE 15.67
Continued

−107	−11	−12	4	0	0	0	0
−1	−6	−2	1	−1	0	0	0
−10	1	4	−1	0	0	0	0
−7	0	3	0	0	0	0	0
0	−1	−1	0	0	0	0	0
2	0	0	0	0	0	0	0
1	0	0	0	0	0	0	0
1	0	0	0	0	0	0	0

(e) Normalized Matrix

0	1	5	6	14	15	27	28
2	4	7	13	16	26	29	42
3	8	12	17	25	30	41	43
9	11	18	24	31	40	44	53
10	19	23	32	39	45	52	54
20	22	33	38	46	51	55	60
21	34	37	47	50	56	59	61
35	36	48	49	57	58	62	63

(f) Zig-zag Reordering Pattern

[−107 −11 −1 −10 −6 −12   4 −2   1 −7   0   0   4   1   0   0
       −1 −1   3 −1   2   1   0 −1   0   0   0   0   0   0
        0   0   0   0   1   EOB]

(g) JPEG Vector

49	52	54	57	67	79	81	76
48	52	56	62	72	79	75	65
68	75	86	100	111	111	96	79
54	62	78	97	110	106	86	67
65	66	73	87	98	96	84	72
77	69	64	67	72	73	72	73
81	74	71	72	72	69	71	75
67	67	73	80	79	70	65	67

(h) Reconstructed Subblock

−1	0	−3	−2	6	−4	2	1
−3	3	0	2	4	−1	0	−5
1	0	−4	−2	−8	0	8	7
−4	3	3	9	−3	2	−5	−8
−2	2	4	−6	−1	11	0	10
8	−1	−9	2	−8	2	−1	−3
3	0	0	2	6	0	−1	−3
−3	−1	1	0	−2	2	−1	3

(i) Errors (Reconstructed Block – Original Block)

When it is time to reconstruct the JPEG coded image, the Huffman-encoded vector is decoded to produce the vector in Figure 15.67(g) again. This vector is expanded back into a matrix and then denormalized by multiplying by the normalization matrix entries instead of dividing by them. The inverse 2D DCT is computed. Finally, the gray scale levels are shifted back to the range 0 to 255 by adding 128. The reconstructed image block appears in Figure 15.67(h), followed by a matrix of errors in Figure 15.67(i) that shows the differences between the reconstructed block and the original. Considering the fact that the vector contains less than two thirds of the numbers in the original block, the errors are remarkably small.

Through the use of the DCT and Huffman encoding, JPEG can achieve compression ratios of 40:1 or even higher with relatively small loss of signal information. The larger the numbers in the normalization matrix, the larger the compression ratio, but the greater the rounding errors that occur. JPEG's major flaw, though, is not rounding but rather the evidence of blocking that it leaves behind in the reconstructed image, a result of its use of 8 × 8 sub-blocks. Nevertheless, JPEG is widely and successfully used, especially on the Internet, where space and time are at a premium.

CHAPTER SUMMARY

Matlab
Support

1. A digital image is made up of pixels. Each pixel in a gray scale image is assigned a gray scale level that records the shade of the pixel.
2. The larger the number of pixels in an image, and the larger the number of gray scale levels that are available, the greater the resolution of the image. In an N-bit image, 2^N possible gray scale levels exist.
3. The histogram for a digital image records the number of pixels at each gray scale level. The gray scales in the histogram may be modified to change the nature of the image. Histogram equalization is a means of improving the contrast in an image.
4. Digital images are added or subtracted pixel by pixel. Image addition may be used in image averaging to reduce noise. Image subtraction may be used to identify differences between images. Following image arithmetic, scaling is frequently necessary to ensure that the resultant image contains legal gray scale levels.
5. Warping changes the shape of an object in a digital image. Morphing changes one object in a digital image into another.
6. Images are filtered using a convolution kernel. Low pass kernels blur an image. High pass filters sharpen an image. Edge filters identify boundaries of objects and changes from dark to light.
7. Dilation adds a layer of pixels to all objects in an image. Erosion removes a layer of pixels from all objects. Dilation followed by erosion tends to repair broken edges. Erosion followed by dilation tends to remove specks of noise.
8. The features of an object—such as its texture, color, or shape—may be used to classify the object.
9. A digital image has a magnitude spectrum and a phase spectrum. The magnitude spectrum for an image gives a sense of the color variations present in the image. The phase spectrum carries the most information about object locations and boundaries.

10. Tomography is the science of studying two-dimensional slices of three-dimensional objects. Tomographical techniques are used in computed tomography (CT) scans and magnetic resonance images (MRIs).

11. Compression reduces the storage needed to record a digital image. When lossless compression techniques are used, the original image is eventually reproduced perfectly. With lossy compression, the re-created file is not identical with the original.

12. Two common compression schemes are run-length encoding and Huffman encoding. Run-length encoding compactly codes repeated occurrences of the same number, while Huffman encoding uses the shortest codes for the most common elements.

13. JPEG, a common lossy image compression standard, uses the discrete cosine transform (DCT), as well as run-length and Huffman encoding.

REVIEW QUESTIONS

15.1 A handful of light-colored marbles is thrown down on a black tablecloth. Each marble has a diameter of 13 mm. An 8-bit digital photograph is taken of a 10 × 15 cm area of the tablecloth and marbles. A histogram for the image, which contains only eight distinct colors, is described in Table 15.5.
   a. How many marbles appear in the photograph?
   b. If the marbles are solid in color, how many different colors are there? How many marbles of each color appear in the image?

TABLE 15.5
Histogram Table for Question 15.1

Gray Scale Level	Number of Pixels
18	83
20	12171
21	1656
25	42
180	541
188	136
209	119
216	252

15.2 The histogram table for a 4-bit digital image is given in Table 15.6. A gray scale value of zero corresponds to black and a gray scale value of 15 corresponds to white.
   a. Use histogram equalization to compute a new set of gray scale levels for this image.
   b. Will the new histogram-equalized image be lighter or darker than the original?

TABLE 15.6
Histogram Table for Question 15.2

Gray Scale Level	Number of Pixels
0	1760
1	351
2	485
3	342
4	1176
5	897
6	5014
7	28
8	195
9	782
10	901
11	8905
12	7883
13	6699
14	705
15	808

15.3 The histogram table for a 3-bit image is given in Table 15.7.
    a. How many pixels are in the image?
    b. Sketch the histogram for the image.
    c. If all the black pixels are changed to white, and if gray scale level 6 is changed to 5, what is the effect on the histogram?

TABLE 15.7
Histogram Table for Question 15.3

Gray Scale Level	Number of Pixels
0	950
1	600
2	350
3	400
4	450
5	500
6	300
7	300

15.4 a. The 8-bit image of a black cross on a white background shown in Figure 15.68(i) is degraded by noise in the other four parts of the figure. Average the four noisy images.
    b. Find the gray scale levels for the averaged image after scaling and rounding.

FIGURE 15.68
Images to be averaged for Question 15.4.

(i) Original Image

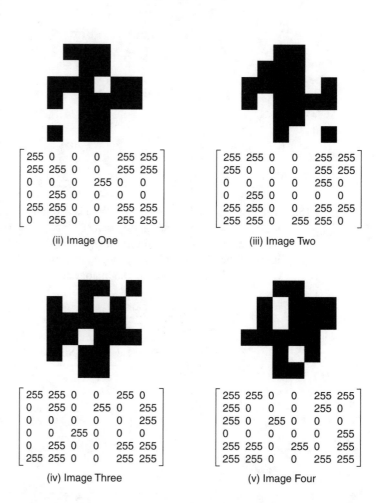

$$\begin{bmatrix} 255 & 0 & 0 & 0 & 255 & 255 \\ 255 & 255 & 0 & 0 & 255 & 255 \\ 0 & 0 & 0 & 255 & 0 & 0 \\ 0 & 255 & 0 & 0 & 0 & 0 \\ 255 & 255 & 0 & 0 & 255 & 255 \\ 0 & 255 & 0 & 0 & 255 & 255 \end{bmatrix}$$

(ii) Image One

$$\begin{bmatrix} 255 & 255 & 0 & 0 & 255 & 255 \\ 255 & 0 & 0 & 0 & 255 & 255 \\ 0 & 0 & 0 & 0 & 255 & 0 \\ 0 & 255 & 0 & 0 & 0 & 0 \\ 255 & 255 & 0 & 0 & 255 & 255 \\ 255 & 255 & 0 & 255 & 255 & 0 \end{bmatrix}$$

(iii) Image Two

$$\begin{bmatrix} 255 & 255 & 0 & 0 & 255 & 0 \\ 0 & 255 & 0 & 255 & 0 & 255 \\ 0 & 0 & 0 & 0 & 0 & 255 \\ 0 & 0 & 255 & 0 & 0 & 0 \\ 0 & 255 & 0 & 0 & 255 & 255 \\ 255 & 255 & 0 & 0 & 255 & 255 \end{bmatrix}$$

(iv) Image Three

$$\begin{bmatrix} 255 & 255 & 0 & 0 & 255 & 255 \\ 255 & 0 & 0 & 0 & 255 & 0 \\ 255 & 0 & 255 & 0 & 0 & 0 \\ 0 & 0 & 0 & 0 & 0 & 255 \\ 255 & 255 & 0 & 255 & 0 & 255 \\ 255 & 255 & 0 & 0 & 255 & 255 \end{bmatrix}$$

(v) Image Four

15.5 Figure 15.69 shows two black and white 8-bit digital photographs of the windows on the side of a building, taken 5 minutes apart. Describe the scaled difference image $A - B$.

FIGURE 15.69
Windows on
the side of a
building for
Question 15.5.

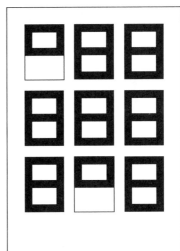

(a) Image A   (b) Image B

15.6 Two $6 \times 6 \times 8$-bit black and white images are shown in Figure 15.70. What is the result, after scaling and rounding, of the image differences:
(**i**) $A - B$
(**ii**) $B - A$

FIGURE 15.70
Images for
Question 15.6.

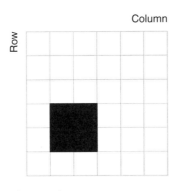

(a) Image A   (b) Image B

15.7 Shape I in Figure 15.71 must be morphed into shape II. Control points are marked in both images. Find the control point locations for the intermediate image that is exactly halfway between the two shapes.

FIGURE 15.71
Shapes for
Question 15.7.

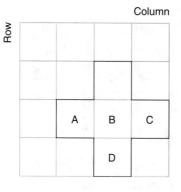

(a) Shape I

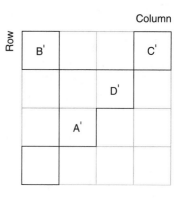

(b) Shape II

15.8 A $512 \times 512 \times 8$-bit image is morphed into another image using control points. The control point locations and gray scale levels for both images are given in Table 15.8. Describe the control points for three intermediate images. Use linear interpolation for locations and weighted averaging for gray scale levels.

TABLE 15.8
Control Points for Question 15.8

Source Control Point	Location	Gray Scale Level	Destination Control Point	Location	Gray Scale Level
A	(200, 35)	40	A′	(75, 40)	188
B	(176, 95)	47	B′	(29, 92)	141
C	(55, 291)	195	C′	(340, 140)	102
D	(55, 323)	203	D′	(331, 381)	70
E	(253, 249)	187	E′	(271, 262)	43
F	(378, 404)	135	F′	(417, 366)	37
G	(380, 341)	105	G′	(380, 205)	185
H	(399, 78)	88	H′	(286, 50)	192

15.9 The left side of the $20 \times 30 \times 8$-bit digital image shown in Figure 15.72 has the gray scale level 200, and the right side has the gray scale level 100. Find the result when the image is filtered by a:
a. 3×3 low pass filter
b. 5×5 low pass filter

FIGURE 15.72

Image for Question 15.9.

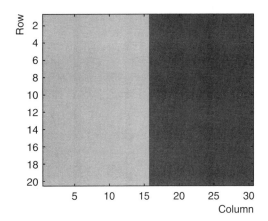

15.10 A 3 × 3 filter is applied to the 8 × 8 image whose gray scale levels are provided in Figure 15.73. Produce the unscaled filtered image for a:

a. High pass filter

b. Vertical Sobel edge filter

c. Horizontal Sobel edge filter

FIGURE 15.73

Digital image for Question 15.10.

20	40	60	80	0	0	0	0
40	40	60	80	0	0	0	0
60	60	60	80	0	0	0	0
80	80	80	80	0	0	0	0
0	0	0	0	0	0	0	0
0	0	0	0	0	80	0	0
0	0	0	0	0	0	0	0
0	0	0	0	0	0	0	0

15.11 For the group of pixels shown in Figure 15.74, perform dilation followed by erosion.

FIGURE 15.74

Pixels for Question 15.11.

15.12 For the group of pixels shown in Figure 15.75, perform erosion followed by dilation.

FIGURE 15.75

Pixels for Question 15.12.

15.13 **a.** Find the total area of the shape in Figure 15.76 using the vertex coordinates given.
b. What is the circularity measure for the shape in (a)?

FIGURE 15.76
Shape for Question 15.13.

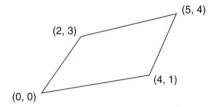

15.14 Write a boundary chain code for the shape shown in Figure 15.77, moving clockwise from "Start."

FIGURE 15.77
Object boundary for
Question 15.14.

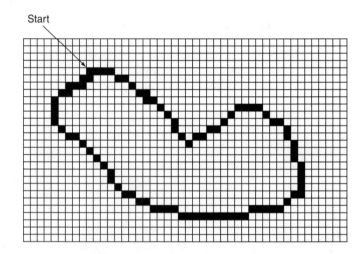

15.15 The boundary chain code for an object in an image is:

1-1-1-1-1-0-0-0-0-0-0-0-0-7-7-7-5-5-5-5-5-5-5-5-5-5-3-3-3-3-3-3-3

a. Draw the object and determine coordinates for its vertices.
b. Compute the perimeter of the object.
c. Compute the area of the object.

15.16 Compute the 2D DFT for the image:

$$x[m, n] = \begin{bmatrix} 4 & 0 \\ 0 & -1 \end{bmatrix}$$

Verify that the result is correct by computing the inverse 2D DFT.

15.17 Predict the shape of the magnitude spectrum of the image shown in Figure 15.78.

FIGURE 15.78

Image for Question 15.17.

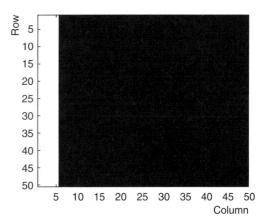

15.18 **a.** Using the Huffman codes in Table 15.4, code the 7-bit ASCII message: "The rain in Spain falls mainly on the plain." Use lowercase letters only and ignore punctuation.

b. What compression ratio is achieved in (a)?

15.19 Compute the 2D DCT for the matrix:

$$x[m, n] = \begin{bmatrix} 4 & 0 \\ 0 & -1 \end{bmatrix}$$

Verify that the result is correct by computing the inverse 2D DCT.

15.20 The DCT of an 8 × 8 subimage is given in Figure 15.79.

a. Compute the normalized matrix using the normalization matrix of Figure 15.67(d).

b. Produce the JPEG vector that corresponds to the normalized matrix.

c. What is the compression ratio for the subimage?

FIGURE 15.79

DCT C[i, k] for Question 15.20.

$$
i \begin{bmatrix}
-572 & -769 & 6 & 170 & -93 & 120 & -36 & -26 \\
52 & -72 & 183 & -66 & 76 & -27 & 30 & -6 \\
-228 & 373 & 75 & -252 & -132 & 12 & -51 & -119 \\
-383 & 87 & 86 & -102 & 22 & 8 & -35 & 50 \\
-93 & 194 & 165 & -27 & -86 & -38 & -45 & -30 \\
-178 & 347 & 49 & -31 & -25 & 38 & -1 & 1 \\
-454 & 57 & -3 & -65 & -52 & -9 & 49 & 3 \\
239 & -12 & -24 & -54 & -53 & 8 & 47 & -48
\end{bmatrix}
$$

k

16

WAVELETS

The once-esoteric topic of wavelets is now part of mainstream DSP technology. Wavelets have in particular found application in signal and image compression. This chapter provides a compact introduction to wavelets and wavelet transforms. The chapter:

- ➤ identifies shortcomings of the DFT
- ➤ shows how wavelets at different scales and translations can overcome these shortcomings
- ➤ defines scaling functions and wavelet functions
- ➤ explains how a father wavelet and a mother wavelet produce a family of wavelets
- ➤ describes how to code a signal using the discrete wavelet transform (DWT)
- ➤ presents a frequency domain view of DWT analysis and synthesis
- ➤ provides a complete example of DWT analysis and synthesis
- ➤ presents examples of two-dimensional DWT analysis
- ➤ illustrates wavelet compression methods

16.1 WAVELET BASICS

The discrete Fourier transform (DFT), studied in Chapter 11, analyzes a signal in terms of its frequency components by finding the signal's magnitude and phase spectra. For example, the DFT (or FFT) produces a single spike in the magnitude spectrum for a pure sine wave, and it produces three spikes for a chord made of three pure sines. The DFT always reports the frequencies present in the portion of the signal that is within the DFT window. As established in Equation (11.4), each point k of an N-point DFT reports a frequency in Hz given by $k\,f_S/N$, where f_S is the sampling frequency. Unfortunately, the DFT cannot identify the times at which various frequency components occur within the window, as Example 16.1 illustrates.

EXAMPLE 16.1

Figure 16.1(a) shows signal A, consisting of 30 seconds of the function

$$x(t) = \frac{\cos(t) + \cos(2t) + \cos(3t)}{3}$$

which contains three separate frequency elements. Figure 16.1(b) shows signal B, which consists of 10 seconds of the low frequency signal, $\cos(t)$, followed by 10 seconds of the midfrequency signal, $\cos(2t)$, followed by 10 seconds of the high frequency signal, $\cos(3t)$. Both signal A and signal B are sampled, at 8 samples/second, and the first 120 points of the DFTs for both signals are shown in Figure 16.1(c) and (d). Despite the fact that the signals are very different, the spectra are very similar. In fact, the transitions between different frequency regions are the only reasons why B's spectrum is less defined than A's.

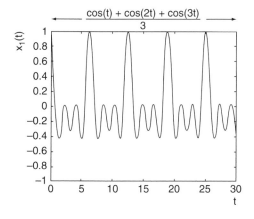

(a) Signal A

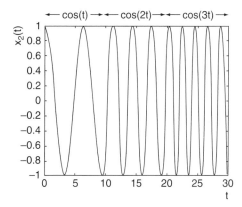

(b) Signal B

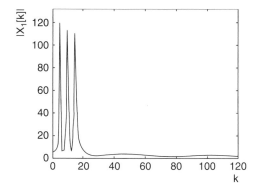

(c) DFT of Signal A

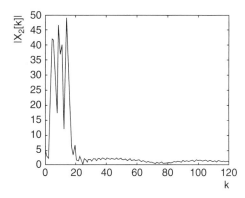

(d) DFT of Signal B

FIGURE 16.1
Signals and spectra for Example 16.1.

Example 16.1 highlights a weakness of the DFT: It fails to distinguish signals whose characteristics change with time, known as nonstationary signals, from those whose characteristics do not change with time, known as stationary signals. Since the DFT is windowed, only the signal's behavior within the window is important. For signals that are stationary within the window, the DFT provides an accurate picture of the frequency content since the same signal behavior persists for the length of the window. For signals with changing behavior, like the signal in Figure 16.1(b), the DFT can report the frequencies but not when they occur. Possible solutions to this problem do exist. For example, several shorter windows could be used, increasing the probability that the signal within each window will be stationary. The spectra obtained from such windows are then presented in time sequence, as Example 16.2 shows.

EXAMPLE 16.2

Instead of the 30 second DFT window used in Example 16.1, three 10-second windows are used on the signal shown in Figure 16.1(b). The results are shown in the three magnitude spectra of Figure 16.2. Since the window lengths are matched perfectly to each portion of

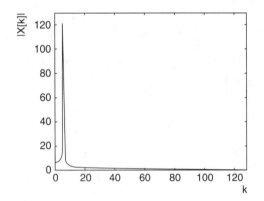

(a) First 10 seconds

(b) Second 10 seconds

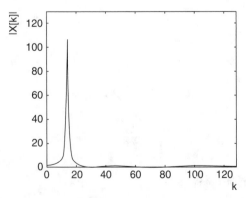

(c) Third 10 seconds

FIGURE 16.2

Spectra for Example 16.2.

the signal, the DFTs shown in Figure 16.2 report single frequency spikes for each 10-second interval. In Figure 16.1(d), the DFT computed for a longer window reported three spikes, obscuring much useful information about the signal.

The results from the three windows in Example 16.2 can be combined into a single **time-frequency diagram**, two views of which are seen in Figure 16.3. The time axis is measured in seconds, while the frequency axis maps to the points of the DFT. In the two-dimensional view, the lightest areas show the strongest responses. The advantage of a time-frequency diagram is that both the frequencies and their locations in the signal can be viewed. At any given instant, a slice parallel to the frequency axis gives a snapshot of the frequency content of the signal at that time. Figure 16.3, for example, shows clearly that frequency is constant in each 10-second segment and that the frequency increases from one segment to the next.

While shorter windows can improve the DFT's ability to analyze nonstationary signals, there are resolution implications. If the sampling rate is left unchanged, a shorter window implies fewer points, which means that the DFT cannot provide as much detail about the true spectrum of a signal. The shorter window means good time resolution, because it provides very local detail, but poor frequency resolution, since the time to monitor signal

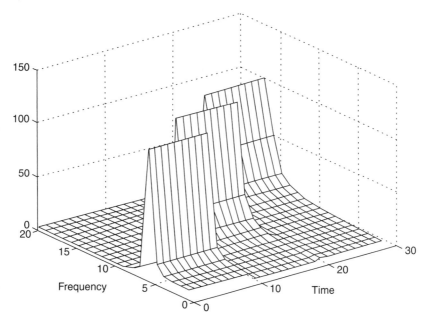

(a) Three-Dimensional View

FIGURE 16.3
Time-frequency diagram.

FIGURE 16.3
Continued

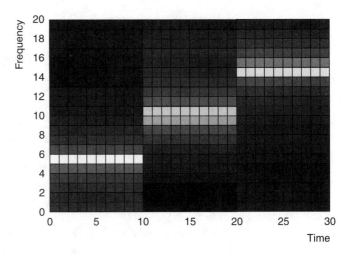

(b) Two-Dimensional View

TABLE 16.1
Time and Frequency Resolution by Window Width

Narrow Window	Good time resolution	Poor frequency resolution
Wide Window	Poor time resolution	Good frequency resolution

characteristics is so short. The only way to improve frequency resolution is to lengthen the window, but this action reduces the time resolution since the DFT cannot pinpoint changes in signal behavior within its window. Thus, good time resolution and good frequency resolution are impossible to achieve at the same time. Table 16.1 summarizes these observations. In addition to the problems associated with choosing an appropriate window length, suitable window boundary positions are difficult to select without a great deal of information about the signal. Thus, attempting to use multiple windows to handle nonstationary signals can be difficult and impractical.

Wavelet transforms are an alternative to short DFTs that are becoming part of mainstream DSP. The most important feature of wavelet transforms is that they analyze different frequency components of a signal with different resolutions. For high frequency elements, good time resolution is used, since signals are changing quickly and it is important to know when high frequency signal elements appear and disappear. This is the equivalent of using a narrow window, which means frequency resolution will be relatively poor. For low frequency signal elements, poor time resolution is acceptable, since signals are changing slowly. This is equivalent to using a wide window, which means a longer collection interval and better frequency resolution, a benefit because so much of the information that makes sounds recognizable occurs in the lower frequencies.

To implement different resolutions at different frequencies requires the notion of functions at different scales. Like scales on maps, small scales show fine details while large scales show gross features only. A scaled version of a function $\psi(t)$ is the function $\psi(t/s)$,

TABLE 16.2
Scaling

$\psi\left(\dfrac{t}{2}\right) = \cos\left(\dfrac{t}{2}\right)$	$s = 2$	expansion $s > 1$	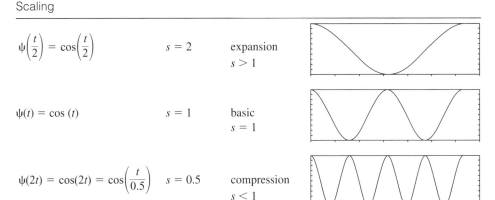
$\psi(t) = \cos(t)$	$s = 1$	basic $s = 1$	
$\psi(2t) = \cos(2t) = \cos\left(\dfrac{t}{0.5}\right)$	$s = 0.5$	compression $s < 1$	

for any scale s. For example, if $\psi(t) = \cos(t)$ represents the basic scale, designated as $s = 1$, then $\psi(t/s) = \cos(t/s)$ is a scaled version of it at some other scale s. When $s > 1$, a function of lower frequency is obtained: This represents an expansion of the basic scale and is capable of describing slowly varying trends in a signal. When $s < 1$, a function of higher frequency is obtained, a compression of the basic scale capable of describing details of more rapid signal changes. Table 16.2 illustrates these cases. Note that scale is inversely proportional to frequency. That is, for some constant α,

$$\text{Scale} = \frac{\alpha}{\text{Frequency}} \qquad (16.1)$$

Table 16.2 and Equation (16.1) provide a connection with the Fourier series, which decomposes a periodic signal by breaking it down into an infinite number of harmonically related sines and cosines. The low frequencies in the signal resonate best with the low frequency, or large scale, sinusoids; and the high frequencies in the signal resonate best with the high frequency, or small scale, sinusoids. It is possible to find these resonances using signals other than sines and cosines. In particular, wavelet functions can serve this purpose. Wavelet functions are localized in frequency in the same way that sinusoids are, but they differ from sinusoids by being localized in time as well. A multitude of wavelet families exists. Each family has a characteristic shape, and the basic scale for each family covers a known, fixed interval of time. The time spans of the other wavelets in the family widen for larger scales and narrow for smaller scales. Thus, wavelet functions can offer either good time resolution or good frequency resolution. For good time resolution, narrow, small scale wavelets are used; for good frequency resolution, wide, large scale wavelets are best. For example, Figure 16.4(a) shows what is called a second order spline wavelet. At the basic scale, $s = 1$, the wavelet function has the value zero outside the range $0 \le t \le 3$, so the function is localized to this range of times. Scaled versions of the second order spline wavelet are shown in Figure 16.4(b) and (c). The wavelet for the scale $s = 1/2$ covers half

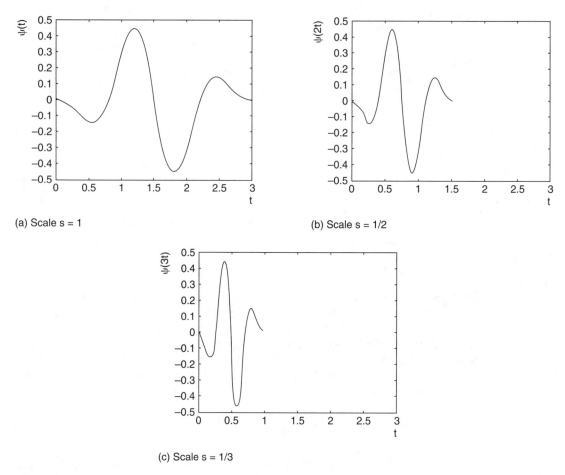

(a) Scale s = 1

(b) Scale s = 1/2

(c) Scale s = 1/3

FIGURE 16.4

Spline wavelet at different scales.

the time of the basic scale, while the wavelet for $s = 1/3$ covers one third of the time. Such scaled copies of the wavelet provide a family of functions with different resonant frequencies. As well as being localized in time, each member of the family is localized to a small range of frequencies.

To determine what frequencies are present in a signal and when they occur, the wavelet functions at each scale must be translated through the signal, to enable comparison with the signal in different time intervals. A scaled and translated version of the wavelet function $\psi(t)$ is the function, $\psi[(t - \tau)/s]$ for any scale s and translation τ. Figure 16.5 shows a sinc signal $x(t)$ superimposed by several translations of each scaled wavelet function. The details of the wavelet transform will be postponed until Section 16.5, but in essence the transform requires multiplying the signal by the wavelet function. Where the wavelet function at a given translation is similar to the signal in frequency, the wavelet transform will be large. Where the

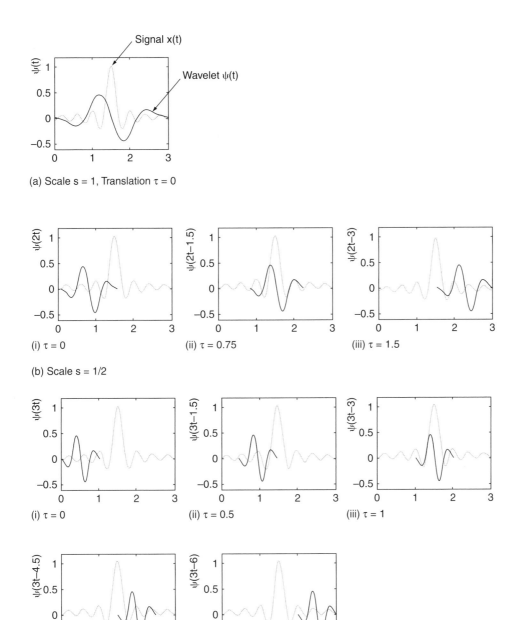

(a) Scale s = 1, Translation τ = 0

(i) τ = 0

(ii) τ = 0.75

(iii) τ = 1.5

(b) Scale s = 1/2

(i) τ = 0

(ii) τ = 0.5

(iii) τ = 1

(iv) τ = 1.5

(v) τ = 2.0

(c) Scale s = 1/3

FIGURE 16.5
Coding a signal with wavelets of different scalings and translations.

wavelet function is dissimilar to the signal, the transform will be small. For the sinc signal shown in Figure 16.5, the wavelet with scale $s = 1/3$ seems to resonate best at the beginning and end of the signal, while the wavelet with scale $s = 1/2$ works best in the middle of the signal. The signal can be coded using these wavelets if it can be decomposed into scaled and translated copies of the basic wavelet function.

A signal is coded through wavelet transforms by comparing the signal to many scalings and translations of a wavelet function. The widest wavelet matches the size of the largest features of interest in the signal. This wavelet responds to the slowest variations in the signal but, because of its width, is not able to identify places where signal behavior changes. This job is left to smaller scale wavelets. These have a shorter temporal range, so their translations can respond differently to different parts of the signal. The narrower the time base for the wavelet, the more accurately it can identify important time boundaries, where signal characteristics change significantly. Since small scale wavelets themselves contain high frequencies, they respond best to high frequency elements in the signal. The wavelet transform results can be presented on a **time-scale diagram**, similar to the time-frequency diagrams in Figure 16.3, with smaller scales corresponding to higher frequencies and larger scales to lower frequencies. The time-scale diagram reports the wavelets that resonate best with the signal throughout its duration. Examples 16.3 and 16.4 present time-scale diagrams for two different signals.

EXAMPLE 16.3

Figure 16.6 shows one sine wave that begins at time 0 and ends at 130 msec, and a second sine wave that begins at 70 msec and ends at 200 msec. The two sine waves overlap and add together for a period of time in the middle. Figure 16.7 shows a time-scale diagram for the signal. The white areas show where the strongest resonances between the signal and wavelets of various scales occur. As expected, a single large scale (low frequency) signal is present at the start, a single small scale (high frequency) signal is present at the end, and

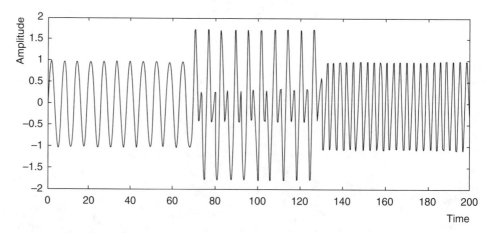

FIGURE 16.6
Signal for Example 16.3.

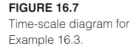

FIGURE 16.7
Time-scale diagram for
Example 16.3.

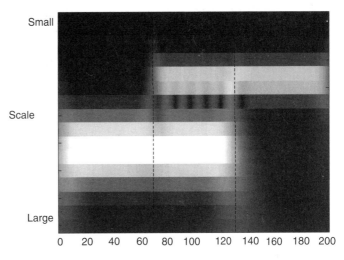

both signals coexist for about 60 msec in the middle. It is the wavelets' frequency localization property that permits them to distinguish the frequency elements in the signal. The time localization property of the wavelets also plays an important role in analyzing the signal. As the wavelet scale, and hence the width of the wavelet, grows smaller, the wavelet transform is able to identify more and more accurately the clear boundaries in the signal's behavior. The exact locations are marked in Figure 16.7 by a dashed line.

EXAMPLE 16.4

Figure 16.8 shows a "chirp" signal. The frequency of this signal varies smoothly from 0 to 5 Hz in 5 seconds. Figure 16.9 shows its time-scale diagram. The largest scales, and hence

FIGURE 16.8
Chirp signal for Example 16.4.

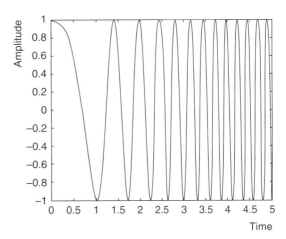

FIGURE 16.9

Time-scale diagram for chirp signal for Example 16.4.

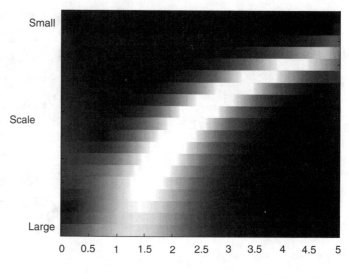

the lowest frequencies, occur at the beginning, while the smallest scales, and therefore the highest frequencies, occur at the end. These features can be confirmed by examining the time signal in Figure 16.8.

16.2 FAMILIES OF WAVELETS

The purpose of the wavelet transform is to decompose a signal into component wavelets. Each component wavelet belongs to a **wavelet family**. Each wavelet family consists of some combination of **scaling functions**, derived from a **father wavelet** $\varphi(t)$, and **wavelet functions**, derived from a **mother wavelet** $\psi(t)$. The father wavelet represents the basic scale for the scaling functions, whereas the mother wavelet fulfills this role for the wavelet functions. All other members of the wavelet family are scalings and translations of either the mother or the father wavelet. Example wavelets for three different wavelet families are shown in Figures 16.10 and 16.11. What all these wavelets have in common is that they are localized in time. They are nonzero only over a finite interval of time, called the **support** of the wavelet. This feature makes wavelets more suitable than sines and cosines for coding physical signals like speech or music. These signals are characterized by abrupt starts and stops, and by rapid short-term variations. To code them using sines and cosines, which go on forever in both directions, is not an efficient process. Wavelets, on the other hand, are functions with speechlike shapes and timing, which makes them highly effective signal coders. For a given application, the most suitable wavelet family can be selected.

Section 16.1 introduced the idea that wavelet functions at multiple scales and translations are necessary to code a signal. The discrete wavelet transform, or DWT, requires

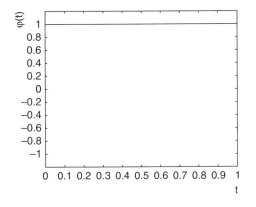

(a) Haar

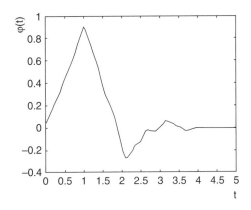

(b) Daubechies-6

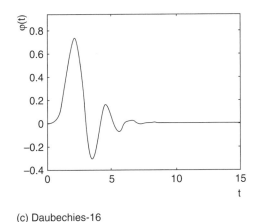

(c) Daubechies-16

FIGURE 16.10

Father wavelets.

that the scales s and translations τ be made discrete. Among the wavelet functions, the basic scale, $s = 1$, is defined by the mother wavelet. All other scales are compressions or expansions of this basic scale. Because powers of 2 allow the most efficient algorithms, scale values are chosen according to

$$s = 2^{-j} \tag{16.2}$$

where j is an integer. For each halving of the scale, the wavelet is squeezed to half its former width; for each doubling of the scale, the wavelet is stretched to twice its original width. Because all wavelets in a family must be able to contribute equally to coding a signal, a wavelet compressed by a factor of 2 must be available twice as often, while a wavelet expanded by a factor of 2 need be available only half as often. At the basic scale, $s = 1$,

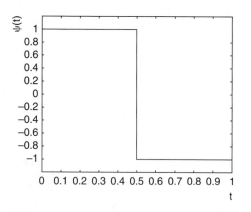

(a) Haar

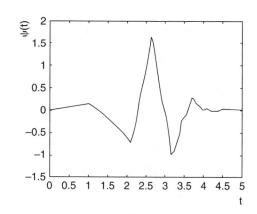

(b) Daubechies-6

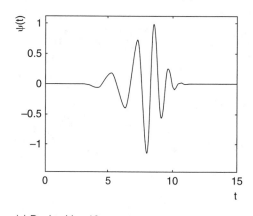

(c) Daubechies-16

FIGURE 16.11

Mother wavelets.

multiples of the translation $\tau = 1$ are used; for $s = 0.5$, multiples of $\tau = 0.5$ are used; and for $s = 2$, multiples of $\tau = 2$ are used. This scheme can be summarized by the formula

$$\tau = k2^{-j} = ks \tag{16.3}$$

where k is an integer, which links the required translation to the scale.

Using the relationships in Equations (16.2) and (16.3), the scaled and translated wavelet function becomes

$$\psi\left(\frac{t - \tau}{s}\right) = \psi\left(\frac{t - ks}{s}\right) = \psi(s^{-1}t - k) = \psi(2^j t - k)$$

These choices for scale s and translation τ have the advantage that the members of the wavelet family generated by $\psi(2^j t - k)$ are all independent of one another and can code signals efficiently. To ensure that the energy is the same for wavelets at all scales, a scaling factor of $2^{j/2}$ must be applied to this function, as explained in Appendix N.1. The expression that generates all scalings and translations of the mother wavelet $\psi(t)$ is, then:

$$\psi_{jk}(t) = 2^{j/2} \psi(2^j t - k) \tag{16.4}$$

where j indexes the scale and k indexes the translation. This equation generates the wavelet functions for a wavelet family, as illustrated in Examples 16.5 and 16.7 for the Haar and Daubechies-4 families. All scalings and translations of the father wavelet $\varphi(t)$ are described by an analogous equation:

$$\varphi_{jk}(t) = 2^{j/2} \varphi(2^j t - k) \tag{16.5}$$

This equation generates the scaling functions for a wavelet family, as illustrated in Example 16.6.

EXAMPLE 16.5

As shown in Figure 16.11(a), the Haar mother wavelet $\psi(t)$ is defined between $t = 0$ and $t = 1$. This wavelet generates the wavelet functions for the Haar family. It represents the basic scale $s = 1$, which is obtained for $j = 0$ in Equation (16.2). At this scale, copies of the wavelet are placed one time unit apart, for $k = 0, 1, 2, \ldots$, since $\tau = k2^{-0} = k$ according to Equation (16.3). Since the mother wavelet has width 1, all copies are placed exactly adjacent to one another. When $j = 1$ is substituted into Equation (16.2), $s = 2^{-1} = 0.5$, so half-scale wavelets are obtained, covering 0.5 seconds. The translations for this scale are given by $\tau = k2^{-1} = 0.5k$. Since this wavelet has width 0.5, half the width of the mother wavelet, translated copies of this wavelet appear exactly adjacent to one another for successive values of k, between $t = 0$ and $t = 0.5$, between $t = 0.5$ and $t = 1$, and so on. Equation (16.4) describes the half-scale wavelet as $\sqrt{2}\psi(2t - k)$, which means that the scaled-down wavelet has an amplitude $\sqrt{2}$ times that of the mother wavelet. Table 16.3 shows the translations for the mother wavelet and three scaled-down versions of it. Figure 16.12 illustrates a few translations for each scale shown in the table. As expected, the width of the wavelet halves and the height is multiplied by $\sqrt{2}$ each time the scale is halved. The amplitude change is due to the scaling factor $2^{j/2}$ in Equation (16.4). For the Haar family, there is no overlap between translated copies of a wavelet at a given scale. This is not usually the case, but merely a side effect of the unit width of the Haar mother wavelet.

TABLE 16.3
Wavelet Function Translations for Haar Family for Example 16.5

Translations		Scale	Width	k							
				0	1	2	3	4	5	6	7
	0	1	1	0	1	2	3	4	5	6	7
j	1	0.5	0.5	0	0.5	1.0	1.5	2.0	2.5	3.0	3.5
	2	0.25	0.25	0	0.25	0.5	0.75	1.0	1.25	1.5	1.75
	3	0.125	0.125	0	0.125	0.25	0.375	0.5	0.625	0.75	0.875

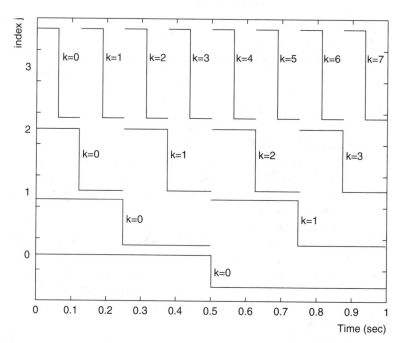

FIGURE 16.12
Family of Haar wavelet functions for Example 16.5.

EXAMPLE 16.6

The father wavelet for the Haar wavelet family is shown in Figure 16.10(a). It produces the scaling functions for the Haar wavelet family according to Equation (16.5). A few translations of the scaling functions for the scales $s = 1$, $s = 0.5$, $s = 0.25$, and $s = 0.125$ are shown in Figure 16.13. The translations for the scaling functions are computed from Equation (16.3). They are identical with the translations for the Haar wavelet functions in the previous example. As noted in that example, translated copies for this wavelet family do

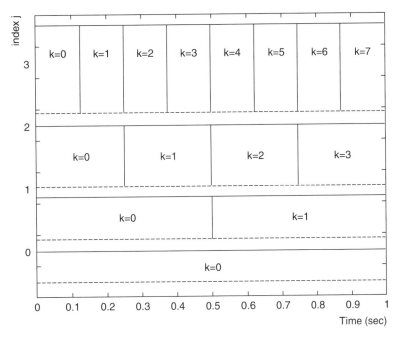

FIGURE 16.13
Family of Haar scaling functions for Example 16.6.

not overlap. The scaling functions at different scales also exhibit the same amplitude changes as the wavelet functions.

EXAMPLE 16.7

The mother wavelet for the Daubechies-4 wavelet family, $\psi(t)$, by definition at the basic scale $s = 1$, lasts from $t = 0$ to $t = 3$ seconds. Scalings and translations of this wavelet produce the wavelet functions for this wavelet family according to Equation (16.4). In Table 16.4, scales for $j = 0, 1$, and 2 are computed from Equation (16.2), and translations for the scales are calculated using Equation (16.3). For instance, $j = 2$ refers to a wavelet at one quarter the scale of the mother wavelet. As such, its width is $0.25(3) = 0.75$. The translations at this scale lie 0.25 time units apart, according to $\tau = k2^{-2} = 0.25k$. Thus, for successive values of k, copies of this wavelet, starting at time zero, lie between $t = 0$ and $t = 0.75$, between $t = 0.25$ and $t = 1.0$, between $t = 0.5$ and $t = 1.25$, and so on. These values are confirmed in Figure 16.14, which illustrates several translations of each scale in Table 16.4. Wavelets $\psi(t)$ for $j = 0$, $\sqrt{2}\,\psi(2t-k)$ for $j = 1$, and $2\psi(4t-k)$ for $j = 2$ are plotted. Notice that the translated copies overlap. The amplitude changes are caused by the $2^{j/2}$ factor in Equation (16.4).

TABLE 16.4

Wavelet Function Translations for the Daubechies-4 Family for Example 16.7

Translations	Scale	Width	k									
			0	**1**	**2**	**3**	**4**	**5**	**6**	**7**	**8**	**9**
0	1	3	0	1	2	3	4	5	6	7	8	9
j **1**	0.5	1.5	0	0.5	1.0	1.5	2.0	2.5	3.0	3.5	4.0	4.5
2	0.25	0.75	0	0.25	0.5	0.75	1.0	1.25	1.5	1.75	2.00	2.25

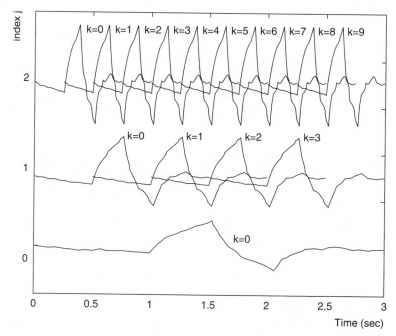

FIGURE 16.14

Family of Daubechies-4 wavelet functions for Example 16.7.

16.3 CODING A SIGNAL

A wavelet family is used to code a signal, that is, to decompose it into scaled and translated copies of a basic function. As discussed in the last section, a wavelet family consists of scaling functions, which are scalings and translations of a father wavelet, and wavelet functions, which are scalings and translations of a mother wavelet. Scaling functions alone are adequate to code a signal completely, but as this section will show, a coding that uses scaling and wavelet functions together is most efficient. To code a signal using scaling functions requires, first, that the characteristics of the family be suitable for the signal being coded and, second, that the smallest scale be capable of coding the finest detail in the signal.

For a function $f(t)$ to be coded in terms of the scaling functions $\varphi_{jk}(t)$ of Equation (16.5) alone implies that the expression

$$f(t) \approx f_j(t) = \sum_{k=-\infty}^{\infty} c_j[k]\varphi_{jk}(t) = \sum_{k=-\infty}^{\infty} c_j[k]2^{j/2}\varphi(2^j t - k) \qquad (16.6)$$

must hold for some set of coefficients $c_j[k]$ and some scale $s = 2^{-j}$. The index j must be chosen large enough to give a scale 2^{-j} small enough to capture all the important signal details. The set of functions $f_j(t)$ that can be perfectly described by the scaling functions at scale j are said to belong to a subspace called S_j. To code a signal in S_j defined on a specified interval of time, all translations of the scaling functions at scale $s = 2^{-j}$ that are defined during the interval of interest are needed.

The Haar family seems a good choice for coding the signal shown in Figure 16.15 because the shapes present in the scaling function are a good match for the blocklike shapes occurring in the signal, although the slopes appear to present a problem. The signal can be coded at any level of detail, determined by the set of scaling functions used. Figure 16.16(a), for example, codes the signal with the coarsest Haar scaling function pictured in Figure 16.13, defined by $j = 0$. Clearly, this is a poor approximation to the signal. Figure 16.16(b) uses all translations of the $j = 1$ scaling function of Figure 16.13. This approximation is not much better. Each subsequent part of Figure 16.16 codes the signal with a finer set of scaling functions from Figure 16.13. As j increases, the scaling functions provide a better and better approximation to the desired signal. In fact, the differences could be further reduced from Figure 16.16(h) by further increasing j to use a scaling function of an even finer scale. A suitable scale can be selected for a signal after deciding how much coding error is acceptable. When no detail of consequence can be gained by adding greater

FIGURE 16.15

Signal to be coded.

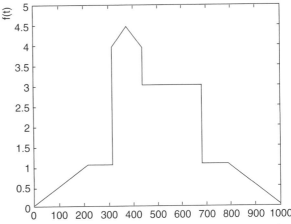

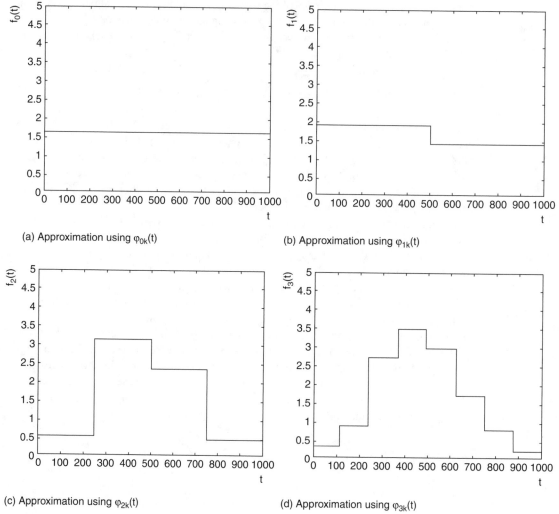

(a) Approximation using $\varphi_{0k}(t)$

(b) Approximation using $\varphi_{1k}(t)$

(c) Approximation using $\varphi_{2k}(t)$

(d) Approximation using $\varphi_{3k}(t)$

FIGURE 16.16

Approximations using scaling functions $\varphi_{jk}(t)$.

detail, the scale is fine enough. The approximation to the signal given in Figure 16.16(h), for example, is given by

$$f(t) \approx f_7(t) = \sum_{k=-\infty}^{\infty} c_7[k]\varphi_{7k}(t) = \sum_{k=-\infty}^{\infty} c_7[k]2^{7/2}\varphi(2^7 t - k)$$

In this case, the function $f(t)$ does not lie in the subspace S_7 because it cannot be represented perfectly by the scaling functions at this scale. Identifying the coefficients like $c_7[k]$ is the task of the discrete wavelet transform, explained in Section 16.5.

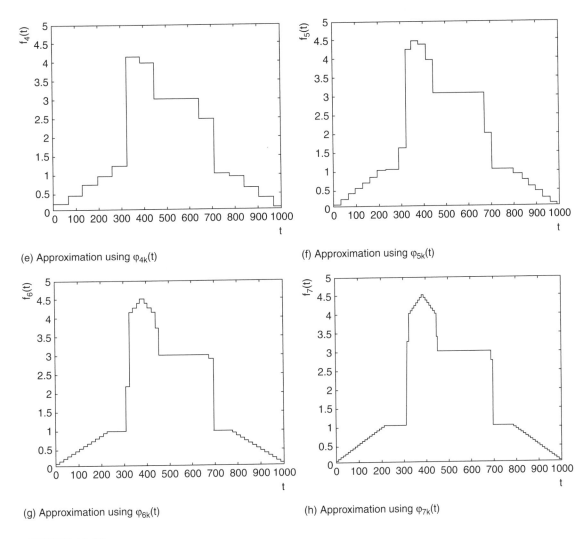

(e) Approximation using $\varphi_{4k}(t)$

(f) Approximation using $\varphi_{5k}(t)$

(g) Approximation using $\varphi_{6k}(t)$

(h) Approximation using $\varphi_{7k}(t)$

FIGURE 16.16

Continued

While scaling functions alone can code a signal to any desired degree of accuracy, as Figure 16.16 suggests, efficiency can be gained by using wavelet functions. To explain this, consider the Haar family, from Figures 16.12 and 16.13, for which the effects are easiest to see. The first two columns in Figure 16.17 show the functions $\varphi(4t-k)$ and $\psi(4t-k)$, which form the basis for the scaling and wavelet functions at the scale $j = 2$. These functions cover the subspaces S_2 and W_2, respectively. Also shown are the functions $\varphi(8t-k)$, which, when scaled appropriately, form the scaling functions for $j = 3$. These functions cover the subspace S_3. From examination of the figures, it is evident that the functions that cover S_2 cannot be combined to produce the functions that cover S_3. The only way to obtain these functions is to combine the functions covering S_2 with the functions covering W_2. For example, by inspection of the first

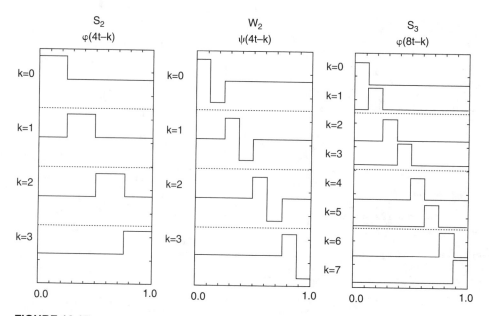

FIGURE 16.17
Contribution of wavelet functions.

row of the figure, $\varphi(8t) = (\varphi(4t) + \psi(4t))/2$ and $\varphi(8t - 1) = (\varphi(4t) - \psi(4t))/2$. Thus, the wavelet functions $\psi(4t-k)$ can describe any signal that can be described by the set of scaling functions $\varphi(8t-k)$ but that cannot be described by the set $\varphi(4t-k)$.

This relationship holds true at other scales as well. In other words, the subspace formed by the wavelet functions covers the "difference" between the subspaces covered by the scaling functions at two adjacent scales. This means that a combination of the scaling functions $\varphi(t-k)$ and the wavelet functions $\psi(t-k)$, for $j = 1$, can describe any signal that can be described by the scaling functions $\varphi(2t-k)$, with $j = 2$. Thus, a function in S_2 may be coded either by using the scaling functions in S_2 or by using a combination of the scaling functions in S_1 and the wavelet functions in W_1. The contribution of W_1 is, in effect, that it fills in the details in S_2 that cannot be provided in S_1. More generally, a function that lies in S_{j+1} can be described by the scaling functions in S_j and the wavelet functions in W_j instead:

$$f(t) \approx f_{j+1}(t) = \sum_{k=-\infty}^{\infty} c_{j+1}[k]\varphi_{(j+1)k}(t) = \sum_{k=-\infty}^{\infty} c_j[k]\varphi_{jk}(t) + \sum_{k=-\infty}^{\infty} d_j[k]\psi_{jk}(t)$$

$$= \sum_{k=-\infty}^{\infty} c_j[k]2^{j/2}\varphi(2^j t - k) + \sum_{k=-\infty}^{\infty} d_j[k]2^{j/2}\psi(2^j t - k) \qquad \textbf{(16.7)}$$

As Equations (16.6) and (16.7) indicate, a function that lies in S_{j+1} can be expressed in terms of scaling functions at scale $s = 2^{-(j+1)}$, or in terms of scaling and wavelet functions at scale $s = 2^{-j}$.

Figure 16.18 shows the wavelet functions that are needed to pick up the "between-scale" details for the first few scales of the Haar family. The subspace S_3, for example, can

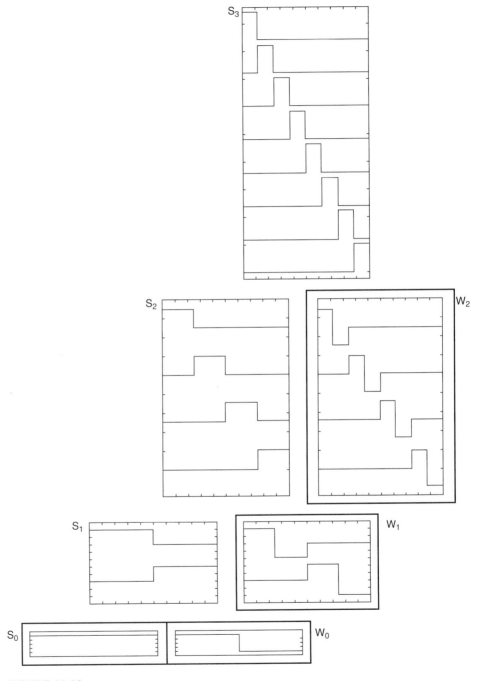

FIGURE 16.18
Wavelet functions and scaling functions for the Haar family.

always be covered by the combination of the subspaces S_2 and W_2, expressed as $S_2 \cup W_2$, and the same rule can be applied for each scaling function subspace. That is,

$$
\begin{aligned}
S_3 &= S_2 \cup W_2 \\
&= (S_1 \cup W_1) \cup W_2 \\
&= ((S_0 \cup W_0) \cup W_1) \cup W_2 \\
&= S_0 \cup W_0 \cup W_1 \cup W_2
\end{aligned}
\tag{16.8}
$$

Equation (16.8) is simply an expression of the fact that a function can be coded either entirely by scaling functions or by a combination of wavelet functions and a large-scale scaling function.[1] Any function that can be coded by using the eight scaling functions in S_3 can alternatively be coded using the scaling function in S_0 and the wavelet functions in W_0, W_1, and W_2, also a group of eight functions in all.

Wavelet functions play the same role in other wavelet families that they do in the Haar family. When a signal is coded with maximal use of wavelets, as suggested by Equation (16.8), Equation (16.7) becomes

$$
\begin{aligned}
f(t) \approx f_M(t) &= \sum_{k=-\infty}^{\infty} c_0[k]\varphi_{0k}(t) + \sum_{j=0}^{M-1} \sum_{k=-\infty}^{\infty} d_j[k]\psi_{jk}(t) \\
&= \sum_{k=-\infty}^{\infty} c_0[k]\varphi(t-k) + \sum_{j=0}^{M-1} \sum_{k=-\infty}^{\infty} d_j[k]2^{j/2}\psi(2^j t - k)
\end{aligned}
\tag{16.9}
$$

where the function $f_M(t)$ is covered by the subspace S_M. The first term relies on the functions $\varphi(t-k)$, which refer to translations of the scaling function at scale $j = 0$, while the functions $2^{j/2}\psi(2^j t - k)$ refer to all scalings and translations of the wavelet functions up to $j = M - 1$. The $c_0[k]$ and $d_j[k]$ are called the wavelet coefficients. Together these coefficients form the discrete wavelet transform of the signal. Section 16.5 develops the techniques necessary to calculate these coefficients. Equation (16.9) provides an efficient means to code a signal. A signal defined in S_j requires wavelet functions up to the scale $s = j - 1$. All translations of the scaling functions and wavelet functions that impact the interval on which the signal is defined must be included.

Figure 16.19 recodes the signal of Figure 16.15, which was coded with scaling functions in Figure 16.16, by using a base scaling function and a family of wavelet functions. The signal is constructed by summing together the scaling function and as many wavelet detail functions as desired. The sum of all eight functions in Figure 16.19 is shown in Figure 16.20. According to Equation (16.9), the function shown in this figure is

$$
f(t) \approx f_7(t) = \sum_{k=-\infty}^{\infty} c_0[k]\varphi(t-k) + \sum_{j=0}^{6} \sum_{k=-\infty}^{\infty} d_j[k]2^{j/2}\psi(2^j t - k)
$$

[1]   Equation (16.8) assumes that the scale given by $j = 0$ is the coarsest scale needed to code the signal. This scale is suitable if the features of interest in the signal are approximately the same length as the support of the $j = 0$ wavelet. If larger scales are needed, negative values of j may be used.

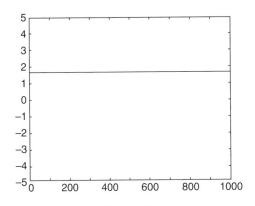

(a) Due to j = 0 Scaling Function

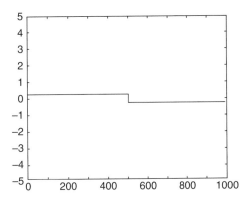

(b) Due to j = 0 Wavelet Function

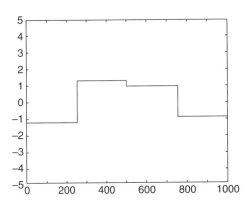

(c) Due to j = 1 Wavelet Functions

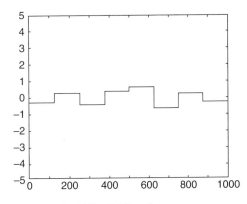

(d) Due to j = 2 Wavelet Functions

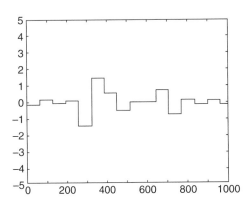

(e) Due to j = 3 Wavelet Functions

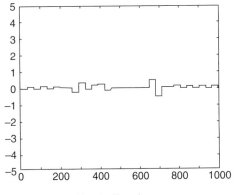

(f) Due to j = 4 Wavelet Functions

FIGURE 16.19

Approximations using wavelet functions $\psi_{jk}(t)$.

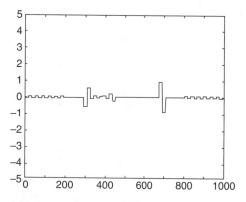

(g) Due to j = 5 Wavelet Functions

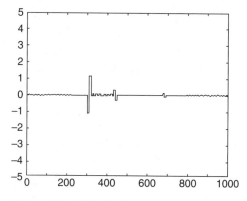

(h) Due to j = 6 Wavelet Functions

FIGURE 16.19

Continued

FIGURE 16.20

Sum of scaling function plus wavelet functions of Figure 16.19.

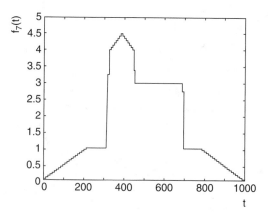

Figure 16.19 clearly shows the contributions that the details make to the overall picture. The wavelet functions for $j = 5$ and $j = 6$, for instance, have an effect on the sum in only two or three important places. Figure 16.19 suggests why wavelet codings are so well suited to compression: A good picture of a signal can be obtained even if the wavelet details beyond a certain j level are dropped. This illustration suggests why wavelet functions code signals so efficiently.

EXAMPLE 16.8

A signal defined between $t = 0$ and $t = 2$ seconds must be coded in subspace S_2 using scaling functions $\varphi(t)$ and wavelet functions $\psi(t)$ from the Daubechies-4 family, as shown in Equation (16.9). The $j = 0$ scaling function for this family is shown in Figure 16.21, and the wavelet functions for $j = 0$, 1, and 2 are shown in Figure 16.14.

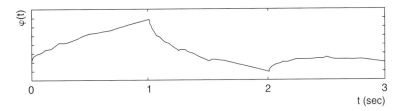

FIGURE 16.21
Daubechies-4 scaling function $\varphi(t)$ for $j = 0$ (father wavelet).

Since the signal is defined between $t = 0$ and 2 seconds, only those scaling functions and wavelet functions that are defined in this range affect the coding. Both the scaling function $\varphi(t)$ and the wavelet function $\psi(t)$ for $j = 0$ are defined between $t = 0$ and $t = 3$ seconds for the Daubechies-4 family, so this scale is sufficiently coarse for the task. For $j = 0$, translations are at $\tau = k2^0 = k$, or 1 second intervals, according to Equation (16.3). The translations $\tau = -2, -1, 0, 1$, and 2 all impact on the range $t = 0$ to 2 seconds on which the signal is defined, as shown in the left column of Table 16.5. The $j = 1$ wavelet function $\psi(2t)$ is defined between $t = 0$ and 1.5 seconds. Its translations are $\tau = k2^{-1} = 0.5k$, or at 0.5 second intervals. The $j = 2$ wavelet function $\psi(4t)$ is defined on $t = 0$ to 0.75, with translations at $\tau = k2^{-2} = 0.25k$, or 0.25 second apart. The translations for each of these scales that impact on the range of times for which the signal is defined are listed in Table 16.5. According to the table, for the Daubechies-4 family, five translations each of $\varphi(t)$ and $\psi(t)$, eight translations of $\psi(2t)$, and fourteen translations of $\psi(4t)$ are needed to code a signal defined between 0 and 2 seconds.

TABLE 16.5
Translations of Scaling Functions and Wavelet Functions Required to Code the Signal in Example 16.8

$j = 0$		$j = 1$		$j = 2$	
τ	Range	τ	Range	τ	Range
-2.00	-2.00 to 1.00	-1.00	-1.00 to 0.50	-0.50	-0.50 to 0.25
-1.00	-1.00 to 2.00	-0.50	-0.50 to 1.00	-0.25	-0.25 to 0.50
0.00	0.00 to 3.00	0.00	0.00 to 1.50	0.00	0.00 to 0.75
1.00	1.00 to 4.00	0.50	0.50 to 2.00	0.25	0.25 to 1.00
2.00	2.00 to 5.00	1.00	1.00 to 2.50	0.50	0.50 to 1.25
		1.50	1.50 to 3.00	0.75	0.75 to 1.50
		2.00	2.00 to 3.50	1.00	1.00 to 1.75
		2.50	2.50 to 4.00	1.25	1.25 to 2.00
				1.50	1.50 to 2.25
				1.75	1.75 to 2.50
				2.00	2.00 to 2.75
				2.25	2.25 to 3.00
				2.50	2.50 to 3.25
				2.75	2.75 to 3.50

16.4 MULTIRESOLUTION ANALYSIS

The development of the previous section omitted one important detail: the calculation of the coefficients $c_0[k]$ and $d_j[k]$ that give the weightings for each wavelet or scaling function component in Equation (16.9). In order to find a method of computing the coefficients at one scale from those at another, it is first necessary to find a relationship between scaling and wavelet functions at different scales. The purpose of this section is to identify these relationships.

An important feature of every scaling function $\varphi(t)$ is that it can be built from translations of double-frequency copies of itself, $\varphi(2t)$, according to the general equation

$$\varphi(t) = \sum_{k=-\infty}^{\infty} \sqrt{2}h_0[k]\varphi(2t - k) \tag{16.10}$$

where the $h_0[k]$ values are called the **scaling function coefficients** and $\sqrt{2}$ is a scaling factor that keeps the energy of scaling functions at every scale the same. Equation (16.10) is called a **multiresolution analysis (MRA)** equation. It expresses the fact that each scaling function in a wavelet family can be written as a weighted sum of scaling functions at the next finer scale. The equation applies for any scale, that is, $\varphi(2t)$ would be given by $\varphi(2t) = \sum_k \sqrt{2}h_0[k]\varphi(4t - k)$. Example 16.9 illustrates how a Haar scaling function can be constructed from higher freque3ncy copies of itself using MRA.

EXAMPLE 16.9

The Haar father wavelet $\varphi(t)$ of Figure 16.10 is repeated in Figure 16.22(a). The function $\varphi(2t)$ has the same shape as $\varphi(t)$ but is compressed to half the width, as shown in (b). Its translated brother $\varphi(2t - 1)$, shown in (c), begins where $2t - 1 = 0$, at $t = 0.5$. This wavelet is shown in Figure 16.22(c). It is easy to see from Figure 16.22 that $\varphi(t) = \varphi(2t) + \varphi(2t - 1)$. Since only $\varphi(2t)$ and $\varphi(2t - 1)$ are needed to construct $\varphi(t)$, only the terms $k = 0$ and $k = 1$ are required in Equation (16.10). The equation

$$\varphi(t) = \sqrt{2}h_0[0]\varphi(2t) + \sqrt{2}h_0[1]\varphi(2t - 1)$$

requires $h_0[0] = 1/\sqrt{2}$ and $h_0[1] = 1/\sqrt{2}$. These coefficients define how Haar scaling functions at different scales are related, according to the MRA equation.

FIGURE 16.22
MRA for Haar scaling functions
for Example 16.9.

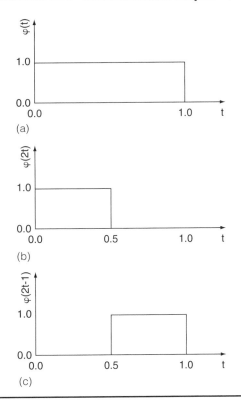

(a)

(b)

(c)

Wavelet functions $\psi(t)$ can also be built from translations of $\varphi(2t)$. The required expression, also an MRA equation, has the same form as Equation (16.10):

$$\psi(t) = \sum_{k=-\infty}^{\infty} \sqrt{2} h_1[k] \varphi(2t - k) \qquad (16.11)$$

where the $h_1[k]$ values are called the **wavelet function coefficients**. This equation expresses the fact that each wavelet function in a wavelet family can be written as a weighted sum of scaling functions at the next finer scale. It applies for any scale. The equation explains why father wavelets and mother wavelets have so much in common. The wavelet functions are determined from scaling functions, which are derived from a father wavelet. In this sense, the father wavelet determines the characteristics of all members of the wavelet family. Example 16.10 illustrates how wavelet functions are constructed from scaling functions for the Haar wavelet family. Other wavelet families also obey Equations (16.10) and (16.11), but the relationships between the scaling and wavelet functions at different scales are not as easy to see as for the Haar family. Example 16.11 provides an illustration for the Daubechies-4 family.

EXAMPLE 16.10

The Haar mother wavelet function is shown in Figure 16.23. Using Figure 16.22, it is easy to see that the mother wavelet can be constructed as:

$$\psi(t) = \varphi(2t) - \varphi(2t - 1)$$

FIGURE 16.23
Haar mother wavelet for Example 16.10.

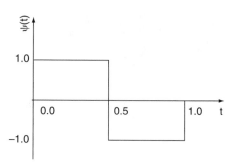

According to Equation (16.11),

$$\psi(t) = \sqrt{2}h_1[0]\varphi(2t) + \sqrt{2}h_1[1]\varphi(2t - 1)$$

and $h_1[0] = 1/\sqrt{2}$ and $h_1[1] = 1/\sqrt{2}$. These coefficients define how Haar wavelet functions are related to Haar scaling functions at other scales, through MRA.

EXAMPLE 16.11

The scaling function coefficients $h_0[k]$ and the wavelet function coefficients $h_1[k]$ for the Daubechies-4 family are nearly impossible to deduce. They are obtained instead using iterative methods outside the scope of this text, and their values are published (for example, see Burrus, Gopinath, Guo, 1998):

$$h_0[0] = 0.4830 \qquad h_0[1] = 0.8365 \qquad h_0[2] = 0.2241 \qquad h_0[3] = -0.1294 \qquad \textbf{(16.12)}$$

$$h_1[0] = -0.1294 \quad h_1[1] = -0.2241 \quad h_1[2] = 0.8365 \qquad h_1[3] = -0.4830$$

Using these values, the MRA equation for the scaling function, Equation (16.10), becomes

$$\varphi(t) = \sum_{k=-\infty}^{\infty} \sqrt{2}h_0[k]\varphi(2t - k)$$

$$= \sqrt{2}(0.4830\varphi(2t) + 0.8365\varphi(2t - 1) + 0.2241\varphi(2t - 2) - 0.1294\varphi(2t - 3))$$

In Figure 16.24, the individual terms of this equation are plotted in (a). The sum of these components gives the father wavelet shown in (b). Furthermore, the wavelet function for the Daubechies-4 family can be constructed, using the MRA equation in Equation (16.11), as:

$$\psi(t) = \sum_{k=-\infty}^{\infty} \sqrt{2}h_1[k]\varphi(2t - k)$$

$$= \sqrt{2}(-0.1294\varphi(2t) - 0.2241\varphi(2t - 1) + 0.8365\varphi(2t - 2) - 0.4830\varphi(2t - 3))$$

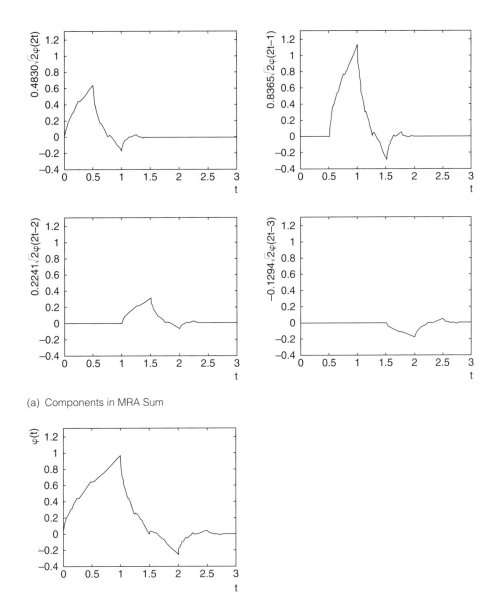

(a) Components in MRA Sum

(b) Sum of Components ≡ Father Wavelet

FIGURE 16.24

Construction of Daubechies-4 father wavelet for Example 16.11.

16.5 THE DISCRETE WAVELET TRANSFORM

16.5.1 Discrete Wavelet Transform Basics

Wavelet analysis is the process of decomposing a signal into its components. The **discrete wavelet transform**, or **DWT**, provides the coefficients $c_j[k]$ and $d_j[k]$ for the decomposition of the signal $f(t)$ into its scaling function and wavelet function components, given in Equation (16.7):

$$f(t) \approx f_{j+1}(t) = \sum_{k=-\infty}^{\infty} c_j[k]2^{j/2}\varphi(2^j t - k) + \sum_{k=-\infty}^{\infty} d_j[k]2^{j/2}\psi(2^j t - k)$$

When the expansion makes maximal use of wavelet functions, only the $j = 0$ scaling function is required, as shown in Equation (16.9):

$$f(t) \approx f_M(t) = \sum_{k=-\infty}^{\infty} c_0[k]\varphi(t - k) + \sum_{j=0}^{M-1} \sum_{k=-\infty}^{\infty} d_j[k]2^{j/2}\psi(2^j t - k)$$

The coefficients $c_j[k]$ and $d_j[k]$ are called the **DWT coefficients**. They express the weightings of the scaling function and wavelet function components that make up a function $f_M(t)$. Because the discrete wavelet transform decomposes a signal into its wavelet components, it might better be described as a discrete wavelet series, since it performs a task similar to that of the discrete Fourier series, but traditionally the term "transform" has been used. **Wavelet synthesis** is the process of combining the components of a signal together again to reproduce the original signal, $f(t)$. The inverse discrete wavelet transform, or **IDWT**, performs this operation.

As shown in Appendix N.2, the analysis equations, which implement the DWT, are:

$$c_j[k] = \sum_{m=-\infty}^{\infty} c_{j+1}[m]h_0[m - 2k] \qquad \textbf{(16.13a)}$$

$$d_j[k] = \sum_{m=-\infty}^{\infty} c_{j+1}[m]h_1[m - 2k] \qquad \textbf{(16.13b)}$$

These equations relate the DWT coefficients at a finer scale, $j + 1$, to the DWT coefficients at a coarser scale, j. The analysis operations are similar to ordinary convolution, defined in Equation (5.2) as:

$$x[k]*h[k] = \sum_{m=-\infty}^{\infty} x[m]h[k - m]$$

The similarity between ordinary convolution and the analysis equations suggests that the scaling function coefficients $h_0[k]$ and the wavelet function coefficients $h_1[k]$ may be viewed as impulse responses of filters. Thus, DWT analysis is a form of filtering. In fact, DWT analysis differs in just two respects from simple convolution: First, the impulse response samples are reversed, $h[m-k]$ instead of $h[k-m]$. Second, the index k is doubled.

The doubling of this index in the analysis equations means that every second convolved sample is dropped, as described in Section 3.2. Equation (16.13) is very efficient to implement. For a signal of length N, the DFT requires N^2 multiplications and N^2 additions, and the FFT requires $N\log_2 N$ of each. The DWT requires only N multiplications and N additions.

The synthesis equation, which is derived in Appendix N.3 and which implements the IDWT, is:

$$c_{j+1}[k] = \sum_{m=-\infty}^{\infty} c_j[m]h_0[k-2m] + \sum_{m=-\infty}^{\infty} d_j[m]h_1[k-2m] \qquad (16.14)$$

This equation relates the DWT coefficients at a coarser scale, j, to the DWT coefficients at a finer scale, $j+1$. In this sense, the synthesis equation performs an inverse DWT (IDWT) function. The terms in the synthesis equation closely resemble ordinary convolution, and therefore imply filtering action. The only difference is that the doubling of the m index has the effect of dropping every other sample of the impulse functions or, equivalently, inserting zeros between the samples in $c_j[m]$ and $d_j[m]$.

As mentioned above, the coefficients $h_0[k]$ and $h_1[k]$ can be seen as the impulse responses for a pair of filters. It turns out that, for all wavelet families, $h_0[k]$ behaves as a low pass filter, while $h_1[k]$ behaves as a high pass filter. The filter shape for each impulse response can be found as usual, by finding the magnitude of the discrete time Fourier transform. As shown in Figure 16.25, the two filters for the Daubechies-4 family are perfectly

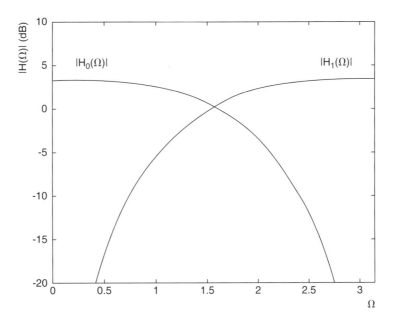

FIGURE 16.25
Low pass and high pass filters for Daubechies-4 family.

complementary. Notice that the gain curves of the filters cross where their gains have dropped by 3 dB from maximum. This means that the filters divide the spectrum into two equal frequency bands. When a signal is filtered, the low pass filter output contains the low frequency elements of the signal, while the high pass filter output contains the high frequency elements. The filters complement one another because the scaling function coefficients $h_0[k]$ and the wavelet function coefficients $h_1[k]$ are related by the equation (Burrus, Gopinath, & Guo, 1998):

$$h_1[k] = (-1)^k h_0[N - 1 - k] \qquad (16.15)$$

where N is the length of the filter, equal to the number of coefficients in $h_0[k]$. Inspection of Example 16.11, for example, shows that the coefficients for the Daubechies-4 family obey this relationship. The $h_1[k]$ coefficients are obtained by reversing the order of the $h_0[k]$ coefficients and changing the sign of every other element. Example 16.12 computes the wavelet function coefficients from the scaling function coefficients for the Daubechies-6 family.

EXAMPLE 16.12

The Daubechies-6 scaling function coefficients are given by (Burrus, Gopinath, & Guo, 1998):

$$h_0[0] = 0.3327 \qquad h_0[1] = 0.8069 \qquad h_0[2] = 0.4599$$

$$h_0[3] = -0.1350 \qquad h_0[4] = -0.0854 \qquad h_0[5] = 0.0352$$

For this family, $N = 6$, so the wavelet function coefficients are, from Equation (16.15):

$$h_1[0] = (-1)^0 h_0[6 - 1 - 0] = 0.0352$$
$$h_1[1] = (-1)^1 h_0[6 - 1 - 1] = 0.0854$$
$$h_1[2] = (-1)^2 h_0[6 - 1 - 2] = -0.1350$$
$$h_1[3] = (-1)^3 h_0[6 - 1 - 3] = -0.4599$$
$$h_1[4] = (-1)^4 h_0[6 - 1 - 4] = 0.8069$$
$$h_1[5] = (-1)^5 h_0[6 - 1 - 5] = -0.3327$$

For many wavelet families, the same wavelet and scaling function filters $h_0[n]$ and $h_1[n]$ are used for both analysis and synthesis. For **biorthogonal wavelet** families, though, the synthesis filters differ from the analysis filters. Instead of via Equation (16.15), the coefficients of the two sets of filters are related through the equations (Burrus, Gopinath, and Guo, 1998):

$$\tilde{h}_0[k] = (-1)^k h_1(N - 1 - k)$$
$$\tilde{h}_1[k] = (-1)^k h_0(N - 1 - k)$$

Before beginning a complete example of wavelet analysis and synthesis using Equations (16.13) and (16.14), it is useful to understand these equations from a frequency point of view. Wavelet analysis depends on the idea of identifying levels of detail in a signal. Section 16.4 introduced the filters $h_0[k]$ and $h_1[k]$. The first impulse function, $h_0[k]$, which according to Equation (16.10) defines the scaling functions, characterizes a filter with a low pass characteristic. The second, $h_1[k]$, defines the wavelet functions according to Equation (16.11). The filter with impulse response $h_1[k]$ has a high pass characteristic. The shapes of both filters for the Daubechies-4 family were presented in Figure 16.25. At the first step of wavelet analysis, the signal to be analyzed is divided into two parts, by filtering it through the low pass and the high pass filters. The high pass filter picks up the small details; the low pass filter picks up everything else. The result, shown in the frequency domain in Figure 16.26(a), is a decomposition of the signal into two halves, a low pass portion defined by translations of a scaling function and a high pass portion defined by translations of a wavelet function. This process continues with the low pass portion of the signal, which is in turn filtered by low and high pass filters. Further detail of the signal is uncovered at each step, and the process of subdividing the lower frequency bands can continue as long as detail remains.

A total of three analysis steps is shown in Figure 16.26(b). As discussed in Section 16.1, good frequency resolution is most important at low frequency. Refiltering the low

FIGURE 16.26

DWT filter shapes.

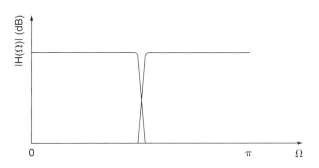

(a) After One Filtering Step

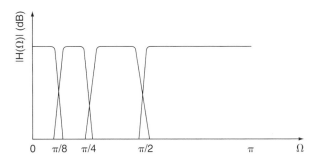

(b) After Three Filtering Steps

pass portions achieves the objective of obtaining the best frequency resolution at the lowest frequencies, while accepting poorer frequency resolution at higher frequencies. The widths of the filters in Figure 16.26(b) show their frequency resolutions: The narrower the filter, the more finely it can discriminate frequencies. Sections 16.5.2 and 16.5.3 provide a frequency domain rationale for the analysis and synthesis equations that appear in Equations (16.13) and (16.14).

16.5.2 A Frequency View of Wavelet Analysis

Analysis is the decomposition of the signal into components that are members of a wavelet family. In the first analysis step, the signal is high pass and low pass filtered. At each subsequent step, the low pass filter output is split into its high and low frequency elements, again through filtering. In the frequency domain, this filtering subdivides bands of frequencies into narrower bands as suggested in Figure 16.26(b). The first low pass filtering step is illustrated in Figure 16.27. The spectrum of the sampled signal is shown in (a), where replicas of the basic spectrum appear at each multiple of the sampling frequency, as discussed in Chapter 2. The signal is low pass filtered by a **quarter-band filter**, shown in (b),

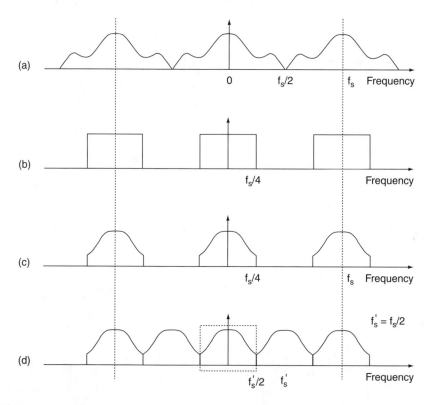

FIGURE 16.27
Low pass filtering followed by downsampling.

which allows a band of frequencies of width $f_S/4$ to pass. This filter is defined by the low pass impulse response $h_0[k]$. The low pass filter shape repeats every f_S Hz, a result of the periodicity of all DTFT magnitude responses. After filtering, multiple copies of the filtered shape appear in the output spectrum, one copy centered at every multiple of f_S. These copies are an unavoidable consequence of sampling and cannot be eradicated until the digital signal is converted back to analog.

As Figure 16.27(c) shows, the low pass filtered output spectrum contains large unused spaces, which means that the sampling frequency is unnecessarily large. To reduce needless extra processing, the sampling rate can be halved, a result accomplished through **downsampling**, or decimation by a factor of 2. As discussed in Section 14.1.2, this is done in the time domain by throwing out every other sample of the signal. It is the reason for the doubling of the k index in Equation (16.13), pointed out in the last section, and has the result in the frequency domain of producing copies centered at every multiple of $f_S/2$ instead of every multiple of f_S. As in Figure 14.3(f), the outcome shown in Figure 16.27(d) is a packed output spectrum that keeps the low pass portion of the original signal's spectrum intact in the baseband, between 0 and $f_S'/2$. Downsampling by a factor of 2 means that the rate of processing will be appropriate to the signal being analyzed, rather than being a factor of 2 too high.

High pass filtering, shown in Figure 16.28, follows the same principles. The spectrum of the sampled signal is again given in (a), and (b) shows the quarter-band high pass filter, defined by the impulse response $h_1[k]$. The filtered output spectrum is shown in (c). As in the low pass case, the sampling frequency is reduced by half, to give spectral copies centered at multiples of $f_S/2$. Figure 16.28(d) illustrates that the downsampled result is a packed spectrum whose baseband contains the high pass portion of the original spectral information. As indicated in Figure 16.26(b), the process of subdivision continues, with the low pass spectrum in Figure 16.27(d) in its turn divided into its low pass and its high pass components. Section 16.5.3 discusses how the subbands are recombined to produce the original spectrum.

16.5.3 A Frequency View of Wavelet Synthesis

Synthesis is the process of reconstituting a signal from its wavelet components. In the frequency domain, this corresponds to recombining the subbands produced during the analysis stage to obtain the original spectrum. Synthesis begins with the packed low pass spectrum of Figure 16.27(d). The first step for synthesis is to "unpack" the packed spectrum by doubling the sampling frequency, through upsampling by a factor of 2. **Upsampling** is the same as zero insertion by a factor of two. As described in Section 14.1.3, it is accomplished in the time domain by inserting a zero between each pair of samples in the signal. This explains the doubling of the m index in Equation (16.14), referred to in the last section. For the low pass spectrum, upsampling means that the subband between $-f_S'$ and f_S' is copied to every multiple of $2f_S'$, or f_S. The resulting spectrum, analogous to Figure 14.6(b), is shown in Figure 16.29(b). No change to the shape of the spectrum occurs, but the change in sampling rate has an important effect on filtering. The upsampled signal is filtered with the quarter-band low pass filter defined by $h_0[k]$ and shown in Figure 16.29(c), producing the spectrum in (d), similar to Figure 14.6(d). Without upsampling, the copies of the low pass

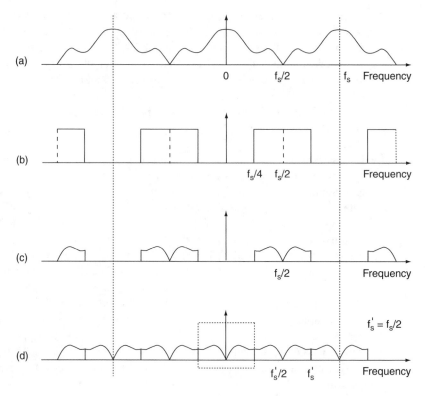

FIGURE 16.28

High pass filtering followed by downsampling.

filter would lie next to one another; with upsampling, they are separated by $f_S/2$. The significance of this becomes apparent when the high pass information is considered.

Upsampling also occurs for the high pass information, as shown in Figure 16.30. The upsampled spectrum in (b) contains copies of the subband between $-f'_S$ and f'_S at every multiple of $2f'_S$, or f_S. As before, no visible change to the spectrum occurs, but the change in sampling rate dramatically affects the output of the high pass filter, defined by $h_1[k]$ and shown in (c). Due to upsampling, copies of the high pass information appear in Figure 16.30(d) separated by $f_S/2$. Without it, no separation between spectral copies would exist. This separation permits the low pass information in Figure 16.29(d) to be summed with the high pass information in Figure 16.30(d) in the last stage of synthesis. The result, shown in Figure 16.31, is the original spectrum of the signal. The use of downsampling and upsampling in wavelet analysis makes wavelet processing a form of multirate DSP.

16.5.4 Calculating the Discrete Wavelet Transform

The frequency view of the DWT presented in the last two sections permits a practical interpretation of the DWT analysis and synthesis equations presented in Equations (16.13)

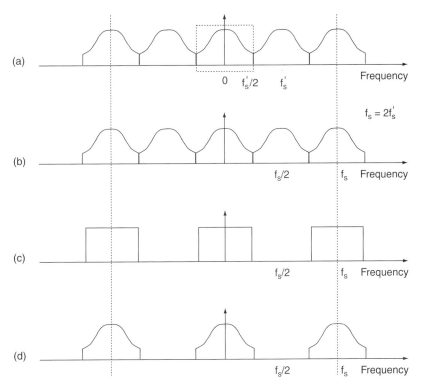

FIGURE 16.29
Upsampling followed by low pass filtering.

and (16.14). The wavelet analysis Equations (16.13a) and (16.13b) perform both the filtering and downsampling described in Section 16.5.2. As Equation (16.13a) shows, filtering is accomplished by convolving the coefficients $c_{j+1}[k]$ with the reversed low pass impulse response $h_0[-k]$, and the result is downsampled to produce $c_j[k]$. In Equation (16.13b), filtering is accomplished through a convolution of the coefficients $c_{j+1}[k]$ and the reversed high pass impulse response $h_1[-k]$, and the filtered output is subsequently downsampled to give $d_j[k]$. In both analysis equations, downsampling, defined in the last section and illustrated in Figure 16.32, halves the sampling frequency by throwing away every other sample. It appears in Equation (16.13a) as the doubled k index, which produces downsampling action. Figure 16.34(a) provides a compact graphical description of the DWT analysis equations. HP refers to convolution with the high pass filter $h_1[-k]$ and LP to convolution with the low pass filter $h_0[-k]$. Downsampling is indicated with the symbol ↓2.

Section 16.5.3 describes synthesis as upsampling followed by filtering, followed by summing the low and high frequency elements. The wavelet synthesis Equation (16.14) performs the filtering task by convolving the impulse responses $h_0[k]$ and $h_1[k]$ with the coefficients $c_j[k]$ and $d_j[k]$. The action of upsampling, defined in the last section, produced by the doubled m index, has the effect of adding a zero between every pair of samples in $c_j[k]$

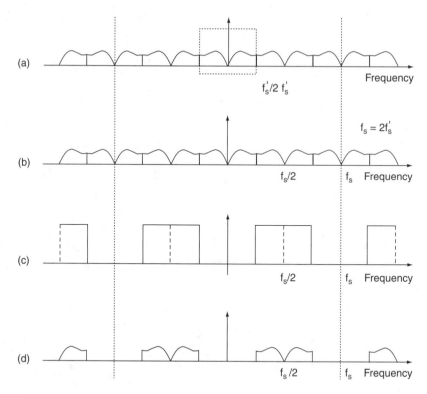

FIGURE 16.30

Upsampling followed by high pass filtering.

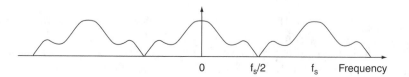

FIGURE 16.31

Synthesizing the original spectrum.

and $d_j[k]$. This is illustrated in Figure 16.33. Figure 16.34(b) shows a graphical view of wavelet synthesis: The coefficients $d_j[k]$ are upsampled and high pass filtered, the coefficients $c_j[k]$ are upsampled and low pass filtered, and these two components are summed to give the coefficients $c_{j+1}[k]$.

While the DWT analysis equations and Figure 16.34(a) link coefficients at one scale with coefficients at the next finer scale, the problem of how to begin the analysis remains. In other words, what set of coefficients $c_j[k]$ are used in the first step? The $c_j[k]$ must al-

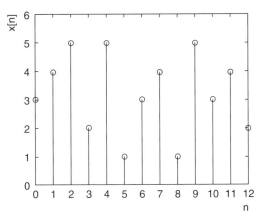

(a) Before Downsampling

(b) After Downsampling

FIGURE 16.32
Time domain view of 2× downsampling.

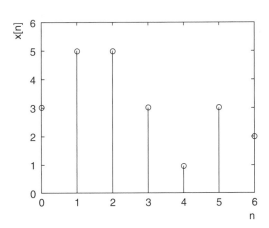

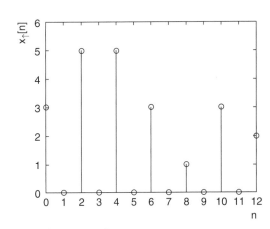

(a) Before Upsampling

(b) After Upsampling

FIGURE 16.33
Time domain view of 2× upsampling.

ways be coefficients that express a function $f(t)$ in terms of scaling functions in subspace S_j, as noted by Equation (16.6):

$$f(t) \approx f_j(t) = \sum_{k=-\infty}^{\infty} c_j[k] 2^{j/2} \varphi(2^j t - k)$$

This equation suggests that the correct procedure (Steacy, 1997) is to express a set of signal samples in terms of some sufficiently small-scale scaling function and then use the coefficients

FIGURE 16.34
DWT analysis and
synthesis.

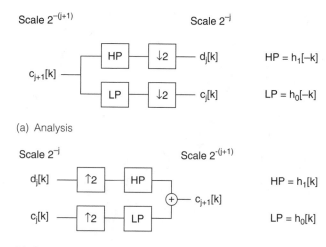

(a) Analysis

(b) Synthesis

$c_j[k]$ as the starting point for DWT analysis. In practice, however, the signal samples themselves, scaled down by $2^{j/2}$, are nearly always used as the first set of coefficients. The motivation for this is that at the finest scales, scaling functions (and wavelet functions) begin to look like impulse functions. Figure 16.35, for example, shows the function $\varphi(8t)$ for the Daubechies-6 family. Like the impulse function, it is narrow in width and close to unit height. A digital signal, which can always be expressed as a sum of impulse functions, can be approximated as a sum of impulse-like functions such as the one shown in this figure. For example, using Equation (16.6), a function in subspace S_3 may be written as

$$f(t) = \cdots + c_3[0]2^{3/2}\varphi(2^3 t) + c_3[1]2^{3/2}\varphi(2^3 t - 1)$$
$$+ c_3[2]2^{3/2}\varphi(2^3 t - 2) + c_3[3]2^{3/2}\varphi(2^3 t - 3) + \cdots$$

whereas, from Chapter 3, any digital signal may be written as the sum of impulse functions as

$$f[n] = \cdots + f[0]\delta[n] + f[1]\delta[n-1] + f[2]\delta[n-2] + f[3]\delta[n-3] + \cdots$$

As the scaling functions become more and more impulse-like at finer scales, it becomes more and more reasonable to equate signal sample values $f[k]$ to $c_j[k] 2^{j/2}$, so that

$$c_j[k] = 2^{-j/2}f[k] \tag{16.16}$$

for samples of a function $f[k]$ in subspace S_j. This approximation can be used in general, with reasonable results: At some fine enough scale given by the index j, the scaling function coefficients can be replaced by the signal samples multiplied by $2^{-j/2}$.

Any number of DWT analysis steps can be performed. Because of downsampling, the number of low pass DWT coefficients $c_j[k]$ is always half that of $c_{j+1}[k]$. Assuming the

FIGURE 16.35

Approaching an impulse function for the Daubechies-6 family.

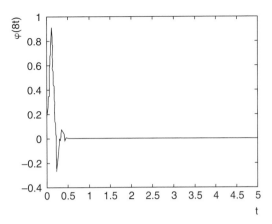

largest scale is $j = 0$, then, a signal length of 2^N samples allows a maximum of N analysis steps. The Nth and final step produces the single DWT coefficient $c_0[k]$. The same observations hold for the high pass DWT coefficients $d_j[k]$. For example, a signal $f(t)$ in S_8 with 256 samples will, during the first analysis step, produce 128 low pass samples, the DWT coefficients $c_7[k]$, and 128 high pass samples, the DWT coefficients $d_7[k]$. At the next step, the low pass samples $c_7[k]$ are further analyzed to produce 64 low pass samples $c_6[k]$ and 64 high pass samples $d_6[k]$. The eighth step takes two low pass samples $c_1[k]$ and produces a single low pass coefficient $c_0[k]$ and a single high pass coefficient $d_0[k]$.

DWT synthesis can begin with the $d_j[k]$ and $c_j[k]$ coefficients at any scale. According to Equation (16.14), the DWT coefficients $c_0[k]$ and $d_0[k]$ can be used to re-create the coefficients $c_1[k]$. Furthermore, combining $c_1[k]$ with $d_1[k]$ gives $c_2[k]$, and combining $c_2[k]$ with $d_2[k]$ gives $c_3[k]$. At each synthesis step, the number of coefficients $c_{j+1}[k]$ doubles from the number in $c_j[k]$, a side effect of upsampling. Since knowledge of $c_0[k], d_0[k], d_1[k]$, and $d_2[k]$ ultimately permits $c_3[k]$ to be computed, this group of coefficients can be used to code any signal that lies in the subspace S_3, as the following example will show.

EXAMPLE 16.13

Eight samples of a signal are collected over the course of 1 second. The sample values are [1 0 −3 2 1 0 1 2]. The signal must be coded using the Haar wavelet family. Compute the DWT coefficients required, and verify that the coding is correct by resynthesizing the signal.

Analysis Phase

The signal covers 1 second. Since the mother and father wavelets for the Haar family cover this same time interval, $j = 0$ will be the coarsest scale needed. Since eight samples are available, a full analysis must include 3 steps, since $2^3 = 8$. The coefficients at the smallest scale, $c_3[k]$, may be approximated, as indicated above, by the samples of the signal multiplied by $2^{-j/2}$, or $2^{-3/2}$. Therefore,

$$c_3[k] = 2^{-3/2} \begin{bmatrix} 1 & 0 & -3 & 2 & 1 & 0 & 1 & 2 \end{bmatrix}$$

$$= \frac{1}{2\sqrt{2}} \begin{bmatrix} 1 & 0 & -3 & 2 & 1 & 0 & 1 & 2 \end{bmatrix}$$

$$= \begin{bmatrix} \dfrac{1}{2\sqrt{2}} & 0 & -\dfrac{3}{2\sqrt{2}} & \dfrac{1}{\sqrt{2}} & \dfrac{1}{2\sqrt{2}} & 0 & \dfrac{1}{2\sqrt{2}} & \dfrac{1}{\sqrt{2}} \end{bmatrix}$$

As Equation (16.13) indicates, only these samples and the Haar filter coefficients, $h_0[k]$ and $h_1[k]$, are necessary to analyze the signal. The analysis tree shown in Figure 16.36 is produced by chaining together three analysis sections from Figure 16.34(a). As derived in Example 16.9, the Haar low pass filter has the impulse response

$$h_0[k] = \begin{bmatrix} \dfrac{1}{\sqrt{2}} & \dfrac{1}{\sqrt{2}} \end{bmatrix}$$

From Example 16.10, the Haar high pass filter has the impulse response

$$h_1[k] = \begin{bmatrix} \dfrac{1}{\sqrt{2}} & \dfrac{-1}{\sqrt{2}} \end{bmatrix}$$

Both of these impulse responses have only two nonzero samples.

The starting set of DWT coefficients is

$$c_3[k] = \begin{bmatrix} \dfrac{1}{2\sqrt{2}} & 0 & -\dfrac{3}{2\sqrt{2}} & \dfrac{1}{\sqrt{2}} & \dfrac{1}{2\sqrt{2}} & 0 & \dfrac{1}{2\sqrt{2}} & \dfrac{1}{\sqrt{2}} \end{bmatrix}$$

Using these coefficients alone, the function $f(t)$ from which the samples are drawn may be expressed, using Equation (16.6) with $j = 3$, as

$$f(t) = \sum_{k=-\infty}^{\infty} c_3[k] 2^{3/2} \varphi(2^3 t - k)$$

$$= \varphi(8t) - 3\varphi(8t - 2) + 2\varphi(8t - 3) + \varphi(8t - 4)$$
$$+ \varphi(8t - 6) + 2\varphi(8t - 7)$$

(16.17)

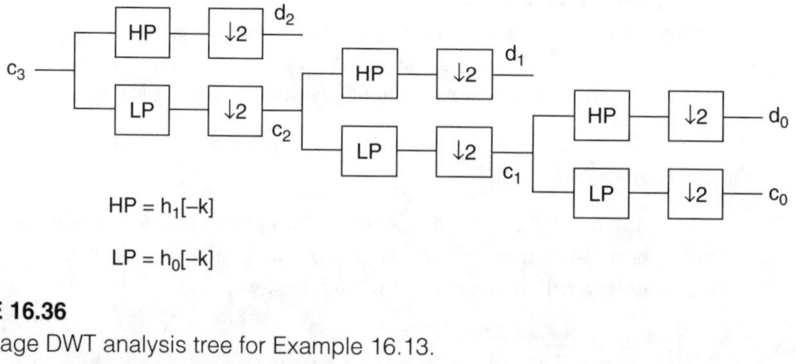

$$\text{HP} = h_1[-k]$$

$$\text{LP} = h_0[-k]$$

FIGURE 16.36
Three-stage DWT analysis tree for Example 16.13.

where $\varphi(t)$ is the father wavelet for the Haar family. In this equation, $f(t)$ has been coded in terms of the scaling functions for $j = 3$ alone. As discussed, efficiency can be gained by coding $f(t)$ instead in terms of scaling and wavelet functions as in Equation (16.7), with $j = 2$.

Beginning with $c_3[k]$, the DWT coefficients $c_2[k]$ at the next coarser scale are calculated using Equation (16.13a):

$$c_2[k] = \sum_{m=-\infty}^{\infty} c_3[m] h_0[m - 2k]$$

Because of downsampling, the number of coefficients $c_j[k]$ is half that of $c_{j+1}[k]$. Since $c_3[k]$ contains eight DWT coefficients, $c_2[k]$ will contain four coefficients, for $k = 0, 1, 2,$ and 3. For the Haar family, the impulse response $h_0[m - 2k]$ is nonzero only where $m - 2k = 0$ or $m - 2k = 1$. For $k = 0$, then, only the $m = 0$ and $m = 1$ terms contribute to the sum. All others are zero. Therefore,

$$c_2[0] = c_3[0]h_0[0] + c_3[1]h_0[1] = \left(\frac{1}{2\sqrt{2}}\right)\left(\frac{1}{\sqrt{2}}\right) + (0)\left(\frac{1}{\sqrt{2}}\right) = \frac{1}{4}$$

Similarly, using Equation (16.13b),

$$d_2[k] = \sum_{m=-\infty}^{\infty} c_3[m] h_1[m - 2k]$$

gives, for $k = 0$,

$$d_2[0] = c_3[0]h_1[0] + c_3[1]h_1[1] = \left(\frac{1}{2\sqrt{2}}\right)\left(\frac{1}{\sqrt{2}}\right) + (0)\left(-\frac{1}{\sqrt{2}}\right) = \frac{1}{4}$$

For $k = 1$, $m - 2k = 0$ occurs for $m = 2$, and $m - 2k = 1$ occurs for $m = 3$, so

$$c_2[1] = c_3[2]h_0[0] + c_3[3]h_0[1] = \left(-\frac{3}{2\sqrt{2}}\right)\left(\frac{1}{\sqrt{2}}\right) + \left(\frac{1}{\sqrt{2}}\right)\left(\frac{1}{\sqrt{2}}\right) = -\frac{1}{4}$$

and

$$d_2[1] = c_3[2]h_1[0] + c_3[3]h_1[1] = \left(-\frac{3}{2\sqrt{2}}\right)\left(\frac{1}{\sqrt{2}}\right) + \left(\frac{1}{\sqrt{2}}\right)\left(-\frac{1}{\sqrt{2}}\right) = -\frac{5}{4}$$

Similar reasoning can be used for the other sample values of $c_2[k]$ and $d_2[k]$, listed in Table 16.7. With both scaling and wavelet function coefficients for $j = 2$ now known, Equation (16.7) expresses the function $f(t)$ as:

$$f(t) = \sum_{k=-\infty}^{\infty} c_2[k] 2^{\frac{2}{2}}\varphi(2^2 t - k) + \sum_{k=-\infty}^{\infty} d_2[k] 2^{\frac{2}{2}}\psi(2^2 t - k)$$

$$= \frac{1}{2}\varphi(4t) - \frac{1}{2}\varphi(4t - 1) + \frac{1}{2}\varphi(4t - 2) + \frac{3}{2}\varphi(4t - 3)$$

$$+ \frac{1}{2}\psi(4t) - \frac{5}{2}\psi(4t - 1) + \frac{1}{2}\psi(4t - 2) - \frac{1}{2}\psi(4t - 3) \quad \textbf{(16.18)}$$

where $\varphi(t)$ and $\psi(t)$ are the father and mother wavelets for the Haar family.

Still another alternative for coding $f(t)$, described in Equation (16.9), makes maximum use of wavelet functions. This coding requires all possible DWT coefficients, which are calculated in the same way as $c_2[k]$ and $d_2[k]$ above. Table 16.6 facilitates these computations by showing the values of $m - 2k$ for all combinations of m and k. Since the Haar filters $h_0[m-2k]$ and $h_1[m-2k]$ are nonzero only when $m - 2k = 0$ or $m - 2k = 1$, only the combinations of m and k that satisfy these equations affect the coefficient calculations. These combinations are shown in bold. Table 16.7 shows all the DWT coefficients computed in the analysis. When the full analysis is complete after three steps, the function $f(t)$ may be expressed according to Equation (16.9) as:

$$f(t) = \sum_{k=-\infty}^{\infty} c_0[k]\varphi(t - k) + \sum_{j=0}^{2}\sum_{k=-\infty}^{\infty} d_j[k]2^{j/2}\psi(2^j t - k)$$

$$= \frac{1}{2}\varphi(t) - \frac{1}{2}\psi(t) + \frac{1}{2}\psi(2t) - \frac{1}{2}\psi(2t - 1)$$

$$+ \frac{1}{2}\psi(4t) - \frac{5}{2}\psi(4t - 1) + \frac{1}{2}\psi(4t - 2) - \frac{1}{2}\psi(4t - 3) \quad \textbf{(16.19)}$$

where $\varphi(t)$ and $\psi(t)$ are the father and mother wavelets for the Haar family. An expansion of this kind can be used to produce a time-scale diagram for a signal, similar to those in Figure 16.7 and Figure 16.9. The size of the coefficient at each time (or translation) and scale determines the brightness of the diagram at each point.

Notice that in each of the expansions in Equations (16.17), (16.18), and (16.19), the total number of coefficients needed to express the function is eight, the same as the number of samples originally collected. This is the case for all wavelet expansions of a sampled signal. Furthermore, all three equations generate the same signal, shown in Figure 16.37.

TABLE 16.6
Wavelet Analysis for Haar Family for Example 16.13

Impulse Response Sample Number $m - 2k$		Scale ($j + 1$) Sample Number m							
		0	1	2	3	4	5	6	7
Scale j	0	0	1	2	3	4	5	6	7
Sample	1	−2	−1	0	1	2	3	4	5
Number	2	−4	−3	−2	−1	0	1	2	3
k	3	−6	−5	−4	−3	−2	−1	0	1

TABLE 16.7
DWT Coefficients for Example 16.13

j	DWT Coefficients
3	$c_3[k] = \left[\dfrac{1}{2\sqrt{2}} \quad 0 \quad -\dfrac{3}{2\sqrt{2}} \quad \dfrac{1}{\sqrt{2}} \quad \dfrac{1}{2\sqrt{2}} \quad 0 \quad \dfrac{1}{2\sqrt{2}} \quad \dfrac{1}{\sqrt{2}}\right]$
2	$c_2[k] = \left[\dfrac{1}{4} \quad -\dfrac{1}{4} \quad \dfrac{1}{4} \quad \dfrac{3}{4}\right]$
	$d_2[k] = \left[\dfrac{1}{4} \quad -\dfrac{5}{4} \quad \dfrac{1}{4} \quad -\dfrac{1}{4}\right]$
1	$c_1[k] = \left[0 \quad \dfrac{1}{\sqrt{2}}\right]$
	$d_1[k] = \left[\dfrac{1}{2\sqrt{2}} \quad -\dfrac{1}{2\sqrt{2}}\right]$
0	$c_0[k] = \left[\dfrac{1}{2}\right]$
	$d_0[k] = \left[-\dfrac{1}{2}\right]$

FIGURE 16.37
Function generated by wavelet expansions for Example 16.13.

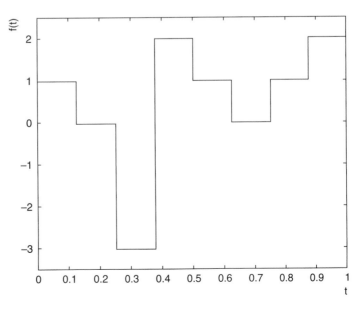

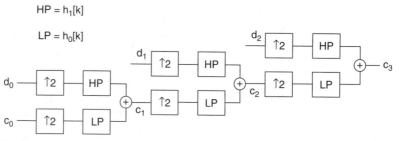

FIGURE 16.38
Three-stage DWT synthesis tree.

Synthesis Phase

Equation (16.14) describes wavelet synthesis. As this equation suggests, not all the coefficients identified in the analysis phase are needed for synthesis. For example, since $c_0[k]$ and $d_0[k]$ can be used to generate $c_1[k]$, $c_1[k]$ does not need to be included. To generate the coefficients in $c_3[k]$, only $c_0[k]$, $d_0[k]$, $d_1[k]$, and $d_2[k]$ are needed. The three-step synthesis tree, produced by chaining three synthesis sections from Figure 16.34(b) together, is shown in Figure 16.38.

At the first stage of synthesis, the coefficients at the next finer scale, $c_1[k]$, are calculated from the coefficients $c_0[k]$ and $d_0[k]$ using Equation (16.14):

$$c_1[k] = \sum_{m=-\infty}^{\infty} c_0[m]h_0[k-2m] + \sum_{m=-\infty}^{\infty} d_0[m]h_1[k-2m]$$

Since $c_0[k]$ and $d_0[k]$ each contain one DWT coefficient, $c_1[k]$ will contain two coefficients. For the Haar family, the impulse responses $h_0[k-2m]$ and $h_1[k-2m]$ are nonzero only where $k - 2m = 0$ or $k - 2m = 1$. Table 16.8 facilitates the calculations by showing $k - 2m$ for all combinations of m and k. Only those that produce a result of 0 or 1 affect the synthesis sum. For example, the $k = 0$ and $k = 1$ coefficients in $c_1[k]$ are computed as:

$$c_1[0] = c_0[0]h_0[0] + d_0[0]h_1[0] = \left(\frac{1}{2}\right)\left(\frac{1}{\sqrt{2}}\right) + \left(-\frac{1}{2}\right)\left(\frac{1}{\sqrt{2}}\right) = 0$$

$$c_1[1] = c_0[0]h_0[1] + d_0[0]h_1[1] = \left(\frac{1}{2}\right)\left(\frac{1}{\sqrt{2}}\right) + \left(-\frac{1}{2}\right)\left(-\frac{1}{\sqrt{2}}\right) = \frac{1}{\sqrt{2}}$$

These results agree with Table 16.7, and other DWT coefficients $c_{j+1}[k]$ can be synthesized in the same way. Finally, the coefficients $c_3[k]$ can be scaled by $2^{3/2}$ to obtain the original signal samples $f[k]$, according to Equation (16.16).

TABLE 16.8
Wavelet Synthesis for Haar Family for Example 16.13

Impulse Response Sample Number $k - 2m$		Scale j Sample Number m			
		0	**1**	**2**	**3**
Scale	**0**	0	−2	−4	−6
$(j + 1)$	**1**	1	−1	−3	−5
Sample	**2**	2	0	−2	−4
Number	**3**	3	1	−1	−3
	4	4	2	0	−2
k	**5**	5	3	1	−1
	6	6	4	2	0
	7	7	5	3	1

The collection of DWT coefficients $c_j[k]$ and $d_j[k]$ required for the synthesis of a signal is frequently referred to as the signal's DWT. The DWT always contains the same number of elements as the signal being transformed. For example, using the assumption embodied by Equation (16.16), 1024 samples of a signal lying in S_{10} provide 1024 DWT coefficients $c_{10}[k] = 2^{-10/2}f[k] = 2^{-5}f[k]$ to start DWT analysis. This set of coefficients is shown in Figure 16.39(a). After one DWT analysis step, all of the information in $c_{10}[k]$ is completely described by 512 low pass DWT coefficients $c_9[k]$ combined with 512 high pass DWT coefficients $d_9[k]$. This DWT is shown in (b). In the next step, $c_9[k]$ is further broken down into 256 values in $c_8[k]$ and 256 values in $d_8[k]$. At this point, 256 coefficients in $c_8[k]$, 256 in $d_8[k]$, and 512 in $d_9[k]$ can be used to synthesize the original signal samples perfectly. This DWT is shown in Figure 16.39(c). After a total of seven DWT analysis steps, the DWT contains the coefficients shown in (d). DWT coefficients are frequently presented

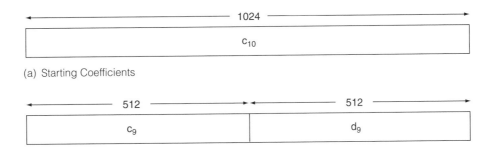

(a) Starting Coefficients

(b) After One DWT Analysis Step

FIGURE 16.39
Layout of DWT coefficients.

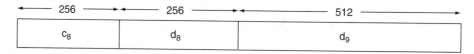

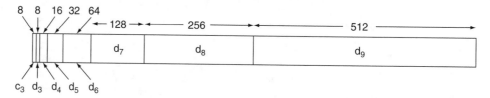

(c) After Two DWT Analysis Steps

(d) After Seven DWT Analysis Steps

FIGURE 16.39
Continued

in the form shown in Figure 16.39. The coefficients for the lowest frequency components appear at the left; those for the highest frequency components appear at the right. Note that the total number of DWT coefficients always equals 1024, the number of samples. For the signal in this illustration, a maximum of 10 steps is possible, since $2^{10} = 1024$.

EXAMPLE 16.14

Write out the DWT for the signal samples in Example 16.13.

The following coefficients form the minimum set needed to synthesize the signal:

$$c_0[k] = \begin{bmatrix} 1 \\ 2 \end{bmatrix}$$

$$d_0[k] = \begin{bmatrix} -\dfrac{1}{2} \end{bmatrix}$$

$$d_1[k] = \begin{bmatrix} \dfrac{1}{2\sqrt{2}} & -\dfrac{1}{2\sqrt{2}} \end{bmatrix}$$

$$d_2[k] = \begin{bmatrix} \dfrac{1}{4} & -\dfrac{5}{4} & \dfrac{1}{4} & -\dfrac{1}{4} \end{bmatrix}$$

The DWT of the signal is given by $[c_0 \, d_0 \, d_1 \, d_2]$, or

$$\left[\underbrace{\frac{1}{2} \quad -\frac{1}{2}}_{c_0 \qquad d_0} \quad \underbrace{\frac{1}{2\sqrt{2}} \quad -\frac{1}{2\sqrt{2}}}_{d_1} \quad \underbrace{\frac{1}{4} \quad -\frac{5}{4} \quad \frac{1}{4} \quad -\frac{1}{4}}_{d_2}\right]$$

It has the same number of points as the original signal.

16.5.5   2D DWT

Two-dimensional signals like digital images require a two-dimensional wavelet treatment. The 2D DWT analyzes the gray scale levels of an image across rows and columns in such a way as to separate horizontal, vertical, and diagonal details. Figure 16.40 suggests how a 2D DWT analysis proceeds. In the first stage, the rows of an $N \times N$ image are filtered using high pass and low pass filters. This filtering is accomplished using 1D convolution with the coefficients $h_0[-k]$ and $h_1[-k]$, as in Figure 16.34(a), because each row of the image is a one-dimensional signal. As usual, downsampling by a factor of 2 removes every other sample in the filtered result, which has the effect of removing every other column of the $N \times N$ block, giving an $N \times (N/2)$ image. In the second stage, 1D convolution with $h_0[-k]$ and $h_1[-k]$ is applied to the columns of the filtered image. Downsampling removes every other sample in each column of the now twice-filtered result, which results in the removal of every other row. Each of the branches in the tree shown in Figure 16.40 therefore produces an $(N/2) \times (N/2)$ image.

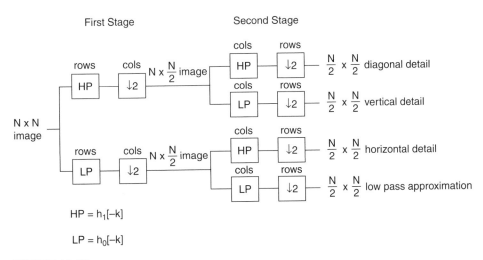

FIGURE 16.40
2D DWT analysis.

Each branch in Figure 16.40 plays a different role in the analysis of an $N \times N$ image. Initial low pass filtering of the rows blurs the gray scale values along each row. When this is followed by low pass filtering along the columns, the result is a low pass approximation to the entire image, shown as the fourth, and bottom, branch of the figure. If, instead, low pass filtering of the rows is followed with high pass filtering of the columns, the changes that occur between rows—that is, the horizontal details—are highlighted, as shown in the third branch of the figure.

In the top branches of the tree in Figure 16.40, the original image is first high pass filtered instead of low pass filtered. Initial high pass filtering of the original rows of the image highlights the changes between elements in any given row. Subsequent low pass filtering of the columns blurs the changes that may occur between the rows, ultimately providing the vertical details only, as noted at the second branch of the tree. When high pass filtering of the rows is followed by high pass filtering of the columns, the changes that are neither horizontal nor vertical are emphasized. This sequence of operations, then, gives the diagonal detail of the original image, as suggested at the top branch in Figure 16.40.

A convenient arrangement of the low pass approximation and image details is shown in Figure 16.41(a). As discussed above, each filtered image contains one half as many rows and one half as many columns as the original. Parts (b) and (c) of the figure show examples of 2D DWTs that use the Daubechies-6 family of wavelets. The image in (b) is made up of primarily horizontal and vertical elements and has essentially no diagonal component. The image in (c) contains strong diagonals and shows only small responses for the horizontal and vertical details. The horizontal, vertical, and diagonal detail portions of the filtering tree are analogous to the set of high pass coefficients $d_j[k]$ for the 1D case. The low pass approximation is analogous to the set of low pass coefficients $c_j[k]$ for the 1D case, so this is the part that is filtered further when additional analysis is desired. As in 1D wavelet analysis, each analysis step extracts a new layer of detail from the filtered image. When the $(N/2)$

| Original Image | Low Pass Approximation | Vertical Detail |
| | Horizontal Detail | Diagonal Detail |

(a)

FIGURE 16.41
DWT image analysis.

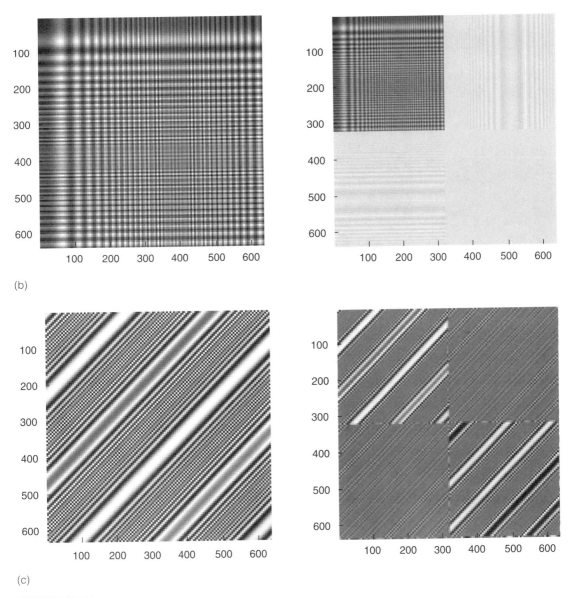

(b)

(c)

FIGURE 16.41

Continued

$\times$ (N/2) low pass approximation is subjected to the same process as the original image, four (N/4) $\times$ (N/4) subimages are produced: a low pass part, and horizontal, vertical, and diagonal detail. Analysis can continue until the subimages obtained contain a single pixel only.

Figure 16.42 shows two steps of a DWT analysis using the Daubechies-8 wavelet family on a 348 $\times$ 348 image. In the second pass, the low pass approximation obtained during

(a) Original Image

(b) First Pass of DWT

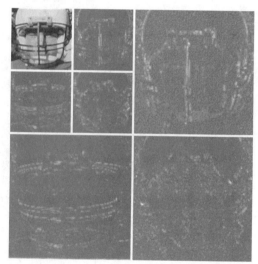

(c) Second Pass of DWT

FIGURE 16.42
Two-stage 2D DWT analysis.

the first pass is itself divided into four subimages, by extracting detail and leaving a new low pass approximation as a residue. The effects of low pass filtering are most easily seen through the loss of detail in the helmet label, the chin strap stripes, and the jersey mesh. The low pass approximation of the second stage can be transformed as well, to produce four new subimages. As mentioned, the process may continue until the subimages contain a single pixel each. In the example of Figure 16.42, almost no diagonal information contributes to

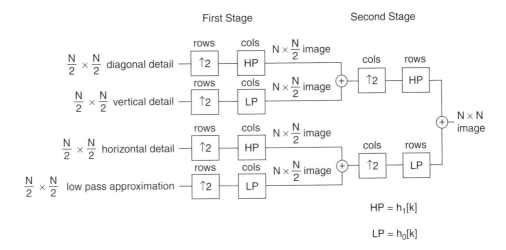

FIGURE 16.43
2D DWT synthesis.

the overall image, which accounts for the relatively diffuse appearance of the regions where diagonal details are reported. The vertical details, at the top right of each analysis square, and the horizontal details, at the bottom left, have a stronger presence.

When a digital image must be reconstructed from its 2D DWT subimages, the details are recombined with the low pass approximation using upsampling and convolution, as shown in Figure 16.43. In the context of 2D DWT synthesis, upsampling refers to the insertion of a zero row after each existing row (first stage) or a zero column after each existing column (second stage). In the first stage, the columns of the upsampled subimages are convolved with the impulse responses $h_0[k]$ and $h_1[k]$; in the second stage, the rows of the upsampled sums are convolved with the same impulse responses.

16.6   TILING THE TIME-SCALE PLANE

Both discrete Fourier transforms and wavelet transforms provide information about a signal's frequency content. Discrete Fourier transforms, because they generate the shape of signal spectra, can provide precise frequency localization, but no time localization within the transform window: The time a frequency element appears or disappears is not registered by the transform. Discrete Fourier transform coverage of the time-scale plane is shown in Figure 16.44, where scale is proportional to the reciprocal of frequency. The width of the bars along the frequency axis corresponds to the frequency spacing of the DFT, or f_S/N, where f_S is the sampling frequency and N is the number of points of the spectrum between 0 and f_S calculated by the DFT. The frequency resolution is the same across all frequencies. The lack of divisions along the time axis indicates that the time resolution is no better than the width of the DFT window. As suggested in Example 16.2, the time resolution is improved as the DFT window narrows. For the same amount of effort, for example, one 32-point DFT could

FIGURE 16.44

Time-scale tiling for DFT with long window.

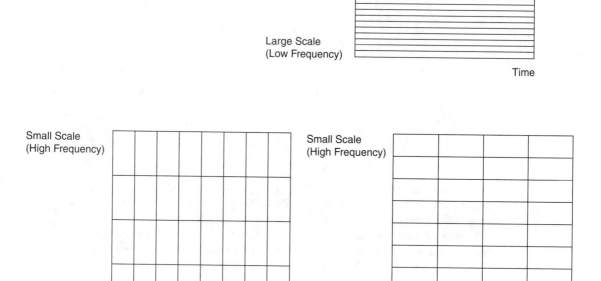

FIGURE 16.45

Time-scale tiling for DFTs with shorter windows.

be replaced with eight 4-point DFTs or four 8-point DFTs. These variations are presented in Figure 16.45. The better the time resolution, the poorer the frequency resolution.

The DFT is not flexible enough to allow different resolutions at different frequencies. As suggested by Figures 16.44 and 16.45, whatever the frequency spacing, it is the same for all frequencies. Flexible management of frequency resolution is the major contribution of the DWT. DWT coverage of the time-scale plane is illustrated in Figure 16.46. The relationships presented in the figure were first discussed in Section 16.1: At high frequencies, the DWT provides poor frequency resolution but good time resolution; at low frequencies, it provides poor time resolution but good frequency resolution. So, the DWT accurately reports the frequencies of low frequency signals using wide time windows and accurately reports the timing for high frequency signals using narrow time windows. On the other hand,

FIGURE 16.46

Time-scale tiling for DWT.

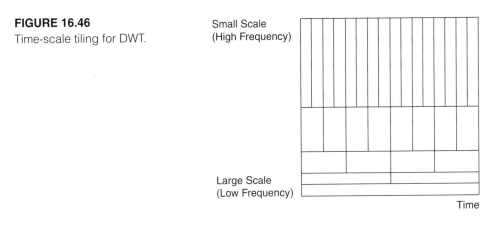

Small Scale
(High Frequency)

Large Scale
(Low Frequency)

Time

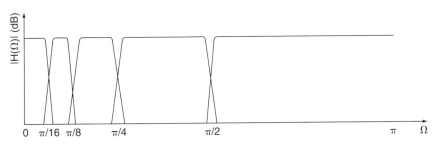

FIGURE 16.47

Vertical slice through time-scale diagram for DWT.

the time resolution is poor at low frequency precisely because the windows are wide, and the frequency resolution is poor at high frequency because the windows are too narrow to identify frequencies accurately. A vertical slice through this time-scale plot at any instant of time produces the band pass filter array in Figure 16.47, familiar in form from Figure 16.26(b).

The DWT coverage of the time-scale plane shown in Figure 16.46 is more sensible in many ways than that of the DFT. Furthermore, wavelet transforms are flexible enough to provide other coverage options as well. Figure 16.48 shows some examples. These variations are called **wavelet packets**. The standard DWT places the narrowest filters at low frequencies where the most important signal details are often located. Wavelet packets can be used to place the narrowest filters in other regions of the frequency spectrum, wherever more detail about the signal is desired. The packet in Figure 16.48(a), for example, provides the best frequency resolution at low frequencies, but resolution at the highest frequencies that is no worse than that of the midrange frequencies. The packet in (b) provides the best frequency resolution at high frequencies and the poorest frequency resolution at the low frequencies.

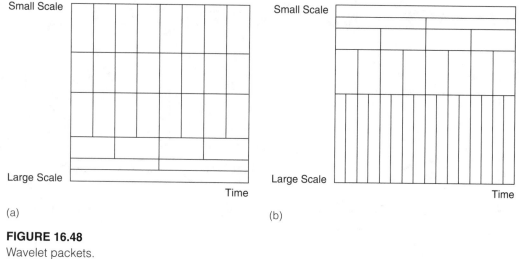

Small Scale

Large Scale

Time

(a)

Small Scale

Large Scale

Time

(b)

FIGURE 16.48
Wavelet packets.

16.7 WAVELET COMPRESSION

16.7.1 Wavelet Compression Basics

Section 15.8 introduced a number of image compression techniques. As indicated there, compression refers to a reduction of the amount of data needed to represent a signal. This is important when the signal must be stored because memory is saved, and also when the signal must be transmitted from one place to another because time is saved. The crudest way to compress a signal coded using a DWT is to remove layers of high frequency detail from the signal by eliminating high frequency coefficients from the DWT. Wavelets are extremely efficient for coding signals. Typically, the sizes of the DWT coefficients drop off very quickly as the scale becomes finer, as seen in Figure 16.19. The coarser scales tend to contain the majority of signal information, with progressively less information at finer scales, captured by the higher frequency coefficients. Omitting these provides a natural possibility for compression: Details can be dropped, but useful approximations to the original signal can be retained. The details are dropped by simply zeroing all the high frequency coefficients above a certain scale. Figure 16.49(a) shows how the high frequency details at the finest scales might be removed from a DWT like Figure 16.39(d), while Figure 16.49(b) shows how progressive layers of detail can be removed from a 2D DWT like the one shown in Figure 16.42(c).

In Figure 16.50 the effects of scale reduction on a 1024-point signal represented by 1024 Daubechies-6 wavelet coefficients are examined. If an accurate portrayal of the rapidly changing signal in Figure 16.50(a) is needed, all 1024 DWT coefficients are required. If some loss of information can be borne, 512 points might be adequate. As more and more high frequency DWT coefficients are removed, the reconstructed signal bears less and less

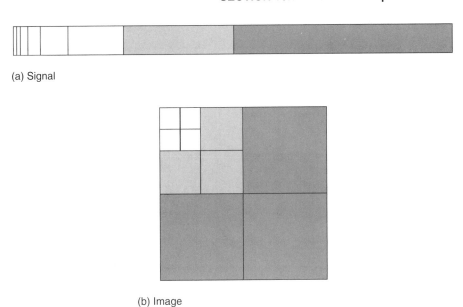

(a) Signal

(b) Image

First Set of DWT
Coefficients Removed

Second Set of DWT
Coefficients Removed

FIGURE 16.49
Compression by scale reduction.

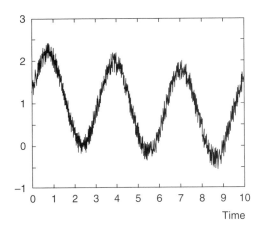

(a) 1024 Points of Original Signal

(b) Reconstructed from 512 Coefficients
(2:1Compression)

FIGURE 16.50
Compression of signal by scale reduction.

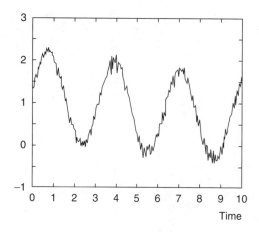

(c) Reconstructed from 256 Coefficients
(4:1 Compression)

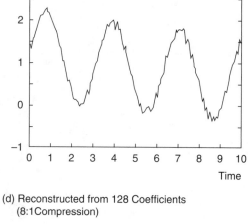

(d) Reconstructed from 128 Coefficients
(8:1 Compression)

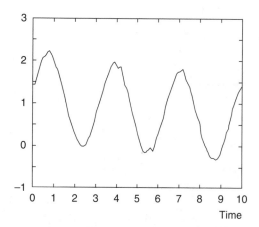

(e) Reconstructed from 64 Coefficients
(16:1 Compression)

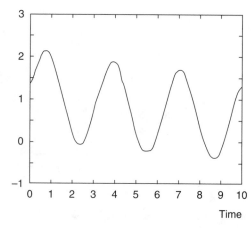

(f) Reconstructed from 32 Coefficients
(32:1 Compression)

FIGURE 16.50

Continued

resemblance to the original signal. On the other hand, Figure 16.50(f) suggests a useful application of scale reduction: The noise in the original signal has been effectively removed. In fact, de-noising signals is an important application of wavelet technology.

Wavelet compression can be applied equally well to images. Figure 16.51(b) shows a reconstruction of the image in (a) from a Daubechies-8 wavelet coding, after one layer of horizontal, vertical, and diagonal detail is removed from a two-stage 2D DWT. Figure 16.51(c) shows the reconstruction after two levels of detail are removed. The original image contains

(a) Original

(b) 4:1 Compression

(c) 16:1 Compression

FIGURE 16.51
Compression of image by scale reduction.

$348 \times 348 = 121,104$ pixels. Part (b) is reconstructed from $174 \times 174 = 30,276$ pixels, for a compression ratio of 121,104:30,276, or 4:1. The image quality is acceptable. In part (c), image degradation is considerably more noticeable. This figure is reconstructed from $87 \times 87 = 7569$ pixels, one sixteenth of the original number of pixels.

Scale reduction is not the most effective way to compress a signal. Efficiency and compressed image quality can be improved through a more selective zeroing of DWT coefficients, called **thresholding**. **Hard thresholding** is the simplest method of removing coefficients. Any coefficient with a magnitude below a specified threshold is set to zero, as Figure 16.52 illustrates. **Soft thresholding** is illustrated in Figure 16.53. In this method,

FIGURE 16.52
Hard thresholding.

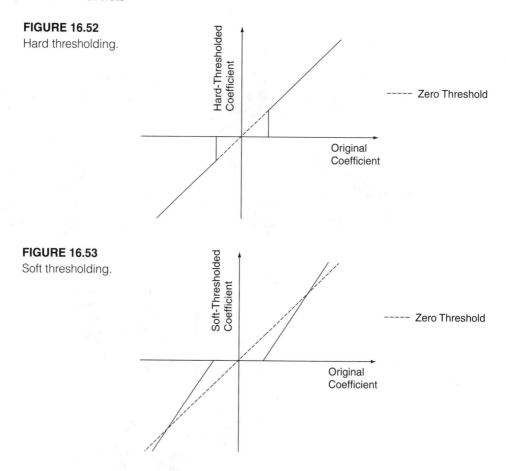

FIGURE 16.53
Soft thresholding.

coefficients below the threshold are still set to zero, but the coefficients above the threshold are affected as well. These shrink in size near the threshold, to avoid the sharp cut-offs that may introduce poor signal behavior.

After some coefficients have been zeroed through thresholding, the remaining DWT coefficients represent the signal with some error. When this smaller number of coefficients is used to reconstruct the signal, some loss of integrity occurs. The degree of acceptable loss is determined by the application. Figure 16.54 shows how different degrees of hard thresholding affect a signal. For similar compression ratios, the reconstructions following hard thresholding retain more fine detail of the original signal than the scale reductions in Figure 16.50, mostly because some coefficients at all scales are retained. Figure 16.55 provides an illustration of hard thresholding for a 2D DWT. The compressed images may be compared to those in Figure 16.51.

Wavelet compression is gaining popularity. Analog Devices, for example, now produces a chip designed for wavelet video compression. Good quality digital video can be reproduced with compression ratios up to 350:1, which means a great deal of video content can be stored in a small space. Applications for the chip include digital camcorders, video

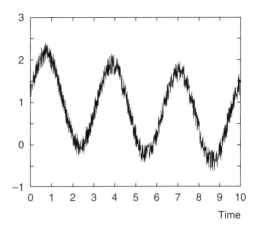

(a) Original Signal

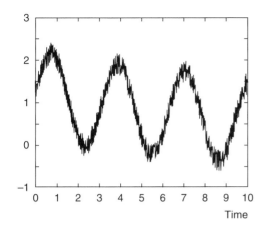

(b) 50% of Coefficients Zeroed (2:1 Compression)

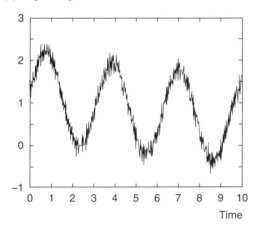

(c) 75% of Coefficients Zeroed (4:1 Compression)

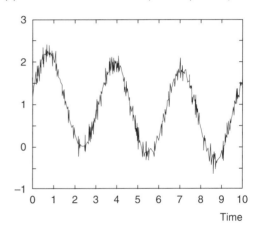

(d) 88% of Coefficients Zeroed (8:1 Compression)

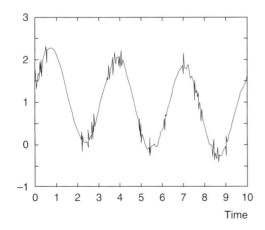

(e) 94% of Coefficients Zeroed (16:1 Compression)

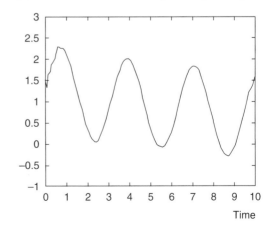

(f) 97% of Coefficients Zeroed (32:1 Compression)

FIGURE 16.54

Compression of signal by hard thresholding.

(a) Original

(b) 4:1 Compression

(c) 16:1 Compression

FIGURE 16.55
Compression of image by hard thresholding.

archiving, and medical imaging. High definition television (HDTV), a standard that will govern the delivery of television programming in the years to come, uses a 16:9 width-to-height aspect ratio for screen images that is better suited to the human field of vision than the conventional 4:3 ratio currently in use. Because HDTV provides 3.5 times as many pixels as standard TV, in addition to multiple digital audio tracks, dramatic compression is absolutely necessary to enable transmission over limited bandwidths. Compression ratios of 200:1, or even 300:1, are basic system requirements. Commercial wavelet technologies were developed too late to compete as the video compression standard, which was selected to be MPEG-2. Wavelets were considered for the audio compression standard, but lost to more established technologies. Dolby Digital was chosen as the North American standard, while

Musicam, a Philips product, was the European choice. On the Internet, wavelet-compressed images accommodate the impatient surfer: A compact low pass approximation can be downloaded rapidly, with detail following thereafter. This way, a decent version of the image from an Internet site is available almost immediately and the image improves with time. The same applies to formats like interlaced GIF and progressive JPEG, but wavelets often produce images with nicer properties. For this reason, JPEG 2000, the newest generation of JPEG compression algorithms, is based on wavelets. Its features include a 20% improvement in compression ratios, better image reproduction, and user-controlled resolution.

16.7.2   The FBI Fingerprint Image Compression Standard

One of the largest applications of wavelet compression to date is the FBI's use of wavelet coding to store digitized fingerprints (Brislawn, 1996). A need was identified as the size of the fingerprint collection, containing over 200 million fingerprint cards collected since 1924, grew to cover over an acre of filing cabinets in the J. Edgar Hoover building in Washington. Direct digitization was impossible: To get adequate resolution, each scanned fingerprint occupied $768 \times 768 = 589,824$ pixels, meaning that a single card required almost 10 Mbytes of data. The JPEG compression standard was considered, but JPEG block operations produced visible straight line artifacts that hampered the efforts of automated systems to track fingerprint ridges. The FBI developed a new standard to deal with the problem, called wavelet scalar quantization (WSQ). The process consists of three steps: DWT decomposition, thresholding, and Huffman encoding (described in Section 15.8). Figure 16.56(a) and (b) show the wavelet and scaling functions used for analysis. Their symmetry has desirable consequences for the coded images. For synthesis, different, biorthogonal, wavelet and scaling functions are used. These functions are shown in Figure 16.56(c) and (d). With WSQ, compression ratios of 18:1 were found to provide satisfactory reconstruction of fine detail, reducing by the same factor the amount of storage needed. Figure 16.57 shows a sample

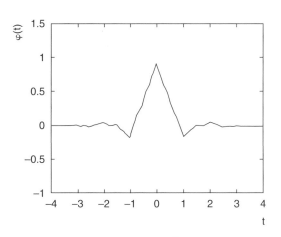

(a) Scaling Function for Analysis $\tilde{h}_0[n]$

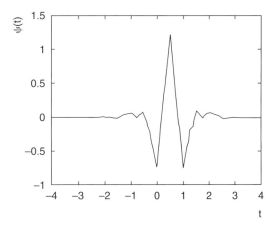

(b) Wavelet Function for Analysis $\tilde{h}_1[n]$

FIGURE 16.56
Cohen-Daubechies-Feauveau family of biorthogonal wavelets.

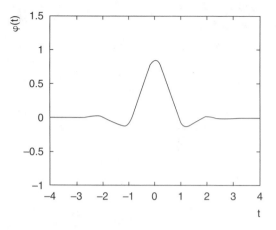

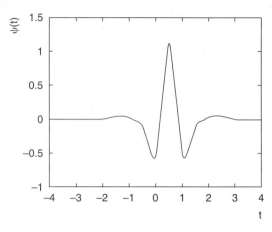

(c) Scaling Function for Synthesis $h_0[n]$

(d) Wavelet Function for Synthesis $h_1[n]$

FIGURE 16.56

Continued

(a) Fingerprint

(b) Wavelet Reconstruction

FIGURE 16.57

Fingerprint compression.

Courtesy Christopher M. Brislawn, Los Alamos
National Laboratory

(c) JPEG Reconstruction

fingerprint along with wavelet and JPEG reconstructions, both using 12.9:1 compression ratios. The wavelet version gives the most faithful rendering. The JPEG version shows blocking artifacts, particularly noticeable when tracing fingerprint ridges at the bottom left and the top right. These artifacts are characteristic of the JPEG compression algorithm.

CHAPTER SUMMARY

Matlab
Support

1. DFTs can identify the frequencies present in a signal, but not the times when they occur. Wavelet transforms can identify both.
2. A wavelet at a particular scale can provide good time resolution or good frequency resolution, but not both.
3. Scale is inversely proportional to frequency.
4. A signal is coded into wavelets at many different scales and translations. Each wavelet is localized not only in frequency but also in time. Large scale wavelets correspond to low frequencies. Small scale wavelets correspond to high frequencies. Time-shifted copies of wavelets at each scale can therefore record not only the frequency but also the timing of signal variations.
5. A time-scale diagram shows the scales that are present in a signal at every point in time.
6. A family of wavelets consists of a father wavelet and its scalings and translations, called scaling functions, and a mother wavelet and its scalings and translations, called wavelet functions. Every wavelet and scaling function can be constructed from the scaling functions at the next finer scale.
7. A signal is coded as the weighted sum of scaling functions and wavelet functions. The weightings are called the discrete wavelet transform (DWT) coefficients.
8. A signal can be coded to any desired level of detail by including smaller-scale wavelets.
9. The first step of a DWT analysis filters a signal into its high frequency elements and its low frequency elements. The second step similarly filters the low frequency portion into its high and low frequency elements. Each subsequent step acts on the low frequency portion of the preceding step, extracting a new layer of high frequency detail. After N steps, the DWT consists of a low pass approximation to the original signal followed by N layers of increasingly high frequency details. Because the DWT has excellent information concentration properties, the DWT coefficients that describe the highest frequency detail in a signal are often very small, and frequently may be eliminated completely without significant signal impairment.
10. One step of a 2D DWT produces a low pass approximation to the original image, plus a set of horizontal, vertical, and diagonal details.
11. Thresholding may be used to eliminate the smallest DWT coefficients in order to compress or de-noise a signal or image.

REVIEW QUESTIONS

16.1 A function $f(t)$ is shown in Figure 16.58.
Sketch:
a. $f(t-1)$
b. $f(2t)$
c. $f(2t-1)$
d. $f(4t)$
e. $f(4t-5)$
f. $f(t/2)$
g. $f(t/2-1)$
h. $f(t/4)$

FIGURE 16.58
Function for Question 16.1.

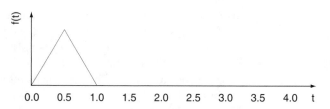

16.2 Sketch the time-scale diagram for:
a. The signal graphed in Figure 16.59

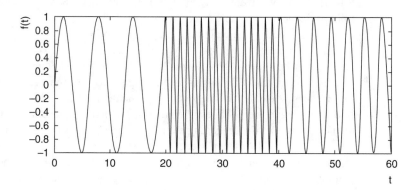

FIGURE 16.59
Signal for Question 16.2(a).

b. $f(t) = \sin(500t) + \sin(5000t)$

16.3 What does the time-scale diagram in Figure 16.60 say about the signal that produced it?

FIGURE 16.60

Time-scale diagram for Question 16.3.

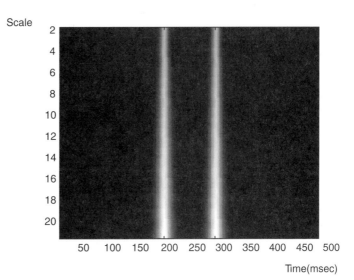

16.4 The Daubechies-8 father wavelet is shown in Figure 16.61. Sketch the scaling functions for the scales $s = 2$, $s = 1$, $s = 0.5$, and $s = 0.25$, with three translations at each scale.

FIGURE 16.61

Daubechies-8 father wavelet for Question 16.4.

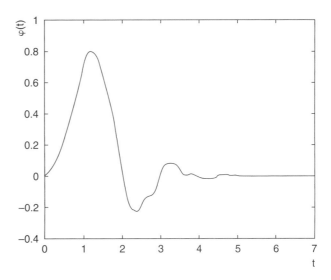

16.5 The Daubechies-20 mother wavelet is shown in Figure 16.62. Sketch the wavelet functions for the scales $s = 2$, $s = 1$, and $s = 0.5$, with two translations at each scale.

FIGURE 16.62

Daubechies-20 mother wavelet for Question 16.5.

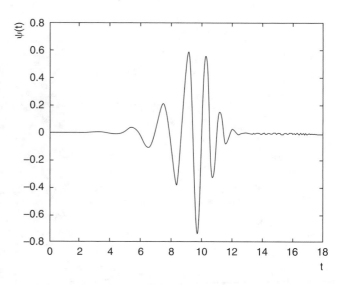

16.6 DWT scaling functions whose father wavelet is shown in Figure 16.63 are used to code a signal in S_2 defined between 0 and 2 seconds. Sketch all the scaling functions that are needed to code the signal using Equation (16.6).

FIGURE 16.63

Father wavelet for Question 16.6.

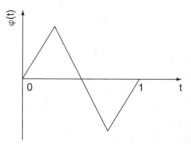

16.7 The father and mother wavelets for a wavelet family are shown in Figure 16.64. Each has a support of 5 seconds. A signal in S_2 is defined between 0 and 3 seconds and is coded using Equation (16.9), with $j = 0$ as the coarsest scale. Sketch the translations of the father wavelet and the scalings and translations of the mother wavelet that are needed to code the signal.

FIGURE 16.64

Father and mother wavelets for Question 16.7.

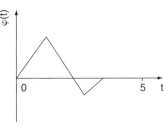

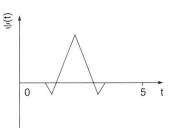

(a) Father Wavelet   (b) Mother Wavelet

16.8 Verify the following for the Haar wavelet family:

a. $\varphi(2t) = \sum_k \sqrt{2}h_0[k]\varphi(4t - k)$

b. $\varphi(2t) = \sum_k \sqrt{2}h_1[k]\varphi(4t - k)$

16.9 The scaling function coefficients for the Battle and Lemarie-4 wavelet family are:

$$h_0[0] = 0.3875 \qquad h_0[1] = 0.6841 \qquad h_0[2] = 0.3875 \qquad h_0[3] = -0.0448$$

a. Find the wavelet function coefficients $h_1[k]$.

b. Draw the filter shapes for the impulse responses $h_0[k]$ and $h_1[k]$.

16.10 A signal $f(t)$ is shown in Figure 16.65. It must be coded using the Haar wavelet family. Find a coding for the signal that relies on:

a. Scaling functions $\varphi(4t)$, $\varphi(4t-1)$, $\varphi(4t-2)$, and $\varphi(4t-3)$

b. The scaling function $\varphi(t)$ and the wavelet functions $\psi(t)$, $\psi(2t)$, and $\psi(2t-1)$

FIGURE 16.65

Signal for Question 16.10.

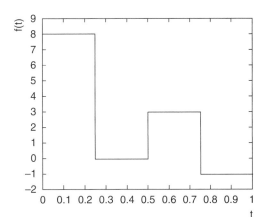

16.11 The Haar wavelet family must be used to code the signal in Figure 16.66. Determine a coding using:

   a. $\varphi(4t)$, $\varphi(4t-1)$, $\varphi(4t-2)$, $\varphi(4t-3)$

   b. $\varphi(2t)$, $\varphi(2t-1)$, $\psi(2t)$, $\psi(2t-1)$

   c. $\varphi(t)$, $\psi(t)$, $\psi(2t)$, $\psi(2t-1)$

FIGURE 16.66

Signal for Question 16.11.

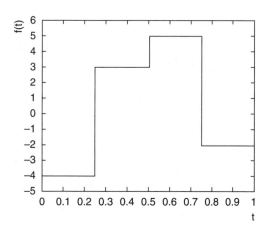

16.12 What DWT analysis tree will produce the filtering effect shown in Figure 16.67?

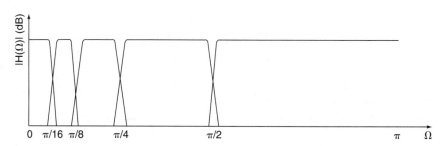

FIGURE 16.67
DWT Filtering for Question 16.12.

16.13 Downsample the signal in Figure 16.68 by a factor of 2.

16.14 Upsample the signal in Figure 16.68 by a factor of 2.

FIGURE 16.68

Digital signal for Question 16.13.

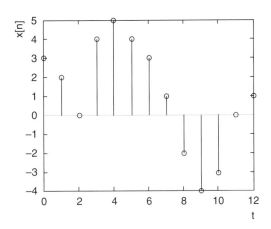

16.15 Use the Haar wavelet family to perform a two-step DWT of the signal:

$$f[k] = [2 \quad 3 \quad -1 \quad -1]$$

Assume that $s = 1$ is the coarsest scale needed, and verify that the DWT is correct by using an inverse DWT.

16.16 The samples of a signal $f[k]$ in subspace S_3 are:

$$f[k] = \left[-4\sqrt{2} \quad 2\sqrt{2} \quad 8\sqrt{2} \quad 4\sqrt{2} \quad 6\sqrt{2} \quad 10\sqrt{2} \quad 0 \quad 2\sqrt{2} \right]$$

a. Find the coefficients $c_3[k]$ to start DWT analysis.
b. Use the analysis equations to perform a three-step Haar analysis of the signal.
c. Write out the 8-point DWT for the signal.

16.17 The three-step DWT of an 8-point signal, computed using Haar filters, is

$$\text{DWT} = \left[4 \quad -2 \quad \sqrt{2} \quad 2\sqrt{2} \quad 1 \quad 1 \quad 2 \quad 1 \right]$$

Use the synthesis equation to reconstruct the original 8-point signal.

16.18 Perform a three-step DWT analysis of the following signal using the Haar wavelet family:

$$f[k] = [1 \quad 2 \quad -1 \quad 3 \quad 0 \quad 5 \quad -1 \quad -2]$$

16.19 The time-scale tiling for a DWT analysis is shown in Figure 16.69.
a. On a single diagram, sketch the set of filters that corresponds to this analysis scheme.
b. At which scale is frequency resolution the best?
c. At which scale is time resolution the best?

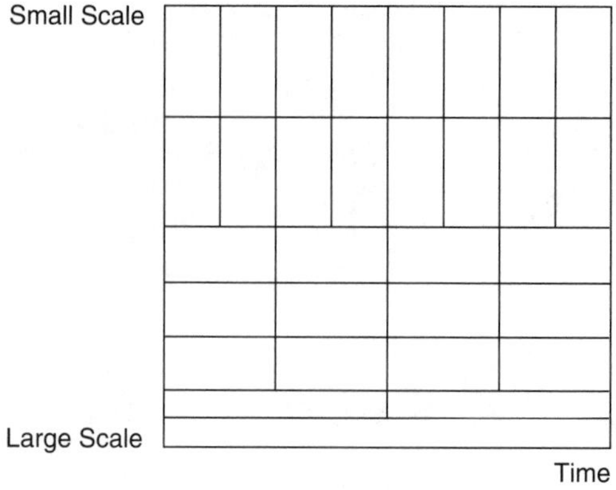

FIGURE 16.69
Time-scale tiling for Question 16.19.

Small Scale

Large Scale

Time

16.20 The DWT for a signal in S_4 is:

$$[\; c_0 \; | \; d_0 \; | \; d_1 \; | \qquad d_2 \qquad | \qquad d_3 \qquad]$$

$$= [\; 4 \; | \; -2 \; | \; 1 \; 2 \; | \; 0.5 \; 0.5 \; 0.2 \; 0.1 \; | \; 0.05 \; 0.1 \; 0.2 \; 0.05 \;]$$

The DWT must be compressed. Rewrite the DWT:
a. After the scale is reduced by 1
b. After the scale is reduced by 2
c. After hard thresholding with a threshold of 0.35
d. After hard thresholding with a threshold of 0.15

16.21 A signal is defined by the samples $f[k] = [\; 1\; 2\; 3\; 4\; 5\; 6\; 7\; 8 \;]$.
a. Compute a three-step DWT for the signal.
b. Hard-threshold the DWT with a threshold of 0.3.
c. Perform a three-step IDWT of the thresholded transform, and reconstruct the original signal samples. How big are the errors?

REFERENCES

Ambardar, Ashok, *Analog and Digital Signal Processing,* Second Edition, Brooks/Cole, Pacific Grove, CA, 1999.

Ambardar, Ashok, and Craig Borghesani, *Mastering DSP Concepts Using MATLAB,* Prentice Hall, Upper Saddle River, NJ, 1998.

Analog Devices, *Analog-Digital Conversion Handbook,* Third Edition, Analog Devices, Norwood, MA, Prentice Hall, Englewood Cliffs, NJ, 1986.

———, "Sigma-Delta Conversion Technology," *DSPatch: The Digital Signal Processing Applications Newsletter,* Winter 1990.

———, *ADSP-2100 Family Assembler Tools & Simulator Manual,* Second Edition, Analog Devices, Norwood, MA, 1994.

———, *ADSP-2100 Family EZ-KIT Lite Reference Manual,* Analog Devices, Norwood, MA, 1995(a).

———, *ADSP-2100 Family User's Manual,* Third Edition, Analog Devices, Norwood, MA, 1995(b).

———, *Practical Analog Design Techniques,* Seminar Notes, Analog Devices, 1995(c).

———, "Analog Devices' Real-Time Music Engine," *www.analog.com/publications/whitepapers/csound.html,* 1996.

———, "ADSP-2141L Safenet DSP Security System on a Chip," *www.analog.com/products/descriptions/ADSP2141L.html.*

———, Analog Devices Announces the First Wavelet Compression IC, [ADV601], *nwd2www1.analog.com/publications/press/products/adv601_052896.html.*

Anderson, Ross, "Perspectives—Automatic Teller Machines," Cambridge University, newsgroup alt.security, December 1992.

———, "Why Cryptosystems Fail," First Conference on Computer and Communications Security '93, Virginia, USA, November 1993.

AT&T, "AT&T Improves Speech Capability," *AT&T Connections, www.att.com/growth/connections/vol98/aso98/speech.html,* August/September/October 1998.

———, "Creating a Network Your Mother Could Talk To," E@AT&T *Magazine,* AT&T, 2000(a).

———, "How May I Help You?" AT&T Press Release, 2000(b).

Braukus, Michael, and Nancy Lovato, "Virtual You—or How to Work in Your 'Jammies,' " Email Release 00-94, Jet Propulsion Laboratory, Pasadena, CA, 16, June 2000.

Berkeley Design Technology, "Processors for DSP: The Options Multiply," *www.bdti.com/articles/ieee/sld001.htm,* Berkeley Design Technology, 1999.

Brislawn, Chris, "The FBI Fingerprint Image Compression Standard," *www.c3.lanl.gov,* July 1996.

Burrus, C. Sidney, Ramesh A. Gopinath, and Haitao Guo, *Introduction to Wavelets and Wavelet Transforms: A Primer,* Prentice Hall, Upper Saddle River, NJ, 1998.

Castleman, Kenneth R., *Digital Image Processing,* Prentice Hall, Englewood Cliffs, NJ, 1996.

Chen, Wei, and Valentin Boriakoff, "Accumulator Length and Quantization Noise," *Embedded Systems Programming,* Vol. 7, No. 9, September 1994, pp. 26–34.

Cole, Ron, Editor-in-Chief, "Survey of the State of the Art in Human Language Technology," National Science Foundation, European Commission, *http://cslu.cse.ogi.edu/HLTsurvey/HLTsurvey.html.*

Cripps, Dale, "HDTV: A Glimpse at The Big Wave," Close-Up with Joe Fedele, HDTV Newsletter, The Wave, *www.fedele.com/website/hdtv/hdtv-wa.htm.*

Danforth, Doug, "Re: Very Simple Speech Recognition Alg. Wanted," Posting to comp.speech newsgroup, November 12, 1992.

Daubechies, Ingrid, *Ten Lectures on Wavelets,* Society for Industrial and Applied Mathematics, Philadelphia, PA, 1992.

Denbigh, Philip, *System Analysis & Signal Processing,* Addison Wesley Longman, Harlow, Essex, England, 1998.

Depalle, Ph., G. Garcia, and X. Rodet, "A Virtual Castrato (!?)," ICMC 94, Aarhus (Danemark), IRCAM—Centre Georges-Pompidou 1994, *http://varese.ircam.fr/articles/textes/Depalle94a/.*

Elmore, Peggy, *Introduction to Geophysics,* NAM CIS D&T, Schlumberger, Canada, May 1982.

Engelfriet, Arnoud, "comp.security.pgp FAQ," *www.pgp.net/pgpnet/pgp-faq,* October 1998.

Folkers, Richard, "Jimmying the Internet," U.S. News Online, *www.usnews.com/usenews/issue/980914/14enco.htm,* September 1998.

Fonte, Gerard, "Breaking Nyquist: Post-Sampling Antialiasing," *Circuit Cellar Ink,* #99, October 1998, pp. 30–34 and pp. 68–73.

Ford, John, *Blackfish Sound—Underwater Communication of Killers Whales in British Columbia,* Vancouver Aquarium, Banff Music, 1992.

Fraunhofer Institut Integnerte Schaltungen, "Real-Time Implementation of High Quality Audio Coding Schemes on DSPs," Fraunhofer Gesellschaft, *www.iis.fhg.de/amm/products/audio_dsp/index.html,* 2000.

Ganssle, Jack G., "An Introduction to Digital Signal Processors," *Embedded Systems Programming,* Vol. 11, No. 5, May 1998.

Gonzalez, Rafael C., and Richard E. Woods, *Digital Image Processing,* Addison-Wesley, Reading, MA, 1992.

Harger, Robert O., *An Introduction to Digital Signal Processing with MathCAD,* PWS Publishing, Boston, MA, 1999.

Heckroth, Jim, "Tutorial on MIDI and Music Synthesis," Crystal Semiconductor Corporation, *www.harmony-central.com/MIDI/Doc/tutorial.html,* The MIDI Manufacturers Association, 1995.

Hewlett-Packard, "Digital Radio Theory and Measurements," Application Note 355A, Hewlett-Packard Company, October 1992.

Higgins, Richard, *Digital Signal Processing in VLSI,* Analog Devices, Prentice Hall, Englewood Cliffs, NJ, 1990.

Hornak, Joseph P., "The Basics of MRI," *www.cis.rit.edu/htbooks/mri/,* 1996–1999.

Hurley, Jim, "Crypto Stats from 'A Tale of Two Cities,'" *www.arachnaut.org/archive/freq.html,* October 1996.

Ifeachor, Emmanuel C., and Barrie W. Jervis, *Digital Signal Processing: A Practical Approach,* Addison-Wesley, Wokingham, England, 1993.

Ilas, Cotistantm, Aurelian Sarca, Radu Giuclea, and Liviu Kreindler, "Using TMS320 Family DSPs in Motion Control Systems," ESIEE, Paris, SPRA 327, Texas Instruments, September 1996.

Ingle, Vinay K., and John G. Proakis, *Digital Signal Processing Laboratory Using the ADSP-2101 Microcomputer,* Analog Devices, Prentice Hall, Englewood Cliffs, NJ, 1991.

Johnson, Curtis D., *Process Control Instrumentation Technology,* Fifth Edition, Prentice Hall, Upper Saddle River, NJ, 1997.

Johnstone, Bob, "Wave of the Future: The Story of the Next Generation in Sound Technology," *www.harmony_central.com/Computer/synth-history.html,* WIRED Online, Ventures USA Ltd., 1994.

Jones, Do-While, "HDTV: The New Digital Direction," *Circuit Cellar Ink,* #86, September 1997, pp. 10–21.

Kak, Avinash C., "Computerized Tomography with X-Ray, Emission, and Ultrasound Sources," Proceedings of the IEEE, Vol. 67, No. 9, 1979, pp. 1245–1272.

Kalcic, Maria T., and Douglas N. Lambert, "Time-Frequency Analysis of Shallow Seismic Imagery," *Sea Technology,* Vol. 40, No. 8, August 1999, pp. 55–60.

Koops, Bert-Joop, "Crypto Law Survey," Version 15.0, *cwis.kub.nl/~frw/people/koops/lawsurvy. htm,* July 1999.

Krutch, Joseph Wood, Editor, *Thoreau: Walden and Other Writings,* Bantam Books, New York, NY, 1962.

Lambert, Douglas N., Donald J. Water, and David C. Young, "Acoustic Sediment Classification Developments," *Sea Technology,* Vol. 40, No. 9, September 1999, pp. 35–41.

Langton, Stewart G., *The Seismic Structure of Tofino Basin and Underlying Accreted Terranes,* Master's Thesis, Department of Physics and Astronomy, University of Victoria, 1995.

Lemaire, Christophe, "Direct and Indirect Out-of-Balance Detection for Future Generation Washing Machines," *www.analog.com/publications/whitepapers/products/dir_indir/,* Appliance Manufacturer Conference & Expo, Nashville, TN, Micromachined Products Division, Analog Devices, September 1999.

Liberty Instruments, "Time, Frequency, Phase and Delay," *www.libinst.com/tpfd.htm,* 1998.

Lim, Jae S., *Two-Dimensional Signal and Image Processing,* Prentice Hall, Englewood Cliffs, NJ, 1990.

Lynn, Paul A., and Wolfgang Fuerst, *Introductory Digital Signal Processing with Computer Applications,* Second Edition, John Wiley & Sons, Chichester, England, 1998.

Lyons, Richard G., *Understanding Digital Signal Processing,* Addison Wesley Longman, Reading, MA, 1997.

Mar, Amy, Editor, *Analog Devices Digital Signal Processing Applications Using the ADSP-2100 Family,* Vol. 1, Prentice Hall, Englewood Cliffs, NJ, 1992.

Markowitz, Judith, "Speech Recognition on the Telephone Network," *Intelligent Systems Report,* Vol. 11, No. 1, January 1994, pp. 1–3.

Marrin, Ken, "CPU and DSP Functionality Converges," *Electronic Systems,* Vol. 38, No. 4, April 1999, pp. 11–14.

Marven, Craig, and Gillian Ewers, *A Simple Approach to Digital Signal Processing,* John Wiley, New York, 1996.

McAlinden, Paul, "DSP Is Cell-Phone Workhorse," *EETimes,* Issue 994, February 23, 1998.

McClellan, James H., C. Sidney Burrus, Alan V. Oppenheim, Thomas W. Parks, Ronald W. Schafer, and Hans W. Schuessler, *Computer-Based Exercises for Signal Processing Using Matlab 5,* Prentice Hall, Upper Saddle River, NJ, 1998.

McClellan, James H., Ronald W. Schafer, and Mark A. Yoder, *DSP First: A Multimedia Approach,* Prentice Hall, Upper Saddle River, NJ, 1998.

Mills, Elinor, "Code Cracked in Record Time," *PC World Communications, www.pcworld.com/pcwtoday/article/0,1510,9413,00.html,* January 1999.

Mlynek, Daniel, and Yusuf Leblebici, "Design of VLSI Systems," *lsiwww.epfl.ch/teaching/webcourse,* October 1998.

Mock, Patrick, "Add DTMF Generation and Decoding to DSP-µP Designs," *EDN,* Vol. 30, March 21, 1985, pp. 205–220.

Morgan, Don, *Practical DSP Modeling, Techniques and Programming in C,* John Wiley, New York, NY, 1994.

National Semiconductor, "An Introduction to the Sampling Theorem," National Semiconductor Application Note 236, pp. 554–565.

Noll, Peter, "Digital Audio Coding Standards," in *Digital Consumer Electronics Handbook,* Ronald K. Jurger, Editor, McGraw Hill, New York, NY, 1997, pp. 8.25–8.50.

Noonan, M., L. Chalupka, M. Viksjo, and D. Perri, "Vocal Development in a Neonatal Orca Whale," Society of Marine Mammalology, Maui, Hawaii, November 1999.

Ohr, Stephan, "Bomb Detector Technology Tops X-Ray's Effectiveness," EDTN Network, *www.edtn.com/news/0199/011199topstory.html,* CMP Media, November 1999.

Oppenheim, Alan V., and Ronald W. Schafer, *Discrete-Time Signal Processing,* Second Edition, Prentice Hall, Upper Saddle River, NJ, 1999.

Orfanidis, Sophocles J., *Introduction to Signal Processing,* Prentice Hall, Upper Sadde River, NJ, 1996.

Papamichalis, Panos, and Jay Reimer, "Implementation of the Data Encryption Standard Using the TMS32010," Application Report: SPRA 130, Digital Signal Processing—Semiconductor Group, Digital Signal Processing Solutions, Texas Instruments, 1989.

Pitas, Ioannis, *Digital Image Processing Algorithms,* Prentice Hall International, Hemel Hempstead, Hertfordshire, England, 1993.

Podanoffsky, Mike, "Compressing Audio and Video Over the Internet," *Circuit Cellar Ink,* #86, September 1997, pp. 22–27.

Pohlmann, Ken C., *Principles of Digital Audio,* Second Edition, SAMS, Prentice-Hall Computer Publishing, 1994.

Proakis, John G., and Dimitris G. Manolakis, *Digital Signal Processing: Principles, Algorithms, and Applications,* Second Edition, Macmillan, New York, NY, 1992.

Quinn, Jack, *Digital Data Communications,* Prentice Hall, Englewood Cliffs, NJ, 1995.

RSA Laboratories, "Frequently-Asked Questions," *www.rsasecurity.com/rsalabs/faq/,* RSA Data Security, 1998.

Sandia National Laboratories, "Scientists Use Digital Paleontology to Produce Voice of Parasaurolophus Dinosaur," Sandia National Laboratories, *www.sandia.gov/media/dino.htm,* 1996.

Scammell, Elsa, "All You Would Like to Know About The Castrati . . . But Not Quite!" *www.cix.co.uk/~velluti/cast.htm,* 1999.

Simone, Luisa, "Image Compression: The Next Wave(let)," *PC Magazine: Trends Online, www. zdnet.com,* April 21, 1999.

Steacy, Robert, "Conversion of Data in Sampled Form to Daubechies Inner Products," Annual Convention of the Canadian Mathematics Society, Victoria, B.C., *http://camel.math.ca/CMS/Events/winter97/w97-abs/node128.html,* December 1997.

Stearns, Samuel D., and Don R. Hush, *Digital Signal Analysis,* Second Edition, Prentice Hall, Englewood Cliffs, NJ, 1990.

Steiglitz, Ken, *A Digital Signal Processing Primer with Applications to Digital Audio and Computer Music,* Addison Wesley, Menlo Park, CA, 1996.

Story, Derrick, "JPEG 2000—More Than New Millenium Buzz," *webreview.com/wr/pub/1999/08/13/feature.index3.html,* August 13, 1999.

Taylor, Insup, and M. Martin Taylor, *Psycholinguistics: Learning and Using Language,* Prentice Hall, Englewood Cliffs, NJ, 1990.

Texas Instruments, "DSP Solutions for Voiceband and ADSL Modems," Application Report SPAA005, Texas Instruments, June 1998.

———, "TMS320C54x DSP Reference Set, Volume 1: CPU and Peripherals," Texas Instruments, Literature Number: SPRU131F, April 1999(a).

———, Analog Seminar Notes, 1999 DSP and Analog Seminar Series, Vancouver, Canada, October 1999(b).

———, DSP Seminar Notes, 1999 DSP and Analog Seminar Series, Vancouver, Canada, October 1999(c).

———, "TMS320C64x Technical Overview," Texas Instruments, Literature Number: SPRU395, February 2000(a).

———, "MPEG-2 Video Decoder: TMS320C62X Implementation," Application Report SPRA649, Texas Instruments, March 2000(b).

———, Seminar Notes, expressDSP Real-Time Software Technology 2000 Seminar Series, Vancouver, Canada, June 2000(c).

———, "TMS320 DSP Product Overview," Texas Instruments.

Tweed, David, "Digital Processing in an Analog World: Basic Issues," *Circuit Cellar Ink,* #99, October 1998, pp. 68–75.

———, "Digital Processing in an Analog World: Technology Choices," *Circuit Cellar Ink,* #100, November 1998, pp. 64–71.

———, "Digital Processing in an Analog World: Dithering Your Conversion," *Circuit Cellar Ink,* #101, December 1998.

Walker, Ross F., "Automated Monitoring of Six Cyanobacterial Taxa from Lake Biwa by Image Processing," SIL 98, 27th Congress of the International Association of Limnology, Dublin, Ireland, August 1998.

———, "Algae Species Classification by Image Processing," Lake Biwa Research Institute, Otsu City, Shiga, Japan, 1999.

Ward, Brice, *Electronic Music Circuit Guidebook,* Tab Books, Blue Ridge Summit, PA, 1975.

Weddell, Steve, "VLIW & TI's New C6x DSP —A Quantum Leap in Performance," Avnet Design, *www.avnet.com.au/AvnetDesign/C6xDSP.htm,* April 14, 1999.

West, Mason, "UT Research Facility Offers Futuristic Computer Graphics," *The Daily Texan,* Texas Student Publications, Associated Press Articles Copyright 1991 – 1997, Associated Press, *http://wwwhost.cc.utexas.edu/cc/vislab/news/DailyTexan_7-16-96.html,* July 1996.

Wolberg, George, *Digital Image Warping,* IEEE Computer Society Press, Los Alamitos, CA, 1990.

Woodard, Jason, "Speech Coding," Department of Electronics & Computer Science, University of Southampton, *http://www-mobile.ecs.soton.ac.uk/speech_codecs/index.html.*

A

THE MATH YOU NEED

A.1 FUNCTIONS

An **analog function** is designated as $x(t)$, where t stands for time. It provides, for every value of time, the amplitude of a continuous-time signal. A **digital function** is designated as $x[n]$, where n stands for the sample number. A digital function is a means of describing the amplitude of a discrete-time signal, a signal defined only at discrete intervals.

An **even function** satisfies the equation $f(t) = f(-t)$, or $x[n] = x[-n]$. An **odd function** satisfies the equation $f(t) = -f(-t)$, or $x[n] = -x[-n]$. Figures A.1 and A.2 show examples of even and odd functions.

A **linear function** refers to a straight line relationship.

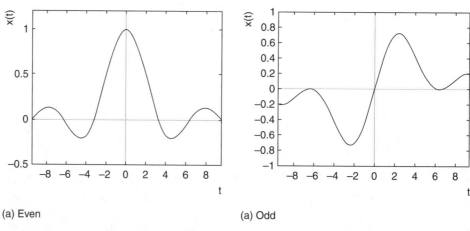

(a) Even   (a) Odd

FIGURE A.1
Even and odd analog functions.

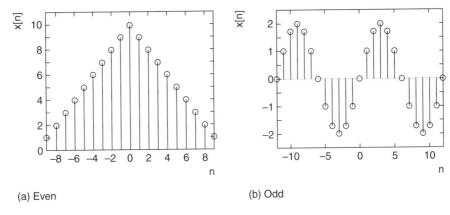

(a) Even   (b) Odd

FIGURE A.2
Even and odd digital functions.

A.2   DEGREES AND RADIANS

Angles are most commonly expressed in **degrees**, with 360° in a full circle. Angles may also be given in radians, with 2π radians in a full circle. Degrees may be converted to radians by multiplying by π radians/180°, and radians may be converted to degrees by multiplying by 180°/π radians. For example, 40° is equivalent to $40(\pi/180) = 2\pi/9$ radians. As another example, $\pi/2$ radians is equivalent to $\pi/2(180/\pi) = 90°$. The units "radians" are frequently shortened to "rads."

A.3   RATIONAL FUNCTIONS

A **rational function** is formed by dividing two functions. For example, the function

$$f(x) = \frac{g(x)}{h(x)}$$

is rational. The **numerator** $g(x)$ is the dividend while the **denominator** $h(x)$ is the divisor.

A.4   LOWEST TERMS

A fraction is expressed in its lowest terms when no common factors are shared by the numerator and denominator. Consider, for instance, the fraction 18/60. The numerator, 18, and the denominator, 60, share a common factor of 2. When this common factor is removed from both, the fraction becomes 9/30. The numerator 9 and the denominator 30 of the new fraction still share a common factor 3. Following its removal, the fraction becomes 3/10. The numerator and the denominator of this fraction do not share any further common factors. Thus, the fraction 18/60, when expressed in lowest terms, is written as the fraction 3/10.

A.5 LOWEST COMMON MULTIPLE

A **prime number** is a number that can be divided evenly only by itself and one. The **lowest common multiple** of a group of integers is the smallest number into which all of the integers in the group divide evenly. It is most easily found by dividing each integer into prime factors and choosing a minimal set. For example, the prime factors of 15, 21, and 28 are

$$15 = 3 \times 5$$

$$21 = 3 \times 7$$

$$28 = 2 \times 2 \times 7$$

The lowest common multiple is $2 \times 2 \times 3 \times 5 \times 7 = 420$.

A.6 RECIPROCALS

The **reciprocal** of a number is equal to one divided by the number. That is, the reciprocal of x is $1/x$.

A.7 LOGARITHMS

Logarithmic functions have the form $y = \log_b(x)$, where b is the base of the logarithm. A base e logarithm is called the natural logarithm, or ln. That is, $\log_e x = \ln x$. When no base is stated, base 10 is assumed. For instance, the logarithm

$$y = \log (100) = 2$$

because the base, 10, must be raised to the power 2 to produce 100. Similarly,

$$y = \log_3 (81) = 4$$

A logarithm with a base other than 10 can be converted to a base-10 logarithm as follows:

$$\log_b y = \frac{\log y}{\log b}$$

Therefore, $\log_2 10 = \log 10 / \log 2 = 3.322$.

Logarithms have some useful properties. For instance,

$$\log_b b^a = a$$

$$\log_b c^a = a \log_b c$$

$$\log_b gh = \log_b g + \log_b h$$

$$\log_b \frac{g}{h} = \log_b g - \log_b h$$

The following examples use these properties:

$$\log (10^5) = 5$$

$$\log (3^4) = 4 \log 3$$

$$\log ((27)(55)) = \log (27) + \log (55)$$

$$\log \left(\frac{34}{71}\right) = \log (34) - \log (71)$$

Finally, a logarithm of the form $\log_b c = a$ can be converted to the form as $c = b^a$, discussed in the next section. This fact is considered the definition of the logarithm.

A.8   POWER AND EXPONENTIAL FUNCTIONS

Power functions have the form b^a. The number b is called the **base**, and the number a is called the **exponent**. For example, 2^x and $(0.5)^{-x}$ are both power functions. Power functions that have a base equal to $e = 2.71828\ldots$, such as e^{4x}, are referred to as **exponential functions**. Special rules are applied to power functions when they are multiplied or divided. Two power functions are multiplied according to the rule

$$b^p \, b^q = b^{p+q}$$

They are divided using the rule

$$\frac{b^p}{b^q} = b^{p-q}$$

That is, in the case of multiplication, exponents are added; in division, exponents are subtracted. For example,

$$e^4 e^{-7.5} = e^{4+(-7.5)} = e^{-3.5}$$

$$\frac{x^{2.6}}{x^{1.4}} = x^{2.6-1.4} = x^{1.2}$$

A negative power is the reciprocal of the same positive power, that is, $z^{-3} = 1/z^3$. A power function of the form $c = b^a$ can be converted to a logarithm of the form $\log_b c = a$. For example, $y = e^{-2x}$ converts to $\ln y = -2x$.

A.9   SINUSOIDAL FUNCTIONS

Analog sinusoidal functions include sine and cosine functions that have the general form:

$$y(t) = A \sin(\omega t + \theta)$$

or

$$y(t) = A \cos(\omega t + \theta)$$

where t stands for time. The sine function $\sin(t)$ has the value zero at time zero, while the cosine function $\cos(t)$ has the value unity at time zero. The amplitude is given by A, which means the maximum of the function is A and the minimum is $-A$. The frequency of the sine wave is given by ω, measured in radians per second. This frequency is related to a frequency f in Hertz through the general relationship

$$\omega = 2\pi f$$

The frequency f in Hz may be converted to a period in seconds through the general equation

$$T = \frac{1}{f}$$

Finally the phase shift of the sinusoid is θ, in radians. When the equation $\omega t + \theta = 0$ is solved for t, the result, $t = -\theta/\omega$, gives the time shift of the sinusoid compared to $\sin(t)$. For example, the sinusoids $y_1(t) = 3\sin(2t + 1)$ and $y_2(t) = \cos(5t)$ are plotted in Figure A.3.

Function values and times may be determined from the function equations. For example, $y_1(1)$ is found by substituting the value $t = 1$ into the function:

$$y_1(1) = 3\sin(2(1) + 1) = 3\sin(3) = 0.4234$$

To find a value of time that produces the function value $y_2(t) = 0.8$ requires the steps:

$$y_2(t) = \cos(5t) = 0.8$$

$$5t = \cos^{-1}0.8$$

$$t = \frac{\cos^{-1}(0.8)}{5} = 0.1287$$

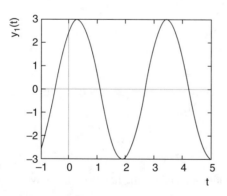

(a) $y_1(t) = 3\sin(2t + 1)$

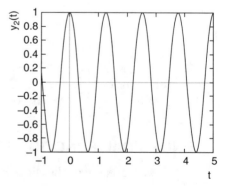

(b) $y_2(t) = \cos(5t)$

FIGURE A.3
Analog sine and cosine functions.

This value of time represents just one of the places where $y_2(t)$ has the value 0.8, as Figure A.3(b) shows, but the inverse cosine function reports its result between 0 and π only.

Digital sinusoids are discussed in Chapter 3.

A.10   DECIBELS

The **gain** of a system is the size of its output divided by the size of its input. For example, if a 1 V peak-to-peak sine wave is applied to a system with a gain of 10, the output will be 10 V peak-to-peak. A **decibel**, or dB, is a unit commonly used to represent system gains. Voltage gains G_V in dB are computed as

$$G_V (\text{dB}) = 20 \log (G_V)$$

where G_V is the linear voltage gain. A gain in dB, which is a logarithmic gain, can be converted back to an ordinary linear gain. Dividing by 20 gives

$$\log G_V = \frac{G_V (\text{dB})}{20}$$

and the definition of the logarithm produces

$$G_V = 10^{G_V(dB)/20}$$

For instance, a gain of 2.5 corresponds to a gain of $G_V(\text{dB}) = 20 \log(2.5) = 7.96$ dB; and a gain of -3.01 dB is the same as a gain of $G_V = 10^{-3.01/20} = 0.7071$. Some common voltage gains are listed in Table A.1. For example, the (logarithmic) voltage gain of -20 dB refers to a linear gain of 0.1. When power gains rather than voltage gains are considered, the equations become:

$$G_P (\text{dB}) = 10 \log (G_P)$$

$$G_P = 10^{G_P(dB)/10}$$

TABLE A.1
Common Decibel Values and Their Linear Equivalents

Logarithmic Gain 20logG$_V$ (dB)	Linear Gain G$_V$
-80	0.0001
-60	0.001
-40	0.01
-20	0.1
0	1
20	10
40	100
60	1000
80	10000

A.11   DECIMAL, BINARY, AND HEXADECIMAL NUMBER SYSTEMS

The **decimal**, or **base 10**, number system counts using the digits 0 through 9. It is the number system most commonly used. The **binary**, or **base 2**, number system is used in all digital formats. It uses the digits 0 and 1 only. The **hexadecimal**, or **base 16**, number system counts 0, 1, 2, 3, 4, 5, 6, 7, 8, 9, A, B, C, D, E, F, for 16 numbers in all. The digits A through F represent the base-10 numbers 10 through 15. Conversions may be made from one number system to any other. The base of a number is sometimes appended to the number as a subscript. Hex numbers are sometimes marked with the annotations $0x$ or h.

Figure A.4 compares three representations for the same number. In the base 10 system, the number is constructed using powers of 10: the decimal number 243 means $2 \times 10^2 + 4 \times 10^1 + 3 \times 10^0$. The base 2 system uses powers of 2. The binary number 11110011 means $1 \times 2^7 + 1 \times 2^6 + 1 \times 2^5 + 1 \times 2^4 + 0 \times 2^3 + 0 \times 2^2 + 1 \times 2^1 + 1 \times 2^0$, which also totals 243, base 10. Similarly, the hexadecimal number F3 means $15 \times 16^1 + 3 \times 16^0 = 243$, which gives the decimal number 243 again. Note that the least significant digit in all number systems always corresponds to groups of one.

When a decimal number is converted to another number system, the largest power in the new number system that is lower than the number must first be identified. For instance, 2^8 is larger than 243, but 2^7 is less. When 2^7 is subtracted from 243, 115 remains. The value 2^6 can be removed from 115 to give 51. Taking away 2^5 and 2^4 leaves 3. It is not possible to subtract 2^3 or 2^2 from this value, but 2^1 and 2^0 can be removed, leaving zero. Thus, 11110011 is a perfect binary representation for the decimal number 243. The conversion to base 16 is similar: 16^2 exceeds 243, but 15 groups of 16^1 can be removed, leaving 3 groups of 16^0. The number 15 is represented with the digit F, giving a hexadecimal representation of F3 for the decimal number 243. Figure A.4 summarizes these findings: $243_{10} = 1111\ 0011_2 = F3_{16}$, where the bases are noted as subscripts. As indicated above, the hexadecimal result may also be written as $0x$F3 or F3h.

A binary or hexadecimal number can be converted to decimal by adding together components for each digit. For example, the binary number $10101 = 1 \times 2^4 + 0 \times 2^3 + 1 \times 2^2 + 0 \times 2^1 + 1 \times 2^0 = 8 + 4 + 1 = 21$ in the decimal number system. The hexadecimal number $2CA5 = 2 \times 16^3 + 12 \times 16^2 + 10 \times 16^1 + 5 \times 16^0 = 8192 + 3072 + 160 + 5 = 11{,}429$ in the decimal system.

Hexadecimal-to-binary and binary-to-hexadecimal conversions can be accomplished easily using the information in Table A.2. Because one hexadecimal digit can represent all the numbers between 0 and 15 in the decimal system, each hex digit can be expressed using exactly four binary digits, which are capable of covering the same decimal range. The

FIGURE A.4
Number places in decimal, binary, and hexadecimal number systems

10^8	10^7	10^6	10^5	10^4	10^3	10^2	10^1	10^0
						2	4	3

2^8	2^7	2^6	2^5	2^4	2^3	2^2	2^1	2^0
	1	1	1	1	0	0	1	1

16^8	16^7	16^6	16^5	16^4	16^3	16^2	16^1	16^0
							F	3

TABLE A.2
Hexadecimal-to-Binary and Binary-to-Hexadecimal Conversion

Decimal	Hexadecimal	Binary
0	0	0000
1	1	0001
2	2	0010
3	3	0011
4	4	0100
5	5	0101
6	6	0110
7	7	0111
8	8	1000
9	9	1001
10	A	1010
11	B	1011
12	C	1100
13	D	1101
14	E	1110
15	F	1111

binary number 1011011001 can be converted to hex by grouping the digits into groups of four, beginning at the right, and zero-padding at the left as required: 0010 1101 1001. The first four digits correspond to the hex digit 2, the second group of four corresponds to the hex digit D, and the last group to the hex digit 9. Thus, binary 1011011001 equals hexadecimal 2D9. Similarly, hex numbers can be converted to binary digit by digit. The hexadecimal number 7F corresponds to the binary number 0111 1111.

A.12 AREA AND PERIMETER

The area of a square is given by x^2, where x is the length of one side. The perimeter of the square is $4x$. The area of a triangle is given by $bh/2$, where b is the length of the base and h is the height of the triangle. The area of a circle with radius r is πr^2. Its perimeter, called the circumference of the circle, is $2\pi r$.

A.13 COMPLEX NUMBERS

A.13.1 j and the Complex Plane

In the realm of complex numbers, j stands for the number $\sqrt{-1}$. It follows that $j^2 = -1$. The concept of j is needed to handle problems like the square root of a negative number. For instance, the square root of -9 is $\pm j3$. The complex plane is a two-dimensional graph, with real numbers along the horizontal axis and imaginary numbers, multiples of j, along

FIGURE A.5

The complex plane.

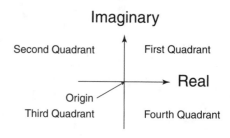

the vertical axis, as shown in Figure A.5. The complex plane is divided into four quadrants. Its center is called the origin.

A.13.2 Rectangular and Polar Form of Complex Numbers

A.13.2.1 Rectangular Form

When a complex number is expressed in **rectangular form**, it is written as

$$z = x + jy$$

where x is the **real part** of the complex number and y is the **imaginary part** of the complex number. For example, the real part of $(13 - j3)$ is 13, and the imaginary part is -3. Figure A.6(a) shows how such a number is plotted on the complex plane: The real part of the number indicates how far left or right of the origin a number lies, and the imaginary part, how far above or below the origin. This is sometimes called a "city block" description of how to move from the origin to a point on the complex plane. When x and y are both positive, the complex number lies in the first quadrant; when x is negative and y is positive, the number lies in the second quadrant; when x and y are both negative, the number lies in the third quadrant; and when x is positive but y is negative, the complex number falls in the fourth quadrant. Parts (b) and (c) of Figure A.6 plot two complex numbers in rectangular form. The **complex conjugate** of a complex number $x + jy$ is $x - jy$. For example, the complex conjugate of $-1 + j5$ is $-1 - j5$.

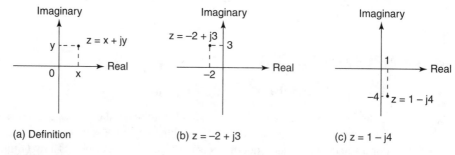

FIGURE A.6

Complex numbers in rectangular form.

Complex numbers in rectangular form are easily added and subtracted: Real parts and imaginary parts are combined separately. For instance,

$$(-1 + j7) + (-9 - j4) = ((-1) + (-9)) + j(7 + (-4)) = -10 + j3$$

$$(-1 + j7) - (-9 - j4) = ((-1) - (-9)) + j(7 - (-4)) = 8 + j11$$

To multiply rectangular complex numbers, real and imaginary parts are multiplied term by term:

$$(-1 + j7)(-9 - j4) = (-1)(-9) + (j7)(-9) + (-1)(-j4) + (j7)(-j4)$$

$$= 9 - j63 + j4 - j^2 28 = 9 - j63 + j4 + 28 = 37 - j59$$

Dividing rectangular form complex numbers requires multiplying top and bottom by the complex conjugate of the denominator:

$$\frac{-1 + j7}{-9 - j4} = \left(\frac{-1 + j7}{-9 - j4}\right)\left(\frac{-9 + j4}{-9 + j4}\right) = \frac{9 - j4 - j63 - 28}{81 - j36 + j36 + 16}$$

$$= \frac{-19 - j67}{97} = \frac{-19}{97} - j\frac{67}{97}$$

As another example,

$$\frac{1}{1 + j\omega} = \left(\frac{1}{1 + j\omega}\right)\left(\frac{1 - j\omega}{1 - j\omega}\right) = \frac{1 - j\omega}{1 + \omega^2} = \frac{1}{1 + \omega^2} - j\frac{\omega}{1 + \omega^2}$$

Many calculators can add, subtract, multiply, and divide complex numbers in rectangular form without requiring the intermediate work shown above.

A complex number in rectangular form, $x + jy$, can be expressed as the ordered pair (x, y) on the complex plane. The distance between two pairs of points (x_1, y_1) and (x_2, y_2) may be found using the distance formula

$$\text{Distance} = \sqrt{(x_2 - x_1)^2 + (y_2 - y_1)^2}$$

A.13.2.2   Polar Form

Polar form is an alternative representation for complex numbers. Polar form uses the distance and angle from the origin to specify a location on the complex plane. The distance from the origin is called the **magnitude** of the complex number, or r. The magnitude can be expressed as $|z|$, where the bars signify that a magnitude is taken. The counterclockwise angle from the positive real axis is called the **phase** of the complex number, or θ. Figure A.7(a) shows how magnitude and phase can locate complex numbers on the complex plane. A complex number in polar form is expressed as

$$z = re^{j\theta} = |z|e^{j\theta}$$

where θ is given in radians. The magnitudes r are always positive. For the complex number $2.5e^{j\pi}$, for instance, the magnitude is 2.5 and the phase is π radians. Figure A.7(b) and (c) provide examples.

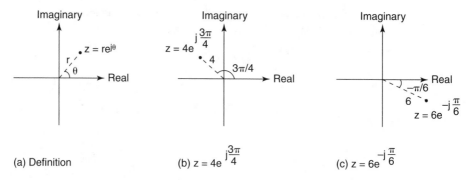

FIGURE A.7
Complex numbers in polar form.

When complex numbers in polar form are multiplied, the magnitudes are multiplied and the phases in the exponents are added. For example,

$$(2e^{-j\frac{\pi}{4}})(5e^{j\frac{\pi}{6}}) = 10e^{j(-\frac{\pi}{4}+\frac{\pi}{6})} = 10e^{-j\frac{2\pi}{24}} = 10e^{-j\frac{\pi}{12}}$$

In division, magnitudes are divided and phases are subtracted:

$$\frac{2e^{-j\frac{\pi}{4}}}{5e^{j\frac{\pi}{6}}} = 0.4e^{j(-\frac{\pi}{4}-\frac{\pi}{6})} = 0.4e^{-j\frac{10\pi}{24}} = 0.4e^{-j\frac{5\pi}{12}}$$

Adding and subtracting complex numbers in polar form is more difficult than multiplying or dividing. These operations are handled most easily by converting the complex numbers into rectangular form before performing the addition or subtraction. This conversion is described in Section A.13.2.4. Alternatively, when polar form numbers are expressed using the short form described in the next section, operations can often be performed on a calculator.

A.13.2.3 Short Form
For convenience, complex numbers in polar form are sometimes written using a short form notation. The complex number $re^{j\theta}$ can be written as $r\underline{|\theta}$. When the short form is used, phase may be quoted in either degrees or radians. For radians, no units are needed, but degrees must be marked with the ° symbol.

Multiplication and division follow the rules identified above for polar form:

$$\left(3\underline{\left|-\frac{\pi}{5}\right.}\right)\left(1.2\underline{\left|\frac{3\pi}{5}\right.}\right) = 3.6\underline{\left|\frac{2\pi}{5}\right.}$$

$$\frac{3\underline{\left|-\frac{\pi}{5}\right.}}{1.2\underline{\left|\frac{3\pi}{5}\right.}} = 2.5\underline{\left|-\frac{4\pi}{5}\right.}$$

When these angles in radians are converted to degrees using the equations in Section A.2, the methods are the same:

$$(3\underline{/-36^\circ})(1.2\underline{/108^\circ}) = 3.6\underline{/72^\circ}$$

$$\frac{3\underline{/-36^\circ}}{1.2\underline{/108^\circ}} = 2.5\underline{/-144^\circ}$$

There are no simple rules that allow addition or subtraction of numbers in this form. However, many calculators allow complex numbers expressed in polar form to be added, subtracted, multiplied, or divided with ease.

A.13.2.4   Conversions Between Rectangular and Polar Form

Figure A.8 combines information from Figure A.6(a) and Figure A.7(a). It shows both rectangular and polar representations for the same complex number z. The cosine of the angle θ may be found as

$$\cos\theta = \frac{\text{real part}}{\text{hypotenuse}} = \frac{x}{r}$$

Similarly,

$$\sin\theta = \frac{\text{imaginary part}}{\text{hypotenuse}} = \frac{y}{r}$$

These two equations can be rearranged to give

$$x = r\cos\theta \qquad\qquad \textbf{(A.1a)}$$

$$y = r\sin\theta \qquad\qquad \textbf{(A.1b)}$$

In other words, $re^{j\theta} = x + jy = r\cos\theta + j\,r\sin\theta$. These equations implement the conversion from polar to rectangular form. For example, the polar complex number $3.6\underline{/72^\circ}$ becomes $1.1125 + j3.4238$ in rectangular form. It should be noted that most calculators can compute the conversion automatically.

FIGURE A.8
Converting between rectangular and polar forms.

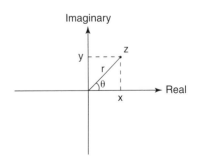

When the real and imaginary parts, x and y, of a complex number z are known, its magnitude and phase can be found using common trigonometric relationships. In particular, the length of the hypotenuse is

$$r = |z| = |x + jy| = \sqrt{(\text{real part})^2 + (\text{imaginary part})^2} = \sqrt{x^2 + y^2} \quad \textbf{(A.2)}$$

and the tangent of the angle θ is

$$\tan \theta = \frac{\text{imaginary part}}{\text{real part}} = \frac{y}{x}$$

which gives

$$\theta = \tan^{-1}\left(\frac{y}{x}\right) \quad \textbf{(A.3)}$$

In other words, $x + jy = re^{j\theta} = \sqrt{x^2 + y^2}\, e^{j\tan^{-1}(y/x)}$. Equation (A.3) gives some ambiguity about phase. If the complex number falls in quadrant two or three, then 180° must be added to the θ value obtained from Equation (A.3). As an example, the rectangular complex number $-1 + j5$, which lies in quadrant two, has magnitude

$$r = \sqrt{(-1)^2 + (5)^2} = 5.099$$

and phase $\theta = \tan^{-1}(5/-1) + 180° = -78.69° + 180° = 101.31°$. As a further example, the quantity

$$z = \frac{1}{1 + j\omega} = \frac{1}{1 + \omega^2} - j\frac{\omega}{1 + \omega^2} = \frac{1}{1 + \omega^2}(1 - j\omega)$$

from Section A.13.2.1 has the magnitude

$$r = \frac{1}{1 + \omega^2}\sqrt{(1)^2 + (-\omega)^2} = \frac{\sqrt{1 + \omega^2}}{1 + \omega^2} = \frac{1}{\sqrt{1 + \omega^2}}$$

and the phase

$$\theta = \tan^{-1}(-\omega) = -\tan^{-1}\omega$$

A few special cases of the conversion of $z = x + jy$ are worth noting: When $x \neq 0$ and $y = 0$, the magnitude is $|x|$ and the phase is either 0° (for $x > 0$) or 180° (for $x < 0$). When $x = 0$ and $y \neq 0$, the magnitude is $|y|$ and the phase is either 90° (for $y > 0$) or $-90°$ (for $y < 0$). As is the case for polar-to-rectangular conversions, most calculators can perform rectangular-to-polar conversion directly.

A.13.2.5   Euler's Identity

Euler's identity relates the complex exponential $e^{j\theta}$ to the trigonometric functions $\sin\theta$ and $\cos\theta$. The identity is

$$e^{j\theta} = \cos\theta + j\sin\theta \qquad\qquad \textbf{(A.4)}$$

This equation is really just an example of a polar-to-rectangular conversion when $r = 1$. For a complex number with $r = 1$, the polar form is $re^{j\theta} = e^{j\theta}$. According to Equation (A.1), the real part of the rectangular equivalent is

$$x = r\cos\theta = \cos\theta$$

and the imaginary part is

$$y = r\sin\theta = \sin\theta$$

Thus,

$$e^{j\theta} = \cos\theta + j\sin\theta$$

Since its magnitude is one, the complex number $e^{j\theta}$ always lies a distance of one from the origin of the complex plane. An alternative form of Euler's identity, for negative angles, is

$$e^{-j\theta} = \cos\theta - j\sin\theta \qquad\qquad \textbf{(A.5)}$$

When Equations (A.4) and (A.5) are added together, the sum gives

$$2\cos\theta = e^{j\theta} + e^{-j\theta}$$

so that

$$\cos\theta = \frac{e^{j\theta} + e^{-j\theta}}{2} \qquad\qquad \textbf{(A.6)}$$

When Equation (A.5) is subtracted from Equation (A.4), the difference is

$$2j\sin\theta = e^{j\theta} - e^{-j\theta}$$

so that

$$\sin\theta = \frac{e^{j\theta} - e^{-j\theta}}{2j} \qquad\qquad \textbf{(A.7)}$$

The Euler forms for $\sin\theta$ and $\cos\theta$ frequently arise in digital signal processing, as do the Euler identities themselves.

A.13.3   The Unit Circle

The unit circle is a circle with radius one on a complex plane, as shown in Figure A.9. Complex numbers with magnitude less than one lie inside the unit circle. The equation

FIGURE A.9
The unit circle.

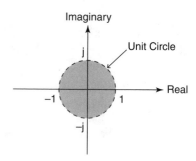

$$|z| < 1$$

describes the shaded area in Figure A.9.

A.14 ABSOLUTE VALUE

The magnitude of a real or integer number is the same as its **absolute value**. The absolute value refers to a number's size without regard to sign. For example, $|-5|$ may be interpreted as the magnitude of a complex number with real part -5 and imaginary part zero, to give magnitude $\sqrt{(-5)^2 + (0)^2} = 5$; or it may be interpreted as an absolute value, to give the result 5. As another example, the expression $|n| \leq 3$ when n is an integer includes all values $-3 \leq n \leq 3$.

A.15 QUADRATIC FORMULA

A **quadratic equation** is a second order polynomial equation like

$$ax^2 + bx + c = 0$$

where a, b, and c are coefficients. The quadratic equation always has two solutions, that is, two values of x that make the equation true. These values of x are called the **roots** of the equation. The **quadratic formula** is a reliable way to find these two roots. The roots are

$$x = \frac{-b \pm \sqrt{b^2 - 4ac}}{2a}$$

When $+$ is used, one root is produced; when $-$ is used, the other root is produced. For example, the quadratic equation $x^2 + 5x + 1 = 0$ has the roots

$$x = \frac{-(5) \pm \sqrt{(5)^2 - 4(1)(1)}}{2(1)} = \frac{-5 \pm \sqrt{25 - 4}}{2} = \frac{-5 \pm \sqrt{21}}{2}$$

$$= \frac{-5 \pm 4.583}{2} = -0.209 \text{ and } -4.791$$

This quadratic equation has $b^2 - 4ac > 0$ and produces two real and distinct roots. As a second example, the quadratic equation $4x^2 + 4x + 1 = 0$ has the roots

$$x = \frac{-(4) \pm \sqrt{(4)^2 - 4(4)(1)}}{2(4)} = \frac{-4 \pm \sqrt{16 - 16}}{8}$$

$$= \frac{-4 \pm 0}{8} = -0.5 \text{ and } -0.5$$

This quadratic equation has $b^2 - 4ac = 0$ and results in two real and equal roots. Finally, the quadratic equation $3x^2 - 2x + 1 = 0$ produces the roots

$$x = \frac{-(-2) \pm \sqrt{(-2)^2 - 4(3)(1)}}{2(3)} = \frac{2 \pm \sqrt{4 - 12}}{6} = \frac{2 \pm \sqrt{-8}}{6}$$

$$= \frac{2 \pm j2.828}{6} = 0.333 \pm j0.471$$

This quadratic equation has $b^2 - 4ac < 0$ and produces a complex conjugate pair of roots.

A.16 $\sum$ SUMS

The Σ sum symbol can be used to write a long sum with similar terms in a compact form. For example, the sum $x + 2x^2 + 3x^3 + 4x^4 + 5x^5$ could be written instead as

$$\sum_{k=0}^{5} kx^k$$

where k is the index of summation. The first term in the expanded sum uses $k = 0$, the second uses $k = 1$, and so on, up to $k = 5$.

Occasionally it is necessary to reexpress a sum by changing its limits. For example, with the substitution $m = k + 1$, a new index is chosen for the sum

$$\sum_{k=0}^{5} kx^k$$

When $k = 0$, $m = 1$; and when $k = 5$, $m = 6$. Furthermore, if $m = k + 1$, then $k = m - 1$, so the sum becomes

$$\sum_{m=1}^{6} (m - 1)x^{m-1}$$

Infinite sums are also possible. For instance,

$$\sum_{n=1}^{\infty} \frac{1}{n\pi} e^{j\frac{2}{n\pi}t} = \frac{1}{\pi} e^{j\frac{2}{\pi}t} + \frac{1}{2\pi} e^{j\frac{2}{2\pi}t} + \frac{1}{3\pi} e^{j\frac{2}{3\pi}t} + \frac{1}{4\pi} e^{j\frac{2}{4\pi}t} + \frac{1}{5\pi} e^{j\frac{2}{5\pi}t} + \cdots$$

In the expansion, the first term is produced with $n = 1$, the second with $n = 2$, and so on. Because the sum is infinite, the final term cannot be specified. An infinite geometric sum is a special class of infinite sum. Each new term in the sum is produced by multiplying the previous term by a **common ratio** r. The first term is normally designated with the letter a. An infinite geometric sum has the form

$$S_\infty = a + ar + ar^2 + ar^3 + ar^4 + ar^5 + \cdots$$

Removing the common factor a from each term,

$$S_\infty = a(1 + r + r^2 + r^3 + r^4 + r^5 + \cdots) \tag{A.8}$$

When $|r| < 1$, this sum has a finite value that can be evaluated in a straightforward way. Multiplying S_∞ by r gives

$$rS_\infty = a(r + r^2 + r^3 + r^4 + r^5 + \cdots) \tag{A.9}$$

Subtracting Equation (A.9) from Equation (A.8) gives

$$(1 - r)\, S_\infty = a$$

Therefore,

$$S_\infty = \frac{a}{1 - r}$$

provided $|r| < 1$. This restriction determines the **region of convergence** for the infinite sum. For example, the sum

$$S_\infty = 5 - e^{-jn} + 0.2e^{-j2n} - 0.04e^{-j3n} + 0.008e^{-j4n} - \cdots$$

has first term $a = 5$ and common ratio $r = -0.2e^{-jn}$. This common ratio has magnitude less than one for all nonzero integer values of n, so the sum converges to

$$S_\infty = \frac{a}{1 - r} = \frac{5}{1 + 0.2e^{-jn}}$$

The Maclaurin series is a special infinite series that allows a function $f(t)$ to be rewritten as a power series. The definition of the Maclaurin series is

$$f(t) = f(0) + f^{i}(0)t + \frac{f^{ii}(0)}{2!}t^2 + \frac{f^{iii}(0)}{3!}t^3 + \frac{f^{iv}(0)}{4!}t^4 + \frac{f^{v}(0)}{5!}t^5 \cdots$$

where the coefficients require the derivatives of $f(t)$ evaluated at $t = 0$. For the function $f(t) = \sin(t), f^{i}(t) = \cos(t), f^{ii}(t) = -\sin(t), f^{iii}(t) = -\cos(t), f^{iv}(t) = \sin(t)$, and $f^{v}(t) = \cos(t)$; so $f(0) = 0, f^{i}(0) = 1, f^{ii}(0) = 0, f^{iii}(0) = -1, f^{iv}(0) = 0$, and $f^{v}(0) = 1$. Thus, the first few terms of the Maclaurin series expansion for $\sin(t)$ are:

$$f(t) = \sin(t) = t - \frac{1}{3!}t^3 + \frac{1}{5!}t^5 - \cdots = t - \frac{t^3}{6} + \frac{t^5}{120} - \cdots$$

B

SIGNAL-TO-NOISE RATIO

The signal-to-quantization noise ratio for the case of quantizing a sine wave may be calculated by finding the ratio of the root-mean-square (rms) signal amplitude to the rms noise amplitude and converting to dB. Quantization errors, the source of quantization noise, are assumed to be random and uniformly distributed between $-Q/2$ and $Q/2$ with zero mean, as shown by the probability distribution in Figure B.1. The mean-square error gives a measure of the power of the quantization noise. In this case, it is the average of the square of the errors:

$$\sigma_e^2 = \int_{-\infty}^{\infty} e^2 P(e) de = \frac{1}{Q} \int_{-Q/2}^{Q/2} e^2 de = \frac{1}{Q} \left. \frac{e^3}{3} \right|_{-Q/2}^{Q/2} = \frac{1}{Q} \left(\frac{Q^3}{24} - \left(-\frac{Q^3}{24} \right) \right) = \frac{Q^2}{12}$$

The root-mean-square noise amplitude, therefore, is

$$\sigma_e = \frac{Q}{\sqrt{12}}$$

FIGURE B.1
Probability distribution of quantization errors.

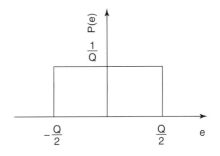

The power of a sine wave with amplitude A and period T is calculated as:

$$\sigma_s^2 = \frac{1}{T}\int_{-T/2}^{T/2}\left(A\sin\frac{2\pi}{T}t\right)^2 dt = \frac{A^2}{T}\int_{-T/2}^{T/2}\sin^2\left(\frac{2\pi}{t}t\right)dt = \frac{A^2}{T}\int_{-T/2}^{T/2}\left(\frac{1}{2} - \frac{1}{2}\cos\frac{4\pi}{T}t\right)dt$$

using the trigonometric identity $\sin^2\theta = (1 - \cos 2\theta)/2$. Since the cosine function is an even function, it integrates to zero. Therefore, the power is

$$\sigma_s^2 = \frac{A^2}{2T}\int_{-T/2}^{T/2}dt = \frac{A^2}{2T}t\Big|_{-T/2}^{T/2} = \frac{A^2}{2T}\left(\frac{T}{2} - \left(-\frac{T}{2}\right)\right) = \frac{A^2}{2}$$

and the root-mean-square signal amplitude is

$$\sigma_s = \frac{A}{\sqrt{2}}$$

Signal-to-noise ratio is defined as $10\log$(signal power/noise power) or, equivalently, $20\log$(signal amplitude/noise amplitude). Using root-mean-square amplitudes for signal and noise, the signal-to-quantization noise ratio (SNR) for a sine wave is

$$\text{SNR} = 20\log\left(\frac{\dfrac{A}{\sqrt{2}}}{\dfrac{Q}{\sqrt{12}}}\right) = 20\log\left(\sqrt{6}\frac{A}{Q}\right)$$

The range of a sine wave is $2A$, so each quantization step Q is, from Equation (2.1),

$$Q = \frac{R}{2^N} = \frac{2A}{2^N}$$

where N is the number of bits used for quantization. Therefore,

$$\text{SNR} = 20\log\left(\sqrt{6}\frac{A}{\left(\dfrac{2A}{2^N}\right)}\right) = 20\log\left(2^N\frac{\sqrt{6}}{2}\right) = 20\log\left(2^N\right) + 20\log\left(\frac{\sqrt{6}}{2}\right)$$

$$= 20N\log(2) + 20\log\left(\frac{\sqrt{6}}{2}\right) = 6.02N\text{dB} + 1.76\text{ dB}$$

C

DIRECT FORM 2 REALIZATION OF RECURSIVE FILTERS

The following steps develop a direct form 2 realization for a recursive filter. Begin with the general recursive difference equation:

$$\sum_{p=0}^{N} a_p y[n-p] = \sum_{k=0}^{M} b_k x[n-k] \tag{C.1}$$

Define a signal $w[n]$ that satisfies the following relationship:

$$\sum_{p=0}^{N} a_p w[n-p] = x[n] \tag{C.2}$$

Substituting this expression into Equation (C.1) gives

$$\sum_{p=0}^{N} a_p y[n-p] = \sum_{k=0}^{M} b_k \sum_{p=0}^{N} a_p w[n-k-p]$$

Interchanging the order of summation:

$$\sum_{p=0}^{N} a_p y[n-p] = \sum_{p=0}^{N} a_p \left(\sum_{k=0}^{M} b_k w[n-k-p] \right)$$

Therefore, from a comparison of the left and right sides of this equation,

$$y[n - p] = \sum_{k=0}^{M} b_k w[n - k - p]$$

or,

$$y[n] = \sum_{k=0}^{M} b_k w[n - k] \tag{C.3}$$

As in Section 4.4, it is possible to assume $a_0 = 1$ without loss of generality. In this case, Equation (C.2) can be rewritten, after replacing the index p with k, as

$$w[n] + \sum_{k=1}^{N} a_k w[n - k] = x[n]$$

or,

$$w[n] = x[n] - \sum_{k=1}^{N} a_k w[n - k] \tag{C.4}$$

Equations (C.3) and (C.4) define the direct form 2 realization of a recursive filter.

D

CONVOLUTION IN THE TIME DOMAIN AND MULTIPLICATION IN THE FREQUENCY DOMAIN

A filter's output can be computed in the time domain using digital convolution, as defined by Equation (5.2):

$$y[n] = x[n] * h[n] = \sum_{k=-\infty}^{\infty} x[k]h[n-k]$$

The impulse response for a causal filter has no samples before zero, which gives

$$y[n] = \sum_{k=0}^{\infty} x[k]h[n-k]$$

This equation may be converted into the z domain by taking the z transform of both sides, using Equation (6.1):

$$Y(z) = \sum_{n=0}^{\infty} \left(\sum_{k=0}^{\infty} x[k]h[n-k] \right) z^{-n}$$

$$= \sum_{n=0}^{\infty} \sum_{k=0}^{\infty} x[k]h[n-k]z^{-(n-k)}z^{-k} = \sum_{k=0}^{\infty} x[k]z^{-k} \sum_{n=0}^{\infty} h[n-k]z^{-(n-k)}$$

For a causal filter, the second summation may begin at $n = k$ instead of $n = 0$:

$$Y(z) = \sum_{k=0}^{\infty} x[k]z^{-k} \sum_{n=k}^{\infty} h[n-k]z^{-(n-k)}$$

Letting $m = n - k$,

$$Y(z) = \left(\sum_{k=0}^{\infty} x[k]z^{-k} \right) \left(\sum_{m=0}^{\infty} h[m]z^{-m} \right)$$

$$= X(z)H(z)$$

It can be shown in a similar manner that

$$y[n] = h[n] * x[n] = \sum_{k=0}^{\infty} h[k]x[n-k]$$

from Equation (5.3) has the z transform

$$Y(z) = H(z)X(z)$$

Since $X(z)H(z) = H(z)X(z)$, the two forms of convolution, in Equations (5.2) and (5.3), must be equivalent.

The above development shows that convolution in the time domain is equivalent to multiplication in the z domain. This is a general property: Convolution in the time domain is equivalent to multiplication in a frequency domain, whether it be the z or the Ω domain:

$$\mathbf{Z}\{a[n] * b[n]\} = A(z)B(z)$$

$$\mathbf{F}\{a[n] * [b[n]\} = A(\Omega)B(\Omega)$$

In fact, the converse is also true: Multiplication in the time domain is equivalent to convolution in a frequency domain, that is:

$$\mathbf{Z}\{a[n]b[n]\} = A(z) * B(z)$$

$$\mathbf{F}\{a[n]b[n]\} = A(\Omega) * B(\Omega)$$

E

SCALING FACTOR IN DISCRETE FOURIER SERIES AND DISCRETE FOURIER TRANSFORM

In Chapter 8, the discrete Fourier series (DFS) expansion was defined as:

$$x[n] = \frac{1}{N}\sum_{k=0}^{N-1} c_k e^{j2\pi\frac{k}{N}n}$$

In Chapter 11, the inverse discrete Fourier transform (IDFT) was defined as:

$$x[n] = \frac{1}{N}\sum_{k=0}^{N-1} X[k] e^{j2\pi\frac{k}{N}n}$$

In both cases, the index n takes values from 0 to $N-1$. This section explains the need for the $1/N$ scaling factor in both the discrete Fourier series and the discrete Fourier transform. Substituting the discrete Fourier series coefficients

$$c_k = \sum_{n'=0}^{N-1} x[n'] e^{-j2\pi\frac{k}{N}n}$$

from Equation (8.3) into the DFS expansion gives:

$$\frac{1}{N}\sum_{k=0}^{N-1} c_k e^{j2\pi\frac{k}{N}n} = \frac{1}{N}\sum_{k=0}^{N-1}\sum_{n'=0}^{N-1} x[n'] e^{-j2\pi\frac{k}{N}n'} e^{j2\pi\frac{k}{N}n}$$

Let $m = n - n'$. Since the index n' goes from 0 to $N-1$, the index m covers a range of N values for each value of n. Since the signal in question is periodic, there is no loss of generality in using values of m from 0 to $N-1$. Since $m = n - n'$, $n' = n - m$. Thus,

$$\frac{1}{N}\sum_{k=0}^{N-1}\sum_{m=0}^{N-1}x[n-m]e^{-j2\pi\frac{k}{N}(n-m)}e^{j2\pi\frac{k}{N}n} = \frac{1}{N}\sum_{k=0}^{N-1}\sum_{m=0}^{N-1}x[n-m]e^{j2\pi\frac{k}{N}m}$$

$$= \frac{1}{N}\sum_{m=0}^{N-1}x[n-m]\sum_{k=0}^{N-1}e^{j2\pi\frac{k}{N}m}$$

Figure E.1 suggests why the sum $\sum_{k=0}^{N-1}e^{j2\pi\frac{k}{N}m}$ is zero as long as $m \neq 0$, by presenting all terms of this sum when $N = 8$ and $m = 1$: The symmetrically arranged terms cancel one another perfectly. When $m = 0$, the sum

$$\sum_{k=0}^{N-1}e^{j2\pi\frac{k}{N}m} = 1 + 1 + 1 + \cdots + 1 = N$$

Thus,

$$\frac{1}{N}\sum_{m=0}^{N-1}x[n-m]\sum_{k=0}^{N-1}e^{j2\pi\frac{k}{N}m} = \frac{1}{N}(x[n])(N) = x[n]$$

FIGURE E.1
Constellation of complex exponentials.

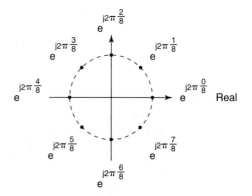

F

INVERSE DISCRETE TIME FOURIER TRANSFORM

The inverse discrete time Fourier transform is defined as

$$x[n] = \frac{1}{2\pi}\int_{2\pi} X(\Omega)e^{jn\Omega}d\Omega$$

where the integral is taken over any 2π interval. When a spectrum $X(\Omega)$ is known, a signal $x[n]$ can be calculated. Similarly, when a frequency response $H(\Omega)$ is known, an impulse response $h[n]$ can be found.

G

IMPULSE RESPONSE OF IDEAL LOW PASS FILTER

An ideal low pass filter has the magnitude response shown in Figure G.1 and zero phase. Using the inverse discrete time Fourier transform, the impulse response for this filter is

$$h_1[n] = \frac{1}{2\pi}\int_{2\pi} H(\Omega)e^{jn\Omega}d\Omega$$

$$= \frac{1}{2\pi}\int_{-\pi}^{\pi} H(\Omega)e^{jn\Omega}d\Omega$$

$$= \frac{1}{2\pi}\int_{-\Omega_1}^{\Omega_1} e^{jn\Omega}d\Omega = \frac{1}{j2\pi n}\left(e^{jn\Omega_1} - e^{-jn\Omega_1}\right)$$

Using Euler's form from Equation (A.7),

$$h_1[n] = = \frac{1}{n\pi}\sin(n\Omega_1)$$

FIGURE G.1
Magnitude response of ideal low pass filter.

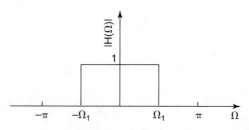

H

SAMPLING PROPERTY

The analog impulse function $\delta(t)$ obeys the following sampling property:

$$f(a) = \int_{-\infty}^{\infty} f(t)\, \delta(t - a)dt$$

When the impulse function occurs within the limits of integration, the integral evaluates to the value of the function $f(t)$ at the location of the impulse, $t = a$ in this case. For example,

$$\int_{-\pi}^{\pi} e^{-t}\delta(t - 1)\, dt = e^{-1}$$

I

SPECTRUM OF DIGITAL COSINE SIGNAL

The easiest way to establish the spectrum of a digital cosine is to begin with the known expression for the spectrum and prove, using the inverse discrete time Fourier transform, that it belongs to the cosine signal. Assume the spectrum of the sampled cosine signal is given by

$$X(\Omega) = \sum_{k=-\infty}^{\infty} A\pi[e^{j\theta}\delta(\Omega-(\Omega_0 + 2\pi k)) + e^{-j\theta}\delta(\Omega+(\Omega_0 + 2\pi k))]$$

as illustrated in Figure I.1. The inverse discrete time Fourier transform is:

$$x[n] = \frac{1}{2\pi}\int_{-\pi}^{\pi} \sum_{k=-\infty}^{\infty} A\pi[e^{j\theta}\delta(\Omega - (\Omega_0 + 2\pi k)) + e^{-j\theta}\delta(\Omega + (\Omega_0 + 2\pi k))]e^{jn\theta}d\Omega$$

Assuming that the frequency of the cosine lies within Nyquist limits, the only elements of the spectrum that lie between $\Omega = -\pi$ and $\Omega = \pi$ occur for $k = 0$. Thus,

$$x[n] = \frac{1}{2\pi}\int_{-\pi}^{\pi} A\pi[e^{j\theta}\delta(\Omega-\Omega_0) + e^{-j\theta}\delta(\Omega+\Omega_0)]e^{jn\Omega}d\Omega$$

Using the sampling property described in Appendix H gives

$$x[n] = \frac{1}{2\pi}\left(A\pi e^{j\theta}e^{jn\Omega_0} + A\pi e^{-j\theta}e^{-jn\Omega_0}\right) = \frac{A}{2}\left(e^{j(n\Omega_0 + \theta)} + e^{-j(n\Omega_0 + \theta)}\right)$$

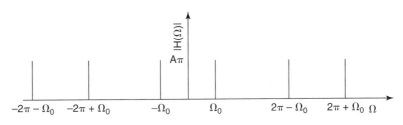

(a) Magnitude Spectrum

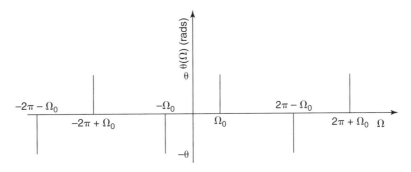

(b) Phase Spectrum

FIGURE I.1
Magnitude and phase spectra of cosine signal.

Using the Euler form for cosine from Equation (A.6), $\cos \theta = \dfrac{e^{j\theta} + e^{-j\theta}}{2}$,

$$x[n] = A \cos(n\Omega_0 + \theta)$$

which is the general form for a digital cosine signal, as expected.

J

SPECTRUM OF IMPULSE TRAIN

The analog impulse train shown in Figure J.1 is a periodic signal. The spectrum for this analog signal is given by

$$c_n = \frac{1}{T}\int_T x(t)e^{-j\frac{2\pi n}{T}t}dt$$

where the c_n are the Fourier coefficients (from the complex exponential Fourier series, not studied in this text). The integral is taken over one period.

FIGURE J.1
Analog impulse train.

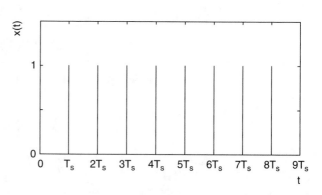

The function $x(t) = \delta(t)$ describes the signal in the interval $-T_S/2$ to $T_S/2$. By the sampling property, the Fourier coefficients become

$$c_n = \frac{1}{T_S} \int_{-T_S/2}^{T_S/2} \delta(t) e^{-j\frac{2\pi n}{T_S}t}\, dt = \frac{1}{T_S}$$

Thus, the coefficients have the same value, $1/T_S$, for each n. The magnitudes are $1/T_S$, and the phases are zero for all values of n. The spacing for the coefficients is the sampling frequency $f_S = 1/T_S$. The magnitude spectrum of the impulse train is shown in Figure J.2.

FIGURE J.2

Magnitude spectrum of analog impulse train.

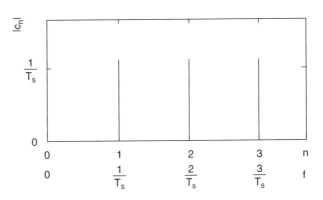

K

SPECTRAL EFFECTS OF SAMPLING

Sampling has a dramatic effect in the frequency domain as well as in the time domain. Sampling in the time domain is equivalent to multiplying an analog signal by an impulse train, as illustrated in Section 11.2. According to Appendix D, multiplication in the time domain is equivalent to convolution in the frequency domain. Thus, multiplying an analog signal by an impulse train in the time domain is the same as convolving the analog signal's spectrum with the spectrum of the impulse train. Figure K.1(a) shows the two-sided spectrum $X(f)$ of a signal, and Figure K.1(b) shows the spectrum $Y(f)$ of an impulse train (see Appendix J). Each impulse in an impulse train in the frequency domain is described by some translation of the function $\delta(f)$. The equation for the entire train, in which pulses are separated by f_S Hz, is $Y(f) = \sum_n \delta(f - nf_s)$. The convolution of two functions $X(f)$ and $Y(f)$ is defined as

$$Z(f) = Y(f) * X(f) = \int_{-\infty}^{\infty} Y(\lambda)X(f - \lambda)d\lambda$$

Therefore, the convolution of a function $X(f)$ with an impulse train is

$$Z(f) = \int_{-\infty}^{\infty} \sum_n \delta(\lambda - nf_s)X(f - \lambda)d\lambda = \sum_n \int_{-\infty}^{\infty} \delta(\lambda - nf_s)X(f - \lambda)d\lambda$$

By the sampling property, introduced in Appendix H,

$$Z(f) = \sum_n X(f - nf_s)$$

This equation describes a function that contains an infinite number of copies of the function $X(f)$, located at integer multiples of f_S, as illustrated in Figure K.1(c). As a result of sampling, copies of the original two-sided signal spectrum appear at each impulse location in the frequency domain. In other words, the sampled spectrum is made up of copies of the original spectrum, one at every multiple of the sampling frequency, 0, $\pm f_S$, $\pm 2f_S$, $\pm 3f_S$, and so on. This result can also be applied when a spectrum is convolved with a single impulse function in the frequency domain: In this case, a copy of the spectrum appears at the location of the impulse.

FIGURE K.1

Effects of sampling in frequency domain.

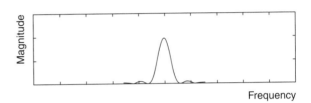

(a) Spectrum of Signal

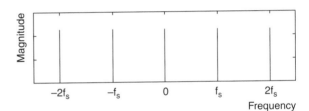

(b) Spectrum of Impulse Train

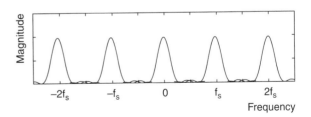

(c) Spectrum of Sampled Signal

L

BUTTERWORTH RECURSIVE FILTER ORDER

The nth order analog Butterworth filter has the filter shape

$$|H(\omega)| = \frac{1}{\sqrt{\left(\dfrac{\omega}{\omega_{p1}}\right)^{2n} + 1}}$$

At the edge of the stop band, where $\omega = \omega_{s1}$, the gain in dB is

$$|H(\omega_{s1})| = \delta_s = \frac{1}{\sqrt{\left(\dfrac{\omega_{s1}}{\omega_{p1}}\right)^{2n} + 1}}$$

This may be solved for n as follows:

$$\delta_s^2 = \frac{1}{\left(\dfrac{\omega_{s1}}{\omega_{p1}}\right)^{2n} + 1}$$

so that

$$\left(\frac{\omega_{s1}}{\omega_{p1}}\right)^{2n} = \frac{1}{\delta_s^2} - 1$$

Taking the log of both sides,

$$\log \left(\frac{\omega_{s1}}{\omega_{p1}} \right)^{2n} = \log \left(\frac{1}{\delta_s^2} - 1 \right)$$

or,

$$2n \log \left(\frac{\omega_{s1}}{\omega_{p1}} \right) = \log \left(\frac{1}{\delta_s^2} - 1 \right)$$

The minimum order for the Butterworth filter is, then,

$$n = \frac{\log \left(\dfrac{1}{\delta_s^2} - 1 \right)}{2 \log \left(\dfrac{\omega_{s1}}{\omega_{p1}} \right)}$$

M

CHEBYSHEV TYPE I RECURSIVE FILTER ORDER

For the nth order Chebyshev Type I filter, the filter shape is given by

$$|H(\omega)| = \frac{1}{\sqrt{1 + \varepsilon^2 C_n^2\left(\dfrac{\omega}{\omega_{p1}}\right)}}$$

where

$$C_n(x) = \begin{cases} \cos\left(n\cos^{-1}(x)\right) & \text{for } |x| \leq 1 \\ \cosh\left(n\cosh^{-1}(x)\right) & \text{for } |x| > 1 \end{cases}$$

and $\cosh(x)$ refers to the hyperbolic cosine function. The required order for the filter can be determined by finding the gains at the edge of the pass and stop bands. At the edge of the pass band, when $\omega = \omega_{p1}$, the function

$$C_n\left(\frac{\omega}{\omega_{p1}}\right) = C_n(1) = 1$$

and

$$|H(\omega_{p1})| = 1 - \delta_p = \frac{1}{\sqrt{1 + \varepsilon^2}}$$

This equation defines the variable ε. For filter design, it is more convenient to solve for ε^2 in terms of δ_p to give

$$\varepsilon = \sqrt{\frac{1}{(1 - \delta_p)^2} - 1}$$

At the edge of the stop band, $\omega = \omega_{s1}$, so

$$|H(\omega_{s1})| = \delta_s = \frac{1}{\sqrt{1 + \varepsilon^2 C_n^2\left(\frac{\omega_{s1}}{\omega_{p1}}\right)}}$$

Since $\omega_{s1} > \omega_{p1}$,

$$\delta_s = \frac{1}{\sqrt{1 + \varepsilon^2\left(\cosh\left(n \cosh^{-1}\left(\frac{\omega_{s1}}{\omega_{p1}}\right)\right)\right)^2}}$$

Rearranging slightly gives

$$\varepsilon^2\left(\cosh\left(n \cosh^{-1}\left(\frac{\omega_{s1}}{\omega_{p1}}\right)\right)\right)^2 = \frac{1}{\delta_s^2} - 1$$

Let $\delta^2 = \dfrac{1}{\delta_s^2} - 1$. This equation defines the variable δ. With this definition,

$$\cosh\left(n \cosh^{-1}\left(\frac{\omega_{s1}}{\omega_{p1}}\right)\right) = \sqrt{\frac{\delta^2}{\varepsilon^2}} = \frac{\delta}{\varepsilon}$$

$$n \cosh^{-1}\left(\frac{\omega_{s1}}{\omega_{p1}}\right) = \cosh^{-1}\left(\frac{\delta}{\varepsilon}\right)$$

Therefore, the minimum order for the Chebyshev Type I filter is

$$n = \frac{\cosh^{-1}\left(\dfrac{\delta}{\varepsilon}\right)}{\cosh^{-1}\left(\dfrac{\omega_{s1}}{\omega_{p1}}\right)}$$

N

WAVELET RESULTS

N.1 WAVELET PRELIMINARIES

An **inner product** of two real functions is defined by the integral

$$\langle x, y \rangle = \int x(t)y(t)dt$$

Two functions $g(t)$ and $g(t-k)$ are **orthogonal** and therefore independent of one another if

$$\int g(t)g(t - k)dt = \begin{cases} A & k = 0 \\ 0 & k \neq 0 \end{cases} \tag{N.1}$$

where A is a constant. The functions are **orthonormal** if Equation (N.1) holds and $A = 1$. The **energy** in a real signal $g(t)$ is given by

$$E = \int g^2(t)dt \tag{N.2}$$

Many wavelet families are designed to be orthonormal; that is, their members are completely independent of one another and the energy of each member of the family is equal to one. If this is the case, the energy of the wavelet $\psi(t)$ at the basic scale is

$$E = \int \psi^2(t)dt = 1 \tag{N.3}$$

Scaling and translation of $\psi(t)$ are given by Equation (16.4). Substituting this equation into Equation (N.2) gives the energy at any other scale:

$$E = \int \psi_{jk}^2(t)dt = \int \left(2^{\frac{j}{2}} \psi(2^j t - k) \right)^2 dt = 2^j \int \psi^2(2^j t - k)dt$$

Setting $t' = 2^j t - k$ gives $t = 2^{-j}(t' + k)$ and $dt = 2^{-j}dt'$, so

$$E = 2^j \int \psi^2(t')2^{-j}dt' = \int \psi^2(t')dt' = 1$$

by Equation (N.3). This development shows the necessity of the $2^{j/2}$ scaling factor that appears in both Equations (16.4) and (16.5).

From Equation (16.10),

$$\varphi(t) = \sum_n \sqrt{2}h_0[n]\varphi(2t - n)$$

Therefore,

$$\varphi(2^j t - k) = \sum_n \sqrt{2}h_0[n]\varphi(2(2^j t - k) - n) = \sum_n \sqrt{2}h_0[n]\varphi(2^{j+1}t - 2k - n)$$

Let $m = 2k + n$, so that $n = m - 2k$. Then

$$\varphi(2^j t - k) = \sum_m \sqrt{2}h_0[m - 2k]\varphi(2^{j+1}t - m) \qquad \textbf{(N.4)}$$

Similarly, from Equation (16.11),

$$\psi(t) = \sum_n \sqrt{2}h_1[n]\varphi(2t - n)$$

so,

$$\psi(2^j t - k) = \sum_m \sqrt{2}h_1[m - 2k]\varphi(2^{j+1}t - m) \qquad \textbf{(N.5)}$$

N.2 WAVELET ANALYSIS EQUATIONS

All translations of the scaling functions $2^{j/2}\varphi(2^j t - k)$ at scale j can describe a group of functions in a subspace called S_j. A function $g(t)$ in this group can be described by a weighted sum of these scaling functions:

$$g(t) = \sum_k c_j[k]2^{j/2}\varphi(2^j t - k)$$

It follows that the group of functions in the subspace S_{j+1} can be described by translations of the scaling functions $2^{(j+1)/2}\varphi(2^{j+1}t - k)$ at scale $j + 1$. A function $f(t)$ in this group can be written as a weighted sum of scaling functions at scale $j + 1$:

$$f(t) = \sum_k c_{j+1}[k]2^{(j+1)/2}\varphi(2^{j+1}t - k) \tag{N.6}$$

As suggested in Section 16.3, the group of functions covered by the subspace S_{j+1} can also be covered by the two subspaces S_j and W_j together, that is

$$S_{j+1} = S_j \cup W_j$$

Therefore, $f(t)$ can also be expressed in terms of scaling functions at scale j and wavelet functions at scale j:

$$f(t) = \sum_k c_j[k]\varphi_{jk}(t) + \sum_k d_j[k]\psi_{jk}(t)$$
$$= \sum_k c_j[k]2^{j/2}\varphi(2^j t - k) + \sum_k d_j[k]2^{j/2}\psi(2^j t - k) \tag{N.7}$$

The analysis equations can be found by using inner products, defined in Appendix N.1. To find $c_j[k]$:

$$c_j[k] = \langle f(t), \varphi_{jk}(t) \rangle = \int f(t)2^{j/2}\varphi(2^j t - k)dt$$

Using Equation (N.4),

$$c_j[k] = \int f(t)2^{j/2}\sum_m \sqrt{2}h_0[m - 2k]\varphi(2^{j+1}t - m)dt$$
$$= \sum_m \int f(t)2^{(j+1)/2}\varphi(2^{j+1}t - m)dt\, h_0[m - 2k]$$
$$= \sum_m \langle f(t), \varphi_{(j+1)m}(t) \rangle h_0[m - 2k]$$

Therefore, the first of the DWT analysis equations is:

$$c_j[k] = \sum_m c_{j+1}[m]h_0[m - 2k] \tag{N.8}$$

The coefficients $d_j[k]$ may be found in a similar manner by using the inner product

$$d_j[k] = \langle f(t), \psi_{jk}(t) \rangle$$

Using Equation (N.5) gives the coefficients

$$d_j[k] = \sum_m c_{j+1}[m]h_1[m - 2k] \tag{N.9}$$

which is the second of the DWT analysis equations.

N.3   WAVELET SYNTHESIS EQUATIONS

According to Equation (N.6), a function in the subspace S_{j+1} may be expressed as

$$f(t) = \sum_k c_{j+1}[k] 2^{(j+1)/2} \varphi(2^{j+1} t - k)$$

but from Equation (N.7), it may also be expressed as

$$f(t) = \sum_k c_j[k] 2^{j/2} \varphi(2^j t - k) + \sum_k d_j[k] 2^{j/2} \psi(2^j t - k)$$

Substituting Equations (N.4) and (N.5) into the latter expression gives

$$f(t) = \sum_k c_j[k] 2^{j/2} \sum_m \sqrt{2} h_0[m - 2k] \varphi(2^{j+1} t - m)$$

$$+ \sum_k d_j[k] 2^{j/2} \sum_m \sqrt{2} h_1[m - 2k] \varphi(2^{j+1} t - m)$$

or,

$$f(t) = \sum_k c_j[k] 2^{(j+1)/2} \sum_m h_0[m - 2k] \varphi(2^{j+1} t - m)$$

$$+ \sum_k d_j[k] 2^{(j+1)/2} \sum_m h_1[m - 2k] \varphi(2^{j+1} t - m) \tag{N.10}$$

The synthesis equations may be found by finding the inner product

$$\langle f(t), \varphi(2^{j+1} t - m) \rangle$$

Using the expression for $f(t)$ in Equation (N.6),

$$\langle f(t), \varphi(2^{j+1} t - m) \rangle = \int \sum_k c_{j+1}[k] 2^{(j+1)/2} \varphi(2^{j+1} t - k) \varphi(2^{j+1} t - m) dt$$

$$= \sum_k c_{j+1}[k] 2^{(j+1)/2} \int \varphi(2^{j+1} t - k) \varphi(2^{j+1} t - m) dt$$

Assuming the scaling functions are orthonormal, the integral has a value of unity when $k = m$, and zero otherwise. Thus,

$$\langle f(t), \varphi(2^{j+1} t - m) \rangle = c_{j+1}[m] 2^{(j+1)/2} \tag{N.11}$$

The same inner product can be evaluated for $f(t)$ in Equation (N.10):

$$\langle f(t), \varphi(2^{j+1}t - m) \rangle = \int \sum_k c_j[k]2^{(j+1)/2} \sum_m h_0[m - 2k]\varphi(2^{j+1}t - m)\varphi(2^{j+1}t - m)dt$$

$$+ \int \sum_k d_j[k]2^{(j+1)/2} \sum_m h_1[m - 2k]\varphi(2^{j+1}t - m)\varphi(2^{j+1}t - m)dt$$

$$= \sum_k c_j[k]2^{(j+1)/2} \sum_m h_0[m - 2k]\int \varphi(2^{j+1}t - m)\varphi(2^{j+1}t - m)dt$$

$$+ \sum_k d_j[k]2^{(j+1)/2} \sum_n h_1[m - 2k]\int \varphi(2^{j+1}t - m)\varphi(2^{j+1}t - m)dt$$

For orthonormal scaling functions, the integrals equal unity. Therefore,

$$\langle f(t), \varphi(2^{j+1}t - m) \rangle = \sum_k c_j[k]2^{(j+1)/2}h_0[m - 2k] + \sum_k d_j[k]2^{(j+1)/2}h_1[m - 2k] \quad \textbf{(N.12)}$$

Combining Equations (N.11) and (N.12) gives

$$c_{j+1}[m]2^{(j+1)/2} = \sum_k c_j[k]2^{(j+1)/2}h_0[m - 2k] + \sum_k d_j[k]2^{(j+1)/2}h_1[m - 2k]$$

or,

$$c_{j+1}[m] = \sum_k c_j[k]h_0[m - 2k] + \sum_k d_j[k]h_1[m - 2k]$$

Interchanging m and k indices, the DWT synthesis equation becomes:

$$c_{j+1}[k] = \sum_m c_j[m]h_0[k - 2m] + \sum_m d_j[m]h_1[k - 2m] \qquad \textbf{(N.13)}$$

INDEX

1.15 number format, 494

A

A/D conversion, 6, 7, 51–52
Acquisition time, 7
Address generators, 500–501
ADSP-2181, 516–518
   supplying coefficients, 527
Aliasing, 7, 32–33, 38–43
Amplitude envelope, 586–587
Analog Devices, 507–509
Analog filter
   compared to digital filter, 104–106
   low pass, 384–386
Analog signal, 2
Analog-to-digital conversion, *See* A/D
      conversion
Antialiasing filter, 34–35
Anticline, 592
Anti-imaging filter, 38, 54
Applications of DSP, 1–2, 16–24
Architecture
   Harvard, 488
   modified Harvard, 488–489
   superscalar, 492
   Von Neumann, 488
Area of object in image, 649–650
Assembly language, 490, 501–503
   declarations and initializations, 533, 535
   for ADSP-2181, 529–537
   MAC instructions, 531–532
   move instructions, 533–534
   multifunction instructions, 533, 535
   program flow control instructions, 533–534
   shifter instructions, 533
Attack, 585–586
Attenuation, 331
Averaging, 127, 157–161, 316–320

B

Band pass filter, 14, 100–102
   FIR design, 353–356, 366–367
   IIR design, 417–419
Band stop filter, 14, 100–101
   FIR design, 364–366
   IIR design, 417–419
Bandwidth, 101, 102
Baseband, 39
Bat echolocation, 474, 476
Best-fit filter design, 417
Big Ben, 475, 476
Bilinear transformation, 386–390
Binary number, 766–767
Bird chirps, 474, 476
Bit, 8
Bit rate, 52
BOPS, 504
Boundary chain code, 650
Boundary effects, 150–151
   for FIR filters, 153
   for IIR filters, 154
   for images, 164
Butterfly, 477
Butterworth filter
   design steps, 399
   filter order, 398, 794–795
   filter shape, 395–398

C

Cache memories, 491
Cascaded filters, 181–184
Castration, 591–592
Causality, 108
CD, 572–573
Cellular telephones, 21–23
Chebyshev Type I filter
   design steps, 407

filter order, 405, 796–797
filter shape, 404–406
Chebyshev type II filter, 386
Chirp, 691–692
Choosing a DSP, 503–507
Circularity, 650
Classification of object in image, 650–654
Clock cycle time, 504
Clock frequency, 504
Closed loop system, 614
Coefficient quantization, 373–375, 426–429
Comb filter, 253–256
Communications, 21–24
Compact disc, *See* CD
Companding, 570–572
Complex exponential, 80–81, 769–771
Complex number, 767–774
    conversion between forms, 771–772
    Euler's identity, 773
    magnitude, 769
    phase, 769
    polar form, 769–771
    rectangular form, 768–769
Composite functions, 89
Compression, 24
    image, 667–673
    lossless, 667
    lossy, 667
    ratio, 667
    wavelet, 738–745
Computed tomography, *See* CT
Continuous word recognition, 575
Controller, 614
Convolution, 145
    2D, 161–165
    boundary effects, 150–151
    connection to difference equation, 155
    graphical method, 148–150
    tabular method, 150–151
    z transform of, 781–782
Convolution kernel, 162
Cross-dissolving, 636
CT, 660–664
Cut-off frequency, 14
Cyanobacteria, 651–652

D

D/A conversion, 9, 52–55
Damped sinusoid, 91
Data Encryption Standard, *See* DES
Data width, 503–504
Daubechies-4 wavelet family, 697–698, 706–707, 710–711
DC component, 293
DCT, 669
Decay, 585, 586
Decibel, 49, 765
Decimation, 564–565
Degree of polynomial, 185
Degrees, 761
DES, 604–612
    triple, 605
DFS, 291–292
    compared to DTFT, 294
    periodic nature, 292
DFT, 434–437
    2D, 480
    frequencies, 444
    interpretation, 443–447
    inverse transform, 435–436
    limitations, 682–685
    magnitude spectrum, 435
    of music, 465–466
    phase spectrum, 435
    relationship to DFS, 463–464
    relationship to DTFT, 436–437
    relationship to Fourier transform, 458–463
    resolution, 444
    smearing, 449–452
    window, 435
    window effects, 464–470
    window length, 452–455
Didjeridoo, 474, 476
Difference equation, 108
    connection to convolution, 155
    converting to frequency response, 234
    converting to transfer function, 176–177
Difference equation diagram, 115
Digital code, 7, 52
Digital filter, *See also* FIR filter, IIR filter, 14, 16
    compared to analog filter, 5, 104–106

Digital frequency, 84
    and analog frequency, 256–257
Digital image processing, *See* image processing
Digital period, 84
Digital playback, 561–562
Digital recording, 561–562
Digital signal
    composite functions, 89
    definition and representation, 2, 7–8, 32, 63
    impulse function, 69
    notation, 65
    power and exponential functions, 79
    sine and cosine functions, 82
    step function, 72
    two-dimensional, 93
Digital signal processing, 1
Digital signature, 604
Digital-to-analog conversion, *See* D/A conversion
Digitizing an image, 619–620
Dilation, 644
Diphone, 591
Direct form 1 realization, 119
Direct form 2 realization, 120, 779–780
    transpose, 121
Direction code, 650
Discrete cosine transform, *See* DCT
Discrete Fourier series, *See* DFS
Discrete Fourier transform, *See* DFT
Discrete time Fourier transform, *See* DTFT
Discrete wavelet transform, *See* DWT
Dither, 568–570, 573
DTFT, 230–232
    and sine wave inputs, 236–239
    compared to DFS, 294
    periodic nature, 233
    plotted against analog frequency, 256–261
    range of digital frequency required, 244
    time-shifting property, 233
DTMF, 100
    decoding tones, 425–426
    generating tones, 422–425
    spectrum of tones, 455–456
Dual tone multifrequency signal, *See* DTMF
DWT, 692–695, 712–713, 729–730
    2D, 731–735

analysis and synthesis example, 723–729
    calculation, 718–723
    coefficients, 729–730
    signal coding, 698–707
Dynamic range, 49–50, 492, 568
Dynamic time warping, 580–581

E
Edge filter for image, 642–643
Eight-to-fourteen (EFM) code, 573
Electrocardiogram (ECG), 456–458
Electroencephalogram (EEG), 5, 19
Elliptic Filter, 386
Encryption, 601–602, 612
    public key, 603
    symmetric key, 603
Equalization of histogram, 627–628
Equiripple FIR filter design, 371–373
Erosion, 644
Euler's identity, 773
Even function, 760
EZ-KIT Lite development board
    assembling and linking programs, 548–549
    set-up and initialization, 537–548
    specifications, 518–519

F
Fast Fourier transform, *See* FFT
Father wavelet, 692
FBI fingerprint image compression, 745–747
Feature identification, 643–650
FFT
    butterfly, 477
    calculation, 476–479
    radix-2 decimation-in-time, 476
    savings over DFT, 479–480
    twiddle factor, 476
    zero padding, 479
Filter, *See also* FIR filter, IIR filter
    antialiasing, 34
    anti-imaging, 38
    moving average, 127, 157–161, 316–320
Filter coefficients, 108
Filter shape, 248
    from pole and zero positions, 261–267

Filtered back projection, 662
Filtering, 99–100, 102
    images, 161–165
    speech, 363
Fingerprint, 746
Finite impulse response filter, *See* FIR filter
Finite word length effects, 115, 527
FIR filter, 127, 315
    arithmetic errors in hardware, 375
    quantization of coefficients, 373–375
    stability, 316
FIR filter program, 553–555
First order filter
    difference equation, 205
    frequency response, 267–268
    impulse response, 205
    pole-zero plot, 205
    steady state output, 205
    transfer function, 204
Fixed point DSP, 493, 503
Floating point DSP, 496, 503
Formant, 576
Fourier coefficients, 292
Fourier transform, 458
Fractional bits, 492
Fractional fixed point number format, 520–523
Frequency domain, 12
Frequency response, 234
    and impulse response, 236
    magnitude response,
    phase response,
Full scale range, 492
Function, 760
Fundamental frequency, 13, 293

G

Gain, 100, 237
    in decibels, 101
Geophone, 599
GFLOPS, 504
Gray scale level, 93, 620

H

Haar wavelet family, 695–697, 699–706,
    708–710

Hardware for DSP, 487–492, 497–501
Harmonic frequency, 13, 293
Harvard architecture, 488
    modified, 488–489
Hash function, 604
HDTV (high definition TV), 744–745
Hexadecimal number, 766–767
High pass filter, 353–356
    FIR design, 16, 640–643
    for image, 417–419
    IIR design, 14, 100, 102
Histogram, 621–622
    equalization, 627–628
    threshold, 625
Horner's method, 502
Huffman encoding, 668–669
Humpback whale song, 473, 476

I

Ideal low pass filter, 326–331, 786
IDFT, 435–436
IDWT, 712
IIR filter, 126, 382–384
    coefficient quantization, 426–429
IIR filter program, 555–557
Image, 36
    morphing, 636–637, 638, 639
    noise reduction by averaging, 631
    scaling, 628
    subtraction, 631–632
    sum, 628–631
    transformation, 632, 635–636
    warping, 636–637
Image processing, 19, 21, 619–673
Improper rational function, 185
Impulse function, 69
Impulse invariance, 413–417
Impulse response, 123
    building general outputs from, 128–131
Infinite impulse response filter, *See* IIR filter
Instruction cycle time, 504
Interpolation, 565–568
    linear, 584, 588–589, 635–636
Inverse DCT, 669–670

Inverse discrete cosine transform, *See* Inverse DCT
Inverse discrete Fourier transform, *See* IDFT
Inverse discrete time Fourier transform, 785
Inverse discrete wavelet transform, *See* IDWT
Inverse z transform, 171, 184–197
    by long division, 188–190
    by partial fraction expansion, 190–197
Isolated word recognition, 575

J
Joint Photographic Experts Group, *See* JPEG
JPEG, 669–673

K
Key splitting, 590
Keyword spotting, 582
Killer whale vocalizations, 473, 476

L
Linear filter effect on sine wave, 238–239
Linearity, 107
Lookup tables, 503
Looping, 586
Low pass filter, 14, 99, 102
    approximating ideal, 326–331, 786
    FIR design guidelines, 342–344
    FIR design steps, 344–345
    for image, 16, 162–165, 637–639
Lowest common multiple, 762

M
MAC, 497–499
Maclaurin series, 584
Magnetic resonance image, *See* MRI
Magnitude response, 240–241
    even nature, 244
Magnitude spectrum, 283
    even nature, 293–294
Mary had a little lamb, 471–472
Masking sound, 574
Matrix, 93
MFLOPS, 504
MIPS, 504

Modems, 23–24
Morphing, 636–637, 638, 639
Mother wavelet, 692
Motor control, 612–616
Motorola, 511–512
Movie projection, 568
Moving average filter, 127, 157–161, 316–320
MP3, 574–575
MPEG Audio Layer 3, *See* MP3
MRA, 708
    frequency view, 715–718
MRI, 665–667
Multifunction instructions, 502
    for ADSP-2181, 533, 535
Multiplier/Accumulator, *See* MAC
Multirate DSP, 565, 718
Multiresolution analysis, *See* MRA
Music synthesis, 583–590

N
Neural network, 653–654
Noise reduction, 631
Noise shaping, 565
Nonperiodic signal, 283
    spectrum, 283–291
Nonrecursive filter, *See also* FIR filter, 109
Number format, 492–497
    for ADSP-2181, 519–523
Nyquist
    frequency, 32
    range, 32
    sampling rate, 32
    sampling theory, 30–32, 38

O
Odd function, 760
Oil wells, 592
Order
    of Butterworth filter, 398
    of Chebyshev Type I filter, 405
    of filter, 42
Overdamped, 613
Overflow and underflow, 495, 497
Oversampling, 42–43, 562–565

P

P wave, 593
Parallel filters, 181, 183–184
Parasaurolophus, 663–664
Partial fraction expansion, 190–197
Pass band, 100
Pass band ripple, 329–330
Perceptual coding, 574–575
Perimeter of object in image, 650
Periodic signal, 291
    spectrum, 291–306
Perplexity, 581
PGP, 603–604
Phase delay, 320
Phase difference, 237
Phase distortion, 320–322
    eliminating, 322–325
Phase response, 240–241
    odd nature, 244
Phase spectrum, 283
    odd nature, 293–294
Phoneme, 576
PID control, 615–616
Pipelining, 490–491
Pitch shifting, 588
Pixel, 93, 620
Poles, 198–201
    effect on system behavior, 219
Pole-zero plot, 199
Polyphony, 585
Position control, 613
Power and exponential functions, 79
Power of digital signal, 578–579
Power series expansion, 502–503, 584
Pretty Good Privacy, *See* PGP
Prewarping equation, 388, 392–395
Private key, 603
Projection-slice tomography, 662
Proper rational function, 185
Prosody, 591
Public key, 603

Q

Quadratic formula, 774–775
Quantization, 9, 44, 54

    bipolar, 48
    bits, 50
    error, 9, 44–45
    level, 7
    noise, 49
    resolution, 44
    unipolar, 44

R

Radians, 761
Rational function, 761
Realization structure, 115, 119, 120, 121
Recursive filter, *See also* IIR filter, 109
Recursive filter
    direct form 1 realization, 119
    direct form 2 realization, 120, 779–780
Reflection coefficient, 595–597
Region of convergence, 171, 776
Registers, 490
    for ADSP-2181, 528–529
Release, 585–586
Repeat instruction, 502
Resolution
    of image, 620–621, 622, 623
    of quantizer, 44
Ringing, 460
Ripple, 329
Roll-off, 100
Run-length encoding, 667–668

S

S wave, 593
Sample and hold, 7, 30, 51, 54
Sampling
    band-limited, 39–41
    frequency, 30
    frequency domain view, 35–38, 792–793
    interval, 30
    period, 30
    rate, 30
    sine waves, 33–34, 38–39
    theorem, 32
Sampling property, 787
Scaling
    for FIR filters, 524–525

for IIR filters, 525–527
Scaling function, 692
   coefficients, 708
Second order filter
   difference equation, 209, 212
   frequency response, 269–273
   steady state output, 217
   transfer function, 209, 212
Seismic cross-section, 597–598, 600
Seismic pulses, 598–599
Sensor, 614–615
   light, 6
   sound, 6
   temperature, 7
Shifter, 499–500
Sign bit, 493
Sign extension, 499
Signal, *See also* analog signal, digital signal
   band-limited, 35
   nonperiodic, 283
   periodic, 291
Signal-to-noise ratio, *See* SNR
Sinc function, 286, 297, 326–327
Sine and cosine functions, 82, 763–765
Sine wave generator program, 549–553
Smoothing filter for images, 162–165
SNR, 50, 777–778
Spectral leakage, 468
Spectrogram, 470–476
Spectrum, 11, 281–283
   imaged frequencies in, 36–37
   of image, 654–660
Speech recognition, 16–18, 575–583
   speaker dependent, 576
   speaker independent, 576
Speech synthesis, 18–19, 590–592
   articulatory synthesizer, 591
   concatenative synthesizer, 591
   formant synthesizer, 591
Speed control, 613
Square wave harmonics, 281–283, 304–305
Stability, 201–204
   analog filter, 389
   digital filter, 388–389
Standard form, 184

Steady state behavior, 151
Step function, 72
Step response, 131
   calculation from impulse response, 131
Stop band, 100
Stop band ripple, 329–330
Strictly proper rational function, 185
Superposition, 107, 112
Superscalar architecture, 492
Sustain, 585–586
System, 3
   DSP, 29–30
   linear, time-invariant, causal, 107–109

T
Template, 576
Texas Instruments, 509–511
Text-to-speech, 590
Thoreau's Walden, 668
Thresholding, 741
   hard, 741–742
   soft, 741–742
Thrinaxodon, 663
Timbre, 590
Time domain, 11
Time invariance, 107
Time shift, 173–174, 233
Time-frequency diagram, 685
Time-scale diagram, 690
Time-scale tiling
   for DFT, 735–736
   for DWT, 736–737
Tomography, 660
Transfer function, 176
   and impulse response, 179–180
   cascade and parallel combinations, 181–184
   converting to difference equation, 177–179
   converting to frequency response, 235
   using to find filter output, 180–181
Transform
   discrete Fourier, 434–435
   discrete time Fourier, 230–231
   discrete wavelet, 692–695
   fast Fourier, 476–479
   Fourier, 458

wavelet, 686
  z, 170–171
Transient behavior, 151
Transition width, 329–330
Twiddle factor, 476
Two's complement, 493

U
Underdamped, 613
Undersampling, 41–42
Unit circle, 201, 773–774

V
Voice, 585
Voice morphing, 592
Voice recognition, *See* speech recognition
Voice synthesis, *See* speech synthesis
Voiced and voiceless sounds, 576
Von Neumann architecture, 488
Vowel sound, 290–291, 303–304

W
Warping, 636–637
Washing machines, 616–617
Wavelet
  analysis, 712–713, 799–800
  biorthogonal, 714
  compression, 738–745
  family, 692
  father, 692
  filters, 713–714
  mother, 692
  packets, 737–738
  scale, 687
  scaling, 686–687
  signal coding, 698–707

support, 692
  synthesis, 712, 713, 801–802
  translation, 688–690
Wavelet function, 692
  coefficients, 709
Wavelet transform, 686
Wavetable synthesis, 585
Wheels turning backward in movies, 43–44
White noise, 281, 458, 459
Window
  Blackman, 338–339
  Hamming, 337–338
  Hanning, 335–337
  Kaiser, 339–341
  rectangular, 332–335

X
X-ray of baggage, 646–647

Y
Youngsters, 659

Z
z domain, 171
  delays, 173–175
z plane, 199
z transform, 170
  region of convergence, 171
  table of transforms, 174
  time-shifting property, 173–175
Zeros, 198–201
  effect on system behavior, 219
Zero insertion, 565–568
Zero order hold, 9, 53, 55
Zero overhead looping, 502
Zero padding, 479